Date Due

NEW MEXICO
EASTERN UNIVERSITY
E N
M U
LIBRARY

Frontispiece: Participants at the Meeting.

Proceedings of an International Meeting, Liverpool, July 1972

Ion Transport in Plants

Edited by

W. P. Anderson

Department of Botany,
University of Liverpool,
Liverpool, England.

1973
ACADEMIC PRESS . London and New York

A subsidiary of Harcourt Brace Jovanovich, Publishers

ACADEMIC PRESS INC. (LONDON) LTD.
24/28 Oval Road,
London NW1

United States Edition published by
ACADEMIC PRESS INC.
111 Fifth Avenue
New York, New York 10003

Library of Congress Catalog Card Number: 73-1473
ISBN: 0-12-058250-3

PRINTED IN GREAT BRITAIN BY
J. W. ARROWSMITH LTD., BRISTOL, ENGLAND

Participants

W. P. ANDERSON, Department of Botany, University of Liverpool, England.

D. A. BAKER, School of Biological Sciences, University of Sussex, Brighton, Sussex, England.

J. BARBER, Department of Botany, Imperial College, University of London, England.

C. E. BARR, Department of Biological Sciences, University College at Brockport, Brockport, New York 14420, U.S.A.

F. W. BENTRUP, Institut für Biologie, der Universität Tübingen, 7400 Tübingen, Germany.

D. J. F. BOWLING, Department of Botany, University of Aberdeen, Scotland.

J. C. COLLINS, Department of Botany, University of Liverpool, Liverpool, England.

W. J. CRAM, School of Biological Sciences, University of Sydney, Sydney 2006, N.S.W., Australia.

J. DAINTY, Department of Botany, University of Toronto, Toronto 181, Canada.

R. F. DAVIS, Department of Botany, Rutgers State University, Newark, New Jersey 07102, U.S.A.

G. DUCET, Physiologie Cellulaire, Centre Universitaire Marseille-Luminy, 13 Marseille (9e), France.

A. ESHEL, Department of Botany, Tel Aviv University, Tel Aviv, Israel.

C. D. FIELD, School of Life Sciences, New South Wales Institute of Technology, Broadway, N.S.W. 2007, Australia.

G. P. FINDLAY, School of Biology, Flinders University, Bedford Park, S. Australia 5042.

T. J. FLOWERS, School of Biological Sciences, University of Sussex, Brighton, England.

v

C. GILLET, Biologie Végétale, Facultés Universitaires de Namur, B-5000 Namur, Belgium.

B. Z. GINZBURG, Department of Botany, Hebrew University, Jerusalem, Israel.

J. L. HALL, School of Biological Sciences, University of Sussex, Brighton, Sussex, England.

R. HELLER, Physiologie Végétale, Sorbonne, Paris (5e), France.

N. HIGINBOTHAM, Department of Botany, Washington State University, Pullman, Washington 99163, U.S.A.

A. E. HILL, Department of Botany, University of Cambridge, Downing St., Cambridge, England.

B. S. HILL, Department of Botany, University of Cambridge, Downing St., Cambridge, England.

H. JAYASURIYA, Department of Botany, University of Cambridge, Downing St., Cambridge, England.

R. L. JEFFERIES, School of Biological Sciences, University of East Anglia, Norwich, England.

D. H. JENNINGS, Department of Botany, University of Liverpool, Liverpool, England.

W. D. JESCHKE, Botanisches Institut, Universität Würzburg, 87 Würzburg, Germany.

R. G. W. JONES, Department of Biochemistry and Soil Science, University College of North Wales, Bangor, Caerns., Wales.

A. KYLIN, Botaniska Institutionen, Universitet Stockholm, Stockholm 50, Sweden.

A. LÄUCHLI, Botanisches Institut der Technischen Hochschule Darmstadt, D-6100 Darmstadt, Germany.

R. A. LEIGH, Department of Biochemistry and Soil Science, University College of North Wales, Bangor, Caerns., Wales.

R. McC. LILLEY, Department of Botany, University of Sheffield, Sheffield, England.

B. C. LOUGHMAN, Department of Agricultural Sciences, University of Oxford, Oxford, England.

U. LÜTTGE, Botanisches Institut der Technischen Hochschule Darmstadt, D-6100 Darmstadt, Germany.

I. R. MACDONALD, Department of Plant Physiology, Macaulay Institute for Soil Research, Aberdeen, Scotland.

A. L. S. MACKLON, Department of Plant Physiology, Macaulay Institute for Soil Research, Aberdeen, Scotland.

E. A. C. MACROBBIE, Department of Botany, University of Cambridge, Downing St., Cambridge, England.

P. MEARES, Department of Chemistry, University of Aberdeen, Aberdeen, Scotland.

A. MEIRI, Institute of Soils and Water, The Volcani Center, Bet Dagan, Israel.

E. MÜLLER, Institut für Biochemie der Pflanzen, Deutsche Akademie der Wissenschaften, Halle, German Democratic Republic.

J. NEUMANN, Department of Botany, Tel Aviv University, Tel Aviv, Israel.

P. NISSEN, Botanisk Laboratorium, Universitet Bergen, Bergen, Norway.

C. K. PALLAGHY, Department of Botany, La Trobe University, Bundoora, Victoria, Australia.

K. R. PAGE, Department of Chemistry, University of Aberdeen, Aberdeen, Scotland.

M. G. PITMAN, School of Biological Sciences, University of Sydney, Sydney 2006, N.S.W., Australia.

R. POOLE, Biology Department, McGill University, P.O. Box 6070, Montreal, Canada.

J. A. RAVEN, Department of Biological Sciences, University of Dundee, Dundee, Scotland.

T. RIPPIN, Department of Biochemistry and Soil Science, University College of North Wales, Bangor, Caerns., Wales.

M. G. T. STONE, A.R.C. Letcombe Laboratory, Wantage, Berkshire, England.

F. A. SMITH, Department of Botany, University of Adelaide, Adelaide, S. Australia 5001.

R. M. SPANSWICK, Division of Biological Sciences, Cornell University, Ithaca, New York 14850, U.S.A.

J. F. SUTCLIFFE, School of Biological Sciences, University of Sussex, Brighton, Sussex, England.

J. F. THAIN, School of Biological Sciences, University of East Anglia, Norwich, England.

M. THELLIER, Laboratoire de Nutrition Minérale, Faculté des Sciences de Rouen, 76-Mont-Saint-Aignan, France.

P. B. H. TINKER, Department of Plant Sciences, University of Leeds, Leeds, England.

M. E. VAN STEVENINCK, Department of Botany, University of Queensland, Brisbane, Queensland 4067, Australia.

R. F. M. VAN STEVENINCK, Department of Botany, University of Queensland, Brisbane, Queensland 4067, Australia.

W. J. VREDENBERG, Centrum voor Plantenfysiologisch Onderzoek, C.P.O., Wageningen, Netherlands.

Y. WAISEL, Department of Botany, Tel Aviv University, Tel Aviv, Israel.

N. A. WALKER, School of Biological Sciences, University of Sydney, Sydney 2006, N.S.W., Australia.

I. C. WEST, Glynn Research Laboratories, Bodmin, Cornwall, England.

R. E. WHITE, Department of Botany, University of Witwatersrand, Johannesburg, South Africa.

Preface

My aim in writing this Preface is to explain to the reader the structure of the Liverpool Workshop on Ion Transport and to try to convey something of the philosophy and intentions in the minds of those who convened the meeting. I leave the published writings and the points of view expressed in the discussion sessions to indicate the scientific mind of the participants.

In the autumn of 1971, when I learned that a number of the Australian ion transport workers would be in Europe the following summer, I thought that it would be a good idea to arrange a meeting to discuss various topics of great current interest. However, the extremely modest original idea grew for one reason or another and the meeting eventually was attended by 61 registered participants from 14 different countries. I wrote in the first instance to twelve well-known authorities in ion transport work and asked for their reaction to the proposal and for suggestions of names of participants and topics for discussion. I received so much enthusiastic encouragement from everyone that it was clear that the time was right to hold such a meeting. Many of the topics suggested themselves, the state of progress being what it is; others were predicated by my and others' selfish interests. However I do believe that the overall coverage of ion transport in this Workshop gives a fairly well-balanced and cohesive view of the state of the game at present.

Each participant was asked to submit his manuscript well in advance of the meeting. The manuscripts were then duplicated in Liverpool, and sent out to all participants, so that everyone attending had a complete set of the proceedings several weeks prior to the meeting. Thus at the actual meeting, participants were in general given only 15 minutes to present their paper, the assumption being that everyone had already read it. The papers were grouped in threes and fours followed by relatively lengthy and free-ranging discussion sessions. Thus the main purpose of the Workshop, to allow a free but formal forum for discussion, was hoped to be achieved. The onus imposed on the Chairmen by this structure was considerable, and I should like to say here that I thought each handled his session beautifully; whatever worth there may be in the discussions for the reader, for those actually participating these sessions were the high points of the Workshop. The Chairmen were largely responsible for this success.

The decision of how best to present a synopsis of the tape-recorded discussion sessions was not easy. In the event, I have written a précis of each discussion session after carefully listening to the tape-recordings. The representation I give of an individual's views has not been agreed by him and I expect to be held solely responsible for errors. I have tried very hard not to misrepresent anyone and to reflect faithfully the feeling as well as the content of each discussion session. The phrases or sentences in quotation marks in these commentaries are verbatim reports from the tapes.

One final thing should be mentioned before I turn to my last paragraph; Professor Jack Dainty was our Honorary Chairman throughout the week of the Liverpool Workshop. All of us who attended the meeting were of a mind that he honoured us in agreeing to fulfil this duty. We hope that he felt we had paid him some little, but sincere tribute too. Many of us can trace our original involvement in plant physiology to him, first in Edinburgh and later in Norwich; many more of us have benefited from his help and advice over the years.

Finally, as always in the last paragraph, I must record my own personal votes of thanks. First and in greatest measure, I wish to offer my heartfelt appreciation of the hard work and understanding of Elsie Allen; she more than anyone bore the burdens, direct and indirect, of this meeting. David Jennings helped every time I turned to him for help, advice or encouragement. Louise Goodwin and Ann Kerrigan both bore up bravely under many a duty or imposition of my far from sunny temper. Tony Bradshaw's generosity ensured the unstinted support of the Botany Department.

* Liverpool, August 1972   W.P.A.

* *Temporary address*:
Department of Botany,
Washington State University,
Pullman, Washington 99163, U.S.A.

Acknowledgements

A meeting such as this can only be organized and brought to completion with the support and goodwill of various organizations and individuals.

We all acknowledge and offer our thanks to the following: the various individual Universities and Institutions which supported the travel to Liverpool of their members; the British Council which gave support to six participants while in England; the Department of Botany, University of Liverpool which subsidized the administrative expenses of the meeting.

We also wish to thank the Local Organizing Committee, D. H. Jennings (Chairman), W. P. Anderson (Secretary) and J. C. Collins for their work in convening the Workshop, and Miss Elsie Allen, Miss Louise Goodwin, Miss Ann Kerrigan and Mr. John Beckett for their willing help in the day-to-day running of the meeting.

Finally, we record our appreciation to Academic Press for expediting the publication of these proceedings.

Contents

Section I

Ultrastructural Localization

Chairman: J. F. Sutcliffe

I.1

Investigation of Ion Transport in Plants by Electron Probe Analysis: Principles and Perspectives

André Läuchli*

Department of Plant Sciences, Texas A & M University, Texas, U.S.A.

I. Introduction

Ultrastructural localization of ions is an area that has aroused considerable attention among transport physiologists in recent years. The aim of ion localization is to measure inorganic solutes qualitatively and quantitatively *in situ* at ultramicroscopic dimensions, thus enabling one to study directly the transport of ions through membranes, in cells, and in tissues and organs of a plant. There are several analytical methods which may be used but electron probe analysis (EPA) is probably the most versatile one.

Electron probe analysis permits an essentially complete elemental analysis of small regions in bulk and sectional specimens. Its versatility has been demonstrated by the fact that all the chemical elements heavier than Be ($Z = 4$) can be easily determined with a sensitivity of 10^{-15}–10^{-16} g, surpassing that of most other analytical methods (Läuchli, 1972b). The spatial resolution of EPA, however, is such that the localization of ions in biological specimens

---

* Present address: Fachbereich Biologie (10) der Technischen Hochschule Darmstadt—Botanik—D-6100 Darmstadt/Germany.

is only accomplished at best to a resolution of approximately 0·1 μm (Hall, 1971; Läuchli, 1972b). Nevertheless, spatial resolution is sufficient to allow compartmentation of ions in plant cells, at least as far as major cell organelles are concerned (cf. Sutfin *et al.*, 1971).

II. Principles of Electron Probe Analysis

The elements in a specimen are determined by irradiating the sample with an electron beam focused on the sample surface and measuring the characteristic X-rays that emerge from the irradiated area. Figure 1 shows

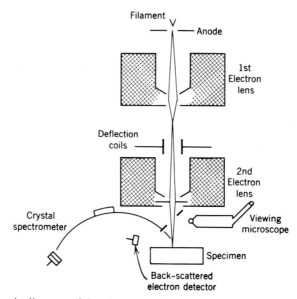

FIG. 1. Schematic diagram of the electron probe analyser (reproduced with permission from Birks, 1971).

the components of an electron probe analyser. They consist of: (i) an electron optical system which produces the electron beam and focuses it to about 0·1–1 μm diameter at the surface of the specimen; (ii) a viewing light microscope for selection of the specimen area to be examined; (iii) electron detectors, used in connection with deflection coils for scanning the beam across the sample surface in a square raster, to obtain electron images of the specimen surface such as that produced by secondary electron emission; (iv) X-ray detection systems to identify the elements by wavelength dispersion with a crystal spectrometer or by energy dispersion with total X-ray detectors. A detailed description of the principles of EPA is given in the monograph by Birks (1971) and the reviews by Hall (1971) and Läuchli (1972b).

A. Wavelength Dispersion

One means of separating the characteristic X-ray lines for the determination of elements is by wavelength dispersion. This is accomplished by diffraction of the X-rays from an analyser crystal in a crystal spectrometer. Several crystals are necessary, each of them covering a range of wavelengths and therefore suitable for the determination of a specific group of elements (cf. Läuchli, 1972b, Table 2). After diffraction from a crystal the intensities of the characteristic X-ray lines are measured by a detector which ordinarily is a gas proportional counter. Two types are in use, the sealed counters and the gas-flow counters. The latter ones have extremely thin windows needed for detecting the long wavelengths of the X-ray lines of the lighter elements ($Z = 5$ to 12).

When the proportional counter is used with a crystal spectrometer, the detector's only function is to monitor the number of X-ray photons per unit of time because the dispersion is done by the crystal. However, when the detector is used without a spectrometer, the characteristic lines must be separated by their energies.

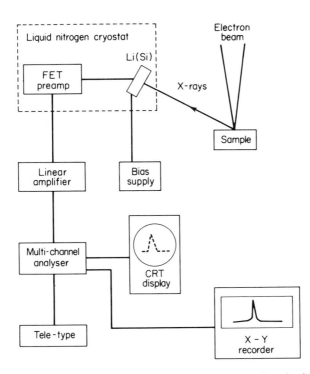

FIG. 2. Block diagram of a solid-state detector system (reproduced with permission from Myklebust and Heinrich, 1971).

B. Energy Dispersion

The method of distinguishing the characteristic X-ray lines according to their energies is known as energy dispersion. This mode of analysis is based on the proportionality between photon energy and pulse amplitude. Owing to rapid advancements in instrumentation energy dispersion has become in many respects competitive with the older method of wavelength dispersion (cf. Frankel and Aitken, 1970; Russ, 1971a).

Solid-state detectors such as the silicon detector are particularly useful for energy dispersion X-ray analysis. A block diagram of a solid-state detector system is shown in Fig. 2. Integral components are the detector which is enclosed in a cryogenic container to damp background noise and the multi-channel analyser. When operated with a silicon detector, this system can be used to measure X-ray photons of energies above 1 keV, allowing analysis of elements with $Z = 11$ (Na) and greater, down to 0·1 % or better; furthermore, it makes possible the total qualitative analysis of a sample in only 1 or 2 min and also the quantitative determination of elements (Myklebust and Heinrich, 1971; Russ, 1971c). The solid-state detector is usually replaced by a gas-flow counter when elements lighter than Na are to be analysed by energy dispersion (Sutfin and Ogilvie, 1971).

C. Analytical Techniques

Electron probe analysis is a method by which qualitative and quantitative information can be obtained. A very useful qualitative technique is the one that utilizes two-dimensional scanning. In the scanning mode, the electron beam is swept back and forth over a square area of the sample by means of deflection coils. When coupled with the detector readout, this allows a two-dimensional X-ray or electron image of the sample to be displayed on an oscilloscope whose flying spot sweeps in synchronization with the scanning electron beam. X-ray images represent qualitative chemical maps of the scanned specimen area while electron images convey information on the topography and mass density of the specimen.

When a static electron beam is focused on the specimen surface, a total qualitative analysis of the irradiated area is obtained by scanning the crystal spectrometers through their wavelength ranges and recording the X-ray spectrum (wavelength scanning). Separation of the characteristic lines by wavelength scanning can be achieved with high resolution, but a complete wavelength scan of an unknown sample usually requires 30–40 min. On the other hand, one may complete a total qualitative analysis with energy dispersion in 1 or 2 min though with somewhat poorer X-ray resolution (cf. Myklebust and Heinrich, 1971).

Two quantitative techniques are used in conjunction with EPA, i.e. point analysis and line scanning. Point analysis requires a static electron beam to be applied to the sample, and the crystal spectrometer or the energy-dispersive detector respectively are set on the most sensitive line or energy of a specific element to be measured. X-ray intensities are obtained from the detector readouts; they are in counts per unit time and represent a relative measure of the amount of an element per sample volume analysed.

Line scanning yields a quantitative picture of the pattern of elemental distribution. A line scan is achieved on any line of traverse across the surface of a specimen when either the sample is moved under the static electron beam or the beam is driven across the stationary sample by means of the beam scanning system. The quantitative data thus obtained are in relative units and reflect profiles of elemental concentration.

The relative quantitative data can be converted to absolute units by standardization with reference standards that are with respect to chemical composition and physical properties as similar to the specimen as possible. Such standards have been produced for use in EPA of animal specimens (Tousimis, 1963; Andersen, 1967; cf. Birks, 1971, p. 79). In contrast, quantitative data from EPA of plant samples were ordinarily expressed in relative units. Humble and Raschke (1971) utilized crystals of K oxalate and KCl to calibrate intensities of X-rays from K in stomata. Additional studies are needed before absolute standardization is established as a routine technique. However, the accuracy of EPA is far greater than with other histochemical methods.

III. Specimen Preparation

When EPA is applied to studies of ion transport in plants, the technique of specimen preparation is of paramount significance. In essence, one must prevent any displacement or leaching of the ions to be detected during the entire preparation procedure. A careful assessment of the various techniques available with extensive literature review was done by Läuchli (1972b); however, only a brief account will be given here.

When dealing with plant tissue, the majority of investigators used the cryostat technique (freeze-drying) to prepare specimens for EPA. The cryostat technique is considered to be reliable as far as retention of ions is concerned, but cellular structures often are preserved poorly in cryostat sections. In addition, "thin" sections are obtained only with great difficulty by this technique. Thin sections, however, are crucial in achieving high resolution in EPA (Läuchli et al., 1970). It is now possible to prepare ultra-thin frozen sections from unfixed plant specimens on an ultra-microtome with freezing attachment (Gahan et al., 1970). Cryo-ultramicrotomy of plant tissues is a very promising approach but has not yet matured to a routine technique.

Freeze-drying of plant specimens without subsequent embedding of the dried material was utilized when ion movements in stomata were studied by EPA (Sawhney and Zelitch, 1969; Humble and Raschke, 1971; Raschke and Fellows, 1971). However, this technique has limited applications because sections cannot be obtained. To prepare sections, freeze-dried plant tissue must be embedded in paraffin or epoxy resin, but marked shrinkage of cell components and displacement of water-soluble compounds may then occur (Fisher, 1972).

An alternative method of dehydrating frozen specimens is by freeze-substitution; embedding is carried out in epoxy resin or paraffin. Ions and water-soluble organic compounds appear to be retained *in situ* (Lüttge and Weigl, 1965; Läuchli *et al.*, 1970; Fisher and Housley, 1972). In experiments on transport of K^+ in plants studied by EPA, Läuchli *et al.* (1970) freeze-substituted the specimens with ether followed by embedding in epoxy resin. The step of washing the dehydrated samples with acetone before infiltration, as described in the original procedure, apparently can be omitted (A. R. Spurr, 1971, unpublished). With this technique thin sections, 1 μm thick, may be prepared anhydrously, thus ensuring good resolution in EPA.

Suitable materials for specimen support and the techniques for vacuum coating the specimens were described by Läuchli (1972b). When K is to be detected by EPA, the samples should be coated with a thin film of Al in order to prevent the volatilization of K due to exposure to the electron beam (Hodson and Marshall, 1971).

IV. Types of Instruments

The conventional electron probe analyser with which most work on plant specimens was conducted is an instrument that was designed to fit the needs of geologists, metallurgists, and physicists. The application of this instrument to the biological sciences came about much later and was hampered by the fact that its resolution is not comparable with that of the electron microscope. Furthermore, it is only very recently that the electron probe analyser has become available with a transmitted electron detector to view thin sections in the transmission mode and simultaneously carry out X-ray analysis. Transmission and scanning electron microscopes are better adapted to biological specimens. Consequently, it is not surprising that there is now a trend to equip both types of electron microscopes with X-ray detectors.

The analytical electron microscope EMMA-4 is a combination of a transmission electron microscope with wavelength- and energy-dispersive detectors (cf. Chandler, 1971). This sophisticated instrument allows X-ray analysis of regions 0·1 μm in diameter when ultra-thin sections are used.

But two-dimensional scanning is not yet available, and the application of the analytical electron microscope is restricted to ultra-thin sections because of its design as a high-resolution electron microscope. The amounts of certain elements, however, may be too low in an ultra-thin section to be measured accurately, and one needs to use slightly thicker sections up to about 1 μm.

The scanning electron microscope with a transmitted electron detector is capable of producing transmission electron images from 1 μm thick sections with good resolution (Matsuo and Suzuki, 1971). Rapid X-ray analysis is accomplished when it is fitted with energy-dispersive detectors (Russ, 1971b). Thurston and Russ (1971) estimated the spatial resolution of X-ray analysis in the scanning electron microscope to be about the same as the thickness of the section, thus in the range of 1 μm to less than 0·1 μm with the detection limits being as little as 10^{-18} g (cf. Russ, 1971d). These projections were actually confirmed for animal specimens (Sutfin et al., 1971 ; Thurston et al., 1972), but corresponding analyses of plant tissues are still lacking. The analytical capabilities of the scanning electron microscope are augmented when a wavelength-dispersive detector is added to the system, since this improves its sensitivity (Kimoto et al., 1970), especially for light elements. Biologists may choose the new method of adapting the scanning electron microscope with energy- and wavelength-dispersive detectors since this system extends EPA to ultramicroscopic dimensions.

The ultimate instrument for elemental analyses in microscopic dimensions may turn out to be the ion microprobe mass analyser in which a high-energy beam of ions is used to irradiate the sample. It allows measurement of isotopes and, thus, analyses not possible previously. Its principles were described by Bayard (1971) and by Andersen and Hinthorne (1972). Very little information exists as to its applications to biological specimens.

V. Application to Ion Transport Studies: Present and Future

The present state of the art of applying conventional EPA to ion transport studies was reviewed by Läuchli (1971, 1972b). Thus far, most studies have

TABLE I. Lateral distribution of potassium in roots of *Zea mays* (data from Läuchli, 1972a)

Lateral distribution[a]					
Ep	Co	En	Pe	Pa	XyP
16	6	10	18	17	39

[a] Electron probe analysis: K (K_a)—average relative X-ray intensities. Ep: epidermis; Co: mid-cortex; En: endodermis; Pe: pericycle; Pa: unspecified stelar parenchyma; XyP: xylem parenchyma.

been on translocation of ions in higher plants, and EPA actually supplemented results obtained by microautoradiography and led to new insights into the pathways and mechanisms of ion translocation (Läuchli, 1972a). Electron probe analysis proved particularly suitable to research on the lateral transport of ions into the xylem of roots. Table I shows the average lateral distribution of K^+ in roots of *Zea mays*, 1 cm from the apex, after absorption of K^+ from a solution of 0.2 mM KCl + 0.5 mM $CaSO_4$ for 4 h (see Läuchli *et al.*, 1971, for further details). Obviously, the cells of the xylem parenchyma adjoining the vessels in the stele accumulated K^+ to a much greater extent than those of any other tissue in these roots and thus may be the sites of ion secretion from the symplasm into the xylem vessels (Läuchli, 1972a).

Another area where experiments involving EPA advanced our physiological knowledge is that of transport of K^+ into and out of cells which control movements through changes in turgor. Examples are the movements of stomata (e.g. Humble and Raschke, 1971) and nyctinasty (Satter *et al.*, 1970).

In recent years, compartmentation of ions in plant cells received widespread attention. The objective is to understand intracellular transport of ions which comprises such processes as membrane transport, ion movement through the cytoplasm, influx and efflux of ions in cell organelles, and so forth. Electron probe analysis has not yet contributed much to this area of ion transport research. The use of the conventional electron probe analyser and of relatively thick samples in the majority of EPA-studies do not yield spatial resolutions sufficiently high to allow an accurate cellular compartmentation. In some of these investigations the emphasis was on distinguishing between the cytoplasmic phase including some organelles on the one hand and the nucleus on the other. Bajaj *et al.* (1971) found that the nuclei in isolated cells of *Phaseolus vulgaris* accumulated K, Ca, and P. Libanati and Tandler (1969) showed that the high concentration of P in the nucleus of plant cells was due principally to an accumulation in the nucleolus. No attempt was made in these experiments to separate the amounts of elements which were in the ionic form from the fraction bound to cellular constituents. The distribution of the ions Na^+, K^+, and Cl^- between the cell wall with the adhering cytoplasm and the vacuole was determined in the mesophyll cells of the halophyte *Suaeda monoica*, using EPA (Waisel and Eshel, 1971). When the plants were grown under saline conditions, the cytoplasm of the mesophyll cells appeared to absorb and retain high quantities of Na^+, whereas the contents of K^+ and Cl^- were low and these ions were distributed more uniformly in the cells.

The experimental combination of EPA with microautoradiography offers the unique possibility of measuring specific activities of labelled ions *in situ*. Läuchli and Lüttge (1968) set out the means by which this aim can be achieved.

Electron probe analysis has broad perspectives in research on compartmentation of ions in cells and on ion translocation. In the past, investigators expressed some disappointment about the gap existing between the prospective resolution of $0.1\,\mu m$ and that actually obtained with biological specimens. Nevertheless, one must bear in mind that thus far we have experimented essentially with relatively thick samples and the conventional electron probe analyser. Technical advancements in the preparation of specimens (e.g. cryo-ultramicrotomy) and in instrumentation (scanning and transmission electron microscope with wavelength- and energy-dispersive detectors) have now provided us with the tools to make EPA a significant method for investigation of ion transport in plants.

References

ANDERSEN, C. A. (1967). *Meth. biochem. Anal.* **15**, 147–270.
ANDERSEN, C. A. and HINTHORNE, J. R. (1972). *Science, N.Y.* **175**, 853–860.
BAJAJ, Y. P. S., RASMUSSEN, H. P. and ADAMS, M. W. (1971). *J. exp. Bot.* **22**, 749–752.
BAYARD, M. (1971). *Am. Lab.* **3**, 15–22.
BIRKS, L. S. (1971). "Electron Probe Microanalysis". 2nd edition. Wiley-Interscience, New York.
CHANDLER, J. A. (1971). *Am. Lab.* **3**, 50–60.
FISHER, D. B. (1972). *Pl. Physiol.* **49**, 161–165.
FISHER, D. B. and HOUSLEY, T. L. (1972). *Pl. Physiol.* **49**, 166–171.
FRANKEL, R. S. and AITKEN, D. W. (1970). *Appl. Spectrosc.* **24**, 557–566.
GAHAN, P. B., GREENOAK, G. C. and JAMES, D. (1970). *Histochemie* **24**, 230–235.
HALL, T. A. (1971). *In* "Physical Techniques in Biological Research", 2nd edition (G. Oster, ed.) Vol. I, Part A, pp. 157–275. Academic Press, New York.
HODSON, S. and MARSHALL, J. (1971). *J. Microsc.* **93**, 49–53.
HUMBLE, G. D. and RASCHKE, K. (1971). *Pl. Physiol.* **48**, 447–453.
KIMOTO, S., HASHIMOTO, H. and TAGATA, S. (1970). Proc. 5th Natl. Conf. Electron Probe Analysis, New York, 1970, pp. 51A–51C. Electron Probe Analysis Society of America.
LÄUCHLI, A. (1971). Proc. 6th Natl. Conf. Electron Probe Analysis, Pittsburgh, 1971, pp. 35A–35B. Electron Probe Analysis Society of America.
LÄUCHLI, A. (1972a). *A. Rev. Pl. Physiol.* **23**, 197–218.
LÄUCHLI, A. (1972b). *In* "Methods of Microautoradiography and Electron Probe Analysis" (U. Lüttge, ed.) pp. 191–236. Springer, Berlin.
LÄUCHLI, A. and LÜTTGE, U. (1968). *Planta* **83**, 80–98.
LÄUCHLI, A., SPURR, A. R. and WITTKOPP, R. W. (1970). *Planta* **95**, 341–350.
LÄUCHLI, A., SPURR, A. R. and EPSTEIN, E. (1971). *Pl. Physiol.* **48**, 118–124.
LIBANATI, C. M. and TANDLER, C. J. (1969). *J. Cell Biol.* **42**, 754–765.
LÜTTGE, U. and WEIGL, J. (1965). *Planta* **64**, 28–36.
MATSUO, T. and SUZUKI, M. (1971). Proc. 29th Ann. EMSA Meet., Boston, 1971, pp. 474–475. Claitor's Publ. Div., Baton Rouge.
MYKLEBUST, R. L. and HEINRICH, K. F. J. (1971). *ASTM Spec. Techn. Publ.* **485**, 232–242.
RASCHKE, K. and FELLOWS, M. P. (1971). *Planta* **101**, 296–316.
RUSS, J. C. (1971a). "Energy Dispersion X-Ray Analysis: X-Ray and Electron Probe Analysis". *ASTM Spec. Techn. Publ.* **485**, 154–179.

RUSS, J. C. (1971c). *ASTM Spec. Techn. Publ.* **485**, 217–231.
RUSS, J. C. (1971d). Proc. 29th Ann. EMSA Meet., Boston, 1971, pp. 54–55. Claitor's Publ. Div., Baton Rouge.
SATTER, R. L., MARINOFF, P. and GALSTON, A. W. (1970). *Am. J. Bot.* **57**, 916–926.
SAWHNEY, B. L. and ZELITCH, I. (1969). *Pl. Physiol.* **44**, 1350–1354.
SUTFIN, L. V. and OGILVIE, R. E. (1971). *ASTM Spec. Techn. Publ.* **485**, 197–216.
SUTFIN, L. V., HOLTROP, M. E. and OGILVIE, R. E. (1971). *Science, N.Y.* **174**, 947–949.
THURSTON, E. L. and RUSS, J. C. (1971). Proc. 4th Ann. Scanning Electron Microscope Symp., Chicago, 1971, pp. 511–516. IIT Research Institute, Chicago.
THURSTON, E. L., RUSS, J. C., TEIGLER, D. J. and ARNOTT, H. J. (1972). Proc. 30th Ann. EMSA Meet., Los Angeles, 1972. Claitor's Publ. Div., Baton Rouge.
TOUSIMIS, A. J. (1963). *ASTM Spec. Techn. Publ.* **349**, 193–206.
WAISEL, Y. and ESHEL, A. (1971). *Experientia* **27**, 230–232.

I.2

Enzyme Localization and Ion Transport

J. L. Hall

*Department of Biological Sciences, University of Sussex,
England*

I. Introduction

There are now methods available for the localization of a wide range of enzymes by microscopical cytochemical techniques, although relatively few of these produce a product which may be studied by both light and electron microscopy. Enzymes which may be localized by these techniques, however, include various phosphatases, esterases, aryl sulphatase, peroxidase, catalase and certain dehydrogenases, and these methods have been utilized in relation to transport studies in plants in a number of ways.

Adenosine triphosphatase (ATPase) activity has been the subject of a number of cytochemical studies since this enzyme has been shown to be closely related to active cation transport in animal cells (Whittam and Wheeler, 1970). The possible role of this activity in ion transport in plant roots will be discussed thoroughly later in this paper. In relation to stomatal movement, Fugino (1967) has reported changes in the ATPase activity of guard cells in the light and dark using cytochemical techniques. Although Willmer and Mansfield (1970) found no changes between light and dark, they reported high levels of ATPase activity in guard cells and this may be associated with the transport of potassium into these cells which occurs during stomatal opening (Humble and Raschke, 1971). Again, largely on the basis of cytochemical studies, another phosphatase, acid phosphatase, has been implicated in the active transport of assimilates in the phloem (Braun and Sauter, 1964; Sauter, 1966); the enzyme is thought to be involved in phosphorylating and dephosphorylating processes associated with the active transport of carbohydrates. Finally, another application of enzyme cytochemistry is illustrated by the work of Yu and Kramer (1967) in which the techniques for the localization of dehydrogenase activity were used to estimate the relative respiratory activity of different tissues of the root in relation to their capacity for ion transport.

II. Problems with Enzyme Cytochemistry

There are a number of problems associated with the reliable localization of enzymes by microscopic techniques and it is perhaps worthwhile to summarize the most important of these in the study of any plant tissue.

A. Fixation

Fixation is normally required for the maintenance of cellular structure, particularly at the ultrastructural level, although fresh and frozen tissue has sometimes been used for light microscopy. Chemical fixation may have profound effects on enzymic activity and it is important to determine biochemically the degree of inactivation produced by fixation. In general, mild glutaraldehyde fixation (1–3 % glutaraldehyde for 1–2 h in the cold) seems satisfactory for a number of enzymes and tissues.

B. Penetration of Staining Medium

Frequently blocks of tissue such as segments of root tips have been used for cytochemical staining but it has now been shown that penetration of plant tissues is slow and leads to differential staining of the tissue block (Hall, 1969; Sexton *et al.*, 1971; Hall and Sexton, 1972). This problem has been illustrated in relation to β-glycerophosphatase and peroxidase activity in which whole root tips and tissue blocks were incubated in the various staining media for periods of up to 2 h. Large areas of the tissue are unstained and there is a differential permeability in different regions of root, the root cap usually being the most rapidly penetrated. The substrates employed in these techniques are normally of quite low molecular weight. For example, 3,3-diaminobenzidine and sodium β-glycerophosphate which are used as substrates for peroxidase and acid phosphatase have molecular weights of 214·3 and 315·1 respectively. Many of the inhibitors, particularly antibiotics, used in ion transport studies are of considerably larger molecular weight and careful consideration should be given to the extent of penetration of these compounds, especially in short uptake experiments. For reliable enzyme cytochemistry thin sections must be used for incubation in the staining medium; 20 μm cryostat- or vibrotome-cut sections have proved to be the most useful in this respect.

C. Leakage of Enzyme Activity

The necessity of using thin sections to overcome the stain penetration problem may result in the loss of some activity from the sections (Sexton *et al.*, 1971). Although pre-fixation in glutaraldehyde reduces leakage from the sections there is still a considerable loss which must be carefully considered in the interpretation of results.

D. Controls

In any cytochemical procedure it is important to run adequate controls to ensure that the staining is a result of enzyme activity, that the enzyme substrate or product have not diffused or been differentially adsorbed, and to determine the specificity of the staining reaction. The use of controls, particularly in relation to phosphatase staining has been discussed thoroughly before (Sexton *et al.*, 1971).

III. ATPase Activity and Cytochemistry in Animal Cells

ATPase activity has been widely studied in relation to ion transport processes and it is convenient that it may be readily assayed biochemically and studied cytochemically by both light and electron microscopy. The interest in ATPase activity stems from work on animal cells which has established a close association between active cation transport utilizing ATP and membrane-bound ATPase activity. This enzyme requires sodium and potassium ions together in addition to magnesium for maximal activity and is inhibited by cardiac glycosides such as ouabain which are very effective inhibitors of cation transport.

The value of enzyme cytochemistry in transport studies is well illustrated by the work of Palade and associates. Marchesi and Palade (1967) used the standard ATPase stain developed by Wachstein and Meisel (1957) to demonstrate that inorganic phosphate is released from ATP on the inside and not the outside of the membranes of mammalian red cell ghosts. In another study, Farquhar and Palade (1966) examined ATPase activity in the epidermis of various amphibians. Staining was found at all cell membranes facing the intercellular spaces and absent from membranes at cell junctions and from those bordering the external medium or the dermis. These findings were interpreted in terms of the presence of sodium pumps along all cell membranes facing the intercellular spaces and result in the movement of sodium across an epithelium from the external medium to the interstitial fluid. A third example is provided by Ryan and Smith (1971) who used cytochemical procedures in the isolation and identification of pinocytotic vesicles from lung tissue. In this tissue the enzyme 5′-nucleotidase which may be localized by a lead phosphate precipitation procedure similar to that used for ATPase staining is restricted to the plasmalemma and pinocytotic vesicles. Homogenates of lung tissue were incubated with the enzyme substrate (AMP) in the presence of lead ions resulting in the precipitation of lead phosphate at the enzymic sites. This led to an increase in density which enabled the vesicles to be separated from the other cellular components by centrifugation and identified in the resulting pellet by the presence of electron-dense lead phosphate deposits.

It should be noted however that some doubt has been raised concerning the validity of the Wachstein–Meisel procedure used in these studies and is based largely on the report that lead ions catalyse the non-enzymatic breakdown of ATP (Moses and Rosenthal, 1968). This finding has been criticized by Novikoff (1970) who reported that little non-enzymatic hydrolysis of the substrate occurred at the temperature and time normally employed for the incubations and concluded that the staining observed represents true enzymic activity. A similar conclusion was reached in relation to the conditions used in plant cytochemical studies (Hall, 1969). Recently an alternative method has been developed for the demonstration of a K-dependent, ouabain-sensitive phosphatase activity which uses strontium as the capture ion for hydrolysed phosphate (Ernst, 1972). This procedure may overcome many of the problems associated with lead ion cytochemistry and it will be interesting to follow the results of its application to both animal and plant tissues.

IV. Plant ATPase Activity

The evidence for a role of ATPase activity in ion transport in higher plant cells is less clear cut. Indeed, the source of energy for active ion transport has not been clearly established. Although it is widely accepted that transport is linked to respiration, this could result from a dependence on ATP or some other high energy intermediate, or be due to a direct coupling to the processes of charge separation and electron flow in the respiratory chain. These possibilities have been discussed recently by Epstein (1972) who concluded that although the argument has not been fully resolved, the weight of evidence favours a phosphorylation mechanism. This conclusion is supported by some recent evidence from Lüttge et al. (1971) who showed that since carrot tissue exhibited no salt respiration in the presence of the uncoupler Cl-CCP, this respiration could not be linked to electron transport. In addition, work on giant algal cells has demonstrated that the transport of sodium and potassium is dependent on ATP although chloride uptake may be linked to electron flow (MacRobbie, 1966; Raven et al., 1969).

The considerable interest and importance of animal transport ATPase activity has stimulated a search for a similar system in higher plant cells but, although a series of papers have been published describing the characteristics of plant ATPase activity, the results have been largely inconclusive.

Plant ATPases show a number of general features in common with the animal enzyme (see Fig. 1a, b). They usually have a pH optimum of about 7–8 and require magnesium or other divalent cations for maximal activity

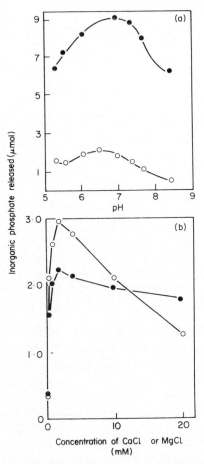

Fig. 1. Effect of pH and divalent cations on the hydrolysis of ATP by cell wall fractions from barley roots. (a) Effect of pH in the presence (●) and absence (○) of $MgCl_2$. (b) Effect of divalent cations in the presence of 2 mM ATP at pH 7·6 and $CaCl_2$ (○) or $MgCl_2$ (●) at the concentrations indicated. Reproduced from Hall and Butt (1969).

(Dodds and Ellis, 1966; Greuner and Neumann, 1966; Hall and Butt, 1969; Sexton and Sutcliffe, 1969; Fisher and Hodges, 1969; Kylin and Gee, 1970). This basic divalent cation-stimulated activity is generally further stimulated by sodium, potassium and other monovalent cations applied singly (Fig. 2). There is little evidence of a synergistic effect between sodium and potassium ions, which is an important characteristic of the animal enzyme, and the

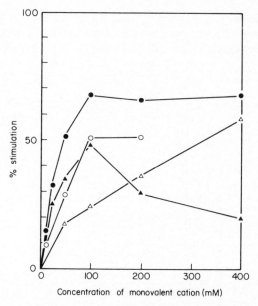

FIG. 2. Effect of monovalent cations on ATP hydrolysis by cell wall suspensions from barley roots. Cell wall suspensions were incubated at pH 9·0 with 2 mM ATP, 2 mM $CaCl_2$ and NaCl (●), KCl (○), RbCl (△) or LiCl (▲) at the concentrations indicated. Results are expressed as percentage stimulation above the activity with no added monovalent cation. Reproduced from Hall (1971a).

activity due to monovalent cations is not inhibited by cardiac glycosides. However, it must be noted that a combined sodium and potassium effect has been reported in some plant systems (Kylin and Gee, 1970; Lai and Thompson, 1971), although the effects are much less marked than found for the animal enzyme and the latter report is unusual in that the maximum sodium-potassium effect is seen in the absence of divalent cations.

Although at first hand these results seem somewhat disappointing there are a number of important features of the plant system which must be considered. Firstly, although monovalent cations have been shown to stimulate a number of enzymes from plant tissues (Evans and Sorger, 1966) it is generally found that, in contrast to ATPase activity, potassium is the most effective ion, the concentration required for maximum activation is lower, and sodium and lithium ions are frequently inhibitory. Secondly, although there is some evidence of a coupled sodium-potassium pump at the plasmalemma of barley roots (Pitman and Saddler, 1967; Jeschke, 1970), this does not appear to be a common feature of ion transport in higher plant cells which, in addition, is largely insensitive to cardiac glycosides (Epstein, 1972). Therefore it is not surprising that a sodium-potassium stimulated ATPase activity that is inhibited by cardiac glycosides has not been found in higher plant tissues. Thirdly, the situation in plants may be confused by the difficulty of preparing suitable membrane material. Generally the ATPase work has been carried out using cell wall fractions or general homogenates or supernatant fractions. Disc electrophoresis has shown that plant ATPase activity exists in multiple forms or isoenzymes (Sexton and Sutcliffe, 1969; Hall, 1971) and both the cell wall fraction (Fig. 3) and the general supernatant

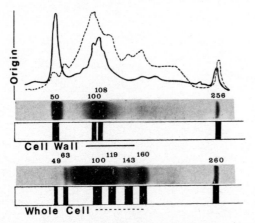

FIG. 3. Electrophoretic separation of ATPase activity from maize roots. Disc electrophoresis of ATPase activity from whole cell homogenates and cell wall preparations showing gels, densitometer tracings and line drawings of gels. The methods were similar to those described by Hall (1971a). Data provided by M. L. Edwards.

are likely to contain a number of these forms. The characteristics of the transport ATPase may be masked by other ATPases which participate in a variety of functions in living cells. The attempt by Lai and Thompson (1971) to prepare purified plasma membrane fractions from plant cells is an important step towards the investigation of transport ATPase activity in higher plants.

Finally, some more positive findings should be mentioned. Fisher *et al.* (1970) have demonstrated that in a number of species and over a range of ion concentrations there is a correlation between ion transport in roots and ion stimulated ATPase activity. Lai and Thompson (1972) have shown a similar correlation between the requirement for the transport of materials from storage cells and the levels of basal and cation-stimulated ATPase activity from *Phaseolus* cotyledons.

V. Cytochemistry of Plant ATPase

The method normally used for the localization of ATPase activity in plant tissues is that described by Wachstein and Meisel (1957) or a modification

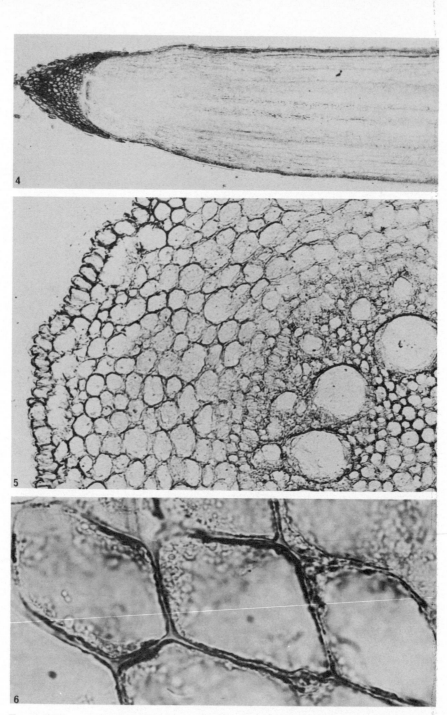

FIGS 4–6. Frozen sections of maize roots stained for ATPase activity. FIG. 4. L.S. showing heavy staining in the root cap and at the root surface. × 36. FIG. 5. T.S. cut 3·5 mm from tip showing high activity at the root and cell surface. × 178. FIG. 6. Root cap cells showing high activity at the cell surface. × 1152. Reproduced from Hall (1969).

of this method. It is clearly desirable to modify the composition of the staining medium in relation to the biochemical characteristics of the enzyme under investigation. Hall (1969, 1970, 1971a, b) has used a medium containing 2 mM ATP, 2 mM $Ca(NO_3)_2$ and 3·6 mM $Pb(NO_3)_2$ buffered at pH 7·0 with 48 mM tris maleate which is based on the characteristics of cell wall ATPase activity. Using this staining medium with cryostat-cut frozen sections high activity was reported at the surface of maize and barley roots (Hall, 1969) although it must be noted that the root surface is highly active for a number of enzymes (Chang and Bandurski, 1964; Hall and Davie, 1971) (Figs 4, 5, 6).

The presence of ATPase activity in cell wall preparations suggested that this was also a property of the cell surface although it was not clear if the ATPase was present in the plasmalemma or cell wall proper. Light microscopy suggested that the enzyme was located in the cell membrane (Fig. 6) and this was confirmed by electron microscopy which showed heavy deposits of lead phosphate in the position of the plasmalemma and plasmodesmata (Figs 7, 9, 10). This staining contrasts with the characteristic staining pattern found for two other surface enzymes, acid phosphatase and peroxidase, which showed highest activity throughout the cell wall and in the inter-cellular regions respectively (Sexton et al., 1971; Hall and Sexton, 1972). This characteristic localization satisfies one requirement of a transport ATPase system which should be located in the plasmalemma. High activity was also observed in the membrane of the central vacuole (Hall, 1971b) although acid phosphatase staining is also found at this site and may be associated with the lysosomal activity of the vacuoles (Hall and Davie, 1971).

In addition to high activity in the cell membrane, ATPase activity was found associated with surface vesicles in maize root cells (Figs 7, 8, 9, 10) and cells of Cucumis roots (Poux, 1967). Similarly stained vesicles have been reported in animal cells (de-Thé, 1968), and it has been suggested that membrane vesiculation or pinocytosis may be involved in active ion transport into animal and plant cells (Bennett, 1956; Sutcliffe, 1962). Membrane vesiculation would carry ions into the cell where the vesicles would either release the ions into the cytoplasm or fuse with the tonoplast, releasing ions into the vacuoles. The energy for the active transport of ions by this process may be provided by the hydrolysis of ATP by an ATPase system located in the vesicular membrane (Hall, 1970). The question of the selectivity of ion accumulation may be explained if this process is simply considered as a mechanism for the transport of specific binding molecules across the cell membrane. Ions bind at specific sites at the outer surface of the plasmalemma and are moved into the cell by membrane vesiculation. Presumably some non-specific transport may occur by this process, particularly at high external ion concentrations.

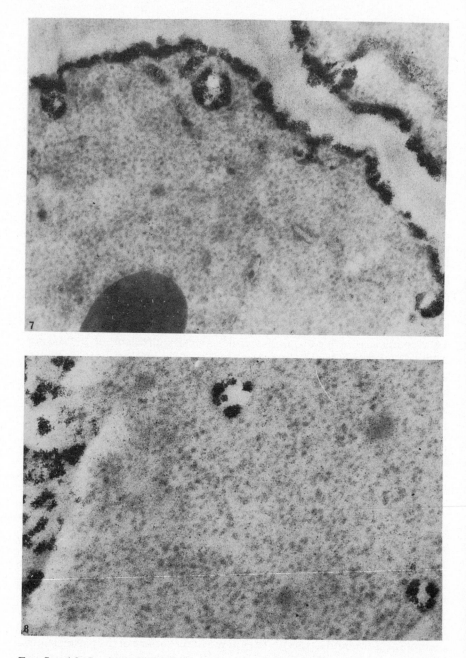

FIGS 7 and 8. Cortical cells of maize roots incubated for ATPase activity showing high staining in the plasmalemma and in small vesicles. FIG. 7. × 51,000. FIG. 8. × 72,000. Reproduced from Hall (1970).

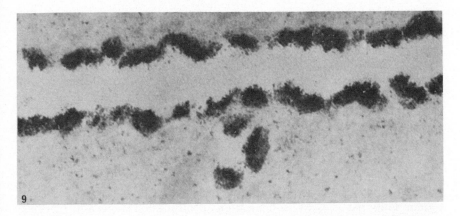

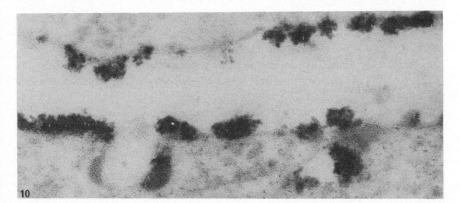

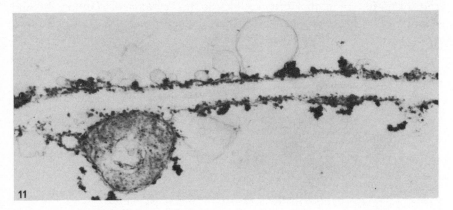

FIGS 9–11. Vesiculation in plant root cells. FIGS 9, 10. Maize cortex cells stained for ATPase showing high activity in the plasmalemma and in surface vesicles. FIG. 9. × 100,000. FIG. 10. × 140,000. FIG. 11. Vacuolated cells from *Suaeda maritima*. × 24,000.

There is now considerable evidence that membrane vesiculation and pinocytosis may be a widely occurring phenomenon in living cells. Microscopic techniques have demonstrated the constant movement, infoldings and invaginations of animal cell membranes and pinocytotic vesicles have been reported in a variety of animal cells (Holter, 1960; Coleman and Finean, 1968). In addition, there have been reports of the isolation of pinocytotic vesicles by centrifugation (Ryan and Smith, 1971).

Electron microscopy has also provided evidence of membrane vesiculation in plant cells (Buvat, 1963; Mahlberg et al., 1971; Roland and Vian, 1971) although it is not possible, of course, to deduce the direction of vesicle movement (Fig. 11). Clear evidence of a pinocytotic mechanism has come from the study of the uptake of large molecules such as ferritin and polystyrene latex particles by isolated plant protoplasts (Mayo and Cocking, 1969; Power and Cocking, 1970). These studies show that pinocytosis is associated with certain active areas of the cell membrane and it is interesting that ATPase staining may also show an uneven pattern at the plasmalemma (Hall, 1970). Other interesting observations have come from the study of plant glands. Gland cells frequently contain numerous microvesicles (Lüttge, 1971) which are sometimes closely associated with the plasmalemma (Thomson and Liu, 1967). Thomson et al. (1969) showed that the salt glands of Tamarix plants that were fed with rubidium contained accumulations of the electron-dense material in the microvacuoles of the secretory cells and postulated that these vacuoles are involved in the secretion of salt from the secretory cells.

These cytochemical observations lend considerable support to certain kinetic studies of ion transport in both algal and higher plant cells which suggest that at least part of the ion uptake into the vacuole is associated with the formation of pinocytotic vesicles in the cytoplasm (Gutknecht and Dainty, 1968; Minchin and Baker, 1969; Pallaghy et al., 1970; MacRobbie, 1971).

Acknowledgement

I should like to thank Dr D. A. Baker and Dr R. Sexton for helpful discussions during the preparation of this manuscript.

References

BENNETT, H. S. (1956). *J. biophys. biochem. Cytol.* **2**, 99–103.
BRAUN, H. J. and SAUTER, J. J. (1964). *Planta* **60**, 543–557.
BUVAT, R. (1963). *Int. Rev. Cytol.* **14**, 41–155.
CHANG, C. W. and BANDURSKI, R. S. (1964). *Pl. Physiol.* **39**, 60–64.
COLEMAN, R. and FINEAN, J. B. (1968). *In* "Comprehensive Biochemistry" (M. Florkin and E. H. Stotz, eds) Vol. 23, p. 99. Elsevier, Amsterdam.
DODDS, J. J. A. and ELLIS, R. J. (1966). *Biochem. J.* **101**, 31P.

EPSTEIN, E. (1972). "Mineral nutrition of plants. Principles and Perspectives". John Wiley, New York.

EVANS, H. J. and SORGER, G. J. (1966). *A. Rev. Pl. Physiol.* **17**, 47–76.

ERNST, S. A. (1972). *J. Histochem. Cytochem.* **20**, 23–38.

FARQUHAR, M. G. and PALADE, G. E. (1966). *J. Cell Biol.* **30**, 359–379.

FISHER, J. and HODGES, T. K. (1969). *Pl. Physiol.* **44**, 385–395.

FISHER, J. D., HANSEN, D. and HODGES, T. K. (1970). *Pl. Physiol.* **46**, 812–814.

FUGINO, M. (1967). *Sci. Bull. Fac. Educ. Nagasaki Univ.* **18**, 1–47.

GREUNER, N. and NEUMANN, J. (1966). *Physiologia Pl.* **19**, 678–682.

GUTKNECHT, J. and DAINTY, J. (1968). *A. Rev. Oceanogr. mar. Biol.* **6**, 163–200.

HALL, J. L. (1969). *Planta* **89**, 254–265.

HALL, J. L. (1970). *Nature, Lond.* **226**, 1253–1254.

HALL, J. L. (1971a). *J. exp. Bot.* **22**, 800–808.

HALL, J. L. (1971b). *J. Microsc.* **93**, 219–225.

HALL, J. L. and DAVIE, C. A. M. (1971). *Ann. Bot.* **35**, 849–855.

HALL, J. L. and SEXTON, R. (1972). *Planta* **108**, 103–120.

HALL, J. L. and BUTT, V. S. (1969). *J. exp. Bot.* **20**, 751–752.

HOLTER, H. (1960). *Int. Rev. Cytol.* **8**, 481–504.

HUMBLE, G. D. and RASCHKE, K. (1971). *Pl. Physiol.* **48**, 447–453.

JESCHKE, W. D. (1970). *Planta* **94**, 240–245.

KYLIN, A. and GEE, R. (1970). *Pl. Physiol.* **45**, 169–172.

LAI, Y. F. and THOMPSON, J. E. (1971). *Biochem. biophys. Acta* **233**, 84–90.

LAI, Y. F. and THOMPSON, J. E. (1972). *Can. J. Bot.* **50**, 327–332.

LÜTTGE, U. (1971). *A. Rev. Pl. Physiol.* **22**, 23–44.

LÜTTGE, U., CRAM, W. J. and LATIES, G. G. (1971). *Z. Pflanzenphysiol.* **64**, 418–426.

MACROBBIE, E. A. C. (1966). *Aust. J. biol. Sci.* **19**, 363–370.

MACROBBIE, E. A. C. (1971). *J. exp. Bot.* **22**, 487–502.

MAHLBERG, P., OLSON, K. and WALKINSHAW, C. (1971). *Am. J. Bot.* **58**, 407–416.

MARCHESI, V. T. and PALADE, G. E. (1967): *J. Cell. Biol.* **35**, 385–404.

MAYO, M. A. and COCKING, E. C. (1969). *Protoplasma* **68**, 223–230.

MINCHIN, F. R. and BAKER, D. A. (1961). *Planta* **89**, 212–223.

MOSES, H. L. and ROSENTHAL, A. S. (1968). *J. Histochem. Cytochem.* **16**, 530–539.

NOVIKOFF, A. B. (1970). *J. Histochem. Cytochem.* **18**, 366–368.

PALLAGHY, C. K., LÜTTGE, U. and VON WIILLERT, K. (1970). *Z. Pflanzenphysiol.* **62**, 51–57.

PITMAN, M. G. and SADDLER, H. D. W. (1967). *Proc. natn. Acad. Sci. U.S.A.* **57**, 44–49.

POWER, J. B. and COCKING, E. C. (1970). *J. exp. Bot.* **21**, 64–70.

POUX, N. (1967). *J. Microsc.* **6**, 1043–1058.

RAVEN, J. A., MACROBBIE, E. A. C. and NEUMANN, J. (1969). *J. exp. Bot.* **63**, 221–235.

ROLAND, J. C. and VIAN, B. (1971). *Protoplasma* **73**, 121–137.

RYAN, J. W. and SMITH, U. (1971). *Biochim. biophys. Acta* **249**, 177–180.

SAUTER, J. J. (1966). *Z. Pflanzenphysiol.* **55**, 349–362.

SEXTON, R. and SUTCLIFFE, J. F. (1969). *Ann. Bot.* **33**, 683–694.

SEXTON, R., CRONSHAW, J. and HALL, J. L. (1971). *Protoplasma* **73**, 417–441.

SUTCLIFFE, J. F. (1962). "Mineral Salts Absorption in Plants". Pergamon Press, Oxford.

DE-THÉ, G. (1968). *In* "The Membranes" (A. J. Dalton and F. Hagueman, eds) p. 121–150. Academic Press, New York and London.

THOMSON, W. W. and LIU, L. L. (1967). *Planta* **73**, 201–220.

THOMSON, W. W., BERRY, W. L. and LUI, L. L. (1969). *Proc. natn. Acad. Sci. U.S.A.* **63**, 310–317.

WACHSTEIN, M. and MEISEL, E. (1957). *Am. J. Clin. Path.* **27**, 13–23.

WILLMER, C. M. and MANSFIELD, T. A. (1970). *New Phytol.* **69**, 983–992.

WHITTAM, R. and WHEELER, K. P. (1970). *A. Rev. Physiol.* **32**, 21–60.

YU, G. H. and KRAMER, P. J. (1967). *Pl. Physiol.* **42**, 985–990.

I.3

Ultrastructural Localization of Ions

R. F. M. Van Steveninck, A. R. F. Chenoweth and M. E. Van Steveninck

Botany Department, University of Queensland, St. Lucia, Australia

I. Introduction

Attempts to provide a direct and visual demonstration of ions in tissues and cells by means of specific reagents which cause *in situ* precipitation of the ions date back more than a hundred years (for a review of early literature see Macallum 1905a). Three reagents gained prominence: sodium cobaltinitrite to show the presence of K^+ (Macallum, 1905b), potassium pyroantimonate to show the presence of Na^+ (Komnick, 1962), and soluble silver salts such as silver nitrate, silver acetate and silver lactate to show the presence of Cl^- (Macallum, 1905a; Komnick, 1962).

During the early 1960s, rapid developments took place in the adaptation of light microscopy methods for use in electron microscopy. Greater resolution in the detection of small deposits can now be achieved but at the same time a much improved preservation of ultrastructural detail is required to make this increased resolution meaningful. It appears that the 1970s will see further major advances in ion detection with the use of electron microscopes with high resolution X-ray energy dispersive analysis facilities (Reuter, 1971), Auger electron microscopy with low energy electron diffraction (LEED) facilities (Chang, 1971) and ion microscopy with mass spectrometer facilities (Muller and Tsong, 1969; Socha, 1971). Although these techniques should eventually offer the ultimate in specificity of detection one should not dismiss the chemical precipitation methods, because of their great sensitivity and resolution, and also because possible improvements in the preparative techniques have not yet been fully exploited. But, a possible lack in specificity dictates that precipitation methods should be used in conjunction with the more recently developed methods of analysis.

The specificity of K^+-pyroantimonate as a reagent for Na^+ has often been questioned (Tandler *et al.*, 1970; Torack and La Valle, 1970; Sumi and Swanson, 1971; Clark and Ackerman, 1971), but in other cases the identity of deposits has been confirmed by electron probe microanalysis (Lane and Martin, 1969). The use of Na^+-cobaltinitrite although applied successfully for quantitative estimates of K^+ in light microscopy (Fisher, 1971) has not been readily adapted to electron microscopy because the reagent interferes with structural preservation. Although, we believe, we may soon succeed in modifying the Macallum method for use in electron microscopy, we will restrict this report to the Ag precipitation method for Cl^- detection as adapted by Komnick (Komnick and Bierther, 1969).

II. Scope, Resolution and Limitations of Ultrastructural Chloride Determinations

A. Scope

Basically, the method involves the fixation (usually for 1 h at $0°C$) of small pieces of tissue in 2% osmium tetroxide (OsO_4) in 0.1 M cacodylate acetate buffer (pH 7.2) to which either 0.5% silver acetate or silver lactate has been added. This is followed by 3 brief rinses in small volumes of 0.1 M cacodylate acetate (pH 7.2). These steps should be performed with a minimum exposure to weak red light. Dehydration is carried out using an acetone series and propylene oxide, before embedding in Araldite or Spurr's medium. As an additional precaution it is advisable to check on the specificity of the Ag deposits by including 0.1 N HNO_3 in the 50% acetone stage of the dehydration in order to remove any non-specific deposits of Ag-phosphate and Ag-carbonate (Komnick and Bierther, 1969). If the size and shape of the AgCl deposits is uniform, it is possible to estimate the amount of Cl^- each deposit represents, and provided sections are of uniform thickness, the approximate concentration of Cl^- in a particular region of a cell can be determined, e.g. a sphere of approximately 25 nm diameter has a volume of $\frac{4}{3} \times \pi \times (0.0125)^3 \, \mu m^3 = 8.2 \times 10^{-6} \, \mu m^3$ $(= V_{AgCl})$. This represents $(V_{AgCl} \times \delta_{AgCl})/MW_{AgCl} = 0.32 \times 10^{-6}$ pequiv Cl^-, in which δ_{AgCl} is the density of AgCl (5.46 g/ml) and MW_{AgCl} its molecular weight ($= 143$).

Assuming a section thickness of 80 nm (uniform "silver" sections) the area of $1 \, \mu m \times 1 \, \mu m$ on an electron micrograph represents $0.08 \, \mu m^3$. The concentration of Cl^- represented by a deposit of 25 nm diameter in $1 \, \mu m^2$ of section will then be $(0.32 \times 10^{-18})/0.08 \times 10^{-15} = 4 \times 10^{-3}$ M or 4 mM. Grain counts in a number of randomly chosen areas of $1 \, \mu m^2$ then allow calculation of approximate overall chloride concentrations.

B. Resolution

Deposits 4–5 nm in diameter were routinely detected in the cytoplasmic phase of leaf mesophyll cells of a river mangrove (*Aegiceras corniculatum* Blanco.) and of a floating hydrophyte (*Nymphoides indica* Mill.). These deposits represent a quantity of 2.56×10^{-9} pequiv Cl⁻ which is only 1550 AgCl molecules. At present it should be possible to resolve deposits of 1 nm diameter if appropriate precautions are taken to avoid solubilization of these deposits during subsequent steps of dehydration and embedding. Thus it should be possible to study mechanisms where clusters of the order of 10–20 Cl⁻ ions are involved. The possibility of achieving this goal is controlled, amongst a number of factors, by the solubility product (K_{AgCl}) and the ionic densities required for the interaction of AgCl molecules to form a core of deposit. The 0·5 % Ag acetate solution used in the fixative represents 30 mM. The solubility product, $K_{AgCl} = 0.2 \times 10^{-10}$ at 2°C, hence precipitation should commence at Cl⁻ concentrations above 7×10^{-7} mM. At this limit of detection, which incidentally is 7 orders of magnitude lower than the average concentrations measured in most plant cells, Cl⁻ ions in random distribution are at an average distance of $1.35\ \mu m$ removed from each other ($\cdot 425$ Cl⁻$/\mu m^3$). Hence, at the concentration when detection should just become possible, considerable diffusion is necessary before sufficient interaction can occur to form an AgCl deposit. On the other hand, Cl⁻, electrostatically associated with a carrier, would most likely be present at a greater density and therefore it seems feasible ultimately to pinpoint Cl⁻ at a carrier site with the minimum of diffusional effects. It is also essential that diffusion should be kept to a minimum for high resolution since the progressive AgCl crystal growth at the expense of the smaller deposits may be expected. It should be worth investigating whether briefest possible periods of fixation, dehydration and embedding make it possible to preserve very small deposits. Abbreviated methods of specimen preparation have been published recently (Bain and Gove, 1971) and appear to give a quality of preservation equal to conventional methods. Alternatively, very small deposits might be preserved using methods involving rapid freezing and cryostat sectioning.

C. Limitations

As has already been indicated diffusional effects are a major problem. Particularly when the size of deposits varies greatly, artifacts are likely to result. Komnick and Bierther (1969) carried out model experiments using gelatin blocks containing NaCl and as a result of these experiments warned against the tendency for existing crystals to grow at the expense of smaller crystals, and the final dissolution of very small AgCl deposits during subsequent steps of specimen preparation. They claimed that if a 1 mm³ block of tissue is rinsed in 6 ml of H_2O at room temperature, all AgCl equivalent to a concentration of 56×10^{-3} M Cl^- would be removed from the tissue. We rinsed about 10 mm^3 tissue at a time in less than 0·6 ml buffer solution very briefly at 2°C, which represents $\frac{1}{700}$ of the solubilizing capacity provided by Komnick and Bierther's example. Although only a relatively small proportion of the AgCl (equivalent to less than 10^{-4} M Cl^-) may be removed from the tissue during this brief rinse, it is possible that the finest deposits of less than 5 nm may disappear as a result of this and subsequent steps.

Another problem is the possibility of formation of non-specific deposits of Ag. Komnick and Bierther (1969) provided evidence of the authenticity of the AgCl deposits by means of electron diffraction patterns, which are difficult to obtain, partly because AgCl rapidly converts to metallic silver in the electron beam. But AgCl melts at a much lower temperature than metallic silver (melting points 455 and 960·5°C respectively) and can thus be recognized (Komnick and Bierther, 1969). These authors also claim that the formation of metallic silver deposits by argentophilic structures is prevented by the presence of a strong oxidant (OsO_4) during the exposure of the tissue to the silver reagent. We carried out preliminary experiments on tissues which were fixed for 2 h and rinsed in buffer overnight to remove all Cl^- and then exposed to either silver acetate alone or a mixture of silver acetate and OsO_4 (Part III, Figs 9 and 10); these experiments showed that Komnick and Bierther's claim is also substantially correct for plant tissues. In the absence of OsO_4, silver acetate gives deposits which are much heavier especially in the chloroplasts than any observed previously, while in the presence of OsO_4 deposits are practically absent. Also the Ag deposits obtained when OsO_4 was not present in the medium have a different appearance, i.e. they lack the angular shape of deposits commonly observed when Cl^- is present and often display a halo of finely divided Ag-grains around them (Part III, Fig. 9).

III. Localization of Cl^- in High- and Low-salt Barley Seedlings

Barley seedlings were raised under sterile conditions in 8 in. long Pyrex tubes (1 in. diameter), the seeds being supported on wire loops immediately above the surface of half-strength Hoagland's solution, with or without the addition of 100 mM NaCl. After 7 days' growth in a Sherer growth cabinet providing 16 h photoperiods at a temperature of $26° \pm 2°C$ (day) and $18° \pm 2°C$ (night) and a light intensity of approximately 9000 lm m^{-2}, fresh weights were recorded, K^+, Na^+ and Cl^- contents were determined and samples taken for electron microscopy from halfway along the length of the primary leaves (after a further period of $12\frac{1}{2}$ hours in the dark, see Van Steveninck and Chenoweth, 1972). From the information provided in Table I it is evident that shoot growth was inhibited by 30 % and root growth

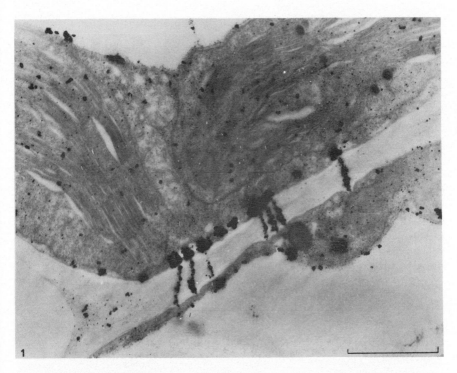

Fig. 1. Portion of leaf parenchyma cells of a high-salt barley seedling. Plasmodesmata are prominent because of heavy silver deposits inside them and in adjacent cytoplasm. Section unstained. Length of bar represents 1 μm.

little affected by the presence of 100 mM NaCl. The Na^+ and Cl^- contents of the shoots had doubled as a result of the high NaCl concentration in the medium. Electron micrographs of high salt plants generally showed a greater amount of AgCl deposited in the cytoplasm and chloroplasts than plants raised under low salt conditions. This was particularly evident in the plasmodesmata which contained very different amounts of deposit (Figs 1 and 2),

TABLE I. Ion concentrations of shoots (mequiv/kg fresh weight) and growth of 10-day barley seedlings raised on half-strength Hoagland solution in the presence or absence of 100 mM NaCl

	Shoot length (cm)	Shoot fresh weight (g)	Root fresh weight (g)	Concentration in shoots (mequiv/kg fresh wt)		
				Na^+	K^+	Cl^-
Control	15·8	0·20	0·084	75	94	8
+NaCl	11·3	0·14	0·080	158	69	18

TABLE II. Concentration of Cl^- in chloroplasts calculated on the basis of size and number of AgCl deposits in thin sections (~ 0.08 μm thick) of mesophyll cells of the primary leaf of barley seedlings (placed in the dark for $12\frac{1}{2}$ hours prior to fixation)

	Cl^- conc (mM)	Areas $(n)^a$
No NaCl	171 ± 4.0^b	9
+ NaCl	333 ± 8.7	9

^{a} n = number of areas of 1 μm^2 counted.
^{b} Standard error of mean.

and also in the quantitative assessment of Cl^- contents of the chloroplasts (Table II). In addition to the prominent deposits in and adjacent to the plasmodesmata, high salt plants occasionally exhibited distinct bands of very dense deposit 0.5–1 μm wide in the cell wall (Figs 3, 4 and 6). Except for those regions next to intercellular spaces where walls of neighbouring cells join (Fig. 5), little or no deposit occurred elsewhere in the cell wall.

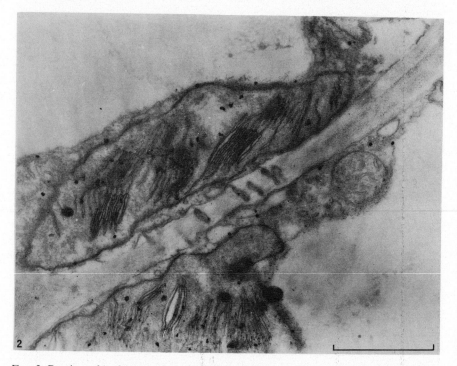

FIG. 2. Portion of leaf parenchyma cells of a low-salt barley seedling. Plasmodesmata contain little or no deposit. Section lightly stained with Pb citrate to assist in the detection of plasmodesmata which are otherwise difficult to find.

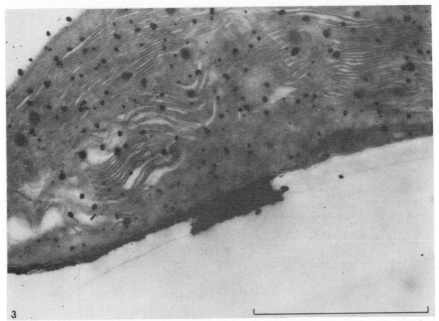

FIG. 3. High-salt barley seedling. Detail of wall region with heavy deposit. Section unstained.

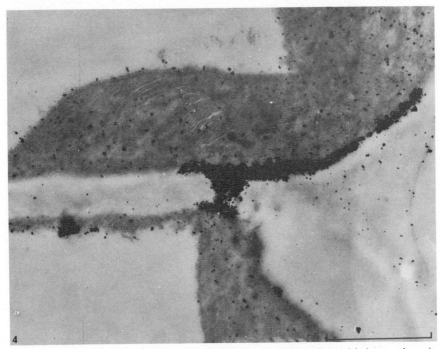

FIG. 4. High-salt barley seedling. Detail of another wall region with heavy deposit. Section unstained.

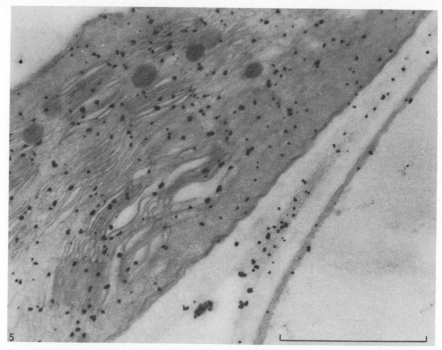

FIG. 5. High-salt barley seedling. Note presence of silver deposits on the cell wall region adjoining an intercellular space. Section unstained.

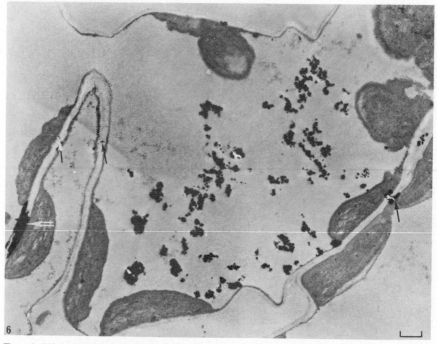

FIG. 6. High-salt barley seedling. Portion of a leaf parenchyma cell showing large central vacuole with dense silver deposits and plasmodesmata (small arrows). Note cell wall region with heavy deposit (large arrow). Section unstained.

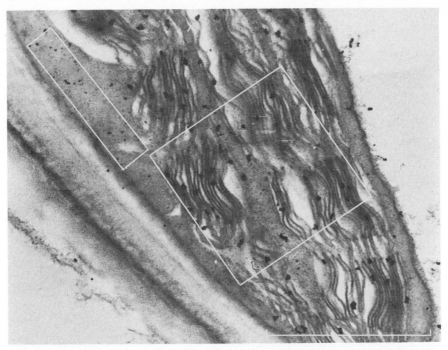

FIG. 7. High-salt barley seedling. Detail of deposits in chloroplast; note their angular shape. The larger deposits are mostly associated with the grana lamellae. The deposits enclosed in the boxed areas of the grana lamellae and of the stroma region represent concentrations of 350 mM Cl⁻ and 65 mM Cl⁻, respectively.

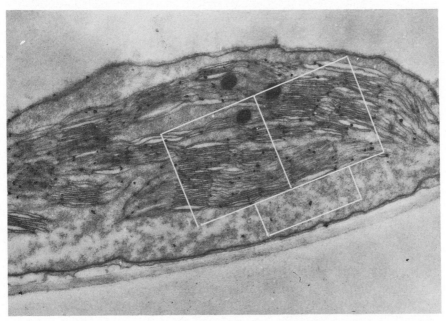

FIG. 8. Low-salt barley seedling. Detail of chloroplast. Deposits in the boxed area of the stroma represent a concentration of 29 mM Cl⁻, while the two areas of grana lamellae represent concentrations of 150 and 196 mM Cl⁻, respectively. Pb stain.

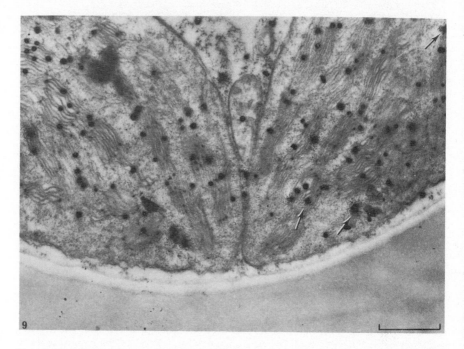

Fig. 9. High-salt barley seedling. Tissue fixed in OsO$_4$, rinsed in buffer overnight and subsequently placed in 0·5% Ag acetate in 0·1 M cacodylate acetate buffer for 2 h. Silver deposits do not have an angular shape and often display a halo of small Ag grains (arrows). Section unstained.

Deposits are absent from most vacuolar sections, but occasionally a cross section will reveal a large amount of deposit (Fig. 6). This non-random distribution pattern may be due to deposits' being swept to one side of the vacuole, probably by the penetrating embedding medium. Since many hundreds of sections can be cut in traversing one leaf parenchyma cell the relatively low occurrence of dense aggregates deposited in the vacuole is easily understood. It is clear that this type of irregular distribution will defy any attempt to assess the deposits in vacuoles quantitatively. The chloroplasts, on the other hand, always showed the presence of relatively intensive but well dispersed deposits of grains of uniform size (Fig. 7). When a distinct non-lamellar region was present, most of the larger grains were associated with the grana lamellae, while the stroma region showed fewer, smaller grains. It was calculated that these deposits would represent a concentration of 350 mM Cl$^-$ for the boxed areas (Fig. 7) with grana and stroma lamellae; 65 mM Cl$^-$ for the area without lamellae in a high salt plant; 150 and 196 mM Cl$^-$ for the two lamellar areas, and 29 mM for the stroma area in a low salt plant (Fig. 8).

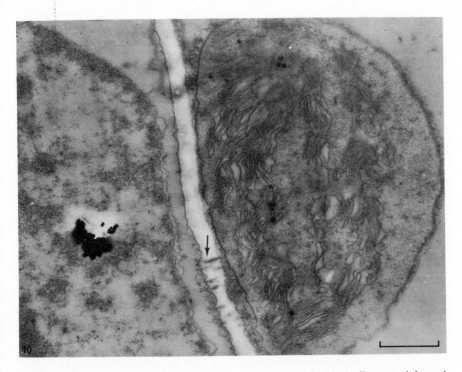

FIG. 10. High-salt barley seedling. Tissue fixed in OsO_4, rinsed in buffer overnight and subsequently exposed to 0·5% Ag acetate in 0·1 M cacodylate acetate + 2% OsO_4 for 2 h. Silver deposits are practically absent in chloroplasts or plasmodesmata (arrow). Occasionally large aggregates of silver were found randomly distributed, this particular one in the nucleus. Section unstained.

A. Removal of Cl^- after Fixation in OsO_4

Leaf tissue fixed for 2 h in OsO_4, was rinsed in relatively large volumes of 0·1 M cacodylate acetate buffer overnight in order to remove any unbound Cl^- from the tissue. Figure 9 shows the result obtained when the tissue was placed in 0·5% silver acetate in 0·1 M cacodylate acetate buffer (ph 7·2) in the absence of OsO_4, and Fig. 10 shows the same tissue exposed to silver acetate in the presence of 2% OsO_4. The numerous deposits shown in Fig. 9 are of a different nature and do not have the angular appearance of the deposits obtained when Cl^- is still present in the tissue (Fig. 8); the deposits often display a halo of finely divided Ag grains around the centre. When OsO_4 was present in the post-treatment with Ag acetate, deposits were practically absent, except for some large randomly distributed aggregates, the presence of which defies explanation at the present time.

IV. Physiological Implications

The presence of silver deposits in the plasmodesmata of high-salt barley seedlings suggests that the principal pathway of transport of Cl^- from cell to cell occurs through the cytoplasmic continuum or symplast (Arisz, 1956). Generally, very little Cl^- seems to be associated with the cell wall, but occasionally certain areas are present in the cell wall of high-salt barley seedlings which show a very high density of silver deposit. Pending confirmation that these large deposits indeed consist of AgCl it might be suggested that these represent special areas of the wall similar to those reported in *Nitella* (Spear *et al.*, 1969) and *Chara* (Smith, 1970). These authors detected distinct acid and alkaline surface areas in cell walls, especially in the light, and suggested that Cl^- influx occurs only, or largely, in the acid-extruding areas. These wall areas showed no evidence of being structurally different from the rest of the wall, but the nature of the deposits indicates that they may have resulted from interaction with substances of a highly argentophilic nature, e.g. acid mucopolysaccharides.

In mature leaf cells, chloroplasts constitute the bulk of the cytoplasmic phase. Larkum's finding (Larkum, 1968; Larkum and Hill, 1970) that chloroplasts constitute a phase of high Cl^- concentration in *Limonium* and *Tolypella* was confirmed for barley chloroplasts and further it was possible to show that the region of the grana lamellae contains a much higher Cl^- concentration than the stroma region. These differences in density of Cl^- might result from the presence of macromolecular components of the membrane system giving rise to a Donnan equilibrium. At this stage, however, it should be mentioned that we have found that plants such as a mangrove (*Aegiceras corniculatum*) and the water plant *Nymphoides indica* contain very few deposits inside the chloroplast (Van Steveninck, unpublished). Hence, evidence to date indicates that high Cl^- content of chloroplasts is not universal. In barley leaf parenchyma cells there is little evidence of Ag deposits in any cytoplasmic vesicles or cisternae of the endoplasmic reticulum, but these cells generally contained little endoplasmic reticulum. Further work on other plant species or other types of cells may provide evidence of transport through a system of cytoplasmic vesicles as suggested by MacRobbie (1969, 1970). It is probable that this method of ultrastructural localization of ions will provide some useful information with respect to gross morphological aspects of the mechanisms of ion transport, e.g. involvement of vesicles, endoplasmic reticulum, pinocytosis. Further, it is hoped that the method can be refined in order to resolve small groups of the order of 10–20 Cl^- ions and thus contribute to molecular aspects of ion transfer such as involvement of carrier molecules or conformational changes.

Acknowledgement

This work was supported by the Australian Research Grants Committee.

References

ARISZ, W. H. (1956). *Protoplasma* **46**, 5–62.
BAIN, J. M. and GOVE, D. W. (1971). *J. Microsc.* **93**, 159–162.
CHANG, C. C. (1971). *Surface Sci.* **25**, 53–79.
CLARK, M. A. and ACKERMAN, G. A. (1971). *J. Histochem. Cytochem.* **19**, 727–737.
FISHER, R. A. (1971). *Pl. Physiol.* **47**, 555–558.
KOMNICK, H. (1962). *Protoplasma* **55**, 414–418.
KOMNICK, H. and BIERTHER, M. (1969). *Histochemie* **18**, 337–362.
LANE, B. P. and MARTIN, E. (1969). *J. Histochem. Cytochem.* **17**, 102–106.
LARKUM, A. W. D. (1968). *Nature, Lond.* **218**, 447–449.
LARKUM, A. W. D. and HILL, A. E. (1970). *Biochem. biophys. Acta* **203**, 133–138.
MACALLUM, A. B. (1905a). *Proc. R. Soc.* **B76**, 217–229.
MACALLUM, A. B. (1905b). *J. Physiol.* **32**, 95–128.
MACROBBIE, E. A. C. (1969). *J. exp. Bot.* **20**, 236–256.
MACROBBIE, E. A. C. (1970). *J. exp. Bot.* **21**, 335–344.
MULLER, E. W. and TSONG, T. T. (1969). "Field Ion Microscopy". Elsevier, New York.
REUTER, W. (1971). *Surface Sci.* **25**, 80–119.
SMITH, F. A. (1970). *New Phytol.* **69**, 903–917.
SPEAR, D. G., BARR, J. N. and BARR, C. E. (1969). *J. gen. Physiol.* **54**, 397–414.
SOCHA, A. J. (1971). *Surface Sci.* **25**, 147–170.
SUMI, S. M. and SWANSON, P. D. (1971). *J. Histochem. Cytochem.* **19**, 605–610.
TANDLER, C. J., LIBANATI, C. M. and SANCHIS, C. A. (1970). *J. Cell Biol.* **45**, 355–366.
TORACK, R. M. and LA VALLE, M. (1970). *J. Histochem. Cytochem.* **18**, 635–643.
VAN STEVENINCK, R. F. M. and CHENOWETH, A. R. F. (1972). *Aust. J. biol. Sci.* **25**, 499–516.

Discussion

The discussion started with a question from Thellier on the resolving power of the electron probe technique. Läuchli replied that resolution was primarily a function of section thickness. He distinguished two categories. (1) When section thickness is of the order of 1 μm and if the instrument is fitted with a transmitted electron detector, resolution is approximately equal to section thickness. Thus improvements in resolution can be achieved by reducing section thickness, but there is a lower limit in that the quantity of the element to be detected may become too small; a working limit is probably of the order of 1 μm. (2) In thicker sections, 20–50 μm, such as might be obtained in a cryostat, it is difficult to set spatial resolution limits.

In reply to a further question Läuchli said that all elements down to beryllium can now be detected. Van Steveninck then asked whether it will shortly be possible to detect H^+ by the electron probe technique. Läuchli pointed out the difficulty in distinguishing H^+ from H atoms present in various biological molecules. There is however a comparatively new technique—the ion microprobe or the ion probe mass analyser—which may be successful. In this system the sample is bombarded with high energy oxygen ions and the emitted radiation is detected in a mass spectrometer. This technique ought to be able not only to detect all the elements, but also to differentiate different isotopes of an element. Läuchli continued that this instrument costs at present about \$600,000 but that the manufacturing company—A.R.L. of California—will allow universities to buy time on one of their instruments. He quoted the work of Galle from Paris who has studied Na^+ in red blood cells with the ion microprobe analyser.

Pallaghy commented on the difficulty of calibration of the electron probe technique, using the example of K^+ estimates in the stomata of different plant species. In tobacco where the stomatal cells contain K^+ as K oxalate, one has to construct quite different calibration curves from those in cases where K^+ is present as KCl. Thus one can only assay K^+ in a tissue in which one knows the K^+ salts present. Läuchli agreed but said that in those situations where only relative comparisons between K^+ in different parts of the same section are required, one can work on the basis of the densitometer traces without knowing the K^+ salts present. Eshel said he wished to stress that it was quantities of ions which were determined by the densitometer profiles, not concentrations; if one wished to convert to concentrations one had either to do it on a dry weight basis or to make some estimate of the water lost during preparation. Again Läuchli agreed and gave some further details. A technique has been developed by T. A. Hall in Cambridge whereby one estimates the so-called "white radiation" from the specimen, i.e. the X-rays scattered non-specifically, and then uses this as a means of estimating mass density. Läuchli stressed that this method is more difficult in plant material because of the highly vacuolated cells. In most preparations the vacuoles are empty and for this reason it is better to use embedded material if one has a technique which prevents ion displacements. In this case the overall density of the specimen is approximately that of the embedding material—one therefore has the advantage of working with much more homogeneous material.

The discussion then turned to Hall's paper. Kylin commented on a phrase in the manuscript, "combined sodium and potassium effects have been reported in plant tissue, but they are in general much less than have been found in animal tissue". He criticized the experimental technique in that he thought it did not sufficiently distinguish between two effects on ATPases: (1) the effects of increasing concentration of either K^+ or Na^+ as the case might be, and (2) the effect of increase of ionic strength *per se* on ATPases. He quoted published work by Hansson and himself which showed responses of Na–K stimulated ATPases to alterations in K^+ and Na^+ concentrations at constant ionic strength. He then took issue with Hall's statement that Na–K effects were found in the absence of divalent cations. The Stockholm results show that in the absence of Mg^{++} there is no activation of the ATPases in sugar beet by Na–K. Kylin finally commented on a further statement by Hall that it might be difficult to detect Na–K activated ATPases in view of the large number of different ATPases in plant cells. In sugar beet preparations with Mg^{++} only, Kylin reported, the specific activity of ATPase is 15 μmol mg protein^{-1} h^{-1}; in the presence of Na^+ or K^+ it is 30 μmol mg protein^{-1} h^{-1}. Thus the Na–K stimulated activity is 15 μmol mg protein^{-1} h^{-1}. In wheat or barley in the presence of Mg^{++} only, the specific activity is 150 μmol mg protein^{-1} h^{-1}, so that in these cases Na–K stimulated activity, if similar to sugar beet, would be difficult to detect.

Hall asked the degree of stimulation in Kylin's preparations. Kylin replied that it was between 100 and 150% of the Mg^{++} only levels in sugar beet. This is only about one quarter of that found in animal tissues but one ought to remember that animal cells are more highly specialized. A plant cell performs a multiplicity of functions; halophytes are probably most specialized in Na–K transport as can be seen by the sugar beet results.

Hall replied that Kylin had made two very important points with which he agreed, the first that the plant cell is much less specialized than the animal cell and the second that there are a large number of different ATPases in the plant cell. He then commented on the necessity of divalent cations; he said that in animal cells Mg^{++} is necessary because the substrate is not ATP but a Mg–ATP complex. He thought that a similar situation pertained in plant cells and that the few cases where one did get Na–K stimulation in the absence of Mg^{++} were exceptions. Kylin agreed but said there was evidence that in certain cases one did get Na–K activation in the absence of Mg^{++}, although in sugar beet where the stimulation was Na–K plus Mg^{++} he thought the substrate was a Na or K–Mg–ATP complex. Hall then returned to Kylin's criticism of experimental design; he thought that the bulk of evidence showed that enzymes were generally relatively insensitive to Na^+ but were of course markedly affected by K^+ concentrations. In most plant systems which were stimulated by K^+ there was no Na^+ effect or a small inhibitory effect. Hall added that perhaps ATPases were distinct in this respect in that there did seem to be a Na^+ stimulation, while other enzymes, e.g. respiratory enzymes, were stimulated by K^+ but inhibited by Na^+.

Lüttge then commented on EM work which shows localization of acid phosphatases in the cell wall protuberances associated with gland cells and with transfer cells of the phloem; it is interesting because these cells are known to have specialized transport function. Hall replied that it should be remembered that acid phosphatases will hydrolyse substrates other than ATP. Pallaghy then commented on the dangers of interpretation of vesicles seen in electron micrographs; he quoted O'Brien who has reported that, by observing cells under the light microscope during the addition of glutaraldehyde and other common fixatives, one could see vesicle formation as the fixative went in. Furthermore, cryostatic media, with which segments are infiltrated before freezing in cryostat preparations, are also known to cause vesicle formation. Hall agreed that it is difficult to interpret vesicles and to know in which direction vesicles were moving before fixation. He said that the correlation which they had found between increase in vesicles per cell and external ion concentration might suggest the vesicles were moving inward. He adduced evidence from Cocking's laboratory on work with latex spheres and ferritin where a vesicular movement from the external medium was reported.

Leigh commented that he had recently been in Nottingham where the current idea was that pinocytotic vesicles budding off from the plasmalemma could not fuse with the tonoplast. Thus vesicles could not be involved in uptake from the external solution to the vacuole. He then asked Hall if he thought the rate of pinocytosis was rapid enough to account for the observed rate of ion uptake. Hall replied that he had no way of telling the rate of pinocytosis in his material. Leigh responded that it might be a property of the plasmalemma which had nothing to do with ion uptake. Hall agreed; he said his contentions were that (a) there is a high vesicular activity at both plasmalemma and tonoplast, and (b) there is high ATPase activity associated with these vesicles. These two factors correlate well with the idea that vesicles are involved with ion transport.

Wyn Jones said that his serious reservation about histochemical localization of ATPase was that one was simply looking at levels; one did not know whether the activity was Mg^{++} dependent, Na–K activated or how it was controlled. Hall agreed and Wyn Jones then asked if it was not surprising that this rather general activity was only found in the vesicles; why is it not found in membranes elsewhere? Hall replied that ATPase activity is found elsewhere but it is greatest in the vesicles. If the incubation time is increased, activity can be found in the membranes of the mitochondria, the nuclear envelope, the endoplasmic reticulum and elsewhere. Wyn Jones asked if the effect of incubation time was really a reflection of activity or whether it was an effect of reagent penetration rate. Hall replied that it is enzyme activity rather than penetration because of the section geometry. West then asked Hall how he distinguished the vesicles from Golgi vesicles associated with cell wall building. Hall said that they were distinguished on the basis of ATPase levels and activities. In the Golgi bodies ATPase activity is low, and further, Golgi vesicles are more electron dense than these vesicles.

MacRobbie commented that in plant cells there are two distinct types of ion transport, a regulatory and an accumulating type. In the first there is Na^+–K^+ exchange, possibly ATP mediated, and K^+–H^+ exchange to maintain cation balance and regulate cell pH; in the second there is accumulation of ions into the vacuole, a transport peculiar to plants. Both these transport systems should have carriers, perhaps ATPases, associated with them.

Läuchli then asked Hall about Fisher's Ph.D. thesis where he had reported his study of ATPase activity versus K^+ concentration in the manner of the Epstein dual isotherm and asked if it were correct that Fisher had detected a low concentration hyperbolic curve and a high concentration hyperbolic curve. Hall replied that he did not think Fisher had tried to make this correlation; all he had done was to correlate rate of K^+ uptake with rate of ATP hydrolysis. One objective had been to test whether there was sufficient ATPase present to account for the rate of K^+ uptake.

Finally Van Steveninck's contribution was discussed. Lüttge showed an autoradiograph of a *Limonium* salt gland with transport through the plasmodesmata between the phloem conducting cells and the companion cells, in confirmation of Van Steveninck's conclusions. Van Steveninck said that this gives useful evidence because the fixative, potassium permanganate, is different from the one he used. Bowling then commented that localization of Cl^- in the plasmodesmata does not necessarily mean that transport is occurring there. He drew the analogy with a hold-up on a motorway; there is a high traffic density but it is stationary. Further, Cl^- at the concentrations employed by Van Steveninck is not a growth requirement in barley; this being so, what is the significance of the high Cl^- concentrations in chloroplasts and cell walls? Van Steveninck agreed with the motorway analogy, but he stressed that there are significantly large concentrations of AgCl precipitate near the plasmodesmata. This may be partly a diffusion artefact in that the deposit is exaggerated, but nevertheless this is the place where precipitation first occurs. Referring to the comment about Cl^- concentration for growth he said that in his low Cl^- plants the leaf Cl^- concentration was about 10 mequiv kg^{-1} fresh weight. In excised barley leaf slices in the light there is a very active Cl^- uptake. Pallaghy said that in maize grown in low Cl^- conditions there are high Cl^- concentrations in the leaf epidermal cells; this Cl^- is vital to stomatal function. Läuchli then commented that he thought one could produce Cl^- deficiency symptoms in barley. Van Steveninck said that his work with high and low Cl^- plants was simply to provide a correlation between the density of precipitate and his notational ideas about the Cl^- concentration in these two classes of plant. Lüttge then made a further comment on the traffic analogy introduced earlier; he agreed that one does not know whether or not the "cars" (ions) were moving prior to localization, but he thinks that it is an indication where the roads are. What one really needs is a time series of localization experiments; it would mean a great deal of work but it might be the only way to proceed. Van Steveninck also returned to the traffic analogy; he said you can have a high local concentration at two places; where there is a hold-up or where there is a process necessary for transport. He gave the example of queuing for passport control at a frontier; Cl^- may have to queue to go across a membrane.

West then asked about Cl^- in the chloroplasts; he said it is known that chloroplasts pump H^+ and Cl^- transport may be coupled. What is the Cl^- permeability of the chloroplast envelope? Van Steveninck said that with intact material it is not possible to do quantitative experiments. However, in the chloroplasts most of the Cl^- was found in the grana with very little in the stroma. It is possible that there are charged macromolecules in the grana which accumulate Cl^- by a Donnan effect. West replied that he did not think it was a Donnan effect but rather an effect of membrane potential. If there is a H^+ pump it would produce a membrane potential of 200–300 mV to accumulate Cl^-. Van Steveninck asked, "What is the inside of the chloroplast?". West said, "From your pictures it looks like the inside of the grana". Van Steveninck said that he was not sure about the structure of the grana; most of the work had been done on preparations without chloroplast envelopes and one must therefore be careful to distinguish inside and outside.

Barber then said that the grana consisted almost entirely of membranes with very little internal volume and he wondered how this volume factor had been taken into account in estimating concentration. Van Steveninck replied that it was difficult but the calculations were based on uniform section thickness which is normally in the range 600–800 Å. Barber said that Cl^- concentration should be based on inter-lamellar volume, not on total section volume. Van Steveninck replied that his sections were almost completely filled with grana lamellae and therefore section thickness is a good estimate of the inter-grana volume. Barber retorted that the inter-lamellar volume in chloroplasts is very small and that he understood Van Steveninck's estimate of 300 mM Cl^- to be calculated over the total section volume; is this is corrected for the inter-lamellar volume fraction the concentration estimate will become very large.

The discussion then moved to a different topic. Waisel showed an electron probe study on *Potamogeton lutens* in which the Cl^- appears to be in the vacuole, with very little in the chloroplasts near the cell wall. In contrast, Lüttge then showed an autoradiograph of ^{36}Cl from a mesophyll cell of *Limonium* with the chloroplasts heavily labelled. Pallaghy commented that non-aqueous chloroplast preparations from maize give rise to two populations, one of high chloride content and the other of low chloride content.

Neumann commented that the large H^+ uptake referred to earlier by West has only been observed in Class II chloroplasts with no intact envelopes. It is hard to monitor these effects in Class I chloroplasts because of the difficulty in getting electron acceptors into the chloroplast. However, what evidence there is suggests that the H^+ fluxes are much smaller. How relevant, therefore, is the large H^+ movement in Class II chloroplasts to the *in vivo* situation?

Finally the transport role of the plasmodesmata was reintroduced. Walker said that in *Chara australis* the Cl^- flux across the tonoplast is 2–5 pmol cm^{-1} s^{-1} whereas the Cl^- flux from one cell to the next is about 4000 pmol cm^{-1} s^{-1}; this flux must be carried by the plasmodesmata between the cells. Walker then said that he was sorry that the traffic analogy had come up. Surely the flux in the plasmodesmata is diffusive; it is somewhat smaller in *Chara* than one can calculate to occur by diffusion unless the pores are blocked. If the pores are blocked one must then postulate "some wretched mechanism" which would give rise to the sort of pictures Van Steveninck had shown.

Section II

Formalism and Membrane Models

Chairman: P. Meares

II.1

Electrokinetic Formulation of Ionic Absorption by Plant Samples

Michel Thellier

Laboratoire de Biologie végétale, Faculté des Sciences de Rouen, Mont-Saint-Aignan, France

I. Introduction

Let us consider the overall absorption process of an ionic species S by a plant sample

$$S \overset{\text{C.D.}}{\rightleftarrows} S_i \tag{1}$$

where S_i holds for the absorbed ion, and C.D. for the overall catalytic device of this absorption mechanism. For such processes, it is possible to measure initial rates of absorption v, for various concentrations of S, by the slope of the first quasi-stationary phase of S accumulation in the plant sample, once the "free space" has been equilibrated.

There is therefore a need for a theoretical relation between the values of the activities (S) and (S_i) in the external and internal media, so that it is possible to summarize the data of absorption experiments by the values obtained from a restricted set of parameters introduced by the theoretical model.

Such a problem, let us remember, had already been encountered in *in vitro* enzymology where the velocity v of a reaction between a substrate S and a product P, as catalysed by an enzyme E

$$S \underset{}{\overset{E}{\rightleftharpoons}} P \tag{2}$$

was expressed against the activities (S) by Michaelis and Menten (1913)

$$v = \frac{V_m(S)}{K_m + (S)} \tag{3}$$

This was done by postulating the existence of a transitory complex ES in the course of the reaction

$$S + E \rightleftharpoons ES \rightarrow E + P \tag{4}$$

and assuming that the formation and dissociation of this complex ES follows classical homogeneous chemical kinetics. Thus the two fundamental parameters were introduced: V_m (maximum velocity at saturation of the enzyme by the substrate) and K_m (first approximation of the inverse of the affinity of the enzyme for the substrate). This turned out to be important for the description of many enzymic reactions.

This type of interpretation we shall call a "mechanistic" one, because it is based on the knowledge of the true mechanism of the reaction process. And whenever it is possible to give such a mechanistic interpretation of a biological system, it is naturally an advantage because there is immediate biological signification for the theoretical parameters.

Unfortunately, there are seldom sufficient data on the catalytic devices of cellular processes to allow a real mechanistic interpretation. This is particularly the case for most of the overall processes, and especially for cellular absorption, where the different active molecules in complicated membranes are generally unknown. The catalytic devices of the *in vivo* process and the exact number and nature of these active molecules are unavailable. In such cases where true mechanistic models cannot be deduced, one will have to accept purely phenomenological models, the analytical form of which will sometimes have to be proposed by analogical reasoning with other domains of biology, chemistry or physics where such analytical forms are already known.

For this reason, however interesting we may have found the formulations involving mechanistic analogies between the Michaelian enzymatic reaction (4) and "vector molecules" in the absorption processes (Epstein and Hagen, 1952), we have tried an alternative formulation, without making any assumptions on the mechanism. We simply write the process as a particular case of a flow and force relation of non-equilibrium thermodynamics (Thellier, 1968, 1970b and 1971). This results in an "electrical" model; the individual ion currents are taken as products of conductances and driving forces, as was suggested by Dainty (1962). Hence the name of "*electrokinetic formulation*" given to our model. It is possible that such relations are linear in the vicinity of thermodynamic equilibrium, but become non-linear in conditions too far from equilibrium (Katchalsky and Curran, 1967).

II. Use of Phenomenological "Flow and Forces" Relations: The "Electrokinetic" Model for Absorption Processes

A. Linear, or Ohmic, Behaviour

1. Theoretical linear relation in $(log\ (S), v)$

Let us consider a certain number of processes, called $1, 2, \ldots \ell$, each associated with a flow J_ℓ and a force X_ℓ. According to non-equilibrium thermodynamics (see for example Katchalsky and Curran, 1967), one can anticipate, at least when not too far from equilibrium conditions, a linear relation between flows and forces. In the simple case, when each flow depends only on its associated force, it can be written

$$J_\ell = L_{\ell\ell}X_\ell \tag{5}$$

where $L_{\ell\ell}$ is a conductance term. This is, for example, the case for Ohm's law, which gives a linear relation between a flow (electric current i) and the associated force (electric potential difference ΔE) with a conductance constant $(1/r)$:

$$i = \Delta E/r \tag{6}$$

In more complicated cases, following Onsager (1931), a given flow J_ℓ will depend linearly not only on its associated force X_ℓ but also on forces associated with other processes, thus leading to

$$
\begin{aligned}
J_1 &= L_{11}X_1 + L_{12}X_2 + \cdots + L_{1\ell}X_\ell + \cdots \\
J_2 &= L_{21}X_1 + L_{22}X_2 + \cdots + L_{2\ell}X_\ell + \cdots \\
&\vdots \\
J_\ell &= L_{\ell 1}X_1 + L_{\ell 2}X_2 + \cdots + L_{\ell\ell}X_\ell + \cdots \\
&\vdots
\end{aligned}
\tag{7}
$$

In all these relations, the Ls are conductance terms; and Onsager's theorem states that whatever ℓ and j, $\ell \neq j$, one verifies

$$L_{\ell j} = L_{j\ell}. \tag{8}$$

This formalism can be applied to metabolic processes, as was shown in particular by Kedem (1961). In the case of an ionic absorption process we can apply it when it is possible to choose external concentrations so that the cells are not very far from equilibrium with the external medium for the ion under study, thus ensuring that the linear laws are valid. Let us then apply it to the ionic absorption summarized by eqn (1).

If this ionic absorption only depends on the electrochemical potentials $\tilde{\mu}_S$ and $\tilde{\mu}_{S_i}$, without being coupled to any cellular process (energetic coupling, metabolic utilization of S, etc.), the force will be written

$$X_1 = \tilde{\mu}_S - \tilde{\mu}_{S_i} \tag{9}$$

or

$$X_1 = 2 \cdot 3 \frac{RT}{n\mathscr{F}} \log \left[B_1(S)/(S_i) \right] \tag{10}$$

where n is the charge of the ion, B_1 a parameter depending in particular on the membrane potential, and R, T and $\mathscr{F}$ have their usual meanings. The velocity v of this absorption process

$$v_1 = 2 \cdot 3 \frac{RT}{n\mathscr{F}} L_1 \log \left[B_1(S)/(S_i) \right] \tag{11}$$

If the ionic absorption is coupled to other cellular processes

$$S_2 \rightleftarrows P_2$$
$$S_3 \rightleftarrows P_3 \tag{12}$$
$$\vdots$$

then, from eqn (7), the velocity $v_{1 \cdot 123...}$ of the absorption process of S as depending on the processes (12) becomes:

$$v_{1 \cdot 123...} = 2 \cdot 3 \frac{RT}{n\mathscr{F}} L_{11} \log \left[B_1(S)/(S_i) \right]$$
$$+ 2 \cdot 3 \frac{RT}{n\mathscr{F}} L_{12} \log \left[B_2(S_2)/(P_2) \right] \tag{13}$$
$$+ \cdots + 2 \cdot 3 \frac{RT}{n\mathscr{F}} L_{1\ell} \log \left[B_\ell(S_\ell)/(P_\ell) \right] + \cdots$$

and this can be written (Thellier et al., 1971)

$$v_{1 \cdot 123...} = 2 \cdot 3 \frac{RT}{n\mathscr{F}} L_{11} \log \left[B_{1 \cdot 123...}(S)/(S_i) \right] \tag{14}$$

with

$$B_{1 \cdot 123...} = B_1 \cdot \left[B_2(S_2)/(P_2) \right]^{L_{12}/L_{11}} \cdots \left[B_\ell(S_\ell)/(P_\ell) \right]^{L_{1\ell}/L_{11}} \cdots \tag{15}$$

It is seen that eqn (14) is of exactly the same form as eqn (11) with only the constant B changed. Therefore, any absorption process such as (1) can be described, in the linear range, by an equation

$$v = 2 \cdot 3 \frac{RT}{n \mathscr{F}} L \log \left[B(S)/(S_i) \right] \tag{16}$$

where B is the parameter that summarizes all couplings of cellular processes coupled to the absorption of (S). B thus appears as the parameter characteristic of the thermodynamic state of the cells in the absorption process. On the other hand the phenomenological parameter L is completely independent of the thermodynamic conditions and characterizes only the membranous devices that catalyse the absorption process.

It must be noted that (S_i), the activity of S inside the cells, is generally unknown so that direct use of eqn (16) is not possible. In this case we can make a simplifying assumption that (S_i) will not vary during the course of the experiment whatever the value of (S), if v is the velocity of the first quasi-stationary state of the absorption process, immediately after the equilibration of the free spaces. Under these conditions, we can summarize $B/(S_i)$ by a single constant π:

$$\pi = B/(S_i) \tag{17}$$

and eqn (16) becomes:

$$v = 2 \cdot 3 \frac{RT}{n \mathscr{F}} L \log \left[\pi(S) \right] \tag{18}$$

or

$$v = 1 \cdot 98 \cdot 10^{-4} (T/n) L \log \left[\pi(S) \right] \tag{19}$$

if v is expressed in amperes(dry wt)$^{-1}$ and L in ohms(dry wt).

Let us also note that we shall refer only to the case of net absorption $(v > 0)$ which implies

$$B(S)/(S_i) = \pi(S) > 1 \tag{20}$$

2. Comparison of the linear model with experiment

It has long been known that experimental points describing an ionic absorption would often give a linear regression curve in a semilogarithmic plot of $\{\log (S), v\}$, (Knauss and Porter, quoted by Sutcliffe, 1962), and we have verified this in many experiments and in recalculations from the literature (Thellier, 1969 and 1970b), at least for monophasic curves of absorption: this gives good agreement between the experimental data and the theoretical form of eqn (19).

There remains, however, the case of the "dual-phase" absorption curves (Fig. 1) as first described by Epstein and Hagen (1952) and almost always found in every case where the range of concentration is sufficiently large.

If such experimental results are plotted using the semilogarithmic coordinates $\{\log(S), v\}$ it is seen (Fig. 2) that they confirm the linear approximation only for the low concentrations (corresponding to the first phase on Fig. 1), whereas they depart markedly from linearity for higher concentrations (second phase on Fig. 1). We should also note that, in the case of macro-nutrient ions, the biological optimum of concentration in the nutrient medium is situated at the beginning of the non-linear phase in $\{\log(S), v\}$.

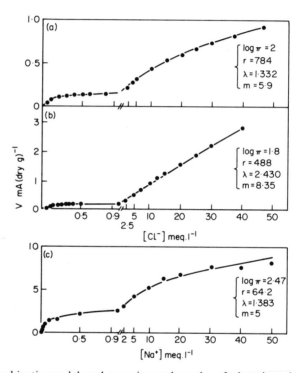

FIG. 1. Electrokinetic model and experimental results of plant ion absorption. The points ● represent the experimental results; the curves were calculated from eqn (25) for values of the parameters indicated in the figure. Experimental data are taken from Torii and Laties 1966 (graphs a and b), and from Rains and Epstein 1967 (graph c). The concentrations are m_N, the velocities v are expressed in $A(g$ of dry tissue$)^{-1}$, π in mN^{-1}, r in $\Omega(g$ of dry tissue$)$, m is a dimensionless constant, and λ is in $V^{(1/m)-1}\Omega^{-1/m}(g$ of dry tissue$)^{-1/m}$.

B. Non-linear, or Non-ohmic, Behaviour

1. Interpretation of the dual phase absorption curves

From the preceding section, we conclude that pure linear flow and force relations for the ionic absorption process, although sufficient so long as the

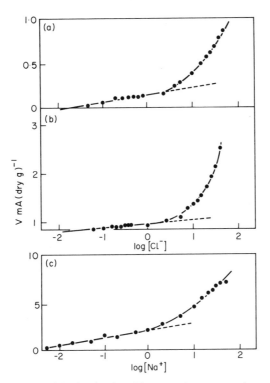

FIG. 2. Use of a semilogarithmic plotting. The experiments are the same as in Fig. 1.

range of studied concentrations is not too high, become insufficient for higher concentrations (that is to say for concentrations far from equilibrium between the cells and the external medium) and that more general non-linear models must then be introduced. This, in fact, is not surprising, being exactly what may be anticipated from non-equilibrium thermodynamics. If we think about it, it is quite logical that pure linear flow and force relations are insufficient to describe an ionic absorption process in the biologically interesting range of concentrations; in fact, one can anticipate that the regulatory processes will be much more efficient if they interact with a non-linear, very rapidly growing, process, rather than with a linear one.

Whatever it is, this flow and force conception of the process makes the dual-phase aspect of the absorption curves appear a mere consequence of the non-equilibrium thermodynamics; the first phase corresponds to a negligible non-linear contribution and the second phase to an important or even dominant non-linear contribution. Naturally, it does not contradict the possible existence of two or more different catalytic devices for the absorption of an ionic species, but it renders unnecessary the hypothesis that a dual-phase absorption curve obligatorily reveals the presence of a dual-catalytic system.

2. Proposal of a non-linear analytical model for ionic absorption processes

Let us consider the non-linear absorption curves, as shown in Fig. 2. As we have said, there are not sufficient microscopic data to allow a mechanistic analytical model of the flow and force relation on the non-linear range. But we can at least try to approximate such a theoretical model with the help of an analogical reasoning with cases in chemistry or physics where it has already been possible to derive non-linear theoretical models.

Figure 3 shows that non-linear flow and force relations are frequently encountered and these are very similar in shape for positive values of flows,

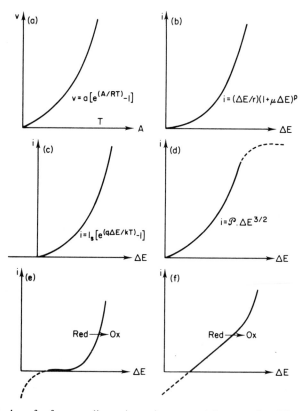

FIG. 3. Reminder of a few non-linear intensity–potential curves for different chemical and physical systems. (a) Relation between the velocity and the affinity for a chemical reaction. (b) Current through a varistant semi-conductor. (c) Current through a junction. (d) Current through a diode. (e) and (f) Oxidation–reduction reactions for slow (e) or rapid (f) systems.

to those encountered with absorption (Fig. 2). There is, for example, the relation between the velocity v and the affinity A of a chemical reaction (De Groot and Mazur, 1962)

$$v = \alpha[e^{(A/RT)} - 1] \tag{21}$$

where α is a parameter of the system; or the current-voltage relations for a varistant semi-conductor (Goffaux, 1957a and b)

$$i = (\Delta E/r)(1 + \mu \, \Delta E)^p; \tag{22}$$

or for a vacuum diode

$$i = I_s(e^{(q\Delta E/kt)} - 1) \tag{23}$$

where q is the charge of the electron and I_s a parameter of the system; and finally for a redox reaction

$$i = \mathscr{P}\Delta E^{3/2} \tag{24}$$

where $\mathscr{P}$ is a parameter (see for example Charlot et al., 1959) containing terms in $\exp(K \, \Delta E)$.

In all these cases we see that non-linear behaviour appears as an exponential or a power function of the force indicating that such a type of law would be a reasonable guess for the non-linear behaviour of absorption. We preferred an electrical model somewhat similar to Goffaux's model (21), both because of the rather wide range of linear relations that is frequently encountered in absorption processes and because experimental (Eley, 1959; Eley et al., 1953) as well as theoretical works (Pullman, B., 1963; Pullman, A., 1964; Pullman and Pullman, 1963; Suard et al., 1961) lead to the hypothesis, for polypeptidic assemblies, of delocalized electronic band structures comparable to those which are encountered, for example, in semi-conductors. We therefore arrived (Thellier, 1971) at a generalizing equation (11)

$$v = 1{\cdot}98 \cdot 10^{-4}(T/n)(1/r) \log[\pi(S)] + \{1 \cdot 98 \cdot 10^{-4}(T/n)\lambda \log[\pi(S)]\}^m \tag{25}$$

with four fundamental parameters r, π, λ and m, the signification of which is discussed later (Section II C).

3. Comparison of the non-linear model (25) with experiment

It is interesting to test the model (25) against experiment, and to try to determine, for each experiment, the best values of the parameters of eqn (25) that will summarize the experimental data.

A graphical method of estimating the parameters ($\log \pi$, r, λ and m) from experimental results has been published previously (Thellier, 1970b; Ayadi and Thellier, 1970). As with any graphical method the results are fairly approximate, but they can be improved by iterative adjustment. A program is being worked out that will allow this to be done automatically. But even with manual adjustment, Figs 1, 2 and 4, which give together the experimental points and the theoretical curves calculated with help of the estimated parameters, show that eqn (25) does constitute a satisfactory regression model for these absorption experiments.

4. Relative importance of linear and non-linear terms: criterion σ

It is easy to see that when (S) is sufficiently small (but obeying relation (20)) so that $\pi(S)$ is near to 1, eqn (25) tends towards the linear ("ohmic") approximation (19); when (S) is sufficiently high, it is the second, non-linear, term of eqn (25) that tends to dominate.

To characterize the passage from ohmic to non-linear behaviour, it is convenient to define a criterion σ as the ratio of the non-linear to the linear term of eqn (25):

$$\sigma = \{1.98 \cdot 10^{-4}(T/n)\lambda \log [\pi(S)]\}^m / \{1.98 \cdot 10^{-4}(T/n)(1/r) \log [\pi(S)]\} \tag{26}$$

or

$$\sigma = \{1.98 \cdot 10^{-4}(T/n)\}^{m-1} r\lambda^m \{\log [\pi(S)]\}^{m-1} \tag{27}$$

The linear behaviour corresponds to values of σ close to 0; and the non-linear appears when σ tends towards and subsequently becomes greater than 1.

C. Signification of the Parameters of Electrokinetic Formulation

1. Introduction of the parameter ε

Numerous processes depend on the absolute temperature T according to an exponential law of $1/T$. It was verified that such was the case for sulphate absorption by Lemna minor (Thellier, 1970a; Thellier et al., 1971) where the phenomenological constant r reasonably obeyed a law

$$r = K e^{(\varepsilon/kT)} \tag{28}$$

or

$$\log r = \log K + \varepsilon/2.3kT \tag{29}$$

thus introducing a new parameter ε (K being a unit-dependent constant and k the Boltzmann constant), and predicting a linear relation between $\log r$ and $1/T$, the slope of which allows ε to be calculated.

A similar treatment can be applied in other cases where temperature behaviour was studied during absorption processes and Table I gives results obtained in that way: it will be remarked that the values of ε thus obtained are of the same order of magnitude as those of real physical semi-conductors.

Moreover, from eqns (19) and (28) it is seen that the velocity of absorption during the linear phase can be written

$$v = K' e^{-(\varepsilon/kT)} \tag{30}$$

TABLE I. Estimation of the parameter ε for different mechanisms of absorption

Reference of the experimental results used in the calculation	Absorbed substance	Plant	ε (eV)
Carter and Lathwell (1967)	$PO_4H_2^-$	Maize roots	0·45
Zsoldos et al. (1968)	Br^-	Sorghum roots	0·90
		Wheat roots	0·80
Zsoldos et al. (1968)	K^+	Sorghum roots	1·10
		Wheat roots	0·70
Thellier et al. (1971)	SO_4^{--}	Lemna minor (whole plants)	0·68
Elchinger (1971)	Arginine	Jerusalem artichoke	0·30

which resembles Arrhenius' equation for enzymic reactions *in vitro*

$$v = K'' e^{-E/NkT} \tag{31}$$

But note that the meaning of the two energy terms (ε and E/N where N is Avogadro number) is not exactly the same: in Arrhenius' classical law, E/N is a molecular energy of activation, whereas here ε cannot be connected to any catalysing molecule in particular, but only to the whole of the cellular device that catalyses the cellular absorption of the studied ion; in this respect, if we persist with the electrokinetic analogy, it plays a role somewhat similar to that played by the forbidden band thickness in the band structure of semi-conductors.

2. Biological signification of the parameters introduced

We have introduced five parameters: π, r, ε, λ and m; the biological signification of which it is interesting to underline.

The parameter π (or better $\log \pi$) is an intensive one, and we have already seen (p. 51) that it summarizes the thermodynamic state of the cells relative to the studied absorption process. On the contrary, the four other parameters (r, ε, λ and m) are completely independent of the thermodynamic conditions and solely characterize the overall cellular device that catalyses the studied absorption process: in the electrokinetic analogy, they are parameters of the semi-conductor, independent of the potential difference which is applied to the semi-conductor. But these four parameters can be themselves classified into two different types.

ε (which is a pure energy) and m (dimensionless number) are intensive parameters that characterize the microstructure of the cellular device which catalyses the studied absorption process, but not the total quantity of this catalytic device: ε and m thus appear as true structural parameters of the catalytic device of absorption (in the electrokinetic analogy, they would depend on the crystalline system of the semi-conductor, on the thickness of the forbidden band, on the levels of impurities, etc., but not on the semi-conductor's total mass).

On the contrary, r and λ (or rather λ^m) are extensive parameters characteristic of the whole of the cellular device that catalyses the absorption process (in the electrokinetic analogy, r, for example, would be an electric resistance).

We therefore anticipate three major types of interactions or regulation mechanisms of the absorption processes. Those affecting $\log \pi$ will correspond to modifications of the thermodynamic couplings; those affecting ε or m will reveal structural modifications of the cellular catalytic devices of absorption; and those affecting r or λ without corresponding modifications of ε or m will be quantitative modifications of this catalytic device corresponding to biosynthesis or destruction of active systems. This is more thoroughly examined in the next paragraph.

III. Application to the Interpretation of Experimental Results

This has already been widely described (Thellier and Ayadi, 1968; Thellier, 1970b; Ayadi et al., 1971; etc.) and we shall only recall here two typical cases: the "competitive" interactions between two similar ionic species from the Mendeleev classification, and the effect of calcium on alkaline ion absorption.

A. "Competitive" Ionic Interactions

Quite often, when one disturbs the absorption of an ionic species (A) by the presence of another ion (B) similar from the Mendeleev classification, a decrease is observed in the velocity of absorption. In that case (Table II) it is generally observed in the linear range of the electrokinetic formulation, that r is not significantly changed, whereas $\log \pi$ is strongly modified: this is a typical "competitive" interaction in the sense of the electrokinetic formalism (Thellier, 1970b).

The interpretation is then as follows: the catalytic device is efficient both for the absorption of A and B; therefore the introduction of B does not modify the properties of this catalytic device, but superimposes a flux of B on the flux of A through the same catalytic device of absorption: this amounts to the introduction of a negative coupling, hence the modification of the thermodynamic parameter $\log \pi$ (p. 50 and 51).

It is interesting to note that Epstein and coworkers arrived long ago at a similar interpretation with the help of their Michaelian formalism of absorption process (Epstein and Leggett, 1954). This will be considered in the general discussion.

TABLE II. Examples of "competitive" ionic interactions during the absorption by *Lemna minor* (modification of log π)

Reference	Absorbed ion	Competitive ion	$\log \pi$ (π mN^{-1})	r (Ω g dry tissue)
Monnier and Thellier (1970)	Phosphate	None	2·58	1800
		Arsenate (1 mN)	1·70	1880
Monnier and Thellier (1970)	Phosphate	None	2·59	1910
		Arsenate (1 mN)	1·76	2280
Ayadi *et al.* (1971)	K$^+$	None	1·91	64
		Rb$^+$ (0·4 mN)	1·07	82
		None	1·10	166
		Rb$^+$ (0·5 mN)	0·76	185
		None	1·23	148
		Rb$^+$ (0·5 mN)	0·70	147
		None	1·05	146
		Rb$^+$ (0·5 mN)	0·58	152
Ayadi *et al.* (1971)	Rb$^+$	None	1·57	93
		K$^+$ (0·4 mN)	1·35	106
		None	1·04	391
		K$^+$ (0·5 mN)	0·75	381
		None	1·13	155
		K$^+$ (0·75 mN)	0·99	163
		None	1·01	97
		K$^+$ (1 mN)	0·62	90

B. Effect of Calcium on the Absorption of Alkaline Cations

Results on the effect of calcium on the absorption of alkaline cations are very conflicting; depending on the conditions and on the plant samples used, the effect is sometimes inhibitory or sometimes activatory. A series of experiments in our laboratory (Thellier and Ayadi, 1971; Ayadi *et al.*, unpublished) on *Lemna minor* potassium absorption has shown that the effect of calcium was generally activatory at low potassium concentrations and inhibitory at high potassium concentrations; this at first seemed rather paradoxical.

Figure 4 gives the interpretation in the electrokinetic formalism, with a slight positive effect of calcium on the thermodynamic parameter log π and a strong effect on all the catalytic parameters (r, ε, λ and m). Thus calcium has a double effect: first it modifies the thermodynamic cellular conditions of alkaline ion absorption (whether it is a direct effect of flux coupling or an

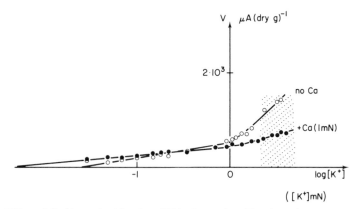

FIG. 4. Effect of Ca^{++} on the kinetics of K^+ absorption. The dotted area represents the domain of optimal biological concentration in a nutrient medium for K^+.

indirect effect on some metabolic coupling); secondly it modifies the catalytic device of alkaline ion absorption through a modification of the structure of this device because ε and m are modified together with r and λ. The first effect is activatory and the second inhibitory.

It is easy to see from Fig. 4 that these two opposite effects are such that the activatory dominates at low potassium concentrations whereas the inhibitory dominates at higher potassium concentrations, thus explaining the apparently paradoxical effect of calcium.

This double effect of calcium tends to flatten the curves $\{\log (K^+), v\}$ and this is especially apparent in the non-linear region (where the biological optimum for potassium concentration in the outside medium is found); with calcium we therefore go from a behaviour pattern where the velocity of potassium absorption (in the biological optimum range) depends very much on potassium concentration, to a behaviour pattern where it is practically independent of this concentration: this is a typical regulatory effect.

Lastly, it should be noticed that this effect of calcium on the structure of the catalytic devices of potassium absorption in plants is consistent with what is known about the effect of calcium on alkaline cation exchange in sensitive animal cells.

IV. General Discussion of Electrokinetic Formulation

A. Non-mechanistic Character of the Electrokinetic Model

Naturally the choice of eqn (25) by analogical reasoning is somewhat arbitrary; for example, we could have tried at least two alternative forms:

$$v = 1.98 \cdot 10^{-4}(T/n)(1/r)\{\log [\pi(S)]\} \{1 + 1.98 \cdot 10^{-4}(T/n)\mu \log [\pi(S)]\}^p \quad (32)$$

with four parameters (π, r, μ and p), nearer to Goffaux's own model (22); or

$$v = I_s\{-1 + e^{\alpha \ln [\pi(S)]}\} \tag{33}$$

with three parameters (π, I_s and α), derived from the model (23), consistent with the fact that quasi-linear relations are often obtained in $\{\log (S),$ $\log (v + c')\}$ coordinates, and with the proposal of Cope and Mandell (quoted by Bookris and Reddy, 1970) for the behaviour of nerve cells.

But it should be noted that all these models, although very different in the mechanistic interpretations from which they originate, have in fact very similar properties if considered as purely formal, phenomenological models: they all admit a linear approximation near equilibrium conditions

$$v = 1\cdot98(T/n)(1/r) \log [\pi(S)] \tag{34}$$

and

$$v = I_s\alpha\ln [\pi(S)] \tag{35}$$

respectively, for eqns (32) and (33); we also find in all cases the thermo-dynamic parameter ($\log \pi$), the intensive catalytic parameter [m for eqn (25), p for eqn (32) and α for eqn (33)] and the extensive parameters [r and λ^m for eqn (25), r and μ^p/r for eqn (32) and I_s for eqn (33)].

Therefore any such model, so long as it can give a good fit with the experimental points, will summarize the experiments by the same three types of parameters—thermodynamic, intensive catalytic or extensive catalytic; in view of the very fundamentally different properties of these different types of parameters, any perturbation found to modify one of them in one of the theoretical models, will be found to modify the same type of parameter in the other models. Thus, from this purely phenomenological viewpoint, the exact form of the theoretical model does not matter much in regard to biological interpretation, so long as it constitutes a good regression model for the experimental points. But this also implies that agreement between any of the models and the experimental points of an absorption experiment must not be taken as evidence for the mechanistic interpretation of absorption: for example, the evidence obtained here (Figs 1, 2, 4) of an agreement between the experimental points and a model derived from semi-conductors does not imply that the catalytic device of absorption is a semi-conductor.

We can also interpret in this way the convergence in the interpretations of experimental results by two very different models, our electrokinetic and the Michaelian of the Epstein school, as indicated above (see III.A). In fact, if we consider Epstein's model from a purely formal point of view as we have done with the electrokinetic model, we see that it summarizes the experiments (in the case of dual absorption curves) with the help of two extensive para-meters V_{m1} and V_{m2} and two intensive ones K_{m1} and K_{m2}. And there again, a given type of perturbation can affect only the same type of parameters, whatever formal model is chosen. From this point of view the two models, Michaelian and electrokinetic, thus appear as differently focused, rather than contradictory.

If we finally prefer our "electrokinetic" formulation of absorption, it is essentially for the two following reasons: (1) It is, at present, impossible to construct a complete mechanistic theory of plant absorption because of the lack of sufficient structural information on the catalytic devices of the process, and we must therefore keep to analogical formal models. In these circumstances, a model derived from flow-and-force general relations might reveal advantages because such formalism is precisely meant for this type of situation and benefits from possibilities of extension. (2) The electrokinetic model distinguishes the thermodynamic aspect of the process (couplings summarized by $\log \pi$) from the catalytic aspect (r, ε, λ and m) and allows the use of the intensive catalytic parameters (m and ε) as true structural (although not geometric) parameters of the catalytic devices of absorption.

B. Utilization and Extension of the Electrokinetic Model

At present the model, as applied to plant absorption, has introduced parameters that are useful for summarizing experimental results and *bear a general significance whatever the unknown supramolecular architectures that catalyse the process are:* according to whether the parameters modified in a perturbation or regulation process are the catalytic ones (intensive or extensive) or "energetic coupling" these perturbations may be classified in three types: structural perturbation of the catalytic device, quantitative modification of the catalytic systems or modification of the energetic couplings. By itself, this classification is useful.

Furthermore, there is no reason why similar reasoning could not lead to useful theoretical models for overall cellular processes other than ionic absorption: transcellular exchanges, respiration, etc.

If, in the study of such phenomena, we encounter a non-linear behaviour different from that found here, we only need to try to find, through other analogical reasoning, an analytical form other than eqn (25) that would give satisfactory agreement with the experimental data; and the new parameters introduced would allow the same type of discussion as above, according to whether they are intensive or extensive.

Acknowledgement

The author acknowledges the financial support of C.N.R.S. through grants C.M.R.S.203, R.C.P.285 and D.G.R.S.T.72.7.0300.

References

AYADI, A. and THELLIER, M. (1970). *C.r. hebd. Séanc. Acad. Sci., Paris* **171**, 1280–1283.
AYADI, A., DEMUYTER, P. and THELLIER, M. (1971). *C.r. hebd. Séanc. Acad. Sci., Paris* **273**, 67–70.
BOCKRIS, J. C. and REDDY, A. K. N. (1970). "Modern Electrochemistry", Vol. 2. Plenum Press, New York.
CARTER, O. and LATHWELL, D. J. (1967). *Pl. Physiol.* **42**, 1407–1412.
CHARLOT, G., BADOZ-LAMBLING, J. and TREMILLON, B. (1959). "Les réactions électrochimiques". p. 29. Masson et Cie, Librairie de l'Académie de Médecine, Paris.
DAINTY, J. (1962). *A. Rev. Pl. Physiol.* **13**, 379–402.
DE GROOT, S. R. and MAZUR, P. (1962). "Nonequilibrium Thermodynamics". North-Holland, Amsterdam.
ELCHINGER, I. (1971). *Third cycle Thesis*, Strasbourg, France.
ELEY, D. D. (1959). *Research* **12**, 293–299.
ELEY, D. D., PARFITT, G. D., PERRY, M. J. and TAYSUM, D. H. (1953). *Trans. Faraday Soc.* **49**, 79–86.
EPSTEIN, E. and HAGEN, C. E. (1952). *Pl. Physiol.* **27**, 457–474.
EPSTEIN, E. and LEGGETT, J. E. (1954). *Am. J. Bot.* **41**, 785–791.
GOFFAUX, R. (1957a). *Rev. gén. Electr.* **66**, 463–472.
GOFFAUX, R. (1957b). *Rev. gén. Electr.* **66**, 569–576.
KATCHALSKY, A. and CURRAN, P. F. (1967). "Nonequilibrium Thermodynamics in Biophysics". Harvard University Press, Cambridge, Mass.
KEDEM, O. (1961). "Membrane Transport and Metabolism", (A. Kleinzeller and A. Kotyk, eds), pp. 87–93. Academic Press, London and New York.
MICHAELIS, L. and MENTEN, M. L. (1913). *Biochem. Z.* **49**, 333–369.
MONNIER, A. and THELLIER, M. (1970). *C.r. hebd. Séanc. Acad. Sci., Paris* **270**, 2178–2181.
ONSAGER, L. (1931). *Phys. Rev.* **37**, 405.
PULLMAN, A. (1964). *Biopolymer Symposia* **1**, 29–33.
PULLMAN, B. (1963). *C.r. hebd. Séanc. Acad. Sci., Paris* **257**, 2488–2491.
PULLMAN, B. and PULLMAN, A. (1963). "Quantum Biochemistry", p. 867. Interscience, New York and London.
RAINS, D. W. and EPSTEIN, E. (1967). *Pl. Physiol.* **42**, 314–318.
SUARD, M., BERTHIER, G. and PULLMAN, B. (1961). *Biochim. biophys. Acta* **52**, 254–265.
SUTCLIFFE, J. F. (1962). "Mineral salts absorption in plants", Pergamon Press, Oxford.
THELLIER, M. (1968). *C.r. hebd. Séanc. Acad. Sci., Paris* **266**, 826–829.
THELLIER, M. (1969). *Bull. Soc. fr. Physiol. vég.* **15**, 127–139.
THELLIER, M. (1970a). *C.r. hebd. Séanc. Acad. Sci., Paris* **270**, 1032–1034.
THELLIER, M. (1970b). *Ann. Bot.* **34**, 983–1009.
THELLIER, M. (1971). *J. theor. Biol.* **31**, 389–393.
THELLIER, M. and AYADI, A. (1968). *C.r. hebd. Séanc. Acad. Sci., Paris* **267**, 1839–1842.
THELLIER, M., THOIRON, B. and THOIRON, A. (1971). *Physiol. Vég.* **9**, 127–139.
TORII, K. and LATIES, G. G. (1966). *Pl. Physiol.* **41**, 863–870.
ZSOLDOS, F., CSEH, E. and BOSZERMENYI, Z. (1968). *Z. Pflanzenphysiol.* **60**, 75–77.

II.2

Solute–Water Interactions in the Teorell Oscillator Membrane Model

K. R. Page and P. Meares

Biophysical Chemistry Unit, Department of Chemistry, University of Aberdeen

I. Introduction

The potential role of electro-osmotic ion–water flux coupling in plant cell membranes is currently attracting attention, especially where connected with the phenomena of excitability. The analogue properties of the Teorell membrane oscillator (Teorell, 1959) are of particular interest in this context, as they demonstrate many of the features observed in excitable tissues. These properties arise from electro-osmotic flux coupling.

The membrane oscillator consists of a broad-pored membrane of low effective internal electric charge separating two solutions of the same electrolyte, each of a different concentration. On passing an electrical current through the membrane an electro-osmotic flux is generated, and this may be directed from the dilute to the concentrated solution. This flow is opposed by a hydrodynamic flow induced by a pressure differential across the membrane. If the current is kept constant, and above a critical density, and the pressure differential allowed to vary it is possible to obtain undamped oscillations of pressure differential and membrane potential. If both the current and the pressure differential are held constant, characteristic stationary states are obtained instead.

A detailed analysis of the operation of this system requires the employment of a membrane of well characterized internal structure. Such a membrane has been lacking until recently and this has limited the scope of the analogue. This paper is based on a re-examination of the stationary states of the Teorell oscillator by Meares and Page (1972) conducted using a new synthetic membrane of especially well-defined pore structure. This work has led to a quantitative understanding of the system, and it is hoped that it will facilitate the application of the analogue to a wider field than has been possible previously.

II. Stationary State Phenomena

The apparatus shown in Fig. 1 was used to study the stationary state phenomena. It represents the membrane separating two compartments; compartment I contains the dilute and compartment II the concentrated

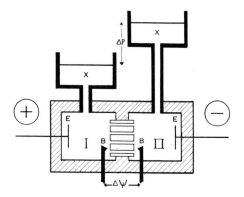

FIG. 1. Outline of membrane cell used for studying stationary state phenomena.

solution, typically 0·01 and 0·1 mol dm^{-3} NaCl. The electrodes E permit the passage of a constant current through the membrane; the potential across the membrane is sensed by probe electrodes connected to the cell interior by means of the salt bridges B. The pressure differential between the compartments is controlled by two constant level reservoirs X. Each of the compartments is stirred by paddles, and the electrolyte is circulated through a closed circuit between the compartment and a one litre capacity reservoir. The entire system is maintained by a thermostat at $25 \pm 0.1°C$.

Figure 2 illustrates the characteristic current versus potential plot obtained with this apparatus. The curve was obtained for a single pressure differential ΔP. Two stable states were observed, a low electrical resistance state R_1 obtained at low current densities, and a high resistance state R_2 obtained at high current densities. In the former the volume flow through the membrane

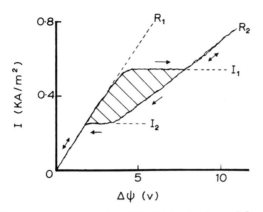

FIG. 2. Current density versus potential plot obtained using a 0·5 μm pore diameter membrane separating 0·1 and 0·01 mol dm^{-3} NaCl and a pressure differential of − 589 Pa. The shaded area is inaccessible under galvanostatic conditions.

is controlled by the pressure differential, and hence flows from compartment II to I filling the membrane with concentrated electrolyte. The opposite situation applies to the latter, electro-osmosis causing a volume flow from compartment I to II. The system shows hysteresis, the transition from R_1 to R_2 occurring at a current density I_1 different from I_2, the current density marking the opposite transition. The difference between I_1 and I_2 increases with increasing ΔP. Under constant current (galvanostatic) conditions the shaded area is inaccessible.

Plots of this type were obtained using "Nuclepore" polycarbonate membranes manufactured by G.E. of America. These are available in a graded series of pore sizes, the pore diameters in our experiments being 0·5, 0·8, 1·0 and 2·0 μm respectively (see Table I). The hydrodynamic permeability of each membrane was measured using the half-time method (Teorell, 1959) and the electro-osmotic permeability using the method of Mackay and Meares (1959). The number of pores per unit area was determined by microscopy, and the membrane thickness estimated using a Mercer–Parnum thickness gauge.

III. The Physical Principles Underlying the Analogue

The stationary-state behaviour of this oscillator has been quantitatively analysed by Meares and Page (1972). Their treatment is based on the analysis of a single pore. The ion fluxes are described using the Nernst–Planck equations with an extra term for convection. The latter term allows for the effect of the barycentric velocity upon the ion fluxes, the entire system being in a membrane fixed frame. This term therefore accounts for ion–water flux coupling.

The barycentric velocity consists of two components, one arising from the pressure differential and calculated using Poisseuille's law, and the other arising from electro-osmosis. The electro-osmotic component is calculated using the Gouy–Chapman theory of the diffuse double layer. The double layer in the pore solutions arises from an effective surface charge on the pore walls. Allowance is made for the observation that with Nuclepore membranes the local surface charge density is proportional to the cube root of the concentration of the solution in the appropriate region of the pore. The surface charge appears to arise from selective ion adsorption on the polycarbonate surface.

Combination of ion flux equations yields the local salt flux and electrical current density, and these are then averaged over the pore cross section. Integration of the averaged equations along the axis of the pore yields the required global equations. The boundary limits of the integration have to allow for the presence of incompletely stirred interfacial layers of solution in contact with the membrane. This is done using the Nernst diffusion layer hypothesis, the effective thickness of the layers being estimated using the silver sheet electrode method of Scattergood and Lightfoot (1968).

The global equations have been found to predict correctly the observed stationary states of the oscillator under a wide variety of conditions. The following analysis of the analogue will therefore utilize these results directly. A detailed study of the validity of the theory may be found in Meares and Page (1972).

The behaviour of the system may be analysed in terms of a feedback loop of the type shown in Fig. 3. The volume flow is a function of the electro-osmotic and hydrodynamic permeabilities of the membrane. For a given

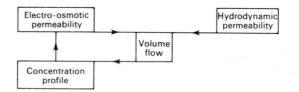

FIG. 3. Feedback loop illustrating the interdependency of the volume flow, concentration profile, and electro-osmotic permeability.

membrane the latter is constant. The electro-osmotic component of the volume flow varies, however, owing to the presence of a variable concentration profile in the membrane pores.

The concentration profile in a given pore is of the form.

$$C_x = A \exp\left(\frac{\bar{v}x}{D_s}\right) + B \tag{1}$$

where A and B are constants for any given membrane and electrolyte concentration pair, and D_s is the salt diffusion coefficient. C_x is the concentration at a position x along the axis of the pore. For large and positive values of $\bar{v}$, the barycentric velocity, eqn 1 shows that the membrane is chiefly filled with dilute solution from compartment I, the profile having a concave exponential form. For large and negative flows the reverse applies, the profile now being convex. These changes in concentration profile underlie the resistance changes from R_1 to R_2.

The effect of the concentration profile is twofold. It governs the thickness of the electrical double layers at the pore wall, and it controls the electrical potential across the pore, the electrical current being kept constant. Both these factors affect the strength of electro-osmosis and hence barycentric velocity so completing the feedback loop. The presence of this loop is necessary for the regenerative properties of the system, and these will be discussed in section V. Figure 4 shows the electrical current density as a

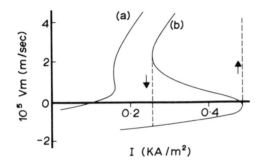

FIG. 4. Current density versus volume flow calculated for a $0.5\ \mu m$ pore diameter membrane separating 0.1 and 0.01 mol dm^{-3} NaCl. Curve (a) is plotted for an applied pressure differential of -196 Pa and curve (b) for a differential of -785 Pa.

function of the volume and pressure differential. The curves in this figure have been calculated from theory for membrane A (see Table I). It will be seen that above a certain pressure differential ΔP the curves have three roots over a range of currents. Two of these barycentric velocities are physically accessible under galvanostatic conditions. The arrows on the highest pressure curve indicate the pathway followed by the system in this situation. The current densities corresponding to the turning points marking the transitions are characterized by I_1 and I_2. These values may be directly compared to the experimentally observed quantities obtained from plots such as shown in Fig. 2.

The relationship between the current and the membrane potential is therefore non-linear as a consequence of electro-osmotic ion–water flux coupling. The non-linearity becomes progressively more marked with increasing pressure differential for a given membrane and concentration pair. Theory also predicts that factors which increase the hydrodynamic permeability relative to the electro-osmotic permeability increase both I_1 and I_2 and their rate of increase with increasing pressure differential. These predictions have always been borne out by experiment as shown in Figs 5 and 6. These figures illustrate the results obtained using membranes whose

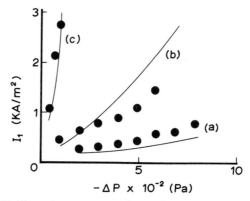

FIG. 5. I_1 versus ΔP. The points represent observed and the lines calculated values for (a) membrane A, (b) membrane B, and (c) membrane C, the membranes separating 0·1 and 0·01 mol dm^{-3} NaCl in each case.

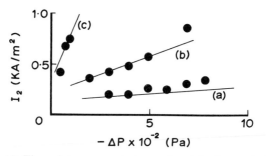

FIG. 6. I_2 versus ΔP. The points represent observed and the lines calculated values for (a) membrane A, (b) membrane B, and (c) membrane C, the membranes separating 0·1 and 0·01 mol dm^{-3} NaCl in each case.

properties are shown in Table I, and 0·1/0·01 mol dm^{-3} solutions of sodium chloride. The factor K in the table is related to the surface charge density σ on the pore walls by eqn 2.

$$\sigma = Kc_s^{\frac{1}{2}} \tag{2}$$

c_s is the concentration of the electrolyte bathing the pore. L_p is the hydro-dynamic permeability.

TABLE I. Membrane properties

Membrane	Pore diameter (μm)	L_p 10^8 m Pa^{-1} s^{-1}	K m cm^{-1} kg$^{-\frac{1}{3}}$
A	0.5	2.35	-9.84
B	1.0	5.62	-5.21
C	2.0	30.8	-3.67

IV. Factors Causing Transitions between Stable States

The theory outlined above permits an analysis of the factors causing transitions between the two stable states of the system. These may be thought of as stimuli inducing excitation. The high concentration side of the membrane (compartment II) will be considered analogous to the interior of a biological cell. The high resistance state will be taken as the resting state, and the low resistance state as the excited state.

A. Current-induced Stimulation

As shown in Fig. 2 the resting state corresponds to a current density greater than I_1. Stimulation is produced by lowering the current density to below I_2, ΔP being kept constant. The membrane is restored to its resting state by allowing the current density to rise above I_1 again. It should be noted that the system returns to the resting state by a different path from that followed by excitation. The system therefore shows hysteresis.

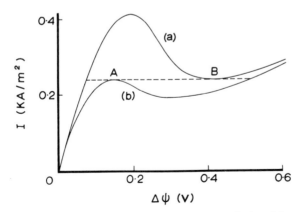

FIG. 7. Current density versus potential difference curves calculated for membrane A separating 0.1 and 0.01 mol dm^{-3} NaCl. Curve (a) plotted for $\Delta P = -686$ Pa and curve (b) for $\Delta P = -392$ Pa. (The potential does not include the potential drop between the probes and membrane.)

B. Pressure-induced Stimulation

Stimulation may be produced by increasing the pressure of the "cell" interior at constant current. This is demonstrated in Fig. 7 which shows two current–voltage plots calculated from theory for membrane A separating $0.1/0.01$ mol dm^{-3} NaCl. For this situation an I_2 value of 0.24 kA m^{-2} is predicted for a ΔP of -686 Pa whilst an I_1 value of the same current density is predicted for a ΔP of -392 Pa. Let the current density be maintained at 0.24 kA m^{-2}. If the pressure differential is less than -392 Pa the system will be in the high resistance state R_2. On raising the pressure the system will remain in this state until a pressure differential of -686 Pa is reached. Excitation will occur at this point, the system moving along the line AB to the low resistance state R_1. On lowering the pressure the system will return along the line AB to the high resistance state when the pressure differential reaches -392 Pa. Hysteresis is observed as the transitions occur at two different pressures.

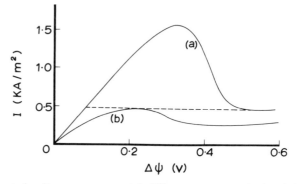

FIG. 8. Current density versus potential difference curves calculated for an elastic membrane with structure similar to membrane A and separating 0.1 and 0.01 mol dm^{-3} NaCl. The curves are plotted for a $\Delta P = -785$ Pa, curve (a) corresponding to a pore radius of 0.28 μm and curve (b) to a pore radius of 0.22 μm. (The potential does not include the potential drop between the probes and membrane.)

C. Stimulation Induced by Mechanical Deformation

Theory predicts that excitation could be produced by stretching an elastic membrane of pore structure similar to that of the Nuclepore membranes. The effect of mechanical deformation cannot be examined with Nuclepore membranes because they are inelastic and so this type of stimulation has not been studied experimentally. Burton (1970) has demonstrated how stress alters the pore size of elastic membranes. Provided the Poisson ratio of the membrane material is relatively high, a small degree of stretch causes a substantial increase in the pore diameter. The possible relevance of this effect as a source of stimulation in biological mechano-elastic transduction justifies a theoretical consideration of elastic parallel pore membranes.

Our theory was found to predict successfully the behaviour of a series of Nuclepore membranes of different pore radii and so permit the effect of varying the pore diameter by stretching a membrane to be predicted.

Figure 8 portrays calculations from theory for an extensible membrane with properties otherwise identical with those of membrane A. Initially the membrane is in the high resistance state R_2 with pore radius less than $0.22 \,\mu$m. On stretching the pores act as foci of stress and their radius increases. When stretching is carried out at constant pressure differential and electric current a transition occurs when the pore radius reaches $0.28 \,\mu$m. The resistance of the system moves along the line AB to the low resistance state R_1. It remains in this state until sufficient stress is removed to restore the radius to $0.22 \,\mu$m. A transition then occurs back to the resting state with resistance R_2.

V. Regenerative Behaviour

When the constant level reservoirs shown in Fig. 1 are replaced by open-topped vertical tubes the analogue becomes capable of regenerative behaviour (Meares and Page 1972). This is a consequence of the feedback loop outlined in section III.

As shown by Fig. 9, if the current density is above a certain critical value undamped oscillations of pressure differential and membrane potential are

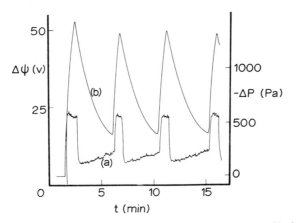

FIG. 9. Undamped oscillations across a membrane with $1 \,\mu$m pores. Variations in the potential $\Delta\psi$ between the probe electrodes are recorded by trace (a), and variations in the pressure differential ΔP are recorded by trace (b). The traces are plotted as a function of the time t. The membrane separated solutions of 0.1 and $0.01 \,\mathrm{mol\,dm^{-3}}$ NaCl, and the cell was fitted with 5 mm diameter tubes. A current density of $1.82 \,\mathrm{kA\,m^{-2}}$ was used.

generated. If the vertical tubes are fitted with an overflow at their top this system may then be used as an analogue for a variety of natural cells (Teorell, 1966). The overflow has two functions. It allows for "hydraulic leakage" and it ensures the "resting" and "excited" states correspond to those used in section IV. The effect of the overflow is illustrated in Fig. 10.

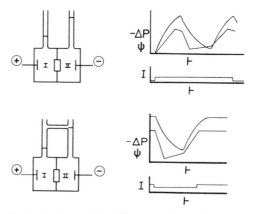

FIG. 10. Regenerative behaviour with different cell configurations. The upper configuration produces oscillatory phenomena corresponding to those shown in Fig. 9. The lower configuration demonstrates the effect of placing an overflow between the two vertical tubes, the low resistance state now corresponding to depolarization of the membrane.

Stimulation of the analogue in this form produces an immediate lowering of the membrane potential. The membrane electrical conductivity rises owing to an outflux of concentrated electrolyte, and the pressure falls on the inside compartment of the membrane cell, that is, there is a loss of turgor. If the stimulus is not removed a train of oscillations will now follow. The system automatically returns to its resting state on removal of the stimulus, leading to a recovery of turgor and resting potential.

These responses have a number of features in common with plant cells capable of excitation. The action potential in the internodal cells of *Chara australis* is closely accompanied by changes in volume flow and turgor analogous to the above sequence (Barry, 1970). Similar sequences have been observed with the motor cells of species of *Mimosa* and *Dionaea* (Sibaoka, 1969). The effects of mechanical deformation discussed in section IV have particular interest in the light of the observations of Benolken and Jacobson (1970) made on *Dionaea muscipula*. Their analysis of the excitatory behaviour of the indented cells forming part of the sensory hairs of this plant indicates permeability changes in the cell membranes as a consequence of mechanical deformation. The sequence of changes leading to excitation bear a remarkable correspondence to the behaviour predicted on the basis of our analogue.

In discussing the analogue role of our membrane it should be appreciated that it is structurally too simple to account for the complete behaviour of any particular biological membrane. Owing to the large size of the pores of a Nuclepore membrane it neglects the extreme ion selectivity found in many natural tissues (Barry, 1970). It also provides no indication of the metabolic origin of the electrical current. The importance of this model is that it demonstrates how electro-osmotic ion-flux coupling can lead to many of those features associated with biological excitability. It illustrates how seemingly complex phenomena may be accounted for without recourse to complicated structural mechanisms. So long as there is a possibility that electro-osmotic mechanisms are present in natural tissues, the effects discussed in this paper must be considered when analysing the transport systems in these tissues.

References

BARRY, P. H. (1970). *J. Membrane Biol.* **3**, 335–371.

BENOLKEN, R. M. and JACOBSON, S. L. (1970), *J. gen. Physiol.* **56**, 64–82.

BURTON, A. C. (1970). "Permeability and Function of Biological Membranes", Vol. 1. North-Holland, Amsterdam.

MACKAY, D. and MEARES, P. (1959). *Trans. Faraday Soc.* **55**, 1221–1238.

MEARES, P. and PAGE, K. R. (1972). *Phil. Trans. R. Soc.* A **272**, 1–46.

SCATTERGOOD, E. M. and LIGHTFOOT, E. N. (1968). *Trans. Faraday Soc.* **64**, 1135–1146.

SIBAOKA, T. (1969). *A. Rev. Pl. Physiol.* **20**, 165–184.

TEORELL, T. (1959). *J. gen. Physiol.* **42**, 831–845.

TEORELL, T. (1966). *Ann. N.Y. Acad. Sci.* **137**, 950–966.

II.3

The Flux-Ratio for Potassium in *Chara corallina*

J. F. Thain

School of Biological Sciences, University of East Anglia

I. Introduction

The Ussing–Teorell flux-ratio equation (Teorell, 1949; Ussing, 1949) has been widely used as a test for the passive independent movement of ions across membranes. According to this criterion, if an ionic species i crosses a membrane without interacting with any other flows and in such a way that the ions themselves do not interact with each other, then the ratio of the unidirectional tracer fluxes for that ionic species should obey the relationship

$$\ln(-j_{i1}/j_{i2}) = [\ln(a_{i1}/a_{i2}) + z_i F(\psi_1 - \psi_2)/RT] \tag{1}$$

where a_{i1} and a_{i2} are the concentrations of i in the bulk solution phases in contact with sides 1 and 2 of the membrane (see Fig. 1); ψ_1 and ψ_2 are the electrical potentials in these phases; z_i, F, R, T are respectively the charge on species i, Faraday's constant, the gas constant and the temperature (K); j_{i1} and j_{i2} are the unidirectional fluxes of i from side 1 to side 2 and from side 2 to side 1 respectively. In practice, concentrations are often used in eqn (1) instead of activities.

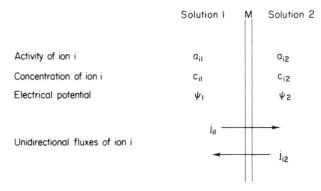

FIG. 1. Diagram of the system.

The value of the unidirectional flux j_{i1} is defined by the relationship

$$j_{i1} = j_{i1}^* \cdot c_{i1}/c_{i1}^* \tag{2}$$

where j_{i1}^* is the measured flux of a tracer isotope of i from side 1 to side 2 when the concentration of this tracer is c_{i1}^* on side 1 and zero (or at least very much less than c_{i1}^*) on side 2. The value of j_{i2} is defined similarly. Equation (1) may be written more simply in the form

$$\ln f_i = X_i/RT \tag{3}$$

where f_i is written for $(-j_{i1}/j_{i2})$ and X_i is equal to the right side of eqn (1) multiplied by RT. X_i is the net thermodynamic driving force acting on species i.

In a number of systems the unidirectional fluxes of an ionic species have been found not to obey this relationship. To accommodate these systems, a factor n, which can be greater or less than unity, is generally incorporated into eqn (3) to give

$$\ln f_i = nX_i/RT \tag{4}$$

Values of n not equal to unity are believed to arise from interactions between the flow of species i and other flows, such as the flow of solvent, flows of other solutes, the "flow" of a (bio)chemical reaction (giving "active" transport), or from interactions between the flows of tracer and non-tracer varieties of species i itself. Interactions of the latter type, often called "isotope interactions", could occur via an exchange-diffusion mechanism, thus giving n less than unity; or they could occur in systems where i crosses the membrane by long, narrow "pores" thus giving n greater than unity (Hodgkin and Keynes, 1955).

In particular, n has been found not to equal unity for the flows of K^+ ion in the coenocytic alga *Chara corallina* (Walker and Hope, 1969) and in

Sepia axons (Hodgkin and Keynes, 1955). In both of these systems the value of n for K^+ was found to be 2·5. (This concurrence is interesting, but may be coincidental.) It has been suggested (Walker and Hope, 1969) that this value of n for K^+ could arise from a K^+/K^+ isotope interaction (perhaps via the "long pore effect") or from a K^+/water interaction. This latter mechanism would operate if a net flow of K^+ in one direction caused a net (electro-osmotic) flow of water in the same direction. This induced flow of water would, in turn, decrease the resistance to the flow of tracer K^+ in the same direction and increase the resistance to the flow of tracer K^+ in the reverse direction. One other suggestion (Coster and George, 1968) is that the flow of K^+ may interact with the flow of another ionic species; the flow of this second ionic species would depend on some of the factors (e.g. the trans-membrane potential difference) which governed the flow of K^+, and would thus result in a relationship of the form of eqn (4).

Interactions between flows are explicitly included in the treatment of transport processes by the methods of non-equilibrium thermodynamics (see e.g. Katchalsky and Curran, 1965). This approach has been used by several authors (Hoshiko and Lindley, 1964; Kedem and Essig, 1965; Coster and George, 1968; Simons, 1969), to obtain equations for the flux-ratio which include separate terms for the different interactions. The aim of the present paper is to carry this approach further in an attempt to find out which of the interactions mentioned above are responsible for the deviation from unity of the value of n for K^+ in *Chara*.

In previous analyses of this problem, Coster and George (1968) ignored the possibility of any interactions other than one between K^+ and some other ion. Simons (1969) argued that the only important interaction was a K^+/K^+ isotope interaction, but his reasons for ignoring other possibilities, such as a K^+/water interaction, were not sufficient.

II. Theoretical Background

According to the theory of non-equilibrium thermodynamics (see, e.g. Katchalsky and Curran, 1965) the flow of one chemical species i crossing a membrane may be affected by the forces acting on all the other species crossing the membrane as well as by the forces acting on species i itself. The relationships between the flows (J) and forces (X) can be expressed either in terms of conductance coefficients (L) or in terms of resistance coefficients (R), as follows

$$J_i = \sum_j L_{ij} X_j \qquad (5a)$$

$$X_i = \sum_j R_{ij} J_j \qquad (5b)$$

where both summations include $i = j$.

Under suitable conditions, which are probably satisfied in most transporting systems (Miller, 1960) the Onsager reciprocal relationship is valid: i.e.

$$L_{ij} = L_{ji} \text{ and } R_{ij} = R_{ji}. \tag{6}$$

When the coupling is such that the flow of j tends to cause a flow of i in the same direction, then L_{ij} is positive and R_{ij} is negative. The straight coefficients L_{ii} and R_{ii} are always positive. It must be pointed out that the forces (X) are the total forces across the whole membrane, and the L and R coefficients are integral coefficients. The situation is much simplified under steady-state conditions when the flows (J) are constant across the membrane, and in the rest of this paper, steady-state conditions will be assumed.

For the present purpose, the most convenient form of the flux-ratio equation derived from non-equilibrium thermodynamics is a modified form of Simons' eqn 18 (Simons, 1969):

$$\ln f_i = \left[X_i - \sum_{j \neq i} R_{ij} J_j - R_{ii*} J_i \right] \Big/ RT \tag{7}$$

Here the first term inside the brackets corresponds to the simple form of the Ussing–Teorell equation (eqn (1) above); the summation allows for the effects of the interactions between i and any other flows in the system; and the third term allows for the effects of isotope interaction. J_i is the net flow of species i. In eqn (7), Simons' separate terms for the components of the thermodynamic driving force on species i have been collected into the term X_i; Simons' terms $\int_0^\delta r_{ij} \cdot J_j \cdot dx$, etc., where r_{ij} is a local resistance coefficient in an element of thickness dx, have been replaced by the terms $R_{ij} J_j$ etc. It is clear that, under steady state conditions

$$R_{ij} = \int_0^\delta r_{ij} \, dx$$

where δ is the thickness of the membrane. Thus, from eqn (7), the flux-ratio equation for K^+ may be written

$$\ln f_K = \left[X_K - R_{KK}^* J_K - R_{KW} J_W - \sum_{j \neq K, W} R_{Kj} J_j \right] \Big/ RT \tag{8}$$

where subscripts K, K*, and W stand for K^+ ion, tracer isotope of K^+, and water respectively. In eqn (8), the term $R_{KW} J_W$, which allows for interaction between K^+ and water, has been separated from the rest of the summation.

III. Discussion

In the following sections, it is assumed that the transport properties of the plasmalemma alone can be discussed separately from those of the combined structural unit consisting of the plasmalemma and cell wall.

In practice, this means that the plasmalemma itself should be the major resistance to the flows of K^+ and water. The plasmalemma is known to be the site of the major resistance to ion movement (Walker, 1960; Findlay and Hope, 1964) and is probably the site of the major resistance to water, at least during bulk flow (Tyree, 1968; Barry and Hope, 1969). The data collected from the literature in order to calculate the values of the required L coefficients are believed to represent the properties of the plasmalemma alone, rather than those of the combination of plasmalemma plus cell wall.

A. Treatment Assuming a Homogeneous Membrane

The relevant data concerning the unidirectional fluxes of K^+ in *Chara corallina* are given in Fig. 2, which is reproduced from the paper of Walker and Hope (1969). In attempting to discover why n does not equal unity for

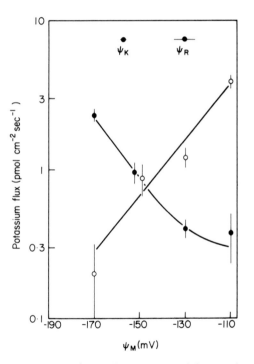

FIG. 2. Unidirectional fluxes of K^+ at different values of the membrane potential (ψ_M) in *Chara corallina*. ψ_K is the Nernst potential for potassium and ψ_R is the normal resting potential. (Reproduced with permission from Walker and Hope, 1969).

the flux-ratio of K^+ in *Chara*, we must consider each of the terms other than X_K in the bracket on the right side of eqn (8).

Simons (1969) has argued that, since a plot of J_K versus $(\ln f_K - X_K/RT)$ is a straight line, then the terms $R_{KW}J_W$ and $\sum R_{Kj}J_j$ must be negligible, so that the only important interaction is the isotope interaction. This argument cannot be conclusive by itself, as it is possible that the flows J_W and J_j may themselves be proportional, or approximately proportional to J_K, either because the flows of water and j are caused by the flow of K^+, or because all these flows may depend on the same driving forces.

In Walker and Hope's experiments, the changes in J_K were caused by imposed changes in the membrane potential. These changes in potential could also cause changes in the fluxes of other ions, such as Na^+; for small changes in potential the changes in J_K and J_{Na} would be proportional to the changes in potential and hence to each other. Under these conditions an interaction between Na^+ and K^+ would not necessarily cause a non-linearity in the plot of J_K versus $(\ln f_K - X_K/RT)$. However, as pointed out by Coster and George (1968) in such a situation it is unlikely that the flux of Na^+ would be zero at the Nernst potential for K^+. Thus if there is any appreciable coupling between the K^+ and Na^+, then the value of the membrane potential for which f_K is unity (the K^+ cross-over potential) should differ significantly from the Nernst potential for K^+. In *Chara*, the Nernst potential for Na^+ is far from the Nernst potential for K^+, while the latter is close to the cross-over potential for K^+. This constitutes strong evidence against the existence of a large contribution to n for K^+ from a K^+/Na^+ interaction, or from any interaction between K^+ and another ionic species.

There is good evidence that, in *Chara*, the transport of K^+ is not linked to any metabolic process; i.e. K^+ is passively distributed (Hope, 1962). The Nernst potential for K^+ is close to the resting potential of the membrane, and the usual inhibitors of active ion transport have very little effect on the fluxes of K^+. Thus it is unlikely that eqn (8) contains a significant term $R_{Kj}J_j$ caused by interaction between K^+ and metabolic processes.

This leaves the possibility of a K^+/K^+ isotope interaction and/or a $K^+/$water interaction. Using his argument outlined above, Simons (1969) argued that there could be no significant contribution to the flux-ratio for K^+ from a $K^+/$water interaction. However, it is with precisely such an interaction, where the water flow is driven by the flow of K^+, that we might expect J_W to be proportional to J_K. Also, a $K^+/$water interaction almost certainly does occur in the plasmalemma of *Chara*; studies of electro-osmosis indicate that the plasmalemma has a significant electro-osmotic coefficient (Barry and Hope, 1969), and K^+ is the major current-carrying ion (Walker and Hope, 1969). Simons' argument is, therefore, not sufficient.

In Walker and Hope's experiments the electro-osmotic flow of water caused by the net flow of K^+ would cause a change, perhaps only very small, in the turgor pressure of the cell, and this change would be in such a direction as to oppose the induced water flow (J_W). The value of the change in turgor required to cause a significant reduction in J_W during the course of a K^+ flux experiment would depend on a number of factors but primarily on the value of L_{WW} for the plasmalemma. It can be shown (see Appendix A) that the change in cell turgor caused by an electro-osmotic flow of water (J_W) would cause that flow to decrease according to the relationship

$$J_W = J'_W \exp(-0.3\,t) \tag{9}$$

where J'_W is the initial value of the electro-osmotic flow and t is the time in seconds from the start of the flow. Thus the value of J_W should drop to 20% of its initial value after only 7 s and to 1% of its initial value after 14 s. As Walker and Hope's K^+ flux experiments usually lasted for 20–30 min., this result suggests that the term $R_{KW}J_W$ in eqn (8) should be very small. Meanwhile, the decrease in J_K produced by the small change in turgor should be only about 0.1% (Appendix A).

This conclusion can be put on a more quantitative basis if the small change in turgor is ignored so that X_W remains zero, and J_W retains its initial value. Under these conditions and for a system where the only interactions are between K^+ and K^* (tracer K^+), K^+ and W (water) and between K^* and water, it can be shown (see Appendix B) that the flux-ratio equation for K^+ becomes

$$\ln f_K = X_K[1 - R_{KW}L_{KW} - R_{KK^*}L_{KK}]/RT \tag{10}$$

where

$$R_{KW} = -L_{KW}/(L_{KK}L_{WW} - L_{KW}^2) \tag{11}$$

Thus

$$(n - 1) = -[R_{KW}L_{KW} + R_{KK^*}L_{KK}] \tag{12}$$

In deriving these relationships it has also been assumed that, because of the very low concentration of K^* present in the system, the effects of the K^+/K^* and K^*/water interactions on J_K and J_W respectively can be ignored.

Sufficient data on transport processes in *Chara* are available in the literature to permit an evaluation of the relevant L coefficients. This evaluation is given in Appendix C, and the values of the various L coefficients are presented in Table I.

With the L values in Table I, it can be calculated from eqns (10) and (11) that the K^+/water interaction makes a contribution of only 10^{-5} to the total difference of 1.5 between the value of n and unity. Furthermore, because the decrease in J_W during the K^+ flux experiment has been ignored, this calculated contribution must be much larger than the true one.

It would appear therefore that the difference between the value of n (i.e. 2.5) and unity for the flux-ratio of K^+ in *Chara* must arise solely from an isotope interaction, perhaps of the "long pore" type.

TABLE I

Values of L coefficients	$mol^2 J^{-1} s^{-1} cm^{-2}$
L_{KK}	3×10^{-15}
L_{KW}	1×10^{-13}
L_{WW}	4×10^{-7}
$L_{K^*K^*}$	9×10^{-24}
L_{KK^*}	$1{\cdot}2 \times 10^{-23}$
L_{K^*W}	7×10^{-21}
$b \cdot L_{WWB}$	?

However, the discussion so far has made the common assumption that the plasmalemma of *Chara* is homogeneous with regard to the flows of K^+ and water; i.e. it has been assumed that all the K^+ and all the water cross the membrane by the same, and only by the same, uniform set of "channels" or "pathways". This assumption has been attacked previously (Dainty *et al.*, 1963), the basis of the attack being the very large difference between the permeability of the membrane to water and its permeability to K^+ (compare the values of L_{WW} and L_{KK} in Table I). Even with allowance for the fact that the values of L_{WW} and L_{KK} are dependent on the concentrations of water and K^+ respectively, the difference between them is still very large. The simplest explanation for the electro-osmosis shown by the cell membrane is that the K^+ and at least some of the water share a common pathway, such as a water-filled pore. It is difficult to see how such a common pathway could have such widely different permeabilities to K^+ and water. Thus we are led to the idea that another set of pathways is available for water, but not for K^+, to cross the membrane; and that the permeability of these pathways to water is greater (perhaps very much greater) than that of the K^+/water pathways. The flux-ratio for K^+ in *Chara* will be analysed in terms of this model of the membrane in the following section.

B. Treatment Assuming a Heterogeneous Membrane

According to the heterogeneous model introduced above, the membrane contains two types of elements arranged in parallel: elements of type A, which occupy a fraction a of the surface area of the membrane, are a major pathway for water movement, but are impermeable to K^+; elements of type B, which occupy a fraction b of the membrane surface, carry all the K^+ flow, and provide a small fraction of the total permeability of the membrane to water. Elements A would be characterized by a water permeability, L_{WWA}; elements B would have a water permeability L_{WWB}, a K^+ permeability L_{KKB}, and an electro-osmotic cross-coefficient L_{KWB}. The values of none of these L coefficients are known, but they are related to the known mean transport coefficients for the whole membrane as follows:

$$L_{KK} = b \cdot L_{KKB}; \qquad K_{KW} = b \cdot L_{KWB} \tag{13a}$$

$$L_{WW} = a \cdot L_{WWA} + b \cdot L_{WWB} \tag{13b}$$

where the term $a \cdot L_{WWA}$ is much larger than $b \cdot L_{WWB}$.

For such a membrane system, the flux-ratio for K^+ would be identical to the flux-ratio for K^+ in the B-type elements alone (Kedem and Essig, 1965), and would be given by an equation of the same form as eqn (7), but containing the transport coefficients for the "B" elements instead of the transport coefficients for the whole membrane: i.e.

$$\ln f_K = [X_K - R_{KK^*B}J_{KB} - R_{KWB}J_{WB}]/RT \tag{14}$$

Here J_{KB} is the net flow of K^+ per unit area of elements B, and J_{WB} is the corresponding flow of water.

The arguments set out in the previous section (A) about the negligible nature of interactions other than the K^+/K^+ isotope interaction and the K^+/water interaction are still valid, so the terms $\sum_{j \neq K^*,W} R_{KjB}J_{jB}$ have been omitted from eqn (14).

Again, as in the case of the homogeneous membrane, the electro-osmotic flow of water through elements B will cause a change in X_W (largely a change in the tugor pressure of the cell) and this change will in turn cause a reverse water flow with the same characteristics as before (eqn 9). However, the total reverse water flow will be distributed between the A and B elements in the ratio $a \cdot L_{WWA}/b \cdot L_{WWB}$. If, as postulated above, this ratio is much bigger than unity, most of this reverse flow will occur via elements A and the water flow through elements B will retain its initial value ($L_{KWB} \cdot X_K$) during the whole K^+ flux experiment.

Under these conditions, eqn (14) can be cast into a form analogous to eqn (10):

$$\ln f_K = X_K[1 - R_{KWB}L_{KWB} - R_{KK^*B}L_{KKB}]/RT \tag{15}$$

so that

$$(n - 1) = -[R_{KWB}L_{KWB} + R_{KK^*B}L_{KKB}] \tag{16}$$

Again, the first term in the brackets arises from a K^+/water interaction, and the second term arises from a K^+/K^+ isotope interaction.

With the help of eqns (13a), the terms in the brackets in eqn (16) become:

$$R_{KWB}L_{KWB} = -L_{KW}^2/[(b \cdot L_{WWB})L_{KK} - L_{KW}^2] \tag{17}$$

and

$$R_{KK^*B}L_{KKB} = \frac{L_{KK}[L_{KW}L_{K^*W} - L_{KK}(b \cdot L_{WWB})]}{L_{K^*K^*}[L_{KK}(b \cdot L_{WWB}) - L_{KW}^2]} \tag{18}$$

Thus in total, the two terms in the brackets in eqn (16) contain six different L coefficients. The values of five of these can be obtained from available experimental data (see Appendix C) and their values have already been given in Table I.

Only the value of $b \cdot L_{WWB}$ cannot be obtained from the experimental data in the literature, and, furthermore, it would appear to be highly unlikely that the value of $b \cdot L_{WWB}$ can be measured. Any experiments aimed at measuring this quantity would require either that the water flow through the B elements should be measured, or that it should be controlled. As the water flow through the B elements is only a part, and probably only a very small part, of the total water flow across the membrane, it would seem to be impossible to satisfy their requirements.

Because the value of $b \cdot L_{WWB}$ is not known, and because both the terms $R_{KWB}L_{WB}$ and $R_{KK^*B}L_{KKB}$ contain this term, it is not possible to calculate the separate contributions from the K^+/K^+ isotope interactions and from the K^+/water interaction to the value of $(n-1)$ for K^+ in *Chara*. It might appear that it should be possible to find, by trial, the value of $b \cdot L_{WWB}$ which gives the required value for $(n-1)$ and then to use this value of $b \cdot L_{WWB}$ to evaluate the separate contributions from K^+/K^+ isotope interaction and the K^+/water interaction. However, it can be shown (Appendix C) that for this system the value of $(n-1)$ is independent of the value of $b \cdot L_{WWB}$.

From Appendix C,

$$(n-1) = L_{KK^*}/L_{K^*K^*} \tag{19}$$

The value of $b \cdot L_{WWB}$ determines the relative values of the contributions from the two types of interaction, but has no effect on the sum of these contributions.

The fact that eqn (19) contains only L_{KK^*} and $L_{K^*K^*}$ does not by itself mean that the K^+/water interaction makes a negligible contribution to the value of n. The L coefficients are not simple quantities and the value of any one may be affected by any of the interactions occurring in the system. The only true measure of the relative importance of the two different interactions are the terms $R_{KWB}L_{KWB}$ and $R_{KK^*B}L_{KKB}$ which appear in eqns (16)–(18).

Although the value of $b \cdot L_{WWB}$ cannot be obtained, we can obtain an estimate of the range of possible values for this quantity. The maximum possible value is obviously the value of L_{WW} for the whole membrane; i.e. $4 \times 10^{-7} \, \text{mol}^2 \, \text{J}^{-1} \, \text{s}^{-1} \, \text{cm}^{-2}$. A minimum possible value can be obtained because it is known from the theory of non-equilibrium thermodynamics (Katchalsky and Curran, 1965, p. 91) that

$$L_{KKB}L_{WWB} \geqslant L_{KWB}^2 \tag{20}$$

Thus, with the help of eqns (13a) we obtain the minimum possible value for $b \cdot L_{WWB}$:

$$b \cdot L_{WWB} \geqslant L_{KW}^2/L_{KK} = 3 \times 10^{-12} \, \text{mol}^2 \, \text{J}^{-1} \, \text{s}^{-1} \, \text{cm}^{-2}$$

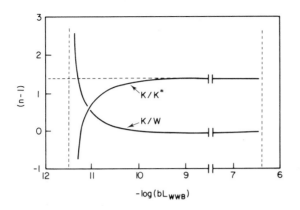

FIG. 3. The contributions from K^+/K^+ isotope interaction (K/K*) and K^+/water interaction (K/W) to the value of $(n-1)$ for K^+ in *Chara corallina*, as functions of the permeability $(b . L_{WWB})$ of the "B" type elements to water.

The variation of the values of the two terms $R_{KWB}L_{KWB}$ and $R_{KK^*B}L_{KKB}$ over the range of possible values of $b . L_{WWB}$ is shown in Fig. 3.

It can be seen that the K^+/K^+ isotope interaction makes the major contribution to $(n-1)$ over most of the range of possible values of $b . L_{WWB}$, while the K^+/water interaction is the major one only if $b . L_{WWB}$ lies between 3×10^{-12} and 8×10^{-12} mol^2 J^{-1} s^{-1} cm^{-2}.

C. Conclusions

A qualitative analysis of the data concerning K^+ fluxes in *Chara corallina* suggests that the only interactions which need be considered in attempts to explain the factor of 2·5 which occurs in the flux-ratio equation are a K^+/water interaction and a K^+/K* isotope interaction. An analysis using transport equations which specifically allow for coupling between flows shows that if the membrane is homogeneous, i.e. if all the water and all the K^+ crossing the membrane use the same single set of uniform pathways— then the K^+/K^+ isotope interaction makes by far the largest contribution to the value of n. If, however, a perhaps more realistic model is used, in which some of the water and all the K^+ use one set of pathways while the rest of the water uses another set, then the values of the contributions to n from the two separate interactions cannot be calculated. Calculations would require a knowledge of the contribution which the K^+/water pathways make to the total permeability of the membrane to water. This information is not available, and is probably unobtainable.

It is, however, possible to define a range of possible values for the contribution of the K^+/water pathways to the total permeability to water, and it is shown that over most of this range the K^+/K^+ isotope interaction makes the major contribution to n. Only over the very lowest part the range of possible values for the water permeability of the K^+/water pathways does the K^+/water interaction make a significant contribution to the value of n.

Speculation about a probable value for $b \cdot L_{WWB}$ is difficult, as this value must depend critically on the as yet unknown mechanism by which water traverses pathways of type B. However Dainty et al. (1963) have argued that these pathways could account for about 10^{-5} of the total permeability of the membrane to water. Their argument was based on a model in which the K^+ and water crossed the common pathways via water-filled pores. The specific conductance of these pores was assumed to be about $10^{-1} \Omega^{-1} cm^{-1}$, which corresponds to a K^+ concentration of about $1 \, mol \, l^{-1}$. This argument therefore leads to a value of about $5 \times 10^{-12} mol^2 J^{-1} s^{-1} cm^{-2}$ for $b \cdot L_{WWB}$, which is very near to the minimum possible value, and which lies in the region of Fig. 3 where the K/water interaction makes the major contribution to n. However the assumed concentration of K^+ in the pores is quite high, and slightly lower but equally acceptable values for the concentration would lead to values of $b \cdot L_{WWB}$ of about $10^{-11} mol^2 J^{-1} s^{-1} cm^{-2}$. This value lies in the region of Fig. 3 where the relative importance of the K/water and K/K* interactions depends greatly on the exact value of $b \cdot L_{WWB}$.

It is sometimes possible to analyse transport coefficients in terms of molecular events occurring in the membrane. This is usually easier by means of an analysis of the values of R coefficients than by means of an analysis of L coefficients. Each L coefficient is a complex quantity whose value may be affected by a number of interactions, whereas each R_{ij} coefficient is a more direct measure of the one interaction between flows i and j.

With the system under consideration here, such an analysis is not practicable. The "homogeneous-membrane" model is not very realistic, and, because the value of $b \cdot L_{WWB}$ is not known, it is not possible to calculate the values of the R coefficients for the B-type elements of the "heterogeneous-membrane" model.

However, one point can be made. If it is true that the major interaction affecting the flow of tracer K^+ is the K^+/K^+ isotope interaction, rather than the K*/water interaction, then it can be shown that, in terms of the "frictional" model of transport (Spiegler 1958),

$$(n - 1) = (K_{KK*}/L_{K*K*}) = F_{K*K}/F_{K*M}$$

where F_{K*K} is the force acting on one mole of tracer K^+ owing to its interaction with non-tracer K^+, and F_{K*M} is the force acting on one mole of tracer K^+ owing to its interaction with the membrane material. From the known value of n ($n = 2 \cdot 5$) it is seen that, in its movement through the membrane, the tracer K^+ interacts more strongly with the non-tracer K^+ than with the membrane material.

From Fig. 3 it can be seen that over most of the range of possible values for $b \cdot L_{WWB}$, the quantity $(-R_{KK*}L_{KK})$ is positive. As L_{KK} must be positive this means that R_{KK*} is negative. This is typical of the type of interaction which involves mutual drag between flows, i.e. where the flow of K^+ in one direction enhances the flow of K^* in the same direction or diminishes the flow of K^* in the reverse direction. However, at very low values of $b \cdot L_{WWB}$ the quantity $(-R_{KK*}L_{KK})$ is negative so that R_{KK*} must be positive. This could arise if the K^+/K^* interaction involved exchange-diffusion. At the very lowest values of $b \cdot L_{WWB}$, the value of $(-R_{KK*}L_{KK})$ appears to be less (more negative) than -1. The physical significance of an interaction of this type between ion flows is not clear; it is possible that these very negative values of $(-R_{KK*}L_{KK})$ are caused by inaccuracy of the values obtained for the L coefficients.

No estimates have been given in Table I for the accuracies of the values quoted there. The accuracy is likely to vary greatly from one coefficient to another, the most accurate probably being the values for L_{WW} and L_{KW} with possible errors of $\pm 10\%$, and the least accurate being the values for L_{KK} and L_{KK*} with possible errors of $\pm 50\%$. Even errors of this magnitude in the values of the L coefficients would not have a significant effect on the general conclusions concerning the relative importance of the K^+/K^+ isotope interaction and the $K^+/$water interaction.

Acknowledgement

The author wishes to thank the editor of the Australian Journal of Biological Sciences for permission to include Fig. 1, which is reproduced from the paper by Walker and Hope (1969).

IV. Appendices

A. The Time-dependence of the Electro-osmotic Flow of Water

For the type of system under discussion, the net flows of K^+, (J_K) and of water (J_W) are given by the relationships

$$J_K = L_{KK}X_K + L_{KW}X_W \qquad (A.1)$$

$$J_W = L_{KW}X_K + L_{WW}X_W \qquad (A.2)$$

The force (X_W) on the water arises mainly from the change (ΔP) in turgor pressure of the cell, so that we can write

$$X_W = -V_W \cdot \Delta P$$

$$= -V_W \cdot e\frac{\Delta V}{V}$$

$$= -(V_W e/V) \int_0^t (J_W A V_W \, dt)$$

$$= -(2V_W^2 e/r) \int_0^t J_W \, dt \qquad (A.3)$$

where V_W is the molar volume of water, e is a constant relating the fractional change in volume of the cell to the change in turgor pressure, V is the cell volume, ΔV is the change in cell volume produced by the flow J_W, A is the surface area of the cell, t is the time from the beginning of the flow J_W, and r is the radius of the (cylindrical) cell.

By combining eqns (A.2) and (A.3) and then intergrating we obtain the expression

$$J_W = (L_{KW}X_K)\exp\left[-(2L_{KW}V_W^2\,et)/r\right]$$
$$= J_W'\exp\left[-2(L_{KW}V_W^2\,et)/r\right] \tag{A.4}$$

where J_W' is the initial value of the electro-osmotic flow. With values of 18 dm^3 for V_W, 0.03 cm for r for a "typical" *Chara cell*, $5 \times 10^7\,Nm^{-2}$ for e (Kelly *et al.*, 1963), and $4 \times 10^{-7}\,mol^2\,J^{-1}\,s^{-1}\,cm^{-2}$ for L_{WW} (Appendix C), eqn (24) becomes

$$J_W = J_W'\exp(-0.3t) \tag{A.5}$$

where t is the time in seconds.

It can be shown in a similar manner, by combining equations (A.1) and (A.3) that the change in J_K due to the change in turgor pressure should be given by the expression

$$J_K = J_K'[1 - L_{KW}^2(1 - \exp(-0.3t)/L_{KK}L_{WW}]$$
$$= J_K'[1 - 10^{-3}(1 - \exp(-0.3t))] \tag{A.6}$$

Thus the maximum reduction in J_K is about 0.1%.

Combination of eqns (A.2) and (A.3) and solution for X_W gives the expression

$$X_W = -(L_{KW}X_K/L_{WW})[1 - \exp(-0.3t)]$$
$$= -2.5 \times 10^{-5}X_K[1 - \exp(-0.3t)] \tag{A.7}$$

where t is in seconds.

This shows that when the membrane potential is held at a value 30 mV from the Nernst potential for K^+, the maximum value of X_W corresponds to a change in turgor pressure of $4 \times 10^3\,Nm^{-2}$ or 4×10^{-2} atmospheres. A typical value for the turgor pressure of a *Chara* cell is 8 atm (Kelly *et al.*, 1963).

B. Derivation of the Flux-ratio Equation (eqn 10)

For a system where the only interactions affecting the flows of K* (tracer isotope of K^+), K^+, and W (water) are a K^+/K^* isotope interaction, a $K^+/$ water interaction and a K*/water interaction, we can write

$$J_K = L_{KK}X_K + L_{KW}X_W$$
$$J_W = L_{KW}X_K + L_{WW}X_W \tag{A.8}$$
$$J_{K^*} = L_{KK^*}X_K + L_{KW^*}X_W + L_{K^*K^*}X_{K^*}$$

and

$$X_K = R_{KK}J_K + R_{KW}J_W$$
$$X_W = R_{KW}J_K + R_{KW}J_W \hspace{2cm} (A.9)$$
$$X_{K^*} = R_{K^*W}J_K + R_{K^*W}J_W + R_{K^*K^*}J_{K^*}$$

Because the concentration and flow of K* are so small, the effects of the K/K* and K*/water interactions have been ignored in the expressions for J_K, J_W, X_K, X_W.

From eqns (A.8) and (A.9) we obtain

$$R_{KW} = -L_{KW}/(L_{KK}L_{WW} - L_{KW}^2)$$
$$R_{K^*W} = (L_{KK^*}L_{KW} - L_{K^*W}L_{KK})/[L_{K^*K^*}(L_{KK}L_{WW} - L_{KW}^2)]$$

and

$$R_{KK^*} = (L_{KW}L_{K^*W} - L_{KK^*}L_{WW})/[L_{K^*K^*}(L_{KK}L_{WW} - L_{KW}^2)]$$

With the assumption, introduced on p. 83, that X_W is zero, eqn (8) can be combined with eqn (A.8) to give the flux-ratio equation

$$\ln f_K = X_K[1 - R_{KW}L_{KW} - R_{KK^*}L_{KK}]/RT \hspace{2cm} (A.10)$$

This equation is probably still valid for the heterogeneous membrane model, in which case eqns (A.8) and (A.9) would refer only to the elements of type B. With this model, it was not assumed that X_W was zero. However the reduction $(L_{WW}X_W)$ in the electro-osmotic water flow through the B elements due to the change in turgor pressure is probably very small compared with the electro-osmotic flow $(L_{KW}X_K)$ itself. Also the decrease $(L_{KW}X_W)$ in the flow of K^+ due to the increase in turgor pressure has been shown (Appendix A) to be small compared with the initial flow $(L_{KK}X_K)$.

If it is assumed that the kinetic behaviour of isotopes is identical, then it must be true that

$$R_{K^*W} = R_{KW}$$

From the relationship given above for R_{K^*W} and R_{KW}, this means that

$$L_{K^*W} = L_{KW}(L_{KK^*} + L_{K^*K^*})/L_{KK}$$

with this relationship and with the relationships given above for R_{KW}, and R_{KK^*}, eqn (A.10) can be converted into the form

$$\ln f_k = (X_K/RT)(1 + L_{KK^*}/L_{K^*K^*})$$

so that

$$n = (1 + L_{KK^*}/L_{K^*K^*})$$

C. Methods of Obtaining Values for L Coefficients

The values of the L coefficients given earlier in Table I were obtained in the following ways.

1. L_{KK}

From eqn (A.8)

$$L_{KK} = (\partial J_K/\partial X_K)_{X_W}$$
$$= (1/F^2)[\partial I_K/\partial(\psi_1 - \psi_2)_{X_W}]$$
$$= (1/F^2)g_K \qquad (A.11)$$

where I_K is the electric current carried by K^+ and g_K is the electrical conductivity due to K^+ ions.

Walker and Hope (1969) give a value of $1.5 \times 10^{-5}\,\Omega^{-1}\,cm^{-2}$ for g_K. However they argue that this value is probably an underestimate while their value of $7 \times 10^{-5}\,\Omega^{-1}\,cm^{-2}$ for L_E, the total membrane conductance is probably an overestimate. They also argue that the K^+ conductance makes the major contribution to the total conductance. A value of $3 \times 10^{-5}\,\Omega^{-1}\,cm^{-2}$ was therefore used to obtain the value of L_{KK} given in Table I.

2. L_{WW}

From eqn (A.8)

$$L_{WW} = (\partial J_W/\partial X_W)_{X_K}$$
$$= (1/V_W^2)(\partial J_V/\partial \Delta\pi)_{X_K}$$
$$= (1/V_W^2)L_p \qquad (A.12)$$

where V_W is the molar volume of water ($18\,cm^3\,mol^{-1}$), J_V is the flow of volume across the membrane, and L_p is the hydraulic permeability coefficient for the membrane. From separate values of L_p for isolated cell wall and for the intact cell, Barry and Hope (1969) calculated a value of 1.4×10^{-5} $cm\,s^{-1}\,atm^{-1}$ (i.e. $1.4 \times 10^{-4}\,cm^4\,s^{-1}\,J^{-1}$) for the L_p of the plasmalemma alone.

3. L_{KW}

From eqn (A.8)

$$L_{KW} = (\partial J_W/\partial X_K)_{X_W}$$
$$= (1/F)[\partial J_W/\partial(\psi_1 - \psi_2)_{X_W}]$$
$$= (1/F)(\partial J_W/\partial I)_{X_W}[\partial I/\partial(\psi_1 - \psi_2)_{X_W}]$$
$$= [1/(V_W F)] . \beta . L_E \qquad (A.13)$$

where I is the total current crossing the membrane, and β is the electroosmotic permeability ($cm^3\,C^{-1}$) of the plasmalemma. From separate values of β for the isolated cell wall and for the whole cell, Barry and Hope (1969) calculated a value of $6 \times 10^{-3}\,cm^3\,C^{-1}$ for β of the plasmalemma. This value for β has also been corrected for the effect of local osmosis. It is arguable that g_K should be used instead of L_E in eqn (A.13). For this reason, and for the reasons given in part 1 of this Appendix, the same value was used for L_E in eqn (A.13) as was used for g_K in eqn (A.11).

4. L_{K^*K}, $L_{K^*K^*}$ and L_{K^*W}

From eqn (A.8),

$$L_{K^*K^*} = (J_{K^*}/X_{K^*}); \quad X_K = X_W = 0 \tag{A.14}$$

Thus the value of $L_{K^*K^*}$ can be obtained by measuring the flux of tracer under a known tracer concentration difference and at the Nernst potential for K^+. X_{K^*} includes contributions from both the tracer concentration gradient and from the membrane potential ($\psi_1 - \psi_2$).

Also if the forces X_K and X_{K^*} are separated into their concentration (or activity) and electrical potential terms, then we get from eqns (A.8)

$$(L_{K^*K^*} + L_{KK^*}) = [\partial J_{K^*}/\partial(\psi_1 - \psi_2)] \tag{A.15}$$

where X_W and all the concentrations are kept constant. Thus the value of $(L_{K^*K^*} + L_{KK^*})$ can be obtained by measuring the change in J_{K^*} caused by a change in the membrane potential.

With the assumption of identical kinetic behavior of isotopes, it can be shown (Kedem and Essig, 1965) that

$$R_{K^*W} = R_{KW} \tag{A.16}$$

This relationship, together with the relationships between the R and L coefficients given in Appendix B, lead to the following equation

$$L_{K^*W} = L_{KW}(L_{KK^*} + L_{K^*K^*})/L_{KK} \tag{A.17}$$

The values of all the L coefficients on the right side of eqn (A.17) have already been obtained, so the value of L_{K^*W} can be calculated.

It must be emphasized that, as the L coefficients are dependent on the concentrations of the various chemical species present, the data used to calculate the L coefficients must all be obtained from experiments having the same concentrations and distributions of the chemical species, and particularly of the tracer isotope. In practice this condition is usually observed, at least approximately, in a series of flux experiments. Also, the values used for J_{K^*} and for the concentrations of tracer in eqns (A.14) and (A.15) must be expressed in moles of tracer and not in c.p.s.

Sufficient data are available (private communication) on the specific activities of bathing solutions, count rates of cell sap and bathing solutions etc used in the experiments of Walker and Hope (1969), together with the results given in their paper, to permit the use of eqns (A.14), (A.15) and (A.17) for the calculation of the values of $L_{K^*K^*}$, L_{K^*K}, L_{K^*W}.

D. Symbols Used

a	fraction of membrane surface occupied by type "A" elements
a_i	thermodynamic activity of chemical species i
b	fraction of membrane surface occupied by type "B" elements
c_i	concentration of chemical species i
e	elasticity constant of cell wall
f_i	flux-ratio of chemical species i
j_i	unidirectional flux of chemical species i
r_{ij}	differential transport coefficient
n	coupling factor in flux-ratio equation
I	electric current
J_i	net flux of chemical species i
$\left.\begin{array}{l} L_{ii} \\ L_{ij} \end{array}\right\}$	transport coefficients
P	turgor pressure of cell
R	gas constant ($8.3 \text{ J mol}^{-1} \text{ K}^{-1}$)
$\left.\begin{array}{l} R_{ii} \\ R_{ij} \end{array}\right\}$	transport coefficients
T	temperature (K)
V_W	molar volume of water
X_i	thermodynamic force acting on species i
β	electro-osmotic permeability
π	osmotic pressure
ψ	electrical potential

References

BARRY, P. H. and HOPE, A. B. (1969). *Biochim. biophys. Acta* **193**, 124–128.

COSTER, H. G. L. and GEORGE, E. P. (1968). *Biophys. J.* **8**, 457–469.

DAINTY, J., CROGHAN, P. C. and FENSOM, D. S. (1963). *Can. J. Bot.* **41**, 953–966.

FINDLAY, G. P. and HOPE, A. B. (1964). *Aust. J. biol. Sci.* **17**, 62–77.

HODGKIN, A. L. and KEYNES, R. D. (1955). *J. Physiol., Lond.* **128**, 61–88.

HOPE, A. B. (1962). *Aust. J. biol. Sci.* **16**, 429–441.

HOSHIKO, T. and LINDLEY, B. D. (1964). *Biochim. biophys. Acta* **79**, 301–317.

KATCHALSKY, A. and CURRAN, P. F. (1965). "Non-equilibrium Thermodynamics in Biophysics". Harvard University Press, Cambridge, Mass.

KEDEM, O. and ESSIG, A. (1965). *J. gen. Physiol.* **48**, 1047–1070.

KELLY, R. B., KOHN, P. G. and DAINTY, J. (1963). *Trans. Proc. bot. Soc. Edinb.* **34**, 373–391.

MILLER, D. G. (1960). *Chem. Rev.* **60**, 15–37.

SIMONS, R. (1969). *Biochim. biophys. Acta* **173**, 34–50.

SPIEGLER, K. S. (1958). *Trans. Faraday Soc.* **54**, 1408–1428.

TEORELL, T. (1949). *Archs Sci. physiol.* **3**, 205–218.

TYREE, M. T. (1968). *Can. J. Bot.* **46**, 317–327.

USSING, H. H. (1949). *Acta physiol. scand.* **19**, 43–56.

WALKER, N. A. (1960). *Aust. J. biol. Sci.* **13**, 468–478.

WALKER, N. A. and HOPE, A. B. (1969). *Aust. J. biol. Sci.* **22**, 1179–1195.

Discussion

This session started by Nissen saying that he had analysed data similar to that shown by Thellier in Figs 1 a–c with different conclusions. He submitted the following written statement:

The data in Fig. 1 can be better represented by single, multiphasic isotherms than by the electrokinetic formulation or the original models:

Figure 1a represents uptake of chloride by root tips of corn as a function of $CaCl_2$ concentration (Fig. 3 in Torii and Laties, 1966). These data are precisely represented by 5 phases of a multiphasic isotherm, as are the corresponding data for proximal sections (Nissen, *Physiol. Plantarum*, in press).

Figure 1b represents uptake of chloride by root tips as a function of NaCl concentration (Fig. 2 in Torii and Laties, 1966). These and the corresponding data for proximal sections are, again, quite precisely represented by 5 phases (in preparation).

Figure 1c represents uptake of sodium by barley roots as a function of NaCl concentration (Fig. 4 in Rains and Epstein, 1967). These and similar data for uptake from solutions of Na_2SO_4 (Fig. 5 in Rains and Epstein, 1967) can be precisely represented by 4 phases of a single, multiphasic mechanism (in preparation).

Discontinuous transitions seem, on the basis of a comprehensive re-examination of concentration-dependence data, an invariant feature of ion uptake in higher plants (Nissen, this volume, *Physiol. Plantarum*, in press, and in preparation). The relevance of models or formulations yielding continuous isotherms over extended concentration ranges may, therefore, be questioned.

Nissen stressed that the discontinuities between the phases on his model are significant. All other models, e.g. Michaelian kinetics or Thellier's electro-kinetics, predict continuous relationships. Thellier replied that his formalism is not an interpretation of a specific set of data, it is rather a thermodynamic language to be used in discussions. There are many reasons for discontinuities or deviations from linearity other than the establishment of a new uptake system. Further the superposition of two hyperbolae in the manner employed by the System I–System II exponents can never give rise to linear Line-weaver–Burke plots. There are interactions between the K_m values and the V_{max} values of the two individual systems.

Walker then asked, "In your equations you have a parameter to the power m which is introduced simply to account for deviations from linearity. Can you justify ascribing properties such as intensive and extensive to this parameter which is introduced rather arbitrarily to account for non-linearity?" Thellier replied indirectly; several of the models he has taken, which predict various sets of experimental data, appear to be physically real systems. In fact they are not. In other words, a model which behaves like a semiconductor, need not be a semiconductor; it simply behaves in a fairly similar fashion. One can derive useful information by examining the extensive–intensive properties of parameters. For example, his interpretation and Epstein's dual isotherm interpretation do provide the same answer on many occasions. On the dual isotherm model, substrate competition appears as variation in K_m, an intensive parameter. In the electrokinetic formalism, competition appears as variation in π, also an intensive parameter. Therefore, although two models are different but both fit the experimental data, what is predicted as variations in extensive parameters on one model will also be variations in extensive parameters on the other model. Finally Thellier again stressed the point that a model may fit data but that is no evidence that the model has any physical, mechanistic reality. Cram then asked what flux Thellier measured in his figures and secondly what predictions he could make from his model. Thellier replied that the fluxes were net fluxes; the formalism makes predictions about the big questions. If the model predicts a variation in an intensive parameter, then one can say with certainty that a structural alteration is involved. But of course it never leads to a mechanistic interpretation.

The discussion then moved to Page and Meares' paper and Dainty asked if the model had been scaled down from pores of 1 μm in diameter and large current densities to the sort of situation likely to be found in biological systems. Page replied that scaling down to pores of 1000 Å would still require current densities of 40–50 μA/cm^2 in order to find the non-linearity; high pressures of 4–5 atm would be developed. Making this adjustment would require a switch from the Gouy–Chapman pore model to the Schmidt, but this in itself would not cause much alteration to the predicted values. If membrane thickness were scaled down, interfacial effects would also become more important.

Field then asked if the Navier–Stokes and Poiseuille laws held at pore diameters of 1000 Å or less. Page said that the little information available suggested that this was the case. However, one might intuitively feel that at pore diameters of 100 Å, different flow mechanisms might take over. Field said that he thought there was evidence that Poisseuille flow held for 50 Å pores, but he knew of no information at smaller diameters. Thain then commented that he thought that Page's work was not a model for action potentials in plant cells, but was rather a demonstration that pressure–electrical interactions were possible. Most biologists would wish to see a similar mechanism for action potentials in both animals and plants. Page replied that in view of the vastly different response times it did not seem to him obvious that similar mechanisms were operating. Since there is not much likelihood of pressure differences in animal cells, any relevance that his analogue might have will be to plant cells.

Thain's paper was next discussed; Walker said that the integration across the membrane of the flux equation given by Thain is formally different from that given by Kedem and Essig. Thain replied that the equations used by him are derived by Simons and are formally identical with that of Kedem and Essig. In the general case the X_is are not multipliers, but in this specific case the water flow is caused by the K^+ flux. The water flux is represented by a $L_{WK}X_K$ term and the J_K by a $L_{KK}X_K$ term, so that the X can be eliminated; thus the equality given by him is true. Dainty then asked if Thain had speculated further on the value of $b . L_{WWB}$; he asked if you assume that K^+ goes through pores several Å in diameter and that Poisseuille flow occurs, what value do you then get? Thain said that he did not see the value of the speculation, but the interesting thing to do would be to get values for the R coefficients and interpret these in molecular terms. Meares commented that if the proportionality between the K^+ flux and the electro-osmotic water flux holds, then if you reverse the current and take the mean of the water fluxes with the current in either direction, this will be a measure of . . . Thain interrupted, "Yes, that is theoretically correct but you run up against a technical problem because you need to measure the K^+ tracer fluxes over a much longer time than you can allow the electro-osmotic water flow to continue".

Smith then asked Thain if he were happy to ignore all the metabolic effects on the K^+ flux. Thain said that the metabolic effect depended on the paper you read, but that it would be interesting to feed in a metabolically coupled component of K^+ flux into the formalism. The problem in doing the sort of thing he has done is that one relies on data collected by different people on different organisms at different times of the year.

Section III

Membrane Resistance and H$^+$ Fluxes

Chairman: U. Lüttge

III.1

Combined Effect of Potassium and Bicarbonate Ions on the Membrane Potential and Electric Conductance of *Nitella flexilis*

C. Gillet and J. Lefebvre*

Département de Biologie Végétale, Facultés Universitaires de Namur Belgium

I. Introduction

The membrane potential of Characean cells becomes more negative when bicarbonate ions are added to the nutrient medium. In this "hyperpolarized" state the potential of *Chara australis*† no longer responds to a change in

---

* Head of Research of the F.N.R.S.
† *Chara australis* R. Brown has been renamed *Chara corallina* Klein ex Wilds, em. (Wood and Imahori, 1965).

the potassium concentration of the external solution, even in the absence of calcium ions (Hope, 1965). The cells can reach the hyperpolarized state in the absence of bicarbonate, provided that the pH of the solution is sufficiently high (Spanswick, 1970). It seems, therefore, that the apparent effect of bicarbonate on membrane potential is an indirect one. It could, in fact be due to an increase of cellular permeability to protons determined by the pH of the external solution.

The electrical conductance of the cell is greatly modified in the presence of bicarbonate, revealing thereby a change in the cell permeability to ions. But the variations may be in opposite directions: a decrease (Hope, 1965) or an increase (Walker, 1962, Spanswick, 1970). In a medium without bicarbonate, Hope and Richards (1971) have shown that the resistance of *Chara corallina* depends very slightly on the pH. Although it has been proved that the cells of Characeae are permeable in HCO_3^- ions (Smith, 1968) and to H^+ ions (Kitasato, 1968), it has not yet been possible to construct a definitive transport model which would account for the electrical modifications observed (Raven, 1970).

In the present communication we show that, contrary to the observations made in the case of *C. corallina* in calcium-free solutions, the membrane potential and the electric conductance of *Nitella flexilis* remains sensitive to K ions when the pH of the solution is increased by addition of bicarbonate or of another buffer. Moreover the hyperpolarization of the cell depends on the K concentration in the external medium.

II. Materials and Methods

A. Materials

Nitella flexilis (L.) Ag. was grown for seven months in the laboratory under normal conditions, in medium II of Forsberg (1965), with a photoperiod of 16 h.

Before the experiment, internodal cells were freed from neighbouring cells and pre-treated for 15 h in a calcium-free mineral medium: 1 mM KCl, 0·1 mM NaCl. Every tested sample contained from 5 to 15 cells. The experiments were performed from February to May inclusive.

B. Experimental Set-up

The membrane potential was measured by means of a microelectrode filled with 3 M KCl inserted into the vacuole and another dipped into the external solution; both were connected, by a bridge of Agar–3M KCl and two calomel batteries, to an electrometer (Keithley 603). A current microelectrode was also inserted into the vacuole and injected, at the moment when resistance was measured, a square wave of positive or negative current.

The chord conductance was measured and if rectification occurred, the averaged value was taken. Since the cells have an average length (1.3 ± 0.2 cm) equal to the space constant of *Nitella flexilis* (1.3 ± 0.4 cm, Volkov and Platonova, 1970), and the current microelectrode is very close to the recording one, the correction which must be made to the measured resistance for dielectric loss is almost nil.

The activity of cytoplasmic chlorides is calculated from the difference in electrochemical potential measured by a microelectrode of Ag introduced into *Nitella* cells and an electrode of Ag–AgCl dipped in the external medium (Lefebvre and Gillet, 1971).

C. Solutions Used

All solutions used had the same ionic strength; concentrations of $(K)_o$ + $(Na)_o$ and $(Cl)_o$ + $(HCO_3)_o$ were equal to 1.1 mM unless specified otherwise.

The variation in the HCO_3 concentration was achieved at the expense of the Cl concentration. The solutions were prepared just before use from stock solutions and at this stage their pH was measured. In some experiments metabolic inhibitors were added to the solutions. INH (isonicotynylhydrazide) and ouabain were used as $1 . 10^{-3}$ M, DNP (2.4) dinitrophenol) and PCMB (*p*-chloromercuribenzoic acid) at $1 . 10^{-4}$ M, DCMU (3',3-4 dichlorophenyl, 1',1' dimethyl urea) at $3 . 10^{-6}$ M.

III. Results

A. Time Course of the Membrane Potential in the Presence of Bicarbonates

When *Nitella flexilis* is bathed in a solution containing bicarbonate the membrane potential can attain equilibrium in 3 different ways. In 85% of the cells, hyperpolarization shows a monotonic rise and the membrane potential becomes stabilized after about 15 min (Fig. 1a). In the others, hyperpolarization reaches a peak after 3–6 min and then increases again (Fig. 1b) or decreases (Fig. 1c) before becoming stabilized. But whatever their behaviour, the final variation of potential (ΔE) is identical for the same external concentration of potassium.

B. Membrane Potential and Resistance Changes in Relation to $(K)_o$ in the Presence of a Constant Concentration of Bicarbonate (0·1 mM)

Two experiments were carried out to test the influence of potassium on the "bicarbonate effect". In the first, for each concentration of potassium a new cell was immersed from the solution without bicarbonate (pH 5·8) into the solution with bicarbonate (pH 7·3). In the second, the same cell was successively bathed with solutions with different $(K)_o$ and then in the same solutions

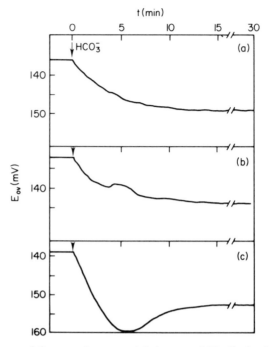

FIG. 1. Time course of the vacuolar potential changes of *Nitella flexilis*. In each case, 0·1 mM bicarbonate was added to the external solution (KCl + NaCl).

plus bicarbonate. The results are roughly similar and have been averaged in Fig. 2a. In the presence of 0·1 mM bicarbonate, the membrane potential is controlled by the external concentration of potassium. Moreover, the hyperpolarization due to the bicarbonate is practically nil in the presence of 1 mM potassium, but it increases with lowering $(K)_o$. Resistance is lower in the presence of bicarbonate, and varies with $(K)_o$ (Fig. 2b).

C. Membrane Potential and Resistance Changes versus $(HCO_3)_o$ in the Presence of Various Constant Concentrations of Potassium

The same cell was bathed successively in solutions of increasing $(HCO_3)_o$, the concentration of the cations (K and Na) remaining constant. Figure 3a shows that the potential varied very little with $(HCO_3)_o$ (or with pH). It was difficult to assess quantitatively the changes in membrane potential because at higher pH the potential becomes unstable in the course of time. The experiment was repeated by changing directly from 0·1 mM to 1 mM HCO_3 solutions. Except for $(K)_o = 0·1$ mM the change in membrane potential does not exceed a few mV. The electric resistance is lowered by increasing the bicarbonate level, especially between pH 7·3 and 8·0 (Fig. 3b).

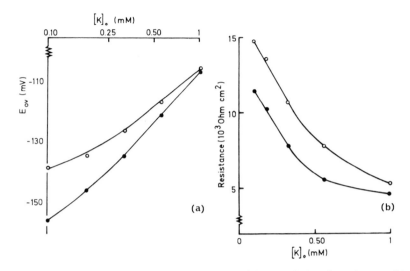

FIG. 2. Experimental values of the vacuolar potential (a) and electric resistance (b) of *Nitella flexilis* as a function of the potassium concentration in an external medium with (●) or without (○) 0·1 mM bicarbonate. In (a) calculated curves from eqn (1) using $\alpha = 0.12$ and $\delta = 10$ in HCO_3-free solutions and $\alpha = 0.012$ and $\delta = 3.3$ in bicarbonate solutions.

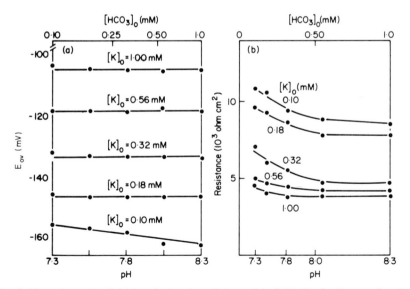

FIG. 3. Vacuolar potential (a) and electric resistance (b) of *Nitella flexilis* as a function of the bicarbonate concentration (or of the pH) in external solutions with different potassium concentrations. The values of the pH are the means of the measures in the different experiments.

D. Reversibility of the Bicarbonate Effect in Presence of $(K)_o = 0.2$ and 1 mM

The change in potential obtained in the presence of 0·1 mM bicarbonate and 0·2 mM potassium is not reversed on reverting to HCO_3-free solutions, as is shown in Table I. (In some experiments, this lack of modification may be prolonged more than one hour.) When the cell is dipped in a solution where $(K)_o = 1$ mM it immediately recovers its initial potential. But if we dip it again in a 0·2 mM solution without bicarbonate we find the value initially observed in presence of bicarbonate. The same type of change is observed for the resistance.

TABLE I. Changes in potential difference and membrane resistance of the same cell before and after having added bicarbonate to the medium

	External medium (mequiv l^{-1})					
	K 1·0 HCO₃ 0	K 0·2 HCO₃ 0	K 0·2 HCO₃ 0·1	K 0·2 HCO₃ 0	K 1·0 HCO₃ 0	K 0·2 HCO₃ 0
E_{ov} (mV)	−105 [−114]	−133 [−143]	−144 [−153]	−143 [−153]	−106 [−114]	−134 [−143]
R_m (kΩ cm^2)	8·5	21·8	15·1	18·3	8·5	22·1

Total cation concentration 1·1 mequiv l^{-1} (K + Na). Means for 15 cells. In brackets: the calculated values of the cytoplasmic potential (E_{oc}) following our model (see discussion).

E. Effect of Other Anions

Table II shows that the hyperpolarization due to 0·1 mM bicarbonate and the lack of potential recovery appear even in the presence of a 50 times greater quantity of chlorides or of nitrates.

TABLE II. Changes in potential difference of the same cell before and after having added bicarbonate to the medium in presence of large concentration of Cl^- or NO_3^-

	External medium (mequiv l^{-1})				
	K 1·0 HCO₃ 0	K 0·2 HCO₃ 0	K 0·2 HCO₃ 0·1	K 0·2 HCO₃ 0	K 1·0 HCO₃ 0
E_{ov} (mV) anion Cl	−99	−116	−124	−120	−95
E_{ov} (mV) anion NO₃	−92	−109	−115	−113	−90

Total cation concentration 6 mequiv l^{-1} (K + Na). Means for 5 cells.

F. Effect of Bicarbonates on the Cytoplasmic Activity of Chlorides

The cytoplasmic activity of chlorides decreases in the presence of bicarbonates and with the augmentation of the pH of the external solution (Table
III). Although the measured activities are less than those previously recorded
(Gillet and Lefebvre, 1971), we again observe the sensitivity of the internal
chlorides to external K concentrations.

TABLE III. Cytoplasmic chloride activity in presence or absence of
bicarbonates

	External medium (mequiv l^{-1})		
	HCO_3 0 pH 5·8	HCO_3 0·1 pH 7·3	HCO_3 1·0 pH 8·3
K 1·0	9·7	9·3	7·3
K 0·2	5·2	4·3	2·1

Total cation concentration 1·1 mequiv l^{-1} (K + Na); HCO_3 sub-
stitutes Cl$^-$; for each test, mean for 5 cells.

G. Comparison with a pH Modification due to the Tricine + KOH Buffer

Without bicarbonate, a hyperpolarization of the cell is obtained by raising
the pH of the external medium provided $(K)_o = 0·1$ mM (Table IV). If
$(K)_o = 1$ mM this effect is not observed. A further addition of bicarbonate
does not modify the potential any further. The resistance is slightly lowered
by high pH.

TABLE IV. Changes in potential difference and membrane resistance of the same cell
by raising the pH in absence or presence of bicarbonate ions

		Without tricine pH 5·8	With tricine pH 7·3	With tricine and HCO_3 pH 7·3
E_{ov} (mV)	$(K)_o = 0·1$ mM	− 137	− 151	− 152
	$(K)_o = 1$ mM	− 104	− 105	− 105
R_m (kΩ cm^2)	$(K)_o = 0·1$ mM	12·5	11·9	10·5
	$(K)_o = 1·0$ mM	5·5	3·4	8·5

Buffer: tricine + KOH, total cation concentration 1·1 mM K + Na. Means of 7 cells.

H. Effect of Metabolic Inhibitors

Several different metabolic inhibitors are used: DCMU, an inhibitor of photosystem II, DNP, an uncoupler of oxidative phosphorylation, INH an inhibitor of the glycolate pathway, ouabain, an inhibitor of the active influx of K, and PCMB as modifying the membrane structure (sulphydryl agent).

Independently of their specific influence on membrane potential, none of the inhibitors tested by us prevents the potential from becoming more negative when bicarbonate ions are added to the medium (Table IV). However, pre-treatment with ouabain reduces hyperpolarization.

TABLE V. Changes in potential difference of a cell pretreated with inhibitors by adding bicarbonate to the medium

	Without HCO_3	With 0·1 mM HCO_3
DCMU	− 136	− 149
DNP	− 78	− 98
INH	− 121	− 141
Ouabain	− 142	− 149
PCMB	− 134	− 150

pH changes: from 5·8 to 7·3 except in presence of DNP (4·5–5·9); Mean potential in KCl 0·2 mM. NaCl 0·9 mM: 135 mV. Means for 8 cells.

IV. Discussion and Conclusions

As Spanswick (1970) has shown for *N. translucens*, our results confirm the indirect effect of bicarbonate ions on the membrane potential and resistance of *N. flexilis*. The observed changes in these two parameters are mainly due to modifications in pH produced by bicarbonates. But in our material, which is pretreated with calcium-free solution, the potential and resistance are also shown to vary with external potassium concentration. Moreover it seems that we do not obtain a hyperpolarized state of *Nitella* in the usual sense of the word, since the membrane potential of the cell is less negative than expected from any of the known ionic gradients (Hope and Richards, 1971).

Our results can be interpreted by assuming that the *Nitella* cell becomes more permeable to K ions when the H concentration is lowered in the external medium. The experimental values of the plasmalemma potential in a solution of pH 5·8 where $[K]_o$ is varying can be fitted with a calculated curve based upon the following Goldman equation

$$E_{oc} = \frac{RT}{F} \ln \frac{[K]_o + \alpha[Na]_o + \delta[H]_o}{[K]_i + \alpha[Na]_i + \delta[H]_i} \tag{1}$$

where R, T, F have their usual meanings; $[\ ]_o$ and $[\ ]_i$ refer to external and internal concentrations; $\alpha = P_{Na}/P_K$ and $\delta = P_H/P_K$; P_K, P_{Na}, P_H are the passive cations permeabilities. The anionic term of the equation has been neglected. As the measured potential is the sum of the plasmalemma and tonoplast potentials we obtain the values for the plasmalemma potential (E_{oc}) by correcting it by 10 mV, this being an average value for the tonoplast potential in *N. flexilis* (Lefebvre and Gillet, 1968). The tonoplast potential is insensitive to external concentration changes. By assuming that $[K]_i = 92$ mequiv l^{-1}, $[Na]_i = 30$ mequiv l^{-1} (mean measured values for K and Na in the cytoplasm of *N. flexilis* from Lefebvre and Gillet, 1968), pH$_i = 5.5$, $\alpha = 0.12$ and $\delta = 10$, there appears to be good agreement between the theoretical and observed values (Table VI). If we introduce these computed values of α and δ into the general equation of the chord conductance (g'_m)

$$g'_m = \frac{F^2 E C_o[1 - \exp(\Delta EF/RT)]}{RT\, \Delta E[1 - \exp(EF/RT)]} \tag{2}$$

where $C = P_K(K_o + \alpha Na_o + \delta H_o)$, it can be observed that the calculated values of P_K for the different tested concentrations of K are nearly constant and are approximately 10^{-5} cm s^{-1}, while $P_H = 10^{-4}$ cm s^{-1} and $P_{Na} = 1.2 \times 10^{-6}$ cm s^{-1}.

These values of α and δ are less satisfactory for cells dipped in the solution of pH 7.3 (bicarbonates added): there is then a considerable deviation between the experimental results and those already calculated (Table VI). By iterative calculations, we have tried to find the values of α and δ which are required to reduce these deviations to a minimum. With $\alpha = 0.012$ and $\delta = 3.3$ we obtain satisfactory agreement between the measurements of the potential and the resistance and the calculated values for each of the K external concentrations. If we assume that P_H remains constant, we observe that P_K becomes three times greater ($P_K \simeq 3 \times 10^{-5}$ cm s^{-1}) and P_{Na} 3.3 times smaller ($P_{Na} \simeq 0.4 \times 10^{-6}$ cm s^{-1}) than before. The correlation between variations of K and Na permeabilities and pH changes, must be compared with the results of Lannoye *et al.* (1970) based upon ion flux measurements. Using the latter values for α and δ, we can predict approximately, for most of the concentrations of potassium, the small variation of potential observed when the pH is increased by addition of more bicarbonate to the medium. However, our model is not consistent with the larger modification of potential when $[K]_o$ is only 0.1 mM; this would need improbable values of α. It seems that under these conditions, the anionic term of the Goldman equation can no longer be neglected. As is seen from Table III, the diminution of $[Cl]_i$ is large enough to influence the potential. Indeed, both a hyperpolarization of the membrane and a decrease in the external K concentration (Lefebvre and Gillet, 1971) favour the diminution of the chloride level in the cytoplasm.

TABLE VI. Comparison between the mean measured values of the plasmalemma potential and the computed values for different potassium concentrations in solution with or without bicarbonates

| [K]$_o$ | E_{oc} (mV) | | | | |
| | pH 5.8 | | pH 7.3 | | |
(mequiv l^{-1})	Experimental	Computed $\alpha = 0.12$ $\delta = 10$	Experimental	Computed $\alpha = 0.12$ $\delta = 10$	Computed $\alpha = 0.012$ $\delta = 3.3$
1	−115	−114	−116	−115	−114
0.56	−126	−126	−131	−127	−130
0.32	−137	−136	−144	−137	−142
0.18	−144	−145	−156	−146	−156
0.10	−148	−151	−166	−153	−169

A closer fit could be obtained by assuming small but finite values for P_{Cl} and $P_{HCO_3^-}$.

The values of the membrane potential observed for different K concentrations when the cells are just removed from the bicarbonate solution (Table I) confirm our model if we suppose that after the cell has been dipped in bicarbonate solution its permeability continues to change over a period of several hours. This is in accordance with the observations of Rent *et al.* (1972). The values computed according to this hypothesis for the plasmalemma potential (normally 10 mV more negative than the membrane potential) are shown in Table I. There is excellent agreement with the observed values. From all these results, it can be suggested that when *Nitella* is bathed with a calcium-free solution, the modification of the potential, caused by changing the pH of the external solution from 5·8 to 8·3, is due to an increased permeability of the membrane to K ions. It is difficult to decide whether this increased permeability is due to a modification of the physical structure of the membrane, which affects the "passive" component of the cation influx (MacRobbie, 1970), or to a modification of the activity of a neutral K pump as the ouabain effect would suggest (Table VI). In this latter case a fraction of the K influx would be linked to H efflux.

The possibility should also be considered that an electrogenic pump transporting H^+ out of the cell, as Kitasato (1968) has suggested, can account for the more negative values of the membrane potential in the presence of bicarbonate. By following an equation adapted from Briggs (1962)

$$E_{oc} = \frac{RT}{F} \ln \frac{[K]_o + \alpha[Na]_o - \dfrac{A_H RT}{P_K EF}\left[1 - \exp\left(\dfrac{EF}{RT}\right)\right]}{[K]_i + [Na]_i}$$

it is possible to calculate the values for A_H/P_K, where A_H is the activity of the H electrogenic pump. Keeping the previous values of α, $[K]_i$ and $[Na]_i$ which fit with the observed membrane potentials in HCO_3^--free solutions, we find that in the presence of 0·1 mM HCO_3, A_H/P_K changes from 0·28 $\times$ 10^{-6} to 0·58 $\times$ 10^{-6} equiv cm^{-3} when $[K]_o$ is changed from 1 to 0·1 mequiv l^{-1}. Taking $P_K = 10^{-5}$ cm s^{-1}, the active efflux would change from about 3 to 6 pequiv cm^{-2} s^{-1}; these values are considerable lower than those proposed by Rent *et al.* (1972). It is not easy to reconcile this last hypothesis with the absence of effects of most of the metabolic inhibitors on the hyperpolarization of the cell. However we cannot discard it *a priori* because, as has already been mentioned by several authors, the action of uncoupling agents is difficult to interpret. This point requires further clarification. Nevertheless it is interesting to note even in this case a relationship between the electrogenic mechanism and the flux of potassium.

Acknowledgements

The authors wish to thank Mrs M. J. Gaspard for technical assistance and the British Council for some financial support.

References

BRIGGS, G. E. (1962). *Proc. R. Soc.* B **156**, 573–577.
FORSBERG, C. (1965). *Physiol. Plantarum* **18**, 275–290.
HOPE, A. B. (1965). *Aust. J. biol. Sci.* **18**, 789–801.
HOPE, A. B. and RICHARDS, J. L. (1971). First European Biophys. Congress, Baden, III, 105.
KITASATO, H. (1968). *J. gen. Physiol.* **52**, 60–87.
LANNOYE, R. J., TARR, S. E. and DAINTY, J. (1970). *J. exp. Bot.* **21**, 543–551.
LEFEBVRE, J. and GILLET, C. (1968). *Bull. Soc. r. bot. Belg.* **102**, 61–78.
LEFEBVRE, J. and GILLET, C. (1971). *Biochim. biophys. Acta* **249**, 556–563.
MACROBBIE, E. A. C. (1970). *Q. Rev. Biophys.* **3**, 251–294.
RAVEN, J. A. (1970). *Biol. Rev.* **45**, 167–221.
RENT, R. K., JOHNSON, R. A. and BARR, C. E. (1972). *J. Membrane Biol.* **7**, 231–244.
SMITH, F. A. (1968). *J. exp. Bot.* **19**, 207–217.
SPANSWICK, R. M. (1970). *J. Membrane Biol.* **2**, 59–70.
VOLKOV, G. A. and PLATONOVA, L. V. (1970). *Biofizika* **15**, 635–642.
WALKER, N. A. (1962). *C.S.I.R.O. Aust. Div. Plant. Ind. A. Rep.* 80.
WOOD, R. D. and IMAHORI, K. (1965). "A revision of the Characeae", J. Cramer, Weinheim.

III.2

Electrogenesis in Photosynthetic Tissues

R. M. Spanswick

Section of Genetics, Development and Physiology
Division of Biological Sciences
Cornell University, U.S.A.

I. Introduction

In the field of ion transport there are several schools of thought, each with its own preconceptions. A gross generalization in the case of studies on plants would be that those who have measured ion fluxes have stressed the relationship of ion movements to metabolism while those who have studied the electrical properties of the membranes have been more concerned with the passive movement of ions. My intention here is to suggest that there is much to be gained from a combination of the two approaches. In particular, I wish to suggest that, in photosynthetic aquatic plants under physiological conditions, it is not possible to measure all fluxes with tracers and the electrical properties of the membranes do not depend only, or even primarily, on the passive movements of ions.

The historical development of ideas concerning the relationship of ion transport in cells and organelles to metabolism has been reviewed by Robertson (1968). Probably the single most important event in this area has been the development of the chemiosmotic hypothesis by Mitchell (1961, 1966). Although this hypothesis has had its greatest influence in the area of phosphorylation, it is gradually becoming clear that it may be of more general significance. Essentially it has two elements: (a) a primary separation of H^+ and OH^- and (b) their recombination through an ATPase or a transport mechanism. A reverse transport of H^+ by the ATPase and an $H^+–Na^+$ exchange system has been used with considerable success to describe ion transport in *Streptococcus faecalis* (Harold and Papineau, 1972). Smith (1970, 1972) has proposed a similar hypothesis to explain chloride transport in *Chara corallina*. Since Smith and his co-workers present their flux studies in detail later in this volume, I shall not deal with these further. Instead, I wish to consider alternative theoretical bases for the description of the electrical properties of plant cell membranes, particularly in relation to the chemiosmotic hypothesis.

In early work with microelectrodes it was assumed that the ion pumps were neutral and the electrical properties of the membrane were due to the passive movements of the ions down the gradients set up by the pumps (Hope and Walker, 1961; Dainty, 1962). Two problems quickly became evident when this hypothesis was applied to the Characeae. First, the electrical conductance of the cell membranes was much greater than that calculated from the passive ion fluxes (MacRobbie, 1962; Williams *et al.*, 1964). Secondly, in the presence of calcium the membrane potential did not behave as a diffusion potential when the ionic concentrations were changed (Kishimoto, 1959; Hope and Walker, 1961; Spanswick *et al.*, 1967). The diffusion potential may be described by the Goldman equation:

$$E = \frac{RT}{F} \ln \frac{P_K K_o^+ + P_{Na} Na_o^+ + P_{Cl} Cl_i^-}{P_K K_i^+ + P_{Na} Na_i^+ + P_{Cl} Cl_o^-} \tag{1}$$

where P_K etc. are the permeability coefficients for K^+ etc. on the inside (i) and outside (o) of the membrane. If the term $P_K K_o^+$ is much larger than the other terms in the numerator, a 10-fold change in K_o^+ will change the membrane potential by 58 mV, assuming P_K remains constant. Except under conditions where the cells have been soaked in 5 mM NaCl and the experiments performed in Ca^{2+}-free solutions, the changes in potential for $10 \times$ changes in Na_o^+ and K_o^+ are about 10 mV (Spanswick *et al.*, 1967). These problems have been difficult to reconcile with what might be called the "classical" theory of ion transport. Walker and Hope (1969) have put forward a vigorous defense of the theory but it cannot account for the effects of light to be described here.

In searching for alternative explanations for the behaviour of the membrane potential and resistance it is necessary to consider that there may be large passive fluxes of the ions normally present in low concentrations (i.e., H^+, OH^- or HCO_3^-) or that there is an electrogenic ion pump present in the membrane.

Apparent evidence for an electrogenic pump came from the hyperpolarization of the membrane potential in *Chara corallina* by HCO_3^- (Hope, 1965). However, similar effects can be produced by solutions containing other buffers at the same pH and HCO_3^- has no effect at constant pH in *Nitella translucens* (Spanswick, 1970a). The effect was therefore attributable to the pH of the solution, a factor that has been investigated in detail by Kitasato (1968). He found that the membrane potential in *Nitella clavata* was apparently controlled by the external H^+ concentration in the pH range 4–6. However, it is generally thought that the pH of the cytoplasm is near neutrality and the membrane potential was too negative to be a simple H^+ diffusion potential. His solution to this dilemma was to suggest that the passive influx of H^+ is balanced by an H^+ extrusion pump which would maintain the pH of the cytoplasm close to neutrality and at the same time polarize the membrane to give the observed membrane potential. With the membrane clamped at the K^+ equilibrium potential, a change in pH from 5 to 6 produced a change in applied current equivalent to a flux of approximately 22 pmol cm^{-2} s^{-1}. Kitasato interpreted this current as a change in the passive influx of H^+ and used it to calculate the contribution of H^+ to the membrane conductance. The H^+ conductance was approximately equal to the observed membrane conductance. Thus he accounted for the main difficulties facing the classical theory by postulating that H^+ is the main carrier of charge across the membrane. Unfortunately, the hypothesis does not account for all the available evidence and it is not consistent with Smith's (1970) hypothesis for chloride transport which requires the membrane to have a low passive permeability to H^+. In an attempt to resolve these difficulties I shall review evidence that confirms the presence of an electrogenic pump in *Nitella translucens* (Spanswick, 1972b), and put forward a hypothesis that attributes the electrical properties of the membrane, including the conductance, to the electrogenic pump itself. A theoretical framework for this interpretation has been provided by Rapoport (1970).

In systems where it is difficult to apply the short-circuit technique, the most reliable indication of the presence of an electrogenic pump is a membrane potential that lies outside the range of possible diffusion potentials given by eqn (1). For the solutions used in studies on the Characeae, the membrane potential is usually more positive than the negative limit set by the K^+ equilibrium potential. By making an appropriate choice of external pH, K^+ concentration and light intensity, it has been possible to achieve a potential that is more negative than E_K in *Nitella translucens*.

In other systems the presence of an electrogenic pump is more obvious. In the marine alga *Valonia ventricosa* the tonoplast potential appears to have an electrogenic component and this is consistent with a large light-stimulated component of the short-circuit current in perfused cells (Gutknecht, 1967). In another marine alga, *Acetabularia mediterranea*, the membrane potential is -170 mV and is outside the range of possible diffusion potentials (Saddler, 1970). Darkness, CCCP, low temperatures and low Cl_o^- all depolarize the cell and reduce the chloride influx. It seems probable that the chloride influx is electrogenic.

Slayman (1965a) showed that the membrane potential of the fungus *Neurospora crassa* was about -200 mV. Again, this is far more negative than the equilibrium potentials for the major ions or H^+. It can be reduced to the level of the diffusion potential by a variety of inhibitors (Slayman, 1956b). In the presence of inhibitors, the decay of the membrane potential is closely correlated with the internal ATP concentration (Slayman *et al.*, 1970). Slayman (1970) considers that the electrogenic pump probably transports H^+.

Higinbotham *et al.* (1970) have provided similar evidence for higher plant tissues. In pea epicotyl tissue, in particular, the membrane potential is more negative than the equilibrium potential for K^+ and its can be reversibly inhibited to a level close to the calculated diffusion potential by 0.1 mM CN^-. Further evidence has been reviewed by Higinbotham (1970).

Light-induced potential changes in green cells have been described by Lüttge and Pallaghy (1969). The membrane potential in *Mnium* was about -200 mV in a solution containing 5 mM KCl $+$ 0.5 mM $CaCl_2$ and, although the absence of Na^+ from the medium makes it difficult to place a limit on the diffusion potential, it seems likely that there may be an electrogenic component. Jeschke (1970) has demonstrated a light-dependent hyperpolarization of about 60 mV in leaf cells of *Elodea densa*, the potential in the light being about -180 mV. In *Elodea canadensis* the membrane potential was -257 mV at an unspecified light intensity (Spanswick, 1972a). This is clearly outside the range of possible diffusion potentials.

A summary of recent work on *Nitella translucens* and *Elodea canadensis* will be presented below and the results will be discussed in terms of the hypotheses outlined above.

II. Methods

The electrical measurements were made using conventional micro-electrode techniques described in detail elsewhere (Spanswick, 1970b, 1972b).

The experimental solutions were similar in composition to the artificial pond water (APW) used previously except that they were buffered with 1 mM MES (2-morpholino ethanesulfonic acid) for APW5 and APW6, the number denoting the pH, 1 mM MOPS (morpholinopropane sulfonic acid) for APW7, 1 mM tricine (tris (hydroxymethyl) methylglycine) for APW8 and 1 mM TAPS (tris (hydroxymethyl) methylaminopropane sulfonic acid) for APW9. The ionic concentrations were: 0.1 mM K^+, 0.1 mM Ca^{2+}, 0.1 mM Mg^{2+}, 1.0 mM Cl^-, and $1.0-1.3$ mM Na^+ depending on the amount of

NaOH required to adjust the pH to the specified value. APW5 also contained 0·2 mM SO_4^{2-}.

Light from an incandescent lamp was passed through glass and water filters. The intensity incident on the bath containing the cell or tissue was 1·0 mW cm^{-2}. *Elodea* leaves were mounted at an angle of 45° to the incident light.

III. Results

A. *Nitella translucens*

In the APW used previously the membrane potential was usually more positive than the K$^+$ equilibrium potential, E_K (Spanswick and Williams, 1964). It was therefore within the range of possible diffusion potentials. To determine whether the potential contained an electrogenic component, a systematic survey of the effects of external K$^+$, pH, light and temperature was conducted to see whether conditions could be established in which the membrane potential was outside the range of possible diffusion potentials (Spanswick, 1972b). The most pertinent results of this work are summarized below:

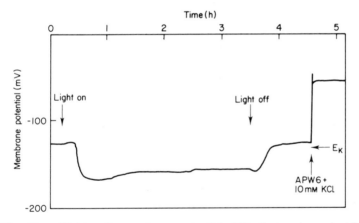

FIG. 1. The effect of light on the membrane potential of *Nitella translucens* in APW6 + 0·4 mM KCl. Also shown is the effect of APW6 + 10 mM KCl and the calculated value of E_K in APW6 + 0·4 mM KCl.

(a) With $K_o^+ = 0·1$ mM, the membrane potential was close to or more positive than E_K in the pH range 5–9. The resistance in the dark was much greater than that in the light.

(b) Increasing the external K$^+$ concentration had little effect on the membrane potential below 1 mM but at higher concentrations the membrane potential suddenly fell to a value close to the calculated value of E_K. This

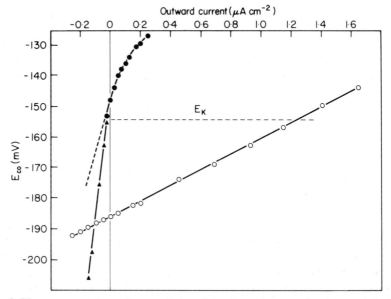

FIG. 2. The current–voltage characteristics of the plasmalemma of an internodal cell of *Nitella translucens* in the light (○) and in the dark before (●) and after (▲) the hyper-polarizing response.

change was accompanied by a sharp decrease in membrane resistance and also by an increase in the membrane fluxes (Spanswick, unpublished). It is probably due to an increase in P_K. Although this sets a limit to the level to which K_o^+ can be raised, it does provide a useful method for estimating E_K for individual cells.

(c) With $K_o^+ = 0.5$ mM, the membrane potential in the light remained at a level considerably more negative than E_K. At pH 5 the potential was close to E_K and changing to a pH higher than 6 usually produced a transient hyper-polarization in the light. In the dark the potential was more positive but the response to changes in pH was less transient.

(d) The effect of light on the membrane potential for a cell in APW6 + 0.4 mM KCl is shown in Fig. 1. Also shown is the potential in APW6 + 10 mM KCl and the calculated value of E_K in APW6 + 0.4 mM KCl. It can be seen that the potential in the dark is close to E_K but is more negative in the light. A summary of the effects of light and temperature on cells in APW6 + 0.4 mM KCl is given in Table I. It can be seen that the potential in the light is much more sensitive to temperature than that in the dark. Light also causes a large reduction in the membrane resistance and the resistance in the light is much more sensitive to temperature than that in the dark.

(e) Figure 2 shows the current–voltage relationship for the plasmalemma of a cell in both the light and the dark. At large inward currents the change in membrane potential is very large due to a change in resistance similar to that observed by Bradley and Williams (1967). The significance of the current required to depolarize the potential in the light to E_K will be dealt with in the Discussion (Section IV of this paper).

(f) Raising the external Na^+ concentration to 10 mM had only a small effect on the membrane potential and resistance at pH 5. Thus the depolarization at low pH cannot be attributed to an increase in P_{Na} as suggested by Lannoye et al. (1970).

TABLE I. Effect of light and dark on the electrical properties of the membrane of Nitella translucens

	Light	Dark
Membrane potential (mV)	-162 ± 3	$-111\cdot5 \pm 2$
E_K (mV)	-117 ± 2	-117 ± 2
Temperature coefficient (mV/°C)	$2\cdot5 \pm 0\cdot5$	$0\cdot97 \pm 0\cdot19$
Membrane resistance (kΩ cm^2)	17 ± 3	153 ± 32
Membrane resistance at 10°C (kΩ cm^2)	35 ± 6	160 ± 44

On the assumption that the light-stimulated hyperpolarization was due to an electrogenic ion pump, the effects of a variety of inhibitors were investigated. Ouabain, an inhibitor of the Na^+–K^+ pump (MacRobbie, 1962) had no effect in the light or the dark. The effects of azide, CN^-, DCMU and CCCP are summarized in Table II. Azide and CN^-, both at 1 mM, depolarize the membrane potential in the light to a value close to the level in the dark and E_K. In the dark the effects are small. DCMU at 5×10^{-6} M had a similar effect but at 10^{-6} M the effect was more variable and the potential in the light was often unaffected. CCCP at 10^{-6} M also depolarizes the cell in the light but the effect is not easily reversible. Again, there is little effect in the dark. This concentration of CCCP halts the cytoplasmic streaming but the other inhibitors do not. 5×10^{-6} M CCCP usually damages the cells irreversibly.

Cl^--free solutions also produced a depolarization in 5 cases out of 7. However, the effect was very slow and therefore cannot be attributed directly to an electrogenic Cl^- pump.

B. Elodea canadensis

The factors affecting the membrane potential of Elodea canadensis have been investigated by Wakefield (1973). At a light intensity of $1\cdot0$ mW cm^{-2}, the membrane potential was -278 ± 5 (23) mV in APW7. The internal K^+ concentration was 91 mM and hence E_K, ignoring activity coefficients, was -172 mV. The Na^+ equilibrium potential was more positive than E_K. It is obvious that the membrane potential in the light cannot be a diffusion

Table II. The effect of inhibitors on the membrane potential of *Nitella translucens*

| | Number of cells | Membrane potential (mV) | | | | | | |
| Inhibitor | | Light | | | Dark | | | |
		APW6 + 0.4 mM KCl	APW6 + 0.4 mM KCl + inhibitor	Recovery in APW6 + 0.4 mM KCl	APW6 + 0.4 mM KCl	APW6 + 0.4 mM KCl + inhibitor	Recovery in APW6 + 0.4 mM KCl	E_K
1 mM CN⁻	7	−177 ± 3	−139 ± 8	−167 ± 9	−134 ± 4	−132 ± 4	−132 ± 3	−128 ± 3.5
1 mM Azide	8	−168 ± 3.5	−105 ± 3	−152 ± 7	−119 ± 3	−100 ± 5	−119 ± 3	−128 ± 2
10⁻⁶ M CCCP	7	−156 ± 5	−128 ± 6	−128 ± 7	−114 ± 3	−107 ± 4	−113 ± 3	−121 ± 2
5.10⁻⁷ M DCMU	7	−168 ± 2.5	−127 ± 3	−171 ± 3	−127 ± 3	−127 ± 4	−126 ± 3	−132 ± 4

TABLE III. Effect of inhibitors on the membrane potential of *Elodea canadensis* at a light intensity of 1.0 mW cm⁻²

| Inhibitor | Concentration (M) | Membrane potential (mV) | | |
		APW7	APW7 + inhibitor	Recovery in APW7
NaN₃	10⁻³	−286 ± 8 (6)	−149 ± 6 (6)	−296 ± 9 (5)
NaCN	10⁻³	−243 ± 6 (5)	−144 ± 5 (5)	−245 ± 15 (2)
CCCP	10⁻⁶	−279 ± 6 (8)	−193 ± 15 (8)	−286 ± 8 (4)
DCMU	10⁻⁶	−275 ± 9 (6)	−195 ± 9 (6)	−265 ± 7 (4)

potential. This was confirmed by experiments in which the external K^+ concentration and pH were changed. The potential was practically independent of pH in the range 5–9 while a 10-fold increase in K_o^+ depolarized the cell by only 24 mV.

The effects of darkness and inhibitors are consistent with the view that the membrane potential is controlled by an electrogenic pump. The time-course of the potential on turning off the light is rather complex (Fig. 3). The initial

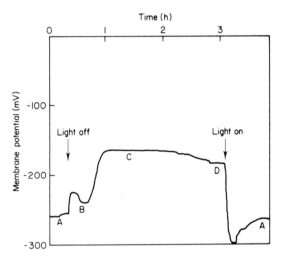

FIG. 3. The effect of light on the membrane potential of *Elodea canadensis*. The average values of the potential at the stages A, B, C and D are given in Table III.

depolarization is followed by a small transient hyperpolarization and then a depolarization to about − 140 mV which lasts for about an hour. There is then a partial recovery. The average values of the potential at the various stages are summarized in Table III, and the effects of inhibitors are summarized in Table IV. As with *Nitella*, azide and CN^- reduced the potential

Table IV. Effect of light and dark on the membrane potential of *Elodea canadensis*

Stage[a]		Membrane potential (mV)
A	Light	− 278 ± 5 (23)
B	Dark (5 min)	
	Min.	− 164 ± 11 (7)
	Max.	− 199 ± 10 (7)
C	Dark (30 min)	− 143 ± 6 (19)
D	Dark (60 min)	− 182 ± 10 (7)
A	Light (recovery)	− 283 ± 8 (9)

[a] See Fig. 3.

to the dark level. CCCP and DCMU reduced the potential to a value that was about the same as that after a long period in the dark (stage D). In separate experiments it was shown that the potential at this stage could be further reduced by azide. The effects of inhibitors at stage C in the dark were small.

IV. Discussion

The values of the membrane potential in the light in *Elodea* can only be accounted for by an electrogenic pump and the effects of darkness and inhibitors are consistent with this explanation. The complicated time-course after turning off the light may prove useful in looking for correlations between the potential and metabolic intermediates. It may also provide some information about the interaction between respiration and photosynthesis. The repolarization in the dark and the partial inhibition to about the same level by DCMU suggest that the substrate for the reaction driving the pump may be common to the two systems.

Although less dramatic and more difficult to obtain, the light-stimulated hyperpolarization in *Nitella* in APW6 + 0·4 mM KCl is also good evidence for an electrogenic pump. Potentials more negative than E_K have been obtained on several occasions by raising the external K^+ concentration (Kishimoto, 1959; Oda, 1962; Hope, 1965; Kitasato, 1968). However, the effects of light have usually been transient (Nishizaki, 1968; Lüttge and Pallaghy, 1969) or within the range of possible diffusion potentials (Nagai and Tazawa, 1962). The greatest number of parallels are to be found in the work of Hope (1965). He showed that in solutions buffered with HCO_3^- the membrane potential of *Chara corallina* was hyperpolarized at a value more negative than $-200\,mV$ and remained so when the external K^+ concentration was raised to 1 mM. Monuron and o-phenanthroline caused partial depolarization and the membrane could also be depolarized by inducing an action potential. A large effect of light on the membrane resistance was also observed by Hope and may account for the wide variation in the values to be found in the literature (Williams et al., 1964).

What are the consequences of this work for the hypotheses currently used to describe the electrical properties of plant cell membranes? Obviously the classical theory cannot account for the membrane potential of *Elodea canadensis* in the light. It would require an internal K^+ concentration of almost 9 M and the measured value is only 0·01 of this. However, Walker and Hope (1969) have adapted the classical theory to explain their work on the Characeae. They suggest that depletion effects at the surface of the plasmalemma give rise to errors in the flux measurements which could account for the discrepancy between the measured and calculated resistances. I have criticized their hypothesis in detail elsewhere (Spanswick, 1972b). Here, I merely wish to point out that the hyperpolarization in *Nitella* in the light would require an approximately 10-fold depletion of the K^+ concentration at the outer surface of the plasmalemma, presumably by an inwardly directed K^+ pump. This seems unlikely in view of the high permeability of the cell wall and would lead to an increase in the membrane resistance rather than the large reduction that is observed.

Kitasato (1968) suggested that the large effect of pH on the membrane potential could be explained if eqn (1) were modified to include terms for the passive diffusion of H^+:

$$E = \frac{RT}{F} \ln \frac{P_K K_o^+ + P_{Na} Na_o^+ + P_H H_o^+ + P_{Cl} Cl_i^-}{P_K K_i^+ + P_{Na} Na_i^+ + P_H H_i^+ + P_{Cl} Cl_o^-} \qquad (2)$$

where P_H is the permeability coefficient for H^+, and H_i^+ and H_o^+ are the internal and external concentrations respectively. With $P_H H_o^+ \gg P_K K_o^+$, the effect of pH on the membrane potential could be explained. This had been rejected earlier on the grounds that the equation would predict a value for the membrane potential that was too positive (Hope, 1965). However, if the passive influx of H^+ is balanced by an electrogenic pump which excretes H^+, as suggested by Kitasato (1968), the membrane will be repolarized and eqn (2) can be modified to take this into account:

$$E = (E_m)_o + \frac{JF}{g_m} \qquad (3)$$

where JF is the current passing through the membrane conductance, g_m, to balance the flux, J, through the pump and $(E_m)_o$ is the potential given by eqn (2) when $J = 0$.

Walker and Hope (1969) suggested that the effect of low pH on the membrane potential could also be explained by (a) a decrease in P_K (b) an increase in P_{Na} (see also Lannoye *et al.*, 1970) or (c) an increase in P_{Cl}. (a) and (c) are inconsistent with estimates of P_K and P_{Cl} made by Kitasato from flux measurements and (b) is inconsistent with the small effects of Na^+ on the membrane potential and resistance of *Nitella translucens* at pH 5. Nevertheless, the large flux of H^+ (22 pmol cm^2 s^{-1}), estimated by Kitasato from the experiment in which he clamped the membrane potential at E_K and changed the pH from 5 to 6, would require a very high value for P_H and there are some difficulties in applying the theory to *Nitella translucens*. The hypothesis is obviously inapplicable to *Elodea* because of the small effect of external pH (Jeschke, 1970). The difficulty in applying it to *Nitella* arises at high pH. Kitasato suggests that the rate of the H^+ pump will decrease at high pH due to a rise in cytoplasmic pH and, since both H_i^+ and H_o^+ will then be small, the membrane potential should then be given by eqn (1). However, although this is true for $K_o^+ = 0.1$ mM, his own data show that increasing K_o^+ to 1 mM at pH 8 results in a depolarization of only 5 mV. Similar observations have been made on *Nitella translucens*. It is also difficult to account for the membrane resistance which remains low in the light at high pH. The decrease in the passive H^+ fluxes should lead to a large increase in resistance but this is not observed in *Nitella translucens* or, apparently, in *Nitella clavata*. This observation cannot be explained by an increase in P_K at high pH since the membrane potential would then be sensitive to K_o^+. It is also difficult to explain the 10-fold difference in resistance in the light and dark in *Nitella translucens* at pH 7, for instance, where the steady values of the membrane potential are similar.

An alternative hypothesis, which is consistent with that of Smith (1970), would be that the electrical properties of the membrane are controlled by an electrogenic pump.

Equation (3) may be used to estimate a minimum value for the flux through the pump from the current–voltage relations of a cell in the light (Fig. 2). When $E = E_K$, $E \approx (E_m)_o$ and the current through the passive channels in the membrane should be zero. The applied current will then be equal to JF, the current through the pump. If $(E_m)_o$ is more positive than E_K, the current through the pump will have been underestimated. For the example in Fig. 2 this current is $1.25\,\mu A$ which is equivalent to a flux of $12\,\mathrm{pmol\,cm^{-2}\,s^{-1}}$. Under most conditions the fluxes of the major ions are an order of magnitude smaller than this. The flux will therefore be tentatively identified with the active H^+ efflux which Spear *et al.* (1969) have estimated to be 5–20 pmol $\mathrm{cm^{-2}\,s^{-1}}$. Strunk (1971) has made a similar estimate from the difference in the currents required to reduce the membrane potential to zero in the light and in darkness plus ouabain, and also from the short-circuit current in perfused cells.

The term J in eqn (3) is usually regarded as being independent of the membrane potential and the term g_m, though referring only to the passive

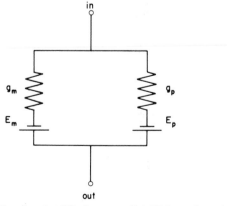

FIG. 4. Equivalent circuit showing the diffusion potential, $(E_m)_o$, and passive conductance, g_m, in parallel with the pump e.m.f., E_p, and conductance, g_p.

diffusion channels in the membrane, is usually identified with the measured membrane resistance. Yet, as long ago as 1964, Finkelstein demonstrated that electrogenic pumps have conductance and Rapoport (1970), using irreversible thermodynamics, has shown that the flux through the pump is potential-dependent. Rapoport applied his theory to the Na^+-K^+ pump, but it is a simple matter to rewrite it for the postulated H^+ pump (Spanswick, 1972b). The free energy change for active transport of H^+ is given by

$$\Delta F_r = \Delta \bar{\mu}_p - v_H \Delta \bar{\mu}_H \qquad (4)$$

where $\Delta \bar{\mu}_p$ represents the free energy change for the non-transported components of the reaction, v_H is the stoichiometric coefficient and $\Delta \bar{\mu}_H$ is the electrochemical potential difference across the membrane for H^+. The rate of the chemical reaction is given by

$$J_r = L_{rr}(-\Delta F_r) \qquad (5)$$

where L_{rr} is a thermodynamic conductance coefficient. The flux through the pump is, of course, equal to $v_H J_r$. It has been shown (Spanswick, 1972b) that it is possible to obtain an expression for the e.m.f. of the electrogenic pump from eqn (5). If the passive permeability coefficients are made zero, the chemical reaction will continue until it is halted by the potential developed across the membrane. This potential, at zero current flow, is equal to the "pump e.m.f.". When $J_r = 0$, $\Delta F_r = 0$. Therefore, from eqn (4):

$$\Delta \bar{\mu}_p = v_H \Delta \bar{\mu}_H$$

$$= v_H \left(RT \ln \frac{H_i^+}{H_o^+} + FE \right)$$

i.e.,

$$E = \frac{\Delta \bar{\mu}_p}{F v_H} - \frac{RT}{F} \ln \frac{H_i^+}{H_o^+}$$

$$= E_p$$

where E_p is the e.m.f. of the pump. It was also suggested (Spanswick, 1972b) that in *Nitella* the conductance of the pump, given by

$$g_p = F^2 L_{rr} v_H^2 \qquad (8)$$

is much larger than the passive conductance, g_m. Since the passive and active channels are in parallel (Fig. 4) this would account for the discrepancy between the measured and calculated conductance and permit description of the membrane potential by eqn (7) since $E \approx E_p$ if $g_p \gg g_m$. Thus the hyperpolarization in the light would be due to a more negative value of $\Delta \bar{\mu}_p$. The effect of external pH could be a direct effect on E_p, and changes in internal pH could still account for the transient response of the potential to changes in external pH. The passive influx of H^+ estimated by Kitasato (1968) from the change in clamping current on changing from pH 5 to pH 6 could be reinterpreted as an increase in the efflux through the pump. In fact, his figure, which shows all the membrane currents directed inwards, is inconsistent with other information in his paper which shows that $E \approx E_K$ at pH 5 when $K_o^+ = 1$ mM. At higher pHs the membrane current should therefore be outwards.

The effect of light on the membrane conductance in *Nitella* (Table I) cannot be explained as easily as the effect on the potential. It can be seen from eqn (8) that there is probably some effect on L_{rr}. The value of L_{rr} cannot be predicted from irreversible thermodynamics. However, in Finkelstein's model, the conductance depends on the concentration of "carriers" in the membrane and it is possible that this changes with the amount of energy available and with temperature.

Rapoport (1970) pointed out that the membrane potential provides a negative feedback loop that regulates the flux through the pump. Thus, when the membrane is hyperpolarized in the light in APW6 + 0·4 mM KCl, the flux through the pump will be smaller than the flux estimated from the current required to reduce the potential to E_K. Thus it is not necessary for the passive fluxes to add up to the "potential flux" through the membrane estimated in this way.

This hypothesis would require some modification for *Elodea* due to the small effects of external pH.

At face value, the effects of azide, CN^- and DCMU may be attributed to their effect on $\Delta\bar{\mu}_p$. However, this will require independent confirmation since the driving reaction has yet to be identified. The reaction could involve an ATPase as in *Streptococcus faecalis* (Harold and Papineau, 1972) and possibly *Neurospora crassa* (Slayman et al., 1970) or perhaps there is a cytochrome system in the membrane as originally suggested by Lundegårdh (1954). Unfortunately, the inhibitors could all be acting at more than one site. This is especially true of uncoupling agents such as CCCP which act as carriers of H^+ across membranes. Kitasato (1968) suggested that the effect of DNP was on the energy supply to the pump and that the membrane depolarized towards the H^+ equilibrium potential. However, he used a solution containing 1 mM K^+ and, at pH 6, the potential reached a steady value close to E_K. Since DNP should increase the permeability to H^+, this provides further evidence of the low permeability to H^+. Under some conditions (Spanswick, 1970a), CCCP initially causes a hyperpolarization. According to the hypothesis presented here (eqn 7), this could be due to an increase in H_i^+ resulting from an increased influx of H^+. The later depolarization and large increase in resistance would be due to an effect on $\Delta\bar{\mu}_p$ as the inhibitor reached the chloroplasts and mitochondria. The increase in resistance could be an effect on L_{rr} similar to darkness.

The effect of Cl^--free solutions is slow and could be an indirect effect on the H^+ pump since $Cl^- - OH^-$ exchange will be halted and this could lead to a rise in cytoplasmic pH with a consequent change in E_p (eqn 7). Although the hypothesis presented above has yet to be rigorously tested, it does appear to resolve some of the difficulties which previously have had to be explained by essentially unverifiable hypotheses such as "filling in pores" and "screening" at the surface of the plasmalemma (Walker and Hope, 1969). It calls for measurements of the cytoplasmic pH, the H^+ fluxes and identification of the reaction driving the H^+ pump.

Acknowledgements

This work was supported by grant number GB-28124 from the National Science Foundation.

References

BRADLEY, J. and WILLIAMS, E. J. (1967). *Biochim. biophys. Acta* **135**, 1078–1080.
DAINTY, J. (1962). *A. Rev. Pl. Physiol.* **13**, 379–402.
FINKELSTEIN, A. (1964). *Biophys. J.* **4**, 421–440.
GUTKNECHT, J. (1967). *J. gen. Physiol.* **50**, 1821–1834.
HAROLD, F. M. and PAPINEAU, D. (1972). *J. Membrane Biol.* **8**, 27–62.
HIGINBOTHAM, N. (1970). *Am. Zoologist* **10**, 393–403.
HIGINBOTHAM, N., GRAVES, J. S. and DAVIS, R. F. (1970). *J. Membrane Biol.* **3**, 210–222.
HOPE, A. B. (1965). *Aust. J. biol. Sci.* **18**, 789–801.
HOPE, A. B. and WALKER, N. A. (1961). *Aust. J. biol. Sci.* **14**, 26–44.
JESCHKE, W. D. (1970). *Z. Pflanzenphysiol.* **62**, 158–172.
KISHIMOTO, U. (1969). *Ann. Rep. Sci. Works Fac. Sci. Osaka Univ.* **7**, 115–146.
KITASATO, H. (1968). *J. gen. Physiol.* **52**, 60–87.
LANNOYE, R. J., TARR, S. E. and DAINTY, J. (1970). *J. exp. Bot.* **21**, 543–551.
LUNDEGÅRDH, H. (1954). *Symp. Soc. exp. Biol.* **8**, 262–296.
LÜTTGE, U. and PALLAGHY, C. K. (1969). *Z. Pflanzenphysiol.* **61**, 58–67.
MACROBBIE, E. A. C. (1962). *J. gen. Physiol.* **45**, 861–878.
MITCHELL, P. (1961). *Nature* **191**, 144–148.
MITCHELL, P. (1966). *Biol. Rev.* **41**, 445–502.
NAGAI, R. and TAZAWA, M. (1962). *Cell Physiol.* **3**, 323–339.
NISHIZAKI, Y. (1968). *Cell Physiol.* **9**, 377–387.
ODA, K. (1962). *Sci. Rep. Tohoku Univ. Ser. IV Biol.* **28**, 1–16.
RAPOPORT, S. I. (1970). *Biophys. J.* **10**, 246–259.
ROBERTSON, R. N. (1968). "Protons, Electrons, Phosphorylation and Active Transport". Cambridge University Press, London.
SADDLER, H. D. W. (1970). *J. gen. Physiol.* **55**, 802–821.
SLAYMAN, C. L. (1965a). *J. gen. Physiol.* **49**, 69–92.
SLAYMAN, C. L. (1965b). *J. gen. Physiol.* **49**, 93–116.
SLAYMAN, C. L. (1970). *Am. Zoologist* **10**, 377–392.
SLAYMAN, C. L., LU, C. Y.-H. and SHANE, L. (1970). *Nature* **226**, 274–276.
SMITH, F. A. (1970). *New Phytol.* **69**, 903–917.
SMITH, F. A. (1972). *New Phytol.* **71**, 595–602.
SPANSWICK, R. M. (1970a). *J. Membrane Biol.* **2**, 59–70.

SPANSWICK, R. M. (1970b). *J. exp. Bot.* **21**, 617–627.
SPANSWICK, R. M. (1972a). *Planta* **102**, 215–227.
SPANSWICK, R. M. (1972b). *Biochim. biophys. Acta* **288**, 73–89.
SPANSWICK, R. M. and WILLIAMS, E. J. (1964). *J. exp. Bot.* **15**, 193–200.
SPANSWICK, R. M., STOLAREK, J. and WILLIAMS, E. J. (1967). *J. exp. Bot.* **18**, 1–16.
SPEAR, D. G., BARR, J. K. and BARR, C. E. (1969). *J. gen. Physiol.* **54**, 397–414.
STRUNK, T. H. (1971). *J. exp. Bot.* **23**, 863–874.
WALKER, N. A. and HOPE, A. B. (1969). *Aust. J. Biol. Sci.* **22**, 1179–1195.
WAKEFIELD, L. E. (1973). M.S. Thesis, Cornell University.
WILLIAMS, E. J., JOHNSTON, R. J. and DAINTY, J. (1964). *J. exp. Bot.* **15**, 1–14.

III.3

The H⁺ Pump in Red Beet

Ronald J. Poole

Biology Department, McGill University
Montreal, Canada

I. Introduction

Although the H^+ pump is currently enjoying great popularity as the presumed missing link in studies of membrane potentials, phosphorylation, cation–anion balance and the like, it remains remarkably elusive to experimental approach. The H^+ pump in red beet may be particularly amenable to investigation because of the ease with which massive H^+ effluxes may be evoked by raising the external pH (Hurd, 1958; Van Stevenink, 1961). An example of this pH effect is shown in Fig. 1.

II. Electrogenicity of H⁺ Transport in Beet

Studies on the membrane potential of beet at high external pH (Poole, 1966) show that the H^+ moves out of the cells against a diffusion gradient, and that this flux is accompanied by a hyperpolarization to more than 60 mV more negative than the K^+ equilibrium potential. It is difficult to account for this potential by the diffusion of any known ion. Moreover, if the metabolic activity of the tissue is varied by washing for various periods before the experiment, it is found that the degree of hyperpolarization is correlated with pump activity, and tends to fall towards the K^+ equilibrium potential as the metabolic activity decreases. It appears, therefore, that the H^+ pump carries

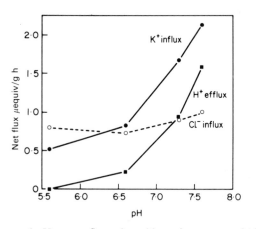

Fig. 1. Effect of external pH on net fluxes in red beet tissue over a 24 h period at 25°C, after washing 4 days at 12°C. The external solution contained 2 mM KCl + 20 mM tris-hydroxymethylaminomethane adjusted to the appropriate pH with H_2SO_4. The H^+ flux was measured by the amount of base (tris) needed to maintain a constant external pH (allowing for the degree of dissociation of tris). Points are the means of 3 replicates, and the standard error does not exceed 10%.

a net charge across the membrane and contributes directly to the membrane potential. This now appears to be typical of H^+ pumps in a number of other systems.

III. Relation to Transport of Other Ions: A Reassessment

As the external pH is increased (Fig. 1) there is a close correlation between the increase in H^+ efflux and the increase in K^+ uptake by beet slices. However, at low pH, when the H^+ efflux falls to zero and the potential falls to the K^+ equilibrium value (Poole, 1966), there may still be an influx of K^+, the magnitude of which depends largely on the Cl^- influx (Pitman, 1964). Thus the K^+ influx appears to be resolved into two components, one of which is sensitive to external pH and balanced by H^+ efflux, while the other is insensitive to external pH and balanced by Cl^- uptake. Measurements of membrane potential seemed to support this notion, since K^+/H^+ exchange at high pH is accompanied by a hyperpolarization, whereas KCl uptake is not (Poole, 1966). It was suggested, therefore, that K^+ may enter the cells by diffusion in response to the potential generated by the H^+ pump, but that KCl uptake may be an independent, electrically neutral process.

Further studies on K^+ uptake in beet have produced evidence which is difficult to fit into the above picture. First, measurements of the K^+ flux ratio across the plasmalemma (Poole, 1969) have not supported the idea that K^+ transport responds freely to the potential during hyperpolarization at high pH. These flux studies suggest that K^+ transport, even at high pH, is largely carrier-mediated. The same can be said for Cl$^-$ uptake, which is an active transport process at the plasmalemma of beet (Gerson and Poole, 1972), and is also not noticeably affected by the hyperpolarization at high pH (Fig. 1). Thus we may have three carrier-mediated processes: H$^+$ efflux, K$^+$ influx and Cl$^-$ influx, which appear to regulate each other by some means other than the membrane potential.

Secondly, the idea that there are two distinct modes of K$^+$ uptake is not supported by studies of Na/K selectivity (Poole, 1971). During washing of beet tissue, a new Na$^+$ transport mechanism develops, and competes with the pre-existing K$^+$ transport mechanism to supply cations to the cell. The resulting change in cation selectivity is observed whether the cation uptake is balanced by H$^+$ efflux or Cl$^-$ influx. In terms of the above hypothesis, a Na$^+$/H$^+$ exchange would compete with a K$^+$/H$^+$ exchange, while at the same time a NaCl pump would compete with a KCl pump. One might hope to find a simpler hypothesis!

A model repeatedly invoked to account for ion interactions during uptake (e.g. Jacobson et al., 1950; Smith, 1970) proposes a link (either electrical or chemical) between K$^+$ influx and H$^+$ efflux, and separate transport of Cl$^-$ together with H$^+$ (or in exchange for OH$^-$) across the plasmalemma. The main purpose of the present paper is to consider the implications of applying this model to transport in beet. A direct (non-electrical) link between K$^+$ and H$^+$ transport is assumed in the following discussion, since this seems at present to offer the simplest model for transport in this tissue.

IV. Cation/Anion Balance and Na/K Selectivity

According to the above model, Cl$^-$ uptake may influence K$^+$ uptake through its effect on cytoplasmic pH. Likewise, Na$^+$ uptake may compete

TABLE I. Effect of anion concentration on cation selectivity in red beet[a]

| External solution (mM) | | | Net uptake (μequiv g^{-1} h^{-1}) | |
NaCl	KCl	CaCl$_2$	Na	K
0·5			3·43	
	0·5			2·03
0·5	0·5		3·78	0·03
0·5		20	3·10	
	0·5	20		2·98
0·5	0·5	20	4·18	3·03

[a] Tissue was washed 6 days at 25°C, and pre-treated for 1 h in the experimental solutions at the same temperature. Uptake was measured over the second hour. Data are means of duplicates. Standard error does not exceed 0·2 μequiv g^{-1} h^{-1}.

with K^+ uptake for cytoplasmic H^+. As Na^+-selective carriers develop with washing of the tissue, they may be expected to compete more successfully for cytoplasmic H^+ (Poole, 1971) since, other things being equal, H^+ binding to the K^+ pump will suffer more severe competition from the cation concentration in the cytoplasm. One might suppose that if the cytoplasmic H^+ supply is increased beyond the saturation point of the Na^+ pump, the inhibition of K^+ uptake by Na^+ would be relieved. Table I shows that addition of $CaCl_2$, which promotes Cl^- uptake (Pitman, 1964) does relieve the inhibition of K^+ uptake by Na^+. Thus both cation–anion interactions and Na^+/K^+ selectivity in beet are readily explained by the model.

V. The Magnitude of the Potential Generated by H^+ Efflux

A re-examination of the electrogenic effect of the H^+ pump in relation to the above model reveals some interesting possibilities. The following experiments use the net uptake of K^+ as a measure of the rate of K^+/H^+ exchange. This is correct for solutions containing K_2SO_4 + $KHCO_3$, since there is clearly a stoichiometric efflux of H^+. During KCl uptake the K^+/H^+ exchange is no longer apparent, but is assumed to take place according to the model. Figure 2 shows the degree of hyperpolarization for various rates of pumping, at constant external K^+ concentration and constant pH. When metabolism is so low as to inhibit the pump, the membrane potential is close to the K^+ equilibrium value. As the rate of pumping increases, the hyperpolarization increases rapidly, but reaches a plateau at which further increases in pump rate produce only modest increases in hyperpolarization. This may have various possible explanations. For instance, there may be a decrease in the net charge carried by the pump at increasing transport rates.

It is known (Poole, 1966) that the degree of hyperpolarization depends on the external pH. However, two factors may be distinguished. These are (1) the large effect of pH on the rate of pumping (Fig. 1) and (2) possible effects of pH on the electrogenicity of the pump. In Fig. 3, all data showing similar uptake rates (between 1·0 and 1·5 μequiv g^{-1} h^{-1}) have been put together to show the effect of external pH at constant external K^+ concentration and constant pump rate. It is clear that hyperpolarization increases with pH, even though the pump rate in all cases was constant at 1·25 $\pm$ 0·05 μequiv g^{-1} h^{-1}. Thus there is an apparent change in the electrogenicity of the pump with pH.

The apparent changes in electrogenicity with increasing pump rate (Fig. 2) and increasing pH (Fig. 3) could perhaps be ascribed to changes in passive permeability. However, a further explanation is required for the original observation (Fig. 1) that K^+ uptake increases greatly with increasing pH. It is suggested, therefore, that the primary effect of high external pH is to remove H^+ from the H^+ efflux pump. The pump could then accept K^+ ions,

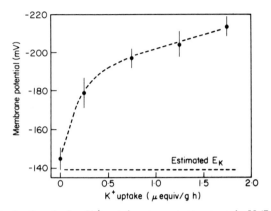

FIG. 2. Hyperpolarization during K^+ uptake at constant external pH (7·2) and constant external K^+ concentration (0·6 mequiv l^{-1}). External solution contained 0·4 mequiv l^{-1} K_2SO_4 or KCl, +0·2 mequiv l^{-1} $KHCO_3$ at 25°C. Various rates of K^+ uptake were obtained by varying the washing pretreatment of the tissue. Readings in which the uptake rates were within a span of 0·5 μequiv $g^{-1} h^{-1}$ were grouped (approx. 10 readings per interval). Standard errors are indicated by vertical lines. Data from Poole (1966, 1969) are included.

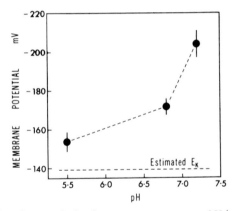

FIG. 3. Effect of pH on hyperpolarization at a constant rate of K^+ uptake (1·25 μequiv $g^{-1} h^{-1}$) and constant external K^+ concentration (0·6 mM). External solution (25°C) contained 0·6 mM KCl or 0·3 mM K_2SO_4 (pH 5·5), or 0·5 mM KCl or 0·25 mM K_2SO_4 + 0·1 mM $KHCO_3$ (pH 6·8), or 0·4 mM KCl or 0·2 mM K_2SO_4 + 0·2 mM $KHCO_3$ (pH 7·2). All results showing a K^+ uptake rate between 1·0 and 1·5 μequiv $g^{-1} h^{-1}$ are included. At each pH the mean uptake rate was 1·25 ± 0·05 μequiv $g^{-1} h^{-1}$.

thus increasing the pump rate, or alternatively, the carrier sites could return empty, thereby generating a potential. The K^+/H^+ exchange pump will thus become electrogenic only when there is a combination of high external pH and low external K^+. Otherwise, the sites will become saturated with K^+ or H^+ and behave like a neutral K^+/H^+ pump.

The changes in electrogenicity shown in Figs 2 and 3 now offer a new explanation for the original observations on the lack of effect of KCl uptake on the potential. It is to be expected, if Cl^- is transported with H^+, and K^+ exchanges against H^+, that KCl uptake will affect the potential in the same way as K^+/H^+ exchange. However, in KCl solutions at low pH, the pump will not show electrogenic properties (Fig. 3), and at high pH there is always enough H^+ efflux to bring the potential to a plateau (Fig. 2), so that any additional hyperpolarization due to KCl uptake will be small. Previous data on the effect of KCl uptake on the potential are therefore inconclusive. In fact, the results of Figs 2 and 3 are typical of uptake from either KCl or K_2SO_4 solutions.

VI. The Number of Binding Sites for K^+ Transport

Although KCl uptake may not, after all, differ from K^+/H^+ exchange as far as the potential is concerned, there is still a component of K^+ uptake which is balanced by chloride and which is not sensitive to reduction of external pH (e.g. Hurd, 1958; and Fig. 1). This might mean that the K^+/H^+ carrier has two or more sites with differing affinities for K^+ or for H^+. A site with relatively low affinity for H^+ on both sides of the membrane would be insensitive to external pH, but may require Cl^- uptake to supply sufficient H^+ on the cytoplasmic side. A site with higher affinity for H^+ would be able to receive H^+ from cytoplasmic organic acids and thus would be independent of Cl^- uptake, but would require a high external pH for its operation. This is, of course, speculation, but it may be that this kind of interpretation will suggest new approaches in experimentation. One obvious question for further investigation is the relation between the pH effect in beet and the well-known dual mechanisms of salt uptake (Epstein, 1966).

References

EPSTEIN, E. (1966). *Nature, Lond.* **212**, 1324–1327.
GERSON, D. F. and POOLE, R. J. (1972). *Pl. Physiol.* **50**, 603–607.
HURD, R. G. (1968). *J. exp. Bot.* **9**, 159–174.
JACOBSON, L., OVERSTREET, R., KING, H. M. and HANDLEY, R. (1950). *Pl. Physiol.* **25**, 639–647.
PITMAN, M. G. (1964). *J. exp. Bot.* **15**, 444–456.
POOLE, R. J. (1966). *J. gen. Physiol.* **49**, 551–563.
POOLE, R. J. (1969). *Pl. Physiol.* **44**, 485–490.
POOLE, R. J. (1971). *Pl. Physiol.* **47**, 735–739.
SMITH, F. A. (1970). *New Phytol.* **69**, 903–917.
VAN STEVENINCK, R. F. M. (1961). *Nature, Lond.* **190**, 1072–1075.

Discussion

Vredenberg asked Gillet if the changes in membrane potential and resistance on the introduction of bicarbonate were related, and whether it might be a pH-induced change as Spanswick had shown. Vredenberg continued that the potential change might be due to a change in the H^+ diffusion potential because of the high H^+ permeability and asked if the observed resistance change could be accommodated on this model. Gillet thought that it could not; the K^+ permeability is higher at high pH and both membrane resistance and potential had been taken into account. Walker then asked if the lower resistance in bicarbonate solution is consistent with an unchanged H^+ permeability and is due only to a change in K^+ permeability. Gillet replied that it was. Smith then commented that bicarbonate effects are now known to be primarily pH effects, but cells do take up bicarbonate. Must one therefore take into account bicarbonate fluxes and OH^- effluxes in the membrane resistance? One might find membrane resistance effects specific for bicarbonate and unrelated to pH changes. Gillet replied that he had no information on this effect; the observed resistance values fit an alteration in K^+ permeability as the only effect.

Vredenberg then asked Spanswick about his assertion that light affects both the electrogenic pump and the membrane resistance: how these effects were distinguished and how the same mechanism could cause an increase in electrogenic pumping and a decrease in membrane resistance. Vredenberg said that in his opinion the primary effect is on resistance and the effect on the electrogenetic pump is secondary. This is important because membrane resistance has been calculated from the I–V curve in the light and has then been referred back to the K^+ diffusion potential in the dark; the implicit assumption is that the K^+ diffusion potential has not been affected by the light–dark switch. Spanswick replied that he knew of no experiment which separately distinguished light effects on potential and resistance. Second, he was not referring back to the K^+ resting potential in the dark, but to an upper limit of any possible diffusion potential which is so arranged as to be more positive than any light potential. It therefore does not matter if there is a light effect on this diffusion potential.

135

Walker then said to Spanswick "you annoyed me by suggesting that the long pore effect is difficult to test; although difficult to test, it does have the advantage of giving predictions. Vredenberg's point to you reaches an interesting situation. If you think that the resistance alteration is due primarily to a change in pumping, and that membrane potential is chiefly due to the pump, then you must offer us experimental evidence. A critical test might be given by your own equation for the pump potential which predicts resistance changes for given changes in potential or vice versa". Spanswick replied that the resistance alteration with pH argues against a high H^+ permeability. He then showed a slide; the membrane resistance in the dark is very high and increases with increasing pH to pH 9 after which it decreases for some unknown reason. In the light the resistance values at the peak potential following a pH change increase only slightly as pH increases; the effect is not very large and is less than would be expected if the membrane were highly H^+ permeable and the conductance were practically all due to the H^+ flux. The plateau potential resistance values decrease slightly as pH increases. This therefore is evidence against a large passive H^+ flux and leads one to an electrogenic pump as the chief component. Walker replied that there was no reason to disbelieve what Spanswick had said, but was there an experimental test? From the equations could one predict a potential given a resistance or vice versa? Spanswick said that he could not predict the resistance but he could show an effect which seemed ambiguous. Sometimes at pH 6 the potential would "come up to a rather positive value"; when acetate was added the membrane would immediately hyperpolarize. The explanation might be the same as in mitochondrial preparations; acetate crosses the membrane, releases H^+ and makes the cytoplasm more acid, thus making the pump more negative in potential. If the membrane potential is due primarily to the pump potential, E_p, then the membrane potential will become more negative. The theory does not give good predictor equations for membrane resistance.

MacRobbie then commented that there seemed to be two possibilities under discussion, the first that H^+ ions affect membrane resistance by passive flow and the second that pH affects the rate of pumping. Bentrup et al.'s paper (III.6) would provide evidence that potential is sensitive to pH in the light but not in the dark. This is consistent with Spanswick's proposal that a pump is involved, but is not consistent with the first proposal unless one assumes alterations to H^+ permeability in the light. MacRobbie asked whether Bentrup's observations in *Vallisneria* were repeated in *Nitella*. Spanswick replied that in *Nitella* potential is also sensitive to pH in the dark but the time course of the effect is quite different. In the dark the effect is usually permanent whereas it is transient in the light. On his model the explanation is that in the light cytoplasmic pH is increased as H^+ ions are pumped out, in turn affecting E_p; a test of his hypothesis would be that cytoplasmic pH should increase in the light. MacRobbie then asked if pH sensitivity were altered in the presence of CCCP which presumably inhibits the electrogenic pump. Spanswick replied that there are difficulties with inhibitor studies; at low pH inhibitors have drastic effects on the cell. But there is some evidence that pH sensitivity in the light is affected by DCMU. MacRobbie then asked, "On your proposal is it fair to say that if we pass current across the membrane and so drive the pump, we ought to detect a consumption of ATP, or reducing power?" Spanswick agreed that it was possible but that it depended on what fraction of the total cell energy consumption was used by the pump; it might be a very small fraction. The flux through the pump in the hyperpolarized state is less than that through the pump in the resting E_K state, because the flux through E_p is potential sensitive. Thus one should find large changes in external pH on passing current through the membrane.

Pallaghy commented that the effect of Ca^{++} on stabilizing membrane potential over large changes in external K^+ concentration is well known. Is it due to a decrease in P_K or to an increase in P_H, or because the Goldman equation can only be used with difficulty in the presence of Ca^{++}? Spanswick replied that K^+ concentration alterations in the presence of Ca^{++} leave the resistance constant up to 1 mM external K^+, where there is a sharp transition to a high K^+ permeability. In the absence of Ca^{++} the resistance changes more smoothly.

Kylin then commented on the difficulties of using DCMU. In photophosphorylation studies there is a sharp discontinuity in the relationship between rate of phosphate assimilation and DCMU concentration at 10^{-6} M DCMU, and a second discontinuity at 6×10^{-6} M; working just at one of these discontinuities may produce irreproducible results. Barr then asked Spanswick about the principal current carrying ion when the cell is hyperpolarized. Is the injected current carried by a H^+ flux? Barr continued that he had attempted the experiment but found no significant pH increase; the small pH increase observed is probably due to the carbonic anhydrase reaction. In this connection the work of Poole is interesting, with the possibility of K^+–H^+ exchange. Is it possible that current injection involves injecting K^+? Spanswick replied by asking, "Does your result argue against passive H^+ movement and are you hyperpolarizing the potential? If you are, the flux through the electrogenic pump will be decreased". Barr replied, "The basic fact is that you do have net current entering the cell; if this is carried by H^+, you must have a pH change in the external solution." Spanswick said that changes in the external solution may not be detectable; one must measure local changes using a pH electrode with a flat end. The problem is that the alkaline and acid bands along the cell confuse the issue. Barr agreed but said that they had passed $0.5\ \mu$A through the cell, i.e. 5 pmol cm^{-2} s^{-1}, and one would imagine this would cause a considerable pH change. Spanswick agreed that it is disturbing that pH changes in the external solution have not been detected.

Poole then said that it would be nice to have some resistance measurements in red beet cells, but that it was not easy in higher plants. Pitman then asked Spanswick about electrogenic potentials; he said basically one is taking the difference between the observed potential and a potential calculated on various assumptions of internal and external conditions, permeabilities etc., and defining this difference as the electrogenic potential. In red beet and also in *Nitella* there is a net flux of ions at the time the potential is measured. This flux of ions through the cell wall may generate a potential which is wrongly ascribed to the electrogenic potential. Such a component would depend on the flux through the wall and therefore would be sensitive to metabolic inhibitors, light, etc. in the same way as is a metabolically driven pump. Poole thought it might be a good explanation in red beet; certainly there are large discrepancies in potential when there is a large net flux of H^+. Spanswick replied that he had made permeability measurements on isolated *Nitella* walls; the Cl^- permeability is of the order $10^{-4}\,\mathrm{cm\,s^{-1}}$, i.e. three orders greater than membrane permeability. Thus the diffusion potential across the *Nitella* wall will be small. Pitman asked whether it depends on the ratio of K^+ to Cl^- in the wall rather than on the permeability of either. Spanswick replied that he thought in the presence of Ca^{++} the wall potential would be negligibly small.

Shone then asked Poole about the effect of Cl^- in his work. He said that it is known that Ca^{++} also affects selectivity; perhaps the $CaCl_2$ concentrations are affecting Na–K selectivity as a result of the Ca^{++}, not the Cl^-. Poole replied that at low concentrations of $CaCl_2$ the inhibition of K^+ uptake is not relieved; in red beet Ca^{++} does not affect selectivity in the way it is known to in barley roots. Sutcliffe commented that in his work on uptake in beet, K^+ is balanced by bicarbonate which is later converted to various organic forms; no efflux of H^+ is required. Poole replied that he thought it difficult to distinguish bicarbonate influx from H^+ efflux.

West then returned to the definition of electrogenic potential; he said it should surely be defined as the difference between the observed potential and the equilibrium electrochemical potential of the pumped ion, not of K^+. If H^+ is pumped one ought to have the equilibrium electrochemical potential of H^+ as the base line. Spanswick then said that any diffusion potential arising in the wall as a result of ion imbalance would be rapidly short-circuited in the Donnan phase which is of low resistance. Walker commented further that the effect of the wall is not to produce a diffusion potential but rather to alter the concentration of the ions at the plasmalemma exterior. Pallaghy then said that his reaction to the whole electrogenic story was that there are so many uncertainties involved in using either the Goldman or Nernst equations that departures from predictions may not be very significant; for example, Lüttge and he had related transients on light switching not to electrogenic pumps but to fluxes of H^+ from the chloroplasts causing transient diffusion potentials across the membranes or the cell walls. He asked if it were reasonable to deduce an electrogenic pump simply on the basis that the observed and calculated potentials are in disagreement. Spanswick replied that it is the least ambiguous way in his opinion. He said that local diffusion potentials of K^+, for example, would cause small changes in potential in *Nitella*, nothing like as large as the observed hyperpolarization.

Van Steveninck then commented on Poole's paper; he said that the tris effect was not simply a H^+-acceptor effect. In freshly cut tissue which has not yet developed a Cl^- uptake, molecules closely similar to tris do not enhance K^+ uptake. He said that he was interested in Poole's conclusion that the K^+ flux has two components, the first 75% which is K^+–H^+ exchange with high H^+ affinity and with concomitant organic acid synthesis, and the second 25% which is also a K^+–H^+ exchange where Cl^- may substitute as the electrically neutralizing ion in aged tissue. Van Steveninck thought that the tris effect as described by him can only be equated with the pH effect when the second site has developed and Cl^- transport is possible.

III.4

The Effect of Light and Darkness in Relation to External pH on Calculated H$^+$ Fluxes in *Nitella*

Danley F. Brown, Thomas E. Ryan and
and C. E. Barr

Department of Biological Sciences
State University of New York
Brockport, U.S.A.

I. Introduction

Some of the points of reference in the investigation of membrane transport are the Goldman (1943) flux equation, the related concept of membrane conductance, and, generally, the pre-eminence of K$^+$, Na$^+$, and Cl$^-$. Kitasato (1968) interrupted the orderly development of research on the Characeae with the idea that, contrary to the common belief, the hydrogen ion was most of the story. His results led him to conclude that active H$^+$ extrusion and the balancing passive H$^+$ influx so dominated the ionic traffic through the membrane as to control the membrane potential, and, in growing cells, provide the means for potassium accumulation.

This hypothesis was supported by the detection of acidic and basic regions appearing in the form of alternating bands along the lengths of the long internodal cells of *Nitella* and *Chara* (Spear *et al.*, 1969; Smith, 1970). Walker and Hope (1969), however, reinvestigated this whole question and emerged with perhaps the most damning evidence against the hypothesis: they looked for and failed to find the net H^+ flux predicted when a current is passed through the cell membrane via an external circuit. A simple calculation reveals, for example, that a hyperpolarizing current of 1 μA should change the pH of 1 ml of unbuffered external solution from 6·0 to 7·0 in less than 2 min if H^+ is, in fact, the main current carrier.

More support for the hypothesis has recently come from the work of Rent et al. (1972) who measured the pH increase when Nitella cells are placed in a solution of pH 4·7. The initial rate of H^+ loss by the solution corresponded to a H^+ influx of about 10^{-11} mol cm^{-2} s^{-1}. This inward current must, of course, be balanced by some other ionic current since the discharging of the membrane capacity accounts for only a negligible fraction of it. As yet the identity of the ion or ions involved in the balancing current is unknown. Lest it be entirely forgotten, another possibility should be mentioned: that under these rather severe conditions some type of organic material may leak from the cells and serve to neutralize the external acidity.

At this juncture it appears that valid testing of the H^+ hypothesis will require careful work with adequate attention being given to the critical points raised by Walker and Hope (1969) and by Spanswick (1970). Despite the doubts concerning its present status, Kitasato's hypothesis has provided a new and vital focal point for current research on the origin of the resting potential.

Our present effort has arisen out of the unexpected finding that darkness induces a depolarization of Nitella cells at pH 5·7 while a hyperpolarization typically occurs at pH 4·7. We wondered if this phenomenon would be consistent with Kitasato's hypothesis and decided to make the measurements necessary for calculating the H^+ fluxes, in much the same way that Kitasato did. The explicit assumption is that the predominance of H^+ is so great as to render the contributions of other ions negligible. The only purpose in making such calculations is to see if reasonable values are obtained, and, if not, to see whether some clue as to the source of the error might be uncovered.

It is only fair to point out that the solution we have been regularly using is not the usual artificial pond water but a high K (1·0 mM), low Ca (0·1 mM) solution. The resting potential might be expected to be rather low, but actually it varies widely over the range -100 mV to -200 mV. Moreover, cells with low potentials can be triggered to a stable hyperpolarized state by a brief exposure to low pH (Spear et al., 1969; Rent et al., 1972). In itself this is of some interest since it directly implicates H^+ in a discrete alteration of the membrane properties of Nitella. The nature of this triggering action has not been fully explored, but it does seem possible that it might be exploited in some way so as to reveal something new about the membrane.

II. Methods

The culturing and experimental methods used throughout this study are identical with those previously described (Rent *et al.*, 1972) except where otherwise indicated. Preconditioning of harvested cells of *Nitella clavata* was under 130 lx cool white fluorescent lighting at 22°C instead of 500 lx. The mean lifespan of cells was considerably increased at this low light intensity. For all experiments cool white light of about 450 lx was used. After the insertion of microelectrodes a period of 15–18 h was allowed for stabilization of the cell before an experiment was begun. The cell was located in a plexiglass trough through which solution slowly flowed.

The main solution, used in both the preconditioning of cells and for the experimental work is designated as K solution and has a pH of 5·7 ± 0·1 when equilibrated with air. It consists of 1·0 mM KCl plus 0·1 mM each of NaCl, CaCl₂ and MgCl₂. For the experimental work at pH 4·7 HCl was used for acidification.

The electrical potential differences between the vacuole and the external solution were measured with the usual Ag/AgCl electrodes with microcapillary salt bridges; the tips were about 10 μm in diameter. For electrical resistances the method of Hogg *et al.* (1968) was employed.

III. Results

A. Survey of Cells: Dependence of Dark Response on pH

In Fig. 1 is presented the clearest type of response of the resting potential of *Nitella* cells to darkness. Darkness induces a depolarization at pH 5·7 whereas at pH 4·7 a hyperpolarization occurs. Figures 2 and 3 include a survey of the dark responses at pH 5·7 of 13 cells, 11 of which were also subjected to the pH 4·7 treatment. At pH 5·7 depolarizations of at least 15 mV

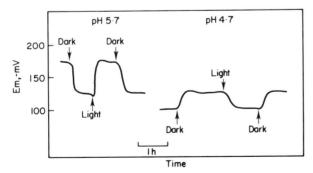

FIG. 1. The influence of pH on the dark response of the resting potential of *Nitella* cells. The post-harvest age of the cell at pH 5·7 was 8 days and that of the cell at pH 4·7 was 5 days. These were the clearest responses obtained.

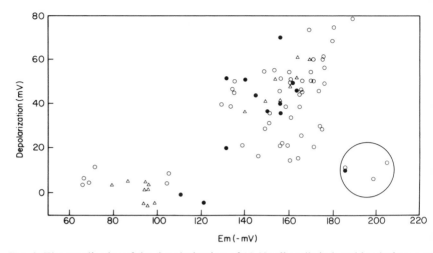

FIG. 2. The amplitudes of the depolarization of 13 *Nitella* cells induced by darkness at pH 5·7. The filled circles represent the values obtained on the initial exposure to darkness, the hollow circles represent subsequent trials, and the triangles represent trials subsequent to treatment at pH 4·7. The circled points are from one cell with a post-harvest age of 7 days. The median resting potential was about −155 mV. The cells varied in post-harvest age from 0·1 to 17 days.

were consistently obtained on cells having resting potentials between −125 and −180 mV, while only small responses occurred for more positive potentials. In only one instance did a cell with a high electronegative potential fail to respond appreciably to darkness. Absence of a dark response at pH 5·7 was more frequent among cells which had been previously exposed to pH 4·7; this would perhaps indicate some damage to the membrane. The median lifespan of cells kept in K solution at pH 4·7 under 450 lx illumination was about 3 days.

Figure 3 shows that hyperpolarizing responses to darkness at pH 4·7 occurred with fair consistency for the resting potential range of −75 to −100 mV. Overall only half the cells behaved in this way, indicating a more select population. The mean hyperpolarization within the responsive potential range was about 18 mV while the mean depolarization at pH 5·7 was 45 mV.*

Of significance is the fact that cells which were not markedly depolarized by the low pH, i.e. with resting potentials more negative than −125 mV, were affected by darkness in the same way as cells within the responsive range at pH 5·7. From this it may be concluded that the direction of the potential displacement depends only on the amplitude of the resting potential and not on pH *per se*.

---

* As Spanswick has pointed out (see Discussion after paper III.6), Figs 2 and 3 clearly show that on the average darkness causes a shift in potential to about −110 mV at both pH 5·7 and pH 4·7. Thus, in the dark E_m is independent of pH and is rather close to E_k.

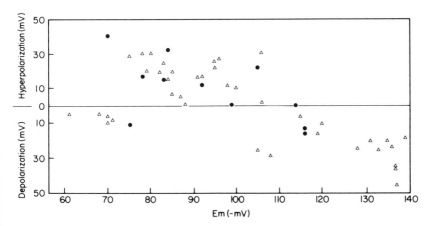

FIG. 3. The amplitudes of the changes in resting potential of 11 *Nitella* cells induced by darkness at pH 4·7. The filled circles represent the values obtained on the initial exposure to darkness and the triangles represent subsequent trials. The median resting potential was about −95 mV. The cells varied in post-harvest age from 0·1 to 12 days.

It is of importance at this point to relate these results to the known acid–base heterogeneity of the cell surface and, in particular, to the actual pH at the outer surface of the cell membrane, for both the acidic and basic zones. Spear *et al.* (1969) found that these two zones had approximately equal areas and that the pH in the bulk solution was altered from 7 to 6 and 8 outside the cell surface of the respective zones. Unfortunately knowledge of the actual pH at the membrane is not available since the cell wall prevents accessibility to the region. It is clear that the overall effect of low pH on the resting potential will represent some sort of summation of the effects on the two zones, and that the relevant pH may be somewhat different from that of the bulk solution.

B. Membrane Properties as Related to pH and Darkness

Figures 4, 5, 6 and 7 consist of the detailed records of the membrane properties of the two cells each subjected to the same 9-h experimental regime on 2 successive days. The amplitude of the resting potential is not well correlated with the resistance and this indicates that at least one other variable must be present. In several instances negative correlations exist between the potential and resistance changes, most notably for the dark/pH 5·7 condition of Fig. 5 in which the resistance increases 15-fold while the potential becomes slightly more electropositive.

The best positive correlations exist for pH 4·7, from which it may be tentatively inferred that the passive properties of the membrane assume increasing importance under these rather unnatural conditions. During the

dark period at pH 4·7 a significant alteration of the membrane properties
appears to occur as evidenced by the much higher resistances of the cells
when returned to light, i.e. as compared with the resistances in light prior to
the dark period. The modification of the membrane behavior seems to be
completely reversed by returning the cells to pH 5·7 solution.

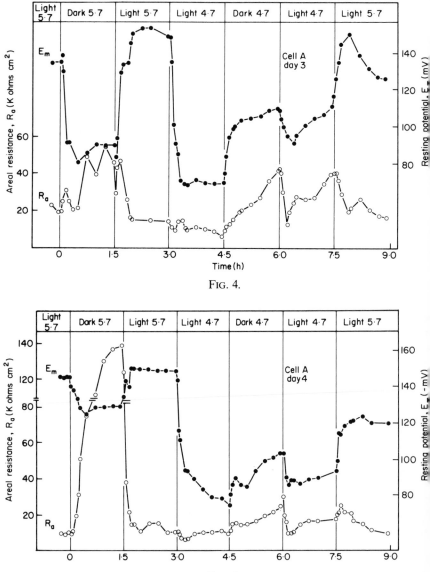

FIG. 4.

FIG. 5.

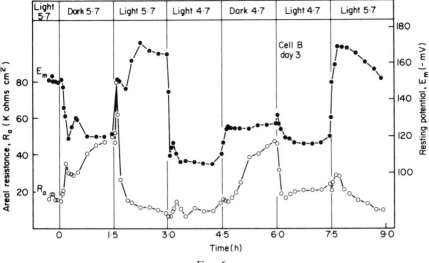

FIG. 6.

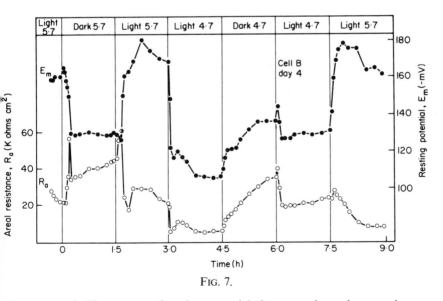

FIG. 7.

FIGS 4, 5, 6, 7. Time course of resting potential changes and membrane resistance changes of two *Nitella* cells under a fixed sequence of light–dark conditions at pH 5·7 and 4·7, as indicated. The cells were kept in the light at pH 5·7 over the 15 h period between the day 3 and day 4 experimental periods. Both cells were harvested 3 days before the beginning of the experiment.

Among the four conditions the highest resistances were for cells in darkness at pH 5·7, with a typical value of about 50 kΩ cm². The lowest resistances occurred for the light/pH 4·7 condition (but only before the dark period). Thus, as previous reports indicate, both light (Hope, 1965; Nishizaki, 1968) and low pH (Kitasato, 1968) bring about a decrease in resistance, the former presumably by a permeability change and the latter by the presence of a higher concentration of current carrier.

C. Calculation of H⁺ Fluxes

The simplest conceptual framework that can be used to analyze these results will include the conventional assumption that measured membrane resistances relate only to the passive movement of ions, and that the amplitude of the membrane potential reflects the relative magnitudes of the electrogenic ion pumping and the ease with which the return current of ions can pass through the membrane.

This analysis will be presented in terms of H^+ alone, i.e. the total ionic traffic through the membrane is to be attributed to H^+, according to the rationale given in the Introduction. The equation to be used in calculating the H^+ fluxes is equivalent to the one used by Kitasato (1968) under similar assumptions.

$$J_H = \frac{E_H - E_m}{\mathscr{F} R_a} \qquad (1)$$

where

J_H = net passive H^+ influx (active H^+ extrusion is equal to this when E_m is steady)

E_m = membrane potential

E_H = H^+ equilibrium potential

= $0.059 \, (pH_i - pH_o)$

$\mathscr{F}$ = Faraday constant

R_a = membrane resistance, Ω cm²

In these calculations it is necessary to introduce approximations for some of the required values: the potential between vacuole and medium is used in place of E_m and, similarly, the resistance between vacuole and medium is used as R_a. In addition the pH of the solution in contact with the cell membrane is assumed to be that of the bulk solution. The pH of the protoplasm is taken to be 6·3 as obtained by Rent et al. (1972).

Table I consists of a compilation of mean values obtained from Figs 4, 5, 6 and 7, and the mean calculated H^+ fluxes. Instead of the resistance values mean conductances are given so that the direct comparison of positive

TABLE Iᵃ. Membrane properties as affected by darkness and pH

	pH 5·7			pH 4·7			pH 5·7
	Light	Dark	Light	Light	Dark	Light	Light
E_m (−mV)	148* ±6	118* ±9	156* ±5	89* ±10	118 ±8	114* ±8	140 ±10
g (μmho cm^{-2})	64 ±12	18* ±4	84 ±18	115* ±17	28 ±4	42 ±7	96 ±13
J_H (10^{-12} mol cm^{-2} s^{-1})	122* ±23	28* ±6	164 ±32	220* ±29	62 ±8	90* ±12	178 ±29

ᵃ The values presented are mean values computed from Figs 4, 5, 6 and 7; they are the values at the end of a given treatment. Variation is expressed as the standard error of the mean.

* Indicates a significant difference at the 0·05 level for the *change* in a membrane property associated with a change in experimental condition, reading from left to right.

parameters may be made. Only the values at the end of each 90 min period are used since the membrane properties appear steadier at this time than during earlier portions of the period.

The calculated H^+ fluxes in light at both pH values average about 150×10^{-12} mol cm^{-2} s^{-1} which is about 4 times the value calculated by Kitasato and 100 times the typical values obtained for K^+ and Cl^- (MacRobbie, 1971). The dark H^+ fluxes are considerably lower as expected. The rather large H^+ flux for the dark/pH 4·7 condition should perhaps be considered separately in view of the obvious after-effects of this treatment: there is perhaps a heavy leakage of K^+ from the cell. As yet no tracer studies have been carried out to examine this point.

The relative effect of darkness on the conductance is similar for the two pH values, i.e. the conductance decreases to about one-fourth the light value. This suggests that one phase of permeability control resides at the inner side of the membrane and that it may be rather independent of the external conditions.

IV. Discussion

A. Calculated H^+ Fluxes and the Concept of Membrane Conductance

The problem of interpreting these results, especially the very high calculated H^+ fluxes, is closely tied to what one's expectations are. Certainly fluxes of 150×10^{-12} mol cm^{-2} s^{-1} are difficult to accept as correct; yet there is no way of rigorously demonstrating that they are wrong. Before proceeding to challenge the idea that the H^+ permeability is extremely high in the Characeae, we suggest that the concept of membrane conductance be subjected to equally stern appraisal.

There appears to be no reasonable basis for assuming that measured membrane conductances are values which relate only to the passive movement of ions through the membrane. It is much easier to defend the idea that all ions within the membrane are influenced by the electric field, and that when the field strength is altered their movements will be modified in some way. Whatever may be the nature of an electrogenic active transport process, it does not seem likely that it would be completely shielded from the effects of an applied potential.

For the sake of presenting a concrete picture, let us assume that in *Nitella* a photosynthetically-produced reductant delivers an H-atom to an H-acceptor molecule at the inner side of the plasmalemma, after which the H is passed along a redox chain to a point midway in the membrane. Here charge separation occurs with the electron returning to the inner side of the membrane via cytochromes while H^+ diffuses through a channel to the outside. This mechanism may be called a redox-driven, pure electrogenic H^+ pump, a mechanism originally proposed by Conway and Brady (1948).

Let us now assume that H^+ is the predominant ion, both actively and passively, as Kitasato suggested. If a small hyperpolarizing current is passed through the membrane, the membrane potential is slightly displaced. As a result H^+ influx should increase, but in addition, the greater field strength will have the effect of opposing the H^+ pump and its rate should be reduced. Thus, the net positive current into the cell actually consists of the difference between the increased H^+ influx and the decreased H^+ extrusion. If they are equally affected by the increment in potential, the calculated conductance will be twice the correct value. One can see that conductance values several times the "true" values might arise in this way.

From this, one is strongly inclined to draw the conclusion that the large values calculated for the H^+ fluxes simply represent a misinterpretation of the measured conductances. Generalized, this type of explanation might account for most of the discrepancies between conductance values and the magnitudes of ionic fluxes as measured isotopically.

H^+ is then left in a rather uncertain state as to its relative importance among the major ions involved in membrane traffic. There is still, of course, a considerable amount of circumstantial evidence indicating a significant role of H^+ in the control of the membrane potential. But in the absence of knowledge of the unidirectional H^+ fluxes and of what membrane conductance means, a quantitative picture is not possible.

V. Concluding Remarks

The idea of a proton pump as the primary electrogenic mechanism in membranes is attractive because of its simplicity; detailed knowledge on the spatial separation of protons and electrons in biological redox processes also makes it defensible, at least in a general way. In the plant kingdom there are numerous examples of net H^+ efflux concurrent with net K^+ uptake, notably in yeast (Conway and Brady, 1950), in *Neurospora* (Slayman and Slayman, 1968), in barley roots (Hiatt, 1967; Pitman, 1970) and in pea leaf fragments (Nobel, 1969). This can be explained, as Kitasato has suggested, in terms of the generation of an electrical potential by active H^+ extrusion which in turn permits K^+ to be accumulated passively.

In the mature cells of the Characeae one would not expect this to occur as these cells are past the stage of growth, and potassium has come to electrochemical equilibrium. Nonetheless, the acidic bands along the cell surface offer some evidence for H^+ extrusion. One wonders if perhaps a similar mosaic of acid and basic patches may be present on a smaller scale in cells of the usual size.

The "primacy of protons" has been an exceedingly fruitful idea in the field of ion transport, but whether H^+ is as predominant as Kitasato proposed seems questionable. The quantitative problems involved here are formidable, both technically and conceptually. One of the most critical needs is to find a way to interpret membrane conductance values. Until a more realistic approach to this problem is made, it does not seem likely that progress can be made toward understanding the origin of the membrane potential.

Acknowledgements

This work was supported by grant GB-18069 from the U.S. National Science Foundation. We gratefully acknowledge the assistance of Mr Joseph DiGregorio, Miss Liz Frank, Mrs Helen Hovey and Mrs Helen Maier.

References

CONWAY, E. J. and BRADY, T. G. (1948). *Nature, Lond.* **162**, 456–457.
CONWAY, E. J. and BRADY, T. G. (1950). *Biochem. J.* **47**, 360–639.
GOLDMAN, D. E. (1954). *J. gen. Physiol.* **27**, 37–60.
HIATT, A. J. (1967). *Pl. Physiol.* **42**, 294–298.
HOGG, J., WILLIAMS, E. J. and JOHNSTON, R. J. (1968). *Biochim. biophys. Acta* **150**, 518–520.
HOPE, A. B. (1965). *Aust. J. biol. Sci.* **18**, 789–801.
KITASATO, H. (1968). *J. gen. Physiol.* **52**, 60–87.
MACROBBIE, E. A. C. (1971). *A. Rev. Pl. Physiol.* **22**, 75–96.
NISHIZAKI, Y. (1968). *Pl. Cell. Physiol.* **9**, 377–387.
NOBEL, P. S. (1969). *Pl. Cell. Physiol.* **10**, 597–605.
PITMAN, M. G. (1970). *Pl. Physiol.* **45**, 787–790.
RENT, R. K., JOHNSON, R. A. and BARR, C. E. (1972). *J. Membrane Biol.* (in press).
SLAYMAN, C. L. and SLAYMAN, C. W. (1968). *J. gen. Physiol.* **52**, 424–443.
SMITH, F. A. (1970). *New Phytol.* **69**, 903–917.
SPANSWICK, R. M. (1970). *J. exp. Bot.* **21**, 617–627.
SPEAR, D. G., BARR, J. K. and BARR, C. E. (1969). *J. gen. Physiol.* **54**, 397–414.
WALKER, N. A. and HOPE, A. B. (1969). *Aust. J. biol. Sci.* **22**, 1179–1195.

III.5

Energy Control of Ion Fluxes in *Nitella* as Measured by Changes in Potential, Resistance and Current–Voltage Characteristics of the Plasmalemma

W. J. Vredenberg

Centre for Plant Physiological Research
Wageningen, The Netherlands

Abbreviations used: CCCP—carbonyl cyanide-*m*-chlorophenyl hydrazone; DCMU—3,4-dichlorophenyl-N,N-dimethyl urea; DCIP—2,6-dichlorophenol-indophenol; DNP—2,4-dinitrophenol.

I. Introduction

The ion transport processes in plant cells are known to be under metabolic control (cf. MacRobbie, 1970; Lüttge *et al.*, 1971; Raven, 1969; Smith, 1968). Electrical measurements on cellular and organellar membranes are a useful means of studying the regulatory mechanisms by which the energy-producing reactions in green cells (photosynthesis and respiration) trigger the energy transductions and ion transport processes at the membranes. These types of experiments have been limited almost exclusively to the measurement of the membrane potential, and changes thereof, occurring upon light- or chemically-induced alterations in energy supply (cf. Bentrup, 1971). Changes in the membrane potential have been interpreted mainly in terms of changes in the passive membrane diffusion potential (Kitasato, 1968; Lüttge and Pallaghy, 1969; Vredenberg, 1969). Relatively little experimental attention has been given so far, at least in plant cells and cell organelles, to a possible primary effect of changes in energy supply on the permeability characteristics of the membranes, and on the active component of the membrane potential, maintained by an electrogenic ion pump. It has been emphasized (cf. Jeschke, 1970; Vredenberg, 1972b; Lüttge et al., 1971), and partial evidence has been presented, that the energy control of the ion transport processes is mediated by an interaction of accumulated energy products with membrane constituents, causing a change in the membrane permeability (—resistance) (Raven, 1968; Jeschke, 1970b; Vredenberg, 1972a, 1972b), and with membrane-bound enzymes catalysing the electrogenic pump(s), causing changes in the potential generated by it (Hope, 1965; Saddler, 1970; Gradman and Bentrup, 1970; Strunk, 1971; Andrianov et al., 1971).

This paper describes the results and interpretations of experiments, in which the effect of light, absorbed by the photosynthetic apparatus, upon the membrane-specific physical determinants of the ion transport processes across the plasmalemma in Nitella translucens, have been studied. These were obtained by using a method which enabled the following electrical membrane parameters and characteristics to be recorded simultaneously as a function of time and of regulatory energy supply: (i) the electrical potential and resistance, and induced changes therein, and (ii) the membrane current–voltage (I–V) relationship, and changes therein.

The results indicate that specific, yet unidentified, energy sources, fed by primary, or associated, photosynthetic reactions control the membrane permeability (—resistance) and an electrogenic ion pump, and consequently the passive and active components of the membrane potential. It is shown that the condition, or what we call the energy state, of the cell in the dark, depends strongly on energy which is generated in prior light periods.

II. Materials and Methods

Freshly grown, internodal cells of Nitella translucens about 5 cm in length and 0·06 cm in diameter (i.e. with a surface area of about 1 cm^2) were used. Culture conditions were identical to those described elsewhere (Vredenberg, 1972a). Usually a cell with its two neighbouring cells was presoaked a few days before the start of the experiments in APW or Ca-APW (1 mM NaCl, 0·1 mM KCl and 0·1 or 1 mM, respectively, CaCl$_2$), enriched with 0·1 mM KHCO$_3$, and exposed to alternating light and dark periods of 12 h each. This pre-treatment was found to yield fairly reproducible conditions of the cell with respect to the responses to be measured. The cell was transferred to the measuring cuvette at the end of a 12 h dark period. In general each experiment was started with the cell suspended in 50 ml of the presoaking medium.

The medium in the measuring cuvette was continuously circulating in one direction along the length of the cell and bubbled with air. The pH of the medium was measured with a glass electrode (e), and recorded on channel 3 of the recorder (Fig. 1). Normally the pH at the start of an experiment was about 6·9. Changes in pH were brought about by adding trace amounts of HCl or tris to the medium. The mixing time for added chemicals was about 1 min. In some cases the medium was withdrawn from the cuvette and replaced by a substituted medium. This procedure also could be completed within 1 min. The experiments were carried out at room temperature.

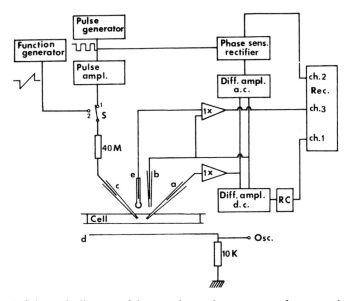

FIG. 1. Schematic diagram of the experimental arrangement for measuring the kinetics of energy-dependent changes in the electrical potential and resistance of plant cell membranes (switch S position 1), and of the membrane current-voltage characteristics (S position 2). The potential is recorded on channel 1 (ch. 1) of the recorder, with a response time of about 1·5 s, selected by the RC circuit in the potential measuring part of the system. The resistance is recorded on ch. 2, with a response time of about 3 s. The recorder deflection on ch. 2 is linearly proportional to resistance values of up to 120 kΩ cm² (Vredenberg, 1972a). The pH of the external medium, probed by electrode (e), is recorded on ch. 3. Further explanations are in the text.

A cell was illuminated perpendicularly upon its surface over its total length by a homogeneous monochromatic light beam (wavelength band around 676 nm, half width 10 nm). Light intensities were measured with a YSI Radiometer, model 65.

The method for measuring energy-dependent changes in the electrical potential across and resistance of the cellular membranes, has been described in a recent paper (Vredenberg, 1972a). Figure 1 shows a diagram of the experimental arrangement. Changes in potential are recorded on channel 1 of the recorder, of which the deflection is proportional to the potential probed by the intracellular capillary electrode (a), with reference to the external electrode (b). Changes in the membrane resistance are recorded, with switch S in position 1, on channel 2 of the recorder, of which the deflection is proportional to the amplitude of the potential modulations caused by square current pulses of 400 ms duration and constant low amplitude (0·01 μA). The current pulses pass across the membrane through the internal current electrode (c) towards the external Ag wire (d). According to conditions, discussed by Hogg et $al.$ (1968), the distance between electrodes a and c is 0·42L, in which L is the half length of the cell. Electrodes a and c were positioned in the vacuole, so that the responses were measured across the cell wall, plasmalemma and tonoplast in series. It has been shown, however, that the responses are mainly, if not exclusively, due to changes occurring at the plasmalemma (Vredenberg, 1971b; 1972a).

The current–voltage characteristics of the plasmalemma were measured by means of a current scanning technique, basically similar to that used by Coster (1965). Switch S is put in position 2, connecting the cell current circuit to the output of a function generator, operating in a single sweep mode, which generates a voltage that changes with time at a uniform rate α. With the 40 MΩ resistor in the current circuit, the current passing across the membranes changes from -0.34 (inward current) to $+0.34$ μA (outward current) at a rate α. In some cases the 40 MΩ resistor was replaced by a 20 MΩ resistor, in order to increase the interval of the current sweep by a factor of 2. Usually the system was operated with $\alpha = 5.6$ nA/s (2 min sweep). In some special cases where a faster recording was necessary, owing to a drift in the membrane rest potential, α was equal to 22·5 nA/s (30 s sweep).

As will be seen in the next section, the current–voltage (I–V) characteristics of the plasmalemma, as measured around the resting potential, are dependent on the condition of the cell. Although the I–V curves measured before and after a change in a particular condition appear to be substantially different in shape, especially around the resting potential, they can be compared quantitatively after special reproduction in one graph. Figure 2 illustrates the considerations which underlie this graphical method. Assume that for a cell in a defined stationary condition, called state S_0, the hypothetical I–V curve, measured over an extended current interval from -0.6 to $+0.6$ μA, is given by curve AB. E_0 is the resting potential in state S_0, with I, V coordinates $(0, -100)$. Part A_0B_0, with $A_0 = (-0.3, -180)$ and $B_0 = (+0.3, -85)$ is the curve measured over the current interval -0.3 to $+0.3$ μA for the cell in state S_0. Let S_1 be the state of the cell in which a $constant$ inward current of 0·1 μA is passed across the membranes, and assume that this current does not cause any chemical or physical change in the membrane

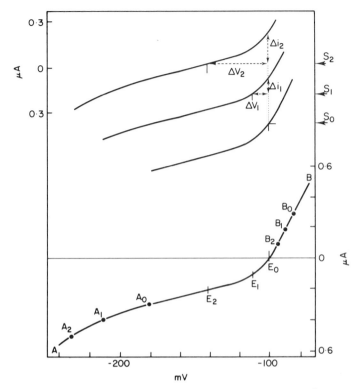

FIG. 2. Lower part: hypothetical membrane I–V curve AB, measured over an extended current interval -0.6 to $+0.6$ μA. A_0B_0, A_1B_1 and A_2B_2 are parts of the I–V curve, measured in a current interval -0.3 to $+0.3$ μA in the absence and presence of *constant* inward currents i_1 (0.1 μA) and i_2 (0.2 μA), respectively. E_0, E_1 and E_2 are the respective dark resting potentials in the absence and presence of these currents, respectively. Upper part: curves A_0B_0, A_1B_1 and A_2B_2, characteristic for the states S_0, S_1, S_2, plotted at equal vertical distances, after translation along a constant voltage coordinate from curve AB. Further explanations are in the text.

and in the cell interior. Then S_1 is characterized by a resting potential E_1 $(-0.1, -110)$. When the I–V curve of the cell in state S_1 would be measured with a current varying from -0.3 to $+0.3$ μA, superimposed on the constant current of 0.1 μA, this curve would be identical to part A_1B_1 of the curve AB, with $A_1 = (-0.4, -210)$ and $B_1 = (+0.2, -90)$. Similarly A_2B_2 is the I–V curve measured for the cell in state S_2, in which a *constant* inward current of 0.2 μA passes across the membranes. In state S_2 the resting potential is E_2 $(-0.2, -140)$. In the upper part of the figure the I–V curves A_0B_0, A_1B_1 and A_2B_2, characteristic for the states S_0, S_1 and S_2, respectively, have been plotted at equal vertical distances, after translating each of them from the curve AB along a vertical line (constant voltage coordinate). The S-marks at the right hand vertical indicate the state of the cell. Each S-mark coincides with the zero current point on the vertical current axis, i.e. the point at which the resting potential is measured. This resting point is also indicated in each curve by the intercept of the small vertical and horizontal bars. Although different in shape around the resting potentials, the parts of the curves that fall within the same potential region are exactly parallel, and when translated, will cover each other and constitute an extended I–V curve A_2B_0 of the membrane. As indicated by the dotted vertical and horizontal lines, the S_1 and S_2-states of the cell can be characterized, with respect to S_0, by the stimulatory membrane currents Δi_1 ($0 \cdot 1\ \mu A$) and Δi_2 ($0 \cdot 2\ \mu A$), respectively, and associated differences in the resting potential ΔV_1 ($10\ mV$) and ΔV_2 ($40\ mV$), respectively.

It has been observed for a variety of experimental conditions, that the I–V curves of *Nitella* cells, measured for each specific condition, but in the absence of an externally applied constant current, are similar to those characteristic for the S_0 and the current-induced S_1 and S_2 states. Thus the I–V curves, measured in a sequence of successive treatments (pre-illumination, changes in pH, etc.), could be reproduced in a similar way as done in the example of the different S-states in the upper part of the figure. Each curve was reproduced at appropriate coordinates on the potential axis. These were determined after translating the curve along the I- and V-axis, such that common parts of the curves were completely, or nearly completely, covering, and consequently after vertical translation, in parallel. In strict analogy with the case discussed above, the different I–V curves can be interpreted as being representative for conditionally-induced states, which are characterized, with reference to a control state, by intrinsic membrane currents Δi, and associated differences in the resting potential ΔV. Usually the control state, S_0, of a cell, responding upon successive variations in a particular condition, is the one characterized by the lowest resting potential.

III. Results and Interpretations

Figure 3 shows the kinetics of the light-induced changes in the membrane potential and resistance, measured in a cell that has been kept in the dark for more than 30 min. Dark-adapted cells usually are in an S_0 state. It has been observed that for cells in an S_0 state, as shown in Fig. 3, the kinetics of the changes in potential and resistance are nearly, if not completely, identical, when light of moderate intensity is given for 1–2 min, and the dark periods between illuminations are not shorter than 10 min.

The I–V characteristics, measured in the dark (full line, d) and in the light (dashed line, l) of the plasmalemma of a cell in an S_0 state are shown in Fig. 4. The dashed curve was measured 2 min after the onset of illumination. As shown in Fig. 3, the potential and resistance of a cell in an S_0 state reach a

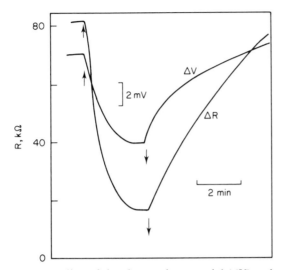

FIG. 3. Simultaneous recording of the changes in potential (ΔV) and resistance (ΔR), upon illumination and darkening of a dark-adapted *Nitella* cell, bathed in Ca-APW enriched with 0·1 mM KHCO$_3$. The dark potential and resistance were -102 mV and 82 kΩ cm^2, respectively. Upward and downward pointing arrows mark the beginning and end, respectively of the illumination period. A downward movement of the potential curve means a membrane depolarization. The intensity of the 676 nm light was about 6·5 nEinstein cm^{-2}s^{-1}.

quasi steady state after 2 min of illumination, i.e. a more depolarized potential and a decreased resistance. Figure 4 shows that the shape of the I–V curve measured in the light is different from the one measured in the dark. The light-induced change was found to be reversible, provided that the illumination period does not last more than 2–3 min (see below). It can be concluded

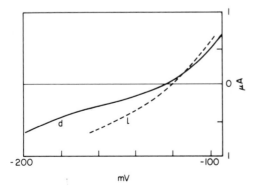

FIG. 4. I–V curves, measured in the dark (full line) and, 2 min after the beginning of an illumination period, in the light (dashed line), of a *Nitella* cell, kept in the dark for more than 30 min. The intensity of the 676 nm light was *ca.* 1 nEinstein cm^{2}s^{-1}.

from the curves that, except in the extreme hyperpolarized region, the differential membrane resistance is substantially decreased in the light. The differential resistance of the membrane at a certain potential is equal to the inverse of the slope of the I–V curve at the point, given by the potential coordinate. Thus, according to the curves of this particular cell, the differential resistance at the resting potential has been changed in the light from 80 to 40 kΩ cm². The resting potential has been changed from -125 to -120 mV. In its hyperpolarized region, for instance at a dark potential of -200 mV, illumination has caused a decrease in the differential resistance from 90 to 80 kΩ cm², and a depolarization of 35 mV.

The effect of prolonged pre-illumination, and of changes in the pH of the external medium on the I–V characteristics of the plasmalemma, measured in the dark, is illustrated in Fig. 5. The condition- or energy state of the cell is

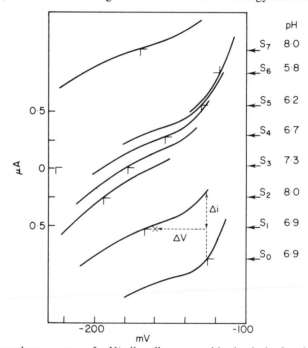

FIG. 5. I–V membrane curves of a *Nitella* cell, measured in the dark after dark adaptation (S_0), 4 min pre-illumination with 1 nEinstein cm^{-2} s^{-1} 676 nm light (S_1), and a change in pH from 6·9 to 8·0 (S_2), from 8·0 to 5·8 through intermittent values (S_2 through S_6), and finally from 5·8 to 8·0 (S_7). The cell was kept in the dark since the pH was changed from 6·9 to 8·0. The S-symbols stand for the energy state of the cell, which apparently has been altered, according to the I–V curves, characteristic for it. The dotted vertical and horizontal lines in the curve of S_1 mark the membrane current (Δi) and membrane hyperpolarization (ΔV), respectively, which are characteristic for the transition of S_0 into S_1. The cross on the horizontal line indicates the membrane potential, actually measured in S_1.

designated by numbered S-symbols. S_0 is the so called low-energy state of the dark-adapted cell at pH 6·9. According to the I–V curve the resting potential in S_0 is $-125\,mV$ and the membrane resistance is about $55\,k\Omega$ cm^2. S_1 is the dark state of the cell, after the medium (pH 6·9) was refreshed and the cell was illuminated with 676 nm light (intensity about 1·0 nEinstein cm^{-2} sec^{-1}) during 5 min. The I–V curve, characteristic for S_1, was measured 10 min after the end of the illumination period. S_1 is characterized by a membrane potential of about $-158\,mV$ (indicated by the cross on the horizontal dotted line in the figure), a dark resistance of about $210\,k\Omega$ cm^2 (at the resting potential), and, with respect to S_0, by an appreciable shift of the I–V curve. According to the method outlined and illustrated in Fig. 2, the I–V curve of S_1 is parallel to the one of S_0, when the resting potential of S_1 is assumed to be $-165\,mV$. Thus apparently pre-illumination has caused a transition of the energy state of the cell from S_0 into S_1, which is reflected by a shift in the membrane I–V curve, a membrane hyperpolarization, and an increase in the membrane resistance. The transition can be interpreted (Fig. 2) in terms of a change in an active membrane current, contributing to the membrane potential, and operating in the dark. Application of the method, illustrated in the previous section (Fig. 2), shows that the current generated in the cell in state S_1, with respect to the one generated in S_0, has increased with an amount of $0.35\,\mu A$ cm^{-2}, or 3.6 pmol cm^{-2} s^{-1}. This analysis shows that the plasmalemma potential is partly maintained by an electrogenic pump which is light-dependent, i.e. stimulated by an energy product formed by a reaction during prolonged illumination, or in successive illumination periods (Vredenberg, 1973). The energy stored in the light is slowly consumed by the pump and/or other energy-dependent cellular dark reactions, causing a slow decrease in the current generated by the pump. This conclusion follows from the observation that a higher energy state of the cell, like S_1, is more or less stationary in the dark for a few minutes, after which it transfers slowly to the S_0 state. This transition, which in general takes more than 30 min for completion, is reflected by a slow membrane depolarization, at a rate of about 1 mV min^{-1}, and a change in the dark resistance.

A change in the pH of the medium from 6·9 to 8·0, and a subsequent short pre-illumination, has caused a dark transition of the cell from S_1 into S_2 (Fig. 5), characterized by the respective I–V curves. This transition apparently is also due to a stimulation of the active membrane current. A subsequent lowering of the pH over small intervals from pH 8·0 to 5·8 is accompanied by a gradual back shift of the membrane I–V characteristics (states S_2 through S_6), a depolarization of the membrane, an increase in the differential resistance in the pH interval between 8·0 and 6·7, and a considerable decrease in the interval between 6·7 and 5·8. The kinetics of the potential and resistance changes are shown in Fig. 6. The pH-dependent changes are reversible, as can be concluded from the responses of the membrane (Figs 5 and 6) upon a final change in pH from 5·8 to 8·0 (S_6 to S_7 transition). The cell could be transferred from S_7 by a short illumination into a dark state S_8 (pH 8·0, not shown), which, according to the I–V characteristics thereof, was approximately similar to S_2. The analysis of the respective I–V curves, characteristic for the states S_2 through S_7, indicates that the membrane current, activated by pre-illumination of the cell, increases in the dark when the pH of the external medium is increased, and vice versa. These changes are accompanied by changes in the membrane potential and resistance. In this particular experiment, in which no intermittent illuminations were applied to the cell in any of the states S_2 through S_7, the membrane resting potential actually measured in any of the states S_3 through S_6 was found to be higher (about 5–15 mV) than that read from the curves in the figure. Presumably the passive membrane diffusion potential increases in the dark, concomitantly with a decrease in the membrane current. An opposite effect was observed when the membrane current was stimulated by pre-illumination (e.g. S_0–S_1 transition in Fig. 5).

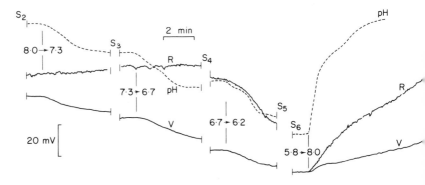

FIG. 6. Kinetics of the potential (V) and resistance (R) upon changes in pH of the external medium (Ca-APW + 0·1 mM $KHCO_3$). The interval between the vertical bars, intercepting the curves, covers a time period of about 3 min, during which the I–V curves (Fig. 5) were measured. It should be noted that the recorder deflection of the resistance signal, as measured in the pH region 8·0–6·7 is not linearly related to the actual resistance, which at these pH values is more than 120 kΩ cm^2 s^{-1}. The resistance measuring system (Fig. 1) is only linearly sensitive to resistances of up to 120 kΩ cm^2 (Vredenberg 1972a).

Figure 7 summarizes the results of an identical experiment carried out with another cell. The procedure was exactly similar to that illustrated in Figs 5 and 6, except that the I–V curves were measured (in the dark) after a short pre-illumination of the cell in any of the energy states, attained after an increase in the pH of the medium. The results were exactly similar to those of Figs 5 and 6, except that the membrane potentials actually measured in

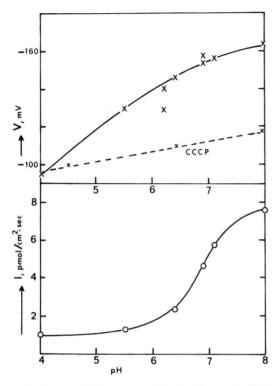

FIG. 7. Membrane dark potential (upper part) in the absence (solid line) and presence (dashed line) of CCCP (1·0 μM), and membrane current (lower part), measured as a function of pH in a *Nitella* cell, pre-illuminated in each pH-induced energy state. In the presence of 1 μM CCCP the membrane current was found to be less than 0·5 pmol cm^2 s^{-1} over the whole pH region (not shown). Values of the membrane current and potential were determined from analyses of I–V curves, similar to the ones shown in Fig. 5, and from the potential recordings (e.g. Fig. 6), respectively. See further the legends of Figs 5 and 6.

this case were found to be lower than the one deduced from the I–V curves. The measured potentials and the membrane currents, which have been read from the I–V curves, according to the method illustrated in Fig. 2, are plotted as a function of the pH. The graph shows that the membrane current is highly dependent on the external pH in the range between 6·0 and 8·0, and at a nearly constant low level in the pH range between pH 4·0 and 6·0. The membrane potential varies between −95 (pH 4·0) and −160 mV (pH 8·0). The pH-dependency below pH 5·5 is much sharper than in the interval between 6·3 and 8·0. The measured points in the curve might suggest, and other experiments agree with this, that the membrane potential is approximately constant in the intermittent pH interval between 5·5 and 6·3. It has been observed that, in contrast to the occurrences at a pH above 5·5 (e.g. Fig. 5), the I–V characteristics of the membrane drastically alter below pH 5·5. The I–V curve measured at pH 4·0 cannot be matched even after extensive translations along the I- and V-axis, with any of the curves measured in the higher pH region. This indicates that, at low pH, changes in the passive permeability of the membrane occur in response to a change in the pH. Thus it seems likely from these observations that the pH control of the membrane potential is triggered by different processes at low and at high pH.

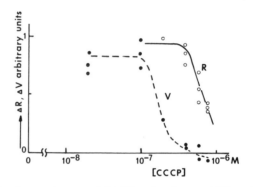

FIG. 8. Steady state changes in potential (ΔV) and resistance (ΔR) on 2 min illumination of a *Nitella* cell in an S_0 state (e.g. Fig. 3), as a function of CCCP concentration in the external medium (APW, pH 6·9). Dark periods between illuminations, during which increasing amounts of CCCP were added, were more than 10 min. Values of ΔV and ΔR are plotted as the fraction of the respective changes in the absence of the uncoupler.

Figure 8 shows the CCCP-inhibition curves of the light-induced changes in potential and resistance, measured in a cell in an S_0 state. It has been discussed (Vredenberg, 1972a), that the selective inhibition of the potential change by low concentrations of CCCP is due to the inhibition of the membrane current, generated by the electrogenic pump. It has been confirmed (Vredenberg, 1973), that the stimulated membrane current generated in a cell in a higher energy state, like S_2 in Fig. 5, is totally inhibited by low concentrations of CCCP. At these inhibitory concentrations of CCCP the membrane potential is hardly sensitive to the pH of the external medium, as shown by the results plotted in Fig. 7.

IV. Discussion

The behaviour of the cell is found to depend on the apparently alterable energy state (S_i) of the cell. It is shown that a change in the energy state of the cell occurs concomitantly with light energy conversion during a prolonged illumination period, and with a change in the pH of the external medium

(Fig. 5). According to the analysis of the I–V curves of the plasmalemma, the alterations in the energy state are caused by changes in the pumping rate of an electrogenic ion pump, contributing to the plasmalemma potential. The polarity of the pump is such that stimulation of its results in membrane hyperpolarization, and a change in the differential membrane resistance. The potential generated by the electrogenic pump in a cell in an S_0 state is relatively small, because of the low current generated by it; the reason is probably that very little or no energy is available for its operation. An S_0 state is conditioned by a relatively long dark treatment of the cell, or by adding low concentrations of CCCP, which apparently uncouples the pump from its energy source (Fig. 8). The current generated by the pump in an S_0 state has been found to be between 0 and 2 pmol cm^{-2} s^{-1} (Vredenberg, 1972a), and can be calculated (e.g. Fig. 3) from the steady state differences in potential and resistance, ΔV and ΔR, respectively, observed in short illumination periods. There is evidence (Vredenberg, 1972a) that in an S_0 state the potential change in the light reflects the change in potential generated by the electrogenic pump, owing to a primary effect of a photosynthetic product, or intermediary or associative reaction, on the plasmalemma, causing a 30–50 % decrease in its resistance in the light. The dark potential and resistance of the plasmalemma of a cell in an S_0 state have been found by us to be -100 to -130 mV and 25 to 80 kΩ cm^2, respectively.

Owing to the membrane hyperpolarization, caused by the stimulation of the electrogenic pump, the differential membrane resistance at the resting potential (-130 to -200 mV) of a cell in a higher energy state (S_i) is considerably increased (up to 200 kΩ cm^2). This change in resistance is exactly determined by the I–V relationship of the membrane, measured in S_0, or in any of the S_i states, and will be more pronounced as the membrane has stronger rectifying properties in the hyperpolarized region. In most cases the resistance decreases in the extreme hyperpolarized region, according to the downward bending of the I–V curve in this region (Coster, 1965; Kishimoto, 1966). It has been observed (Vredenberg, 1972a) that the kinetics of the potential and resistance changes occurring in prolonged illumination, during which the cell is transferred into a higher energy state, are different. This difference can be explained in terms of the time–energy response of the electrogenic pump, and the particular I–V relationship of the membrane in the dynamic hyperpolarized state in light and darkness.

These experimental results indicate at least a dual effect of photosynthetic energy conversion on the transport-determining electrical membrane parameters:

(1) A short term (1–2 min) reversible light effect causes the differential membrane resistance in the depolarized region (i.e. characterized by a steep and linear I–V relationship) to decrease at a relatively high rate. It is suggested that this decrease occurs in association with a chemical or ionic change

in the cytoplasm, due to a rapid light-induced translocation of a reaction product (intermediate), or ions, across the chloroplast envelope. As a direct consequence of the resistance decrease, the active component of the membrane potential, generated by the electrogenic pump, decreases (membrane depolarization).

(2) A long term (2–20 min) light effect causes an increase in the active component of the membrane dark potential, generated by an electrogenic membrane current pump (membrane hyperpolarization). The reaction which triggers the electrogenic pump apparently proceeds at a low rate, both in the light and in the dark. Probably the triggering effector is a high energy product, or intermediate of the photosynthetic reactions, which is in exchange with the chloroplast inner phase, and released into a cytoplasmic pool. A role for the high-potential phosphorylated compound that is translocated across the chloroplast envelope (Heber and Santarius, 1970) is suggested. The finding that phlorizin does not affect the long term effect (nor the short term effect) suggests that ATP formed in photosynthetic phosphorylation is not primarily involved.

(3) The membrane diffusion potential, as a passive component of the membrane potential may alter in direct association with primary light-driven ion translocations at the chloroplast membrane, or in association with the activation of the electrogenic pump, provided that the plasmalemma is selectively permeable for at least one of the ions involved in both cases.

The results, which are schematically summarized in Fig. 9, suggest that the light-dependent processes, mentioned under (1) and (2) contribute to a considerable degree to the potential changes generally observed in *Nitella* and other algal and green plant cells (Hope, 1965; Nishizaki, 1968; Lüttge

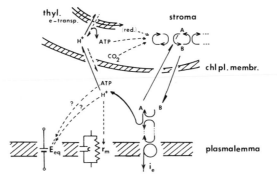

Fig. 9. Simplified scheme of a possible mode of linkage between photosynthetic reactions in the chloroplast and the ion flux determining parameters of the plasmalemma. The electrogenic pump (i_e) is powered by a high-energy intermediate, or product, which is released into the cytoplasm via a translocator shuttle, in which A and B are transferable transport metabolites (cf. Heber and Santarius, 1970). E_{eq} is the membrane potential, measured when $i_e = 0$. Further explanations are in the test.

and Pallaghy, 1969; Spanswick, 1970; Jeschke, 1970a; Andrianov *et al.*, 1971; Vredenberg, 1969, 1972b). The light-induced change in potential, observed in the presence of DCMU and DCPIP (Vredenberg, 1971a), at which no change in resistance occurs (Vredenberg, 1972b), may reflect the exclusive change in the proton diffusion potential.

In view of results reported by others (Kitasato, 1968; Spear *et al.*, 1969; Strunk, 1971), one is tempted to suggest that the electrogenic pump activates and balances the efflux and influx, respectively, of hydrogen ions across the plasmalemma. Kitasato (1968) has concluded from the pH response of the plasmalemma potential, that the passive component of it behaves as a proton diffusion potential. This conclusion has been made under the assumption that the rate of active H^+-transport by the pump remains constant when there is a change in the external pH. Accordingly he has calculated a high membrane permeability coefficient for the protons. This high proton permeability coefficient would account nicely for the discrepancy which has been reported in the literature (cf. Williams *et al.*, 1964) between the membrane conductance (resistance), measured electrically, and determined from flux measurements of K^+, Na^+ and Cl^- ions. Although our results, with respect to the pH response of the plasmalemma potential, are qualitatively similar to those of Kitasato (1968), and to those given by Andrianov *et al.* (1968) and by Hope and Richards (1971), they do not permit Kitasato's conclusions. Our experiments indicate that the membrane resistance changes with an alteration in the energy state of the cell, i.e. increases concomitantly with an increase in the activity of the electrogenic pump; this brings the membrane in a resting hyperpolarized state, at which the differential resistance can be as high as 200 kΩ cm^2, or even greater (e.g. S_2 in Fig. 5). The discrepancy between the resistance measured electrically, and that deduced from flux measurements, which is of the order of 100 kΩ cm^2 (Walker and Hope, 1969), in fact becomes small in *Nitella* cells in a higher energy state. Thus the discrepancy may appear not to exist when both measurements are made in the same cell in a defined stationary energy state. The observation that the ion flux, activated, or balanced for, by the electrogenic pump increases from about 1·0 to 7·7 pmol cm^{-2} s^{-1} when the pH changes from 5·5 to 8·0 (Fig. 7), indicates that the potential change which is associated with it contributes to the pH-dependent change in the membrane potential, at least in this pH region. At pH below 5·5 the pump appears to generate a constant ion flux. However, the membrane permeability characteristics have been observed to change at these low pH values. This may be due to pH-dependent changes in the cation permeability coefficients, as suggested by Walker and Hope (1969) and by Lannoye *et al.* (1970). These effects, which were not accounted for by Kitasato, would considerably influence the quantitative results and conclusions with respect to the proton fluxes across, and proton permeability of the plasmalemma. Moreover our finding that the plasmalemma potential is hardly pH-sensitive in the presence of low concentrations of CCCP (Fig. 7), at which the pump is inhibited (Fig. 8), is inconsistent with Kitasato's conclusion that the plasmalemma behaves as a proton diffusion barrier. His own experiment with DNP (e.g. Kitasato, 1968, Fig. 12) in the presence of which the membrane potential does not markedly change upon a change in pH from 5 to 6, also seems to conflict with his reasoning. We have also found that the active membrane pump is inhibited by 0·2 mM DNP. Ouabain was found by us to be ineffective. Strunk (1971) has given evidence for an ouabain-sensitive electrogenic pump in *Nitella*.

The identity of the primary energy sources in plant cells triggering the membrane transport processes, and the nature of the electrogenic ion pump(s) and membrane reactive sites fed by these sources under various conditions, are still unknown. Further experiments are needed to clarify certain discrepancies between results obtained by various authors. Our results indicate that a probable cause of variable results, even with the same plant species, is the fact that, despite equal growth and experimental conditions, a cell can be in completely different energy states, in which the responses are different from each other. It is necessary that in critical experiments these energy states are controlled and characterized. As shown in the present paper, the membrane I–V characteristics, measured under each condition, can be fruitfully used as descriptive means.

Acknowledgements

This research was partly supported by the Netherlands Foundation for Biophysics (Stichting voor Biofysica), financed by the Netherlands Organization for the Advancement of Pure Research (Z.W.O.). The skilful technical and experimental assistance of Mr W. J. M. Tonk is acknowledged. Thanks are also due to Mr W. R. R. ten Broeke (ITAL, Wageningen) for constructing the single-sweep function generator, used in the current-scanning technique of the I–V membrane characteristics.

References

Andrianov, V. K., Vorob'eva, I. A. and Kurella, G. A. (1968). *Biofizika* (USSR, Engl. transl.) 13, 396–398.
Andrianov, V. K., Bulychev, A. A., Kurella, G. A. and Litvin, F. F. (1971). *Biofizika* (USSR) 16, 1031–1036.
Bentrup, F. W. (1971). *Fortschr. Bot., Berl.* 33, 51–61.
Coster, H. G. L. (1965). *Biophys. J.* 5, 669–686.
Gradman, D. and Bentrup, F. W. (1970). *Naturwiss.* 57, 46–47.
Heber, U. and Santarius, K. A. (1970). *Z. Naturforsch.* 25b, 718–728.
Hogg, J., Williams, E. J. and Johnston, R. J. (1968). *Biochim. biophys. Acta* 150, 518–520.
Hope, A. B. (1965). *Aust. J. biol. Sci.* 18, 789–801.
Hope, A. B. and Richards, J. L. (1971). Paper presented at the 1st Europ. Biophys. Congr., Baden, Austria.

JESCHKE, W. D. (1970a). Z. Pflanzenphysiol. **62**, 158–172.
JESCHKE, W. D. (1970b). Planta **91**, 111–128.
KISHIMOTO, U. (1966). Pl. Cell Physiol. **7**, 429–439.
KITASATO, H. (1968). J. gen. Physiol. **52**, 60–87.
LANNOYE, R. J., TARR, S. E. and DAINTY, J. (1970). J. exp. Bot. **21**, 543–551.
LÜTTGE, U. and PALLAGHY, C. K. (1969). Z. Pflanzenphysiol. **61**, 58–67.
LÜTTGE, U., BALL, E. and VON WILLERT, K. (1971). Z. Pflanzenphysiol. **65**, 336–350.
MACROBBIE, E. A. C. (1970). Q. Rev. Biophys. **3**, 251–294.
NISHIZAKI, Y. (1968). Pl. Cell Physiol. **9**, 377–387.
RAVEN, J. (1968). J. exp. Bot. **59**, 233–253.
RAVEN, J. (1969). New Phytol. **68**, 45–62.
SADDLER, H. D. W. (1970). J. gen. Physiol. **55**, 802–821.
SMITH, F. A. (1968). J. exp. Bot. **19**, 442–451.
SPANSWICK, R. M. (1970). J. Membrane Biol. **2**, 59–70.
SPEAR, D. G., BARR, J. K. and BARR, C. E. (1969). J. gen. Physiol. **54**, 397–414.
STRUNK, T. H. (1971). J. exp. Bot. **22**, 863–874.
VREDENBERG, W. J. (1969). Biochem. biophys. Res. Commun. **37**, 785–792.
VREDENBERG, W. J. (1971a). Biochem. biophys. Res. Comm. **42**, 111–118.
VREDENBERG, W. J. (1971b). In Proc. Ist Europ. Biophys. Congr. (E. Broda, A. Locker and H. Springer-Lederer, eds) Vol. III, pp. 435–439. Verlag der Wiener Medizinischen Akademie, Vienna.
VREDENBERG, W. J. (1972a). Biochim. biophys. Acta **274**, 505–514.
VREDENBERG, W. J. (1972b). In Proc. IInd Int. Congr. Photosynth. Res. (G. Forti, M. Avron and A. Melandri, eds) Vol. II, pp. 1049–1056. Dr. W. Junk N.V., Publ., The Hague.
VREDENBERG, W. J. (1973). Biochim. biophys. Acta **298**, 354–368.
WALKER, N. A. and HOPE, A. B. (1969). Aust. J. biol. Sci. **22**, 1179–1195.
WILLIAMS, E. J., JOHNSTON, R. J. and DAINTY, J. (1964). J. exp. Bot. **15**, 1–14.

III.6

The Membrane Potential of *Vallisneria* Leaf Cells: Evidence for Light-dependent Proton Permeability Changes

Friedrich W. Bentrup, Hans J. Gratz and Hans Unbehauen

Biophysical Laboratory, Institute of Biology
University of Tübingen
Germany

I. Introduction

For many plant cells, particularly those of higher plants, the membrane potential has been shown to depend substantially upon the photosynthetic or respiratory energy flow (cf. reviews by MacRobbie, 1970; Bentrup, 1971). In a few instances it seems quite clear that it consists of two components, a diffusion potential and an additive electrogenic mechanism where metabolic energy is directly transduced into a potential difference, i.e. in *Neurospora* (cf. Slayman *et al.*, 1970), and *Acetabularia* (Gradmann, 1970, Saddler, 1970).

Except for *Acetabularia*, however, the electrogenically transported ion species has not been identified, but the proton is a frequently suspected candidate (cf. Kitasato, 1968). Electrogenic proton extrusion, however requires a reasonable proton permeability, P_H, of the plasmalemma. Then the diffusion potential, given by the constant field equation, must include the term $P_H \cdot [H^+]$.

We started to investigate *Vallisneria* in more detail when it turned out that this plant is intermediate in behaviour between *Nitella* and *Elodea* In the light and also in the dark the membrane of *Nitella* can be regarded as a hydrogen electrode (Kitasato, 1968; Hope and Richards, 1971), and that of *Elodea*, in turn, mainly as a potassium electrode (Jeschke, 1970). However, *Vallisneria* resembles *Nitella* in the light, but *Elodea* in the dark. Further-more, studying the ionic relations of *Vallisneria* might help the analysis of two already well-studied cellular functions of this aquatic plant, viz. sym-plasmic transport (cf. Arisz, 1969) and cyclosis, especially chloroplast orientation (cf. Seitz, 1971).

II. Materials and Methods

A. Material and Test Solution

Leaves from *Vallisneria spiralis* from a greenhouse pond at our Institute were brought into the following test solution (mmol):

0·1 KCl	5·9 Na₂HPO₄ . 2 H₂O
10·0 NaCl	13·9 NaH₂PO₄ . H₂O
0·1 CaCl₂ . 2 H₂O	pH 6·5

This solution facilitated isoosmolar changes of $[K^+]_0$ between 0·1 and 10 mmol, as well as changes in pH between 4·5 and 7·5. The cells were also tested in a medium with 50 mmol total osmolarity; here they behaved less respon-sively.

B. Intracellular Potassium and Sodium

Leaves from the test medium were broken by freezing to $-80°C$ and centrifuged at 40,000 g for 10 min. The supernatant cell sap was mineralized and assayed by flame photometry (Bentrup *et al.*, unpublished). The values were 80·0 mmol potassium and 33·0 mmol sodium for freshly collected plants.

C. Electrophysiological Measurements

Our set-up employed conventional Ag/AgCl-electrodes with 0·5M KCl-filled micropipettes of about 1 μm tip diameter. Electrodes with tip potentials

arger than 10 mV were discarded. The electrodes were connected to a unit-
gain amplifier with 10^{12} Ω input impedance to feed a chart pen recorder with
?50 mV s^{-1} follow-up time. The fastest recorded potential change was 5 mV
$^{-1}$ (Fig. 5). A quartz iodine lamp delivered about 20,000 erg cm^{-2} s^{-1}
)f white light between 400 and 700 nm through a Schott KG 1 heat absorbing
filter. The measurements employed a Plexiglas vessel which could be perfused
)y the medium to give $>90\%$ change within a minute, while the response
vas continuously recorded from a single cell. The micro-electrode was in-
erted by a manipulator into individual cells of 10 cm long leaf pieces. This
)rocedure was observed by a stereomicroscope; though the location of the
·lectrode tip could not be observed each time, it probably always stayed in
he vacuole of the epidermal cells which contain most of the chloroplasts of
he leaf. The reported data each reflect the response of a representative cell
it 25°C. A cell was regarded as intact if it responded at the end of a set of
·xperiments to "light-on" and "light-off" signals in the regular way (Fig. 1).

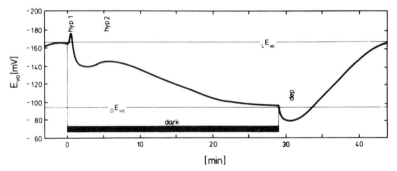

FIG. 1. Membrane potential, E_{vo}, of a *Vallisneria* leaf cell as a function of a light/dark
·egime. $_LE_{vo}$ and $_DE_{vo}$ denote the stationary values in the light and dark, respectively.
n this paper the transient hyperpolarizations upon "light-off" will be referred to as
hyp 1 and *hyp 2*, respectively. Similarly, *dep* refers to the depolarization after "light-on".

III. Results

A. Control of the Membrane Potential by Light

In the light *Vallisneria* leaf cells display a stationary membrane potential,
E_{vo}, usually between -170 and -230 mV. In the dark, the cells keep an
·qually steady value, $_DE_{vo}$, between -90 and -110 mV. The cells respond
o a light–dark regime in a characteristic fashion shown in Fig. 1. Following
a "light-off" signal the membrane at first hyperpolarizes with a latency of
)·8 s by about 15 mV (*hyp 1*), followed by a rapid depolarization and a second
.ransient hyperpolarization (*hyp 2*), then eventually approaching the dark
·evel, $_DE_{vo}$. "Light-on" elicits only one transient, a depolarization (*dep*) by
.0–15 mV, followed by the steady hyperpolarization toward $_LE_{vo}$.

B. Effect of Changes in External K^+ and pH

To characterize the two steady potential levels, $_LE_{vo}$ and $_DE_{vo}$, we changed $[K^+]_o$ of the medium from 0·1 to 10 mmol replacing K^+ with Na^+. The response is shown in Fig. 2. The experimental points are matched by theoretical curves (see section IV—Discussion). The response to changes in the

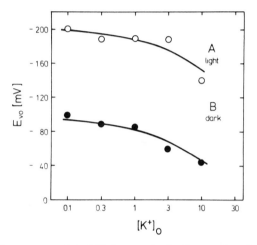

Fig. 2. The membrane potential of *Vallisneria* leaf cells as a function of the medium potassium concentration, $[K^+]_o$. A: values for $_LE_{vo}$ (○); the curve reflects eqn (2) inserting $P_H/P_K/P_{Na} = 200/1/0·1$ and $F(J/g_m) = -105$ mV. B: values for $_DE_{vo}$ (●); the curve represents eqn (2), using $P_{Na}/P_K = 0·1$.

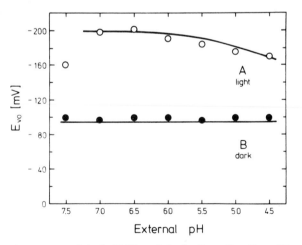

Fig. 3. The membrane potential of a *Vallisneria* leaf cell as a function of the external pH. Experimental and theoretical values of $_LE_{vo}$ and $_DE_{vo}$, respectively, were obtained as indicated in Fig. 2.

pH of the medium is given by Fig. 3. Here only $_LE_{vo}$ responds to changes in pH and this again is fitted by a theoretical curve in the range from 7·0 to 4·5. The response closely resembles that of *Nitella*, including the depolarization for pH > 7·0 (Hope and Richards, 1971). But in contrast to *Nitella*, *Vallisneria* displays no measurable response to pH in the dark.

C. Effect of Metabolic Inhibitors

1. DCMU

Since the presented data suggest that $_LE_{vo}$, at least, is partly of electrogenic origin, inhibitors of metabolic energy production should immediately affect E_{vo}. This holds for DCMU (3-(3,4-dichlorophenyl)-1,1-dimethylurea) as shown by Fig. 4. Though 10^{-6} or 3×10^{-6} mol DCMU apparently do not influence the stationary potentials, the inhibitor abolishes the transient potentials, *hyp 1* and *2*; it also prolongs significantly the *dep* response (cf. Fig. 1). Hence DCMU allows one to differentiate between the mechanisms controlling the stationary and transient potentials, respectively.

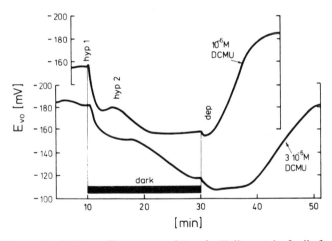

FIG. 4. "Light-on" and "light-off" response of E_{vo} of a *Vallisneria* leaf cell after addition of DCMU at time zero. For *hyp 1*, *hyp 2*, and *dep* see Fig. 1.

2. DNP

Firstly, dinitro-phenol might affect the assumed proton permeability. Secondly, 10^{-5} mol DNP presumably blocks oxidative phosphorylation, though not photophosphorylation, hence should resemble DCMU in not affecting $_LE_{vo}$. Figure 5 supports both ideas. $_LE_{vo}$ is not affected but the transient responses appear considerably changed; like DCMU, DNP abolished the *hyp 1* transient almost completely, but *hyp 2* is enhanced, and the *dep* response appears unchanged (cf. Fig. 1). After a short dark period of 15–30 s the membrane depolarizes and immediately repolarizes without the light-triggered *dep* transient of the control (Fig. 1).

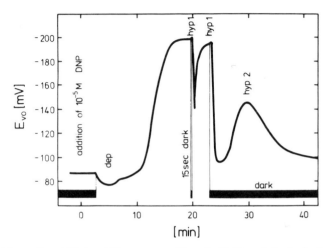

FIG. 5. "Light-on" and "light-off" response of E_{vo} of a *Vallisneria* leaf cell after addition of 10^{-5} mol DNP at time zero. For *hyp 1*, *hyp 2*, and *dep* see Figs 1 and 4.

Addition of 3×10^{-5} mol DNP also eliminates the *hyp 2* response, and the membrane depolarizes with a mean latency of 0·74 s at the highest measured rate of 5 mV s^{-1}. But in the presence of 10^{-4} mol DNP the cell fails to repolarize in the light; instead it exhibits a graded transient hyperpolarization of only 10 mV with the shortest measured latency of 0.45 s (data not presented). In summary, DNP acts differently on the two stationary potentials: at low concentration it affects neither of them, while in 10^{-4} mol DNP $_LE_{vo}$ shifts towards $_DE_{vo}$. Like DCMU, it selectively affects particular transients, but also creates a new one.

IV. Discussion

A. Stationary Membrane Potentials

1. The constant field approach

Vallisneria leaf cells show two stationary potentials, the light potential, $_LE_{vo}$, at about -200 mV, and the dark potential, $_DE_{vo}$, at about -90 mV (Fig. 1). Our data are fully compatible with and comprehensively explained by the following interpretation. $_LE_{vo}$ is composed of a diffusion equilibrium potential, E_{eq}, and an additive electrogenic mechanism, E_X:

$$_LE_{vo} = E_{eq} + E_X \tag{1}$$

The results (Figs 2, 3) allow one to express $_LE_{vo}$ by the constant field equation for K$^+$, Na$^+$, H$^+$, and the respective permeability coefficients, defined here as the ratios $\alpha = P_{Na}/P_K$ and $\beta = P_H/P_K$. Furthermore $E_X = F(J/g_m)$; with F = Faraday constant J = electrogenic ion flux, mol cm^{-2} s^{-1}, and g_m = chord membrane conductance, Ω^{-1} cm^{-2} (cf. Kitasato, 1968). Hence eqn (1) will be given by:

$$_LE_{vo} = \frac{RT}{F} \ln \frac{K_o^+ + \alpha . Na_o^+ + \beta . H_o^+}{K_i^+ + \alpha . Na_i^+ + \beta . H_i^+} + F\frac{J}{g_m} \qquad (2)$$

(R and T have their usual meaning, and the subscripts i and o refer to the intracellular and external ion concentrations, respectively).

E_{eq} can be calculated from the known values of Na$_o^+$, K$_o^+$, H$_o^+$, K$_i^+$, and Na$_i^+$; H$_i^+$ will be taken as 3×10^{-6} mol (cf. Kitasato, 1968). If we select for the relative permeabilities the ratio $P_H/P_K/P_{Na} = 200/1/0·1$, E_{eq} is given by the drawn curves A of Figs. 2 and 3, which fit the experimental data satisfactorily. This fit implies that in both cases $E_X = -105$ mV. While E_{eq} is not very sensitive to an error in H$_i$ by a factor of, say, 3, it alters considerably the experimental data if α and/or β is changed by a factor of between 0·7 and 1·3.

Hence we state that the light potential of *Vallisneria* can quantitatively be accounted for by the diffusion potential, E_{eq}, dominated by the proton permeability, i.e. $P_H \gg P_K$, P_{Na}, and an additional electrogenic term, $E_X = -105$ mV.

The dark potential, $_DE_{vo}$, can correctly be described by eqn (2) substituting $\beta = 0$ and $E_X = 0$. Equation (2) then yields the curves B in Figs 2 and 3.

Altogether the description of the experimental data by eqn (2) implies that the proton conductance is restricted to the light potential; in the dark it drops at least by a factor of 10^3 relative to P_K and P_{Na}.

The interesting question remains open, whether P_H is a function of $_LE_{vo}$, thus resembling excitability as we have suggested for *Acetabularia* (cf. Gradmann, 1970), or solely of the electrogenic flow J. In any event, we at present consider a change of P_H to be the most convenient way to explain our results (see below).

2. The electrogenic mechanism, E_X

What evidence can be inferred for $E_X = -105$ mV to be truly electrogenic? Present knowledge clearly favours electrogenesis, because (a) E_{vo} relies upon immediate presence of metabolic energy (Figs 1, 4, 5); (b) $_LE_{vo}$ ranges well beyond any conceivable diffusion potential. Firstly, E_K never exceeds about -170 mV. $_LE_{vo}$ being due to E_H, on the other hand, would imply an intracellular pH as low as 1·5; in *Nitella* this hypothetical value is only 2·5 (cf. Kitasato, 1968; Hope and Richards, 1971).

The size of the electrogenic flux, J, can crudely be estimated substituting the plasmalemma slope conductance of *Elodea* (Spanswick, 1972) as the best available guess for the chord conductance, g_m, of *Vallisneria*. Equation (2) then yields the large flux, $J = 400$ pmol cm^{-2} s^{-1}. Assuming g_m to be mainly g_H, at flux equilibrium this flux would be a passive proton influx balanced by the electrogenic flux. This, in turn, could be proton export, or, for instance, chloride import causing the necessary proton extrusion via electric coupling. Both mechanisms seem as plausible as more elaborate ones (e.g. that proposed by Smith, 1970). Substituting chloride for sulphate, Gradmann (1970) showed that in *Acetabularia* the electrogenic flux depends upon the presence of chloride in the medium, while Jeschke (1970) found no effect in *Elodea*, hence pointing to proton extrusion. Similarly, *Vallisneria* (and *Riccia*, see below) attain the light potential level, though chloride has been substituted by sulphate. Hence proton extrusion, rather than chloride import, is indicated for these plants also.

3. Metabolic control of $_LE_{vo}$

Following current reasoning, the lack of effect of DCMU (Fig. 4) and of DNP (Fig. 5) on $_LE_{vo}$ would indicate that the membrane potential of *Vallisneria* can be driven exclusively by photophosphorylation. It seems unlikely, however, that $_LE_{vo}$ is really sustained only via ATP consumption by E_X. Firstly, contrary to our observation, E_X should also operate in the dark as a result of oxidative phosphorylation, as does the electrogenic chloride pump of *Acetabularia* (Gradmann, 1970; Saddler, 1970). It will be argued below that in the dark E_X is indeed zero because J is zero. Secondly, we have not yet established that under our experimental conditions cyclic photophosphorylation occurs at all. Furthermore, steady-state consumption of chloroplast ATP requires oxidation of pyridine nucleotides in the cytoplasm which cannot easily be accounted for (Heber and Santarius, 1970; Krause, 1971).

Light-dependent chloride uptake into *Vallisneria* does not depend upon the presence of O_2 or CO_2; this also points to photophosphorylation as its primary energy source (Lookeren Campagne, 1957). In *Elodea*, on the other hand, the light-dependent chloride influx both at the plasmalemma and the membrane potential, also require photosystem II to be operative (Jeschke, 1970, 1971).

Our finding confirms the observation by Lüttge and Pallaghy (1969) that the steady-state light potentials of *Atriplex spongiosa* and *Mnium* leaf cells are not affected by DCMU, while the respective transients clearly are. Our suspicion that in *Vallisneria* and other species $_LE_{vo}$ might be controlled by both photosystems, depending upon the conditions, is substantiated by Throm (1971), who measured action spectra of both photosystems for the light-induced changes of the membrane potential of *Griffithsia setacea* under different conditions. He concluded that the membrane potential of this red alga is not coupled unconditionally to either single one of the two photosystems.

B. Proton Permeability Changes

Inspection of eqn (2) suggests that transient potentials may result from variation in one or several of the parameters, P_H, H_i, J, and g_m; here $g_m(P_H)$ seems at least to be a plausible candidate. Evidence has been presented above that, assuming P_K and P_{Na} to be light-insensitive, P_H, hence g_m, is lower in the dark by orders of magnitude. This is quite contrary to the observed change of E_X in a light–dark regime (cf. eqn 2). Therefore we conclude that in the dark J is in fact zero. In accordance with this, E_{vo} was never found to attain a steady level below $_D E_{vo}$.

In *Acetabularia*, the membrane conductance likewise is up to a hundred times higher when E_X, i.e. J_{Cl}, operates compared to when it does not (Gradmann, 1970; Saddler 1970, 1971). In fact, in a situation identical with the *dep* transient of *Vallisneria*, the membrane conductance of *Acetabularia mediterranea* first increases for the outward current (via the electrogenic flux?), the rectification ratio being as high as 2 (Klemke and Bentrup, in preparation). At any rate, in *Acetabularia* considerable changes in P_{Cl} occur, while P_H seems negligibly small, because the resting potential of *Acetabularia crenulata* is completely insensitive to changes in external pH between 4 and 9 (Gradmann, 1970); but in *A. mediterranea* the excitability, i.e. the changes in P_{Cl}, depends upon the pH (Kurre and Bentrup, unpublished).

DNP, at the concentrations used here (Fig. 5), might act primarily by raising the proton permeability of the plasmalemma, though experimental evidence apparently exists only for organelle membranes (cf. Karlish *et al.*, 1969). While DNP *per se* cannot short-circuit the pH-gradient across the thylakoid because only P_H, not the permeability of the counter ion, P_K, is increased (Grünhagen and Witt, 1970), it might prevent the proposed P_H of the plasmalemma (and/or tonoplast) from decreasing after the onset of dark. Thus the observed rapid drop of E_{vo} of 3 mV s^{-1} (Fig. 5) as compared with 3 mV min^{-1} (Fig. 1) occurs, because J is short-circuited through a persistently high g_m (cf. eqn 2).

A variable proton permeability has been found also in leaf cells of the aquatic moss, *Riccia fluitans* (Felle and Bentrup, in preparation). Here again, $_L E_{vo}$ consists of E_{eq} featuring $P_H \gg P_K$, P_{Na}, P_{Cl}, plus E_X which still operates, if the external chloride is replaced with sulphate; E_X is completely eliminated by DCMU, its magnitude depending on the composition of the medium, or at least upon $[K^+]_o$. The dark potential is dominated by potassium, i.e. $P_{Na}/P_K = 0.02$. The transient behaviour fairly closely resembles that of *Vallisneria*.

C. Origin of the Transient Potentials

Only brief speculation is permitted by the few data presented so far. It seems clear that the specified transients are controlled by both photosystems I and II, because they are very sensitive to the presence of DCMU (Fig. 4). We do not even know all the elements participating in the potential control. A minimum mode of control would involve counteracting parallel mechanisms having variable time constants at their potential controlling level. Evidence for such a system has been provided in the light control of the resting potential of *Acetabularia* (Hansen and Gradmann, 1971).

Somewhat tentatively and arbitrarily, we shall select just one mode of potential control based on eqn (2) for each, the *dep* and the *hyp 1*, transient. In the following, J shall be analogous to a capacitor, for instance the cytoplasmic ATP pool. In *Elodea*, this pool displays light-dependent changes with charging and decharging times in the order of a minute (Santarius and Heber, 1965; Heber and Santarius, 1970).

(a) The *dep* response (Fig. 1) arises, if "light-on", firstly causes J to hyperpolarize the plasmalemma, and secondly causes P_H to increase; consequently E_{v_o} will temporarily swing towards E_H which is below $_DE_{v_o}$, if the cytoplasmic pH is below 5·0 under those particular conditions. (b) The *hyp 1* transient appears, if "light-off" triggers the decrease of P_H, while J still operates by consumption of the ATP pool. Hence E_X transiently hyperpolarizes the plasmalemma. DCMU might abolish this transient, because it affects the ATP pool.

V. Conclusions

We are quite aware that our study on the ionic relations of *Vallisneria* leaf cells still lacks many essential experiments; their outcome might not only modify the given interpretation, but also call for a different theoretical approach. Ultimately, our interest is focused on questions like this: Is P_H controlled by photosynthetically driven proton fluxes through the chloroplast envelope? Or is it controlled, rather, by the membrane potential, i.e. induced by some kinds of electrostrictive changes of the plasmalemma function?

Moreover, could studying *Vallisneria* in this way help in the analysis of its much-studied symplasmic transport (cf. Arisz, 1969)? One could easily argue that a differentially active E_X in symplasmically, hence electrically, coupled cells will establish a metabolically controlled potential gradient within the leaf. The significance of this old transport hypothesis, adapted for *Vallisneria* by Boij (1971), seems questionable, however, because electrical coupling between *Elodea* leaf cells is surprisingly moderate (Spanswick, 1972), and between coleoptile cells of *Avena* it could not be detected at all (Goldsmith *et al.*, 1972). Goldsmith and co-workers conclude, therefore, that polar transport of molecules like auxin is not symplasmic, but transmembrane and through the intercellular space. Incidentally, an auxin-activated ATP-powered proton extrusion at the plasmalemma has been postulated as the primary event of cell wall extension in *Helianthus* hypocotyls (Hager *et al.*, 1971).

It seems that polar functions of plants rely rather upon *cellular* gradients. Could not these partly employ metabolically controlled membrane potentials? In fact, a spatial difference of E_X, i.e. of the electrogenic chloride pump, has been measured on *Acetabularia* stalk segments; its significance is that polar growth is initiated where the membrane potential, specifically E_X, is highest (Novak and Bentrup, 1972).

In *Vallisneria*, the photodinetic orientation of the chloroplasts has recently been shown to reflect an ATP gradient (Seitz, 1970, 1971). Since in the leaf cell light intensity differences range between 1:10 and 1:100, a respective intracellular potential gradient due to a differential E_X can be inferred to exist. The neighbouring mechanisms, E_X and cyclosis, might compete for ATP or some other form of energy, leading to the observed stationary chloroplast distribution pattern which is known to disappear in the dark as does E_X; or they might interact through current loops and local permeability changes which are known to affect cyclosis in Characeae (Barry, 1968; Pickard, 1969). Another conceivable relationship is indicated by the fact that a particular photodinetic photoreceptor (probably a flavine) is located at or near the plasmalemma (Haupt, 1968). These remarks are only to point out interesting relationships, let alone possible causal ones, between different protoplasmic functions of this plant.

Acknowledgements

We are grateful for the efficient technical assistance of Miss B. Hanneck. This work was supported by the Deutsche Forschungsgemeinschaft.

References

ARISZ, W. H. (1969). *Acta bot. Neerl.* **18**, 14–38.
BARRY, W. H. (1968). *J. cell. Physiol.* **72**, 153–160.
BENTRUP, F. W. (1971). *In* "Fortschritte der Botanik" (H. Ellenberg *et al.*, eds), pp. 51–61.
BOIJ, H. L. (1971). Proc. 1st Europ. Biophys. Congr. (Baden/Vienna) Vol. III, pp. 125–129.
GOLDSMITH, M. H. M., FERNÁNDEZ, H. R. and GOLDSMITH, T. H. (1972). *Planta* **102**, 302–323.
GRADMANN, D. (1970). *Planta* **93**, 323–353.
GRÜNHAGEN, H. H. and WITT, H. T. (1970). *Z. Naturforsch.* **25b**, 373–386.
HAGER, A., MENZEL, H. and KRAUSS, A. (1971). *Planta* **100**, 47–75.
HANSEN, U. P. and GRADMANN, D. (1971). *Pl. Cell Physiol.* **12**, 335–348.
HAUPT, W. (1968). *Biol. Rundschau* **6**, 121–136.
HEBER, U. and SANTARIUS, K. A. (1970). *Z. Naturforsch.* **25b**, 718–728.
HOPE, A. B. and RICHARDS, J. L. (1971). Proc. 1st Europ. Biophys. Congr. (Baden/Vienna) Vol. III, p. 105.

JESCHKE, W. D. (1970). Z. Pflanzenphysiol. **62**, 158–172.
JESCHKE, W. D. (1971). Proc. 1st Europ. Biophys. Congr. (Baden/Vienna) Vol. III, pp. 111–118.
KARLISH, S. J. D., SHAVIT, S. and AVRON, M. (1969). Eur. J. Biochem. **9**, 291–298.
KITASATO, H. (1968). J. gen. Physiol. **52**, 60–87.
KRAUSE, G. H. (1971). Z. Pflanzenphysiol. **65**, 13–23.
LOOKEREN CAMPAGNE, R. N. VAN (1957). Acta bot. Neerl. **6**, 543–582.
LÜTTGE, U. and PALLAGHY, CH. K. (1969). Z. Pflanzenphysiol. **61**, 58–67.
MACROBBIE, E. A. C. (1970). Q. Rev. Biophys. **3**, 251–294.
NOVAK, B., and BENTRUP, F. W. (1972). Planta **108**, 227–244.
PICKARD, W. F. (1969). Can. J. Bot. **47**, 1233–1240.
SADDLER, H. D. W. (1970). J. gen. Physiol. **55**, 802–821.
SADDLER, H. D. W. (1971). J. Membrane Biol. **5**, 250–260.
SANTARIUS, K. A. and HEBER, U. (165). Biochim. biophys. Acta **102**, 39–54.
SEITZ, K. (1970). Z. Pflanzenphysiol. **63**, 401–407.
SEITZ, K. (1971). Z. Pflanzenphysiol. **64**, 241–256.
SLAYMAN, C. L., LU, CH. Y.-H. and SHANE, L. (1970). Nature, Lond., **226**, 274–276.
SMITH, F. A. (1970). New Phytol. **69**, 903–917.
SPANSWICK, R. M. (1972). Planta **102**, 215–227.
THROM, G. (1971). Z. Pflanzenphysiol. **65**, 389–403.

Discussion

Findlay asked Barr about the relationship between pH and membrane potential over an extended pH range. Barr said he had no further information but from the data in the paper the light potential is -150 mV at pH 5·8 and -85 mV at pH 4·7; above pH 5·7 the potential does not vary strongly with pH. Findlay then asked if the potential is pH insensitive in the dark, and Barr replied that it is. Vredenberg then commented that the pH sensitivity of membrane potential is only observed when the electrogenic pump is operating; in the dark it is not operating or only at a low rate. Therefore membrane potential is not highly pH sensitive in the dark. Spanswick then said that he did not entirely agree with Vredenberg's statement but went on to ask Barr about Fig. 3 in his paper; how are the data interpreted on the Kitasato model and on the equivalent circuit model as proposed by Spanswick? In the first diffusion potential and electrogenic potential are in series while in the second they are in parallel. The second model may more simply explain the dark effects in that they seem to depend on the previous state of the cell in the light. In Barr's Fig. 3, Spanswick continued, the switch from hyperpolarization to depolarization occurs at approximately 100–110 mV, which is roughly the K^+ diffusion potential. Thus, depending on whether E_p is greater or less than E_K in the light, switching to dark can cause a hyperpolarization or depolarization of the membrane potential. Barr agreed and said that the average dark potential is -118 mV, close to the E_K value. Spanswick said, "Then instead of interpreting this as switching off a current through passive channels, I would interpret it in terms of variation of g_M and g_P in the equivalent circuit." Barr said, "It's like shutting off the pump and allowing the membrane to go to E_K." Spanswick pointed out that shutting off the pump is equivalent to decreasing g_P. Barr asked if it becomes a purely passive system in the dark, and Spanswick replied, "Not purely but chiefly".

Field then asked if a large electric field developed across a thin membrane might alter the conductivity of the membrane, as in the Wien effect in semi-conductors. Can one not interpret the I–V curve in this way? Vredenberg agreed; he said that the I–V curves show that as the activity of the pump increases, the slope conductance of the membrane decreases; in the hyperpolarized state when the pump is active, the slope conductance is lower than in the depolarized state. Field retorted, "What I really want to know is why it is necessary to have an electrogenic pump. Can one not interpret the I–V curve shifts in terms of physical polarization of the membrane by a pH gradient-induced electric field?" Vredenberg replied, "In the hyperpolarized state something is maintaining a current across the membrane".

Meares then injected a cautionary note prompted by Barr's suggestion that different regions of the membrane in Characeae have different properties. He said that a mosaic membrane will not tend to linear behaviour even at vanishingly small forces. Thus it is not a valid procedure to assume that departures from equations based on linearity imply some mechanism or other. The theory of mosaic membranes was first introduced by Kedem and Katchalsky; varying concentrations on either side of such a membrane will not result in the potential alterations one predicts from the theory of a single function membrane. Further changes in the conductivities in the various regions of the mosaic, as one might cause by pH alterations, will again cause potential alterations across the membrane because of changes in the inter-action between the different regions of the mosaic. Light-induced changes may also arise from permeability changes, thus setting up all sorts of feedback interactions within the system. No one has formally added a simple barrier such as a cell wall, in series with a complicated active mosaic barrier; although it is possible to do it in principle, it would be mathematically very complicated.

Turning to Vredenberg's paper, Spanswick showed a slide of a cell in the dark at constant current. The membrane potential changes in two stages; there is an initial rapid response followed by a slow increase in hyperpolariza-tion. This effect was first observed by Kishimoto in high ionic strength solu-tions and was called by him the slow hyperpolarizing response; this response occurs in the dark but not in the light, according to Spanswick. Spanswick said this comment was relevant to Vredenberg's technique because Kishi-moto has shown that the time lag between the first response and the second decreases with frequency of current pulsing. After several pulses one cannot distinguish the two stages; thus one can over-estimate membrane resistance. In the light the I–V relationship is highly linear and there is no hyperpolariz-ing response; in the dark it is linear up to a point and above this threshold it deviates from linearity. Spanswick then claimed that Vredenberg had measured the hyperpolarizing response and so had poor estimates of mem-brane resistance.

Vredenberg replied that he too had observed the hyperpolarizing response described by Kishimoto and that his experiments are so designed that it does not affect the measurements. The biphasic response is not present because the K^+ concentration is low. Spanswick replied that there are in fact two hyperpolarizing responses: the first where the membrane is essentially impermeable to K^+ and the second where the external K^+ concentration exceeds 1 mM. In this latter case by passing a large current one can obtain a hyperpolarizing response. Vredenberg said that in his experiments he injected 0·3 μA, positive and negative, and found no hyperpolarizing response.

Gillet then asked what light intensity was used and whether there was no temperature effect. Vredenberg said that he used monochromatic light at 670 nm with an intensity range 1–10 kerg cm^{-2} s^{-1}. Walker then asked if Vredenberg's Fig. 4 clearly established by experiment that the short time, 2 min response was a change in resistance at constant pump rate. It is not clear whether this is interpretation or whether it is experimentally proven. Furthermore, Walker stated, in Fig. 5 various I–V curves which are said to be merely translated, do not in fact superimpose on translation. To what extent are they translations and to what extent do they vary in shape? Since the shape of the curve depends on Spanswick's model on the various parameters of the pump, or on the old-fashioned model on the various concentrations and permeabilities of the Goldman equation, it seems surprising that the shape of the I–V curves does not alter under these varying conditions. Vredenberg replied that the data in Fig. 4 and Fig. 3 were from different experiments. The curves are measured in the light after 2 min at which time the resistance and potential attain steady values. In Fig. 4 the electrogenic pump is assumed to be steady otherwise there would be a hyperpolarization. In Fig. 3 after the initial depolarization there is a gradual hyperpolarization of the membrane. It is not likely that in a dark-adapted cell there are sufficient energy reserves to activate the pump further. Vredenberg did not reply directly to the question about superposition of the various I–V curves but said that differences were caused by variations in P_K and P_{Na} due to pH variation and other factors.

Barr then commented on Spanswick's suggestion on his paper; a best straight line in his Fig. 2 or 3 extrapolated to zero de- or hyperpolarization gives an intercept value of -115 mV. Thus in the dark the membrane potential is close to the K^+ equilibrium potential and it might be interesting to see whether the membrane resistance values correspond to the K^+ conductance values. Spanswick then showed a slide of membrane resistance and membrane potential of *Nitella* in the dark; between 0·1 and 0·5 mM external K^+ the potential does not vary much and the K^+ conductance is low. Thus the membrane potential is not K^+ dependent in *Nitella*; there is a sharp transition at 1·0 mM external K^+; above this concentration K^+ does control membrane potential and K^+ conductance is high.

Walker then returned to Vredenberg's analysis; he asked how one can test the assumptions of the analysis because he thinks it is an important method of approach. Vredenberg said that measurements of resistance and potential were not by themselves sufficient to determine the energy state of a cell. One should always look at the I–V curves and then one can decide what energy state the cell is in. He has defined energy state purely on the basis of the I–V curve; examining the kinetics of potential change is not sufficient because in a rectifying membrane a potential change is always associated with a resistance change. However, in long term illumination experiments the kinetics of potential change are quite different from those of resistance change. If one takes a dark adapted cell in a certain energy state and illuminates it one finds a resistance change of 60–80 %. If one takes a long illuminated cell, switches the light off and then on again, there is no resistance change. Thus if one works with a cell which has been recently long illuminated, one will get a false impression of what is going on; the resistance will not change much. Without an I–V curve the interpretation can be erroneous; only the I–V curve gives information about the activity of the electrogenic pump. On illuminating a dark adapted cell, Vredenberg continued, both potential and resistance change and the value of the current can be calculated directly as their quotient. In the presence of 0·4 μM CCCP, one finds that the resistance change is still large but that the potential change has decreased; the resistance change is only inhibited at much higher CCCP concentrations. Thus the resistance and potential changes may be due to different processes and perhaps also the long and short term responses are due to separate processes.

Vredenberg then discussed the scheme on which he thought these data could best be understood. Associated with electron transport in the thylakoid membranes there is a proton influx into the grana lamellae, which may cause a H^+ efflux across the chloroplast envelope. Heber has shown that intermediates of the Calvin cycle form a shuttle across the chloroplast envelope, for example there is a PGA–NADPH shuttle, a succinate–oxaloacetate–malate shuttle, etc. These shuttles produce the H^+ concentrations in the cytoplasm. Now at least two processes must be assumed, Vredenberg continued, the first responsible for the membrane potential and the second for the membrane resistance; possibly a third process influences the diffusion potential. The H^+ concentrations in the cytoplasm may affect the membrane or possibly high energy intermediates of the photosynthetic cycle activate the electrogenic pump. A suggestion for further work is an examination of which high energy compounds power the electrogenic pump. It is insensitive to ouabain and to both Dio-9 and fluorescein, which may indicate that ATP is not the primary energy source. However these inhibitors may not penetrate to the sites concerned; perhaps MacRobbie and Raven disagree. It should be fruitful to measure I–V curves in the presence of inhibitors.

Vredenberg then continued by commenting on Bentrup's paper and said he would like to see I–V curves on *Vallisneria* or *Elodea* to see whether the hyperpolarization Bentrup finds is associated with increased electrogenic pumping. Bentrup replied that as far as he knew the pump is operating at a constant rate; it does not vary with pH. Vredenberg asked if he was not assuming that the pump rate is constant and then is altering the permeabilities to fit the observed curve. Bentrup agreed that he was. Vredenberg then asked if the membrane resistance will not be affected when the H^+ permeability changes by the large amount Bentrup suggests.

The discussion finally returned to Vredenberg's paper by Findlay asking for information on the effect of external K^+ variation. He thought the analysis would carry more weight if it could be shown that under conditions which are known to affect the I–V curve, displacements remained unaltered. Vredenberg replied that his K^+ experiments were very preliminary, but there were difficulties because of the hyperpolarizing response at high K^+. His analysis can only be fruitful when used on membranes with rectifying properties, and at high K^+ these properties are lost. In high K^+ the short term responses are present but the long term hyperpolarizing response is much decreased.

Section IV

H^+ Fluxes in Cells and Organelles

Chairman: N. Higinbotham

IV.1

Millisecond Delayed Light as an Indicator of the Electrical and Permeability Properties of the Thylakoid Membranes

J. Barber

*Botany Department, Imperial College, London
England*

I. Introduction

It is becoming increasingly clear that photochemical and energy conversion processes of photosynthesis are dependent on the properties of the chloroplast membranes and the establishment of ionic and electrical gradients across these membranes. Murata *et al.* (1970) have suggested that light-induced net movement of ions across chloroplast membranes may regulate energy transfer between the two photosystems. Mitchell (1966) has argued that an electrochemical potential gradient acting on protons across the thylakoids is the high energy intermediate between electron transfer and photophosphorylation. Moreover, there is the likelihood that light-driven ion fluxes, closely associated with primary electron flow, may control or influence both active and passive ion movements across the plasmalemma and tonoplast of plant cells. Nevertheless, our knowledge of the ionic permeability properties of chloroplast membranes and their ability to maintain electrical and ionic gradients, is meagre.

To date the bulk of the documented work deals with the ability of broken chloroplasts to pump H^+ into the thylakoid interiors (Jagendorf and Uribe 1966). The existence of this proton pump, which seems to be directly linked with photoinduced electron flow (Schwartz, 1971), has given strong support to Mitchell's chemiosmotic hypothesis (1966). With regard to the possibility of changes in cytoplasmic pH being linked to active ion transport, it is worth noting that the above light-induced pH increase is not observed in a suspending medium containing chloroplasts which retain their outer membrane systems (Heber and Krause, 1971). The nature of the coions which move with H^+ is not clear. Evidence for both net Cl^- influx (Deamer and Packer, 1969) and net K^+ efflux (Dilley and Vernon, 1965) have been presented.

As yet little is known about the passive permeability properties of chloroplast membranes. According to Winocur et al. (1968) and Rottenberg et al. (1972) chloroplasts are very permeable to Na^+, K^+ and Cl^-. Their results contrast with the earlier work by Saltman et al. (1963) who concluded that chloroplasts exchange these ions very slowly. Nevertheless, it is usually accepted by most workers that isolated chloroplasts, not retaining their outer membrane, are "leaky" to small monovalent ions (Walker and Crofts, 1970).

Direct and meaningful measurement of electrical gradients in chloroplasts, especially across the thylakoid membranes, is a formidable, if not impossible, task. Only recently have such measurements been attempted (Bulychev et al., 1972). Junge and Witt (1968) have used an indirect method to measure light-induced electrical potentials across the thylakoids. They have argued that a light-induced absorbancy change at 515 nm, due to a shift in the chlorophyll b and carotenoid spectra, results from electrochroism or a "Stark" effect. They have estimated that potentials of about 100 mV (inside positive) are developed across continuously illuminated thylakoids.

This paper will present and discuss experimental evidence that millisecond delayed light emission can also be used as an indirect method for measuring electrical gradients and ionic fluxes across chloroplast membranes.

Delayed light was discovered by Strehler and Arnold (1951) and is a long-lived emission from plants having a spectrum corresponding to chlorophyll a fluorescence. This delayed fluorescence can often persist for minutes after terminating the illumination and seems to result from a back reaction of the primary photoproducts of photosystem II (Z^+ and Q^-).

$$Z\,Chl\,Q \overset{h\nu}{\rightleftharpoons} Z\,Chl^*\,Q \rightleftharpoons Z^+\,Chl\,Q^-$$

As Fig. 1 shows, when spinach chloroplasts are subjected to a KCl gradient seconds after terminating the illumination the intensity of emission increases. Barber and Kraan (1970) interpreted this stimulation as being the result of an electrical potential developed as the ions diffused across the thylakoid membrane.

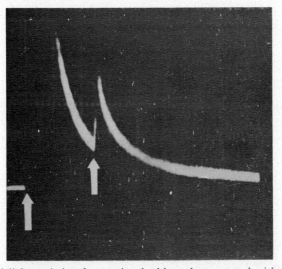

FIG. 1. Delayed light emission from spinach chloroplasts treated with 0·5 μM valino-mycin and its stimulation by injecting 50 mM KCl. The decay was measured using a double shutter system operated mechanically with electromagnets. The first arrow indicates the opening of the photomultiplier shutter approximately 50 ms after ter-minating the illumination. The sensitivity of the oscilloscope was set so that the signal was off scale for the first second of the decay. The KCl injection was made after about 2·5 s dark time. The grid spacing corresponds to 1 s.

Such an argument suggests that Z^+ and Q^- are specifically orientated in the membrane and that an electrical gradient of the correct polarity can decrease the activation energy (E) for lifting the electron from the meta-stable state to the first singlet of chlorophyll (see Crofts *et al.*, 1971),

i.e.
$$L = \phi N f \exp - \left(\frac{E - F \Delta\psi/J}{kT} \right) \tag{1}$$

where L is the intensity of delayed fluorescence, ϕ is the prompt fluorescence yield, N is the number of trapped electrons, f is a frequency factor, $\Delta\psi$ is the membrane potential, k is the Boltzmann constant and the other symbols have their usual meanings.

From studies of the stimulation of 10 s delayed light by alkali metal salts of strong acids, Barber and Varley (1971a, 1972) were able to present good evidence for the diffusion potential model and the exponential relationship between L and $\Delta\psi$.

Unlike the slower components of delayed fluorescence, emission in the one millisecond region of the decay is very sensitive to uncouplers and seems to be dependent on the high energy state of the chloroplasts (Mayne, 1967). This part of the decay is also stimulated by gradients of certain salts (Barber and Varley, 1971a, b; Barber, 1972 a, b) and I shall present some recent studies into this phenomenon and discuss them in relation to the above concepts.

II. Materials and Methods

Delayed light emission from osmotically broken chloroplasts in the region of one millisecond of the decay, was continuously monitored using a rotating sector phosphoroscope shown diagrammatically in Fig. 2. The illumination

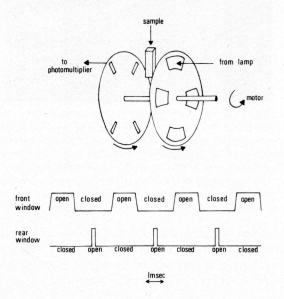

FIG. 2. Phosphoroscope used for measuring one millisecond delayed light emission. The time sequence for illumination and measuring are shown.

and viewing times were 1·5 and 0·2 ms, respectively, with four measurements per cycle. During the illumination period the chloroplasts were exposed to blue light transmitted through a filter combination consisting of a Balzer Calflex C and a 2 mm Schott BG38 giving an intensity of 2·7 × 10^4 erg cm^{-2} s^{-1} at the cuvette. The delayed light was detected with an EMI 9659 B photomultiplier protected by a 2 mm Schott RG 665 filter. The current pulses from the photomultiplier were passed through a diode pump circuit with a time constant of 0·1 s and the signals recorded on a Honeywell chart recorder. Details of chloroplast isolation and experimental procedures have been reported elsewhere (Barber, 1972a, b) but where necessary relevant information has been included in the figure legends.

III. Results

A. KCl-induced Transients and Effect of Valinomycin

Figure 3 shows the general characteristics of millisecond-delayed light signals induced by injecting 50 mM KCl into suspensions of broken chloroplasts which had or had not been treated with valinomycin. Similar transients of varying magnitudes and decays could be detected with other alkali metal salts but not with salts of divalent cations (Barber and Varley, 1971b). As the traces in Fig. 3 show, the relative size of the KCl-induced signals

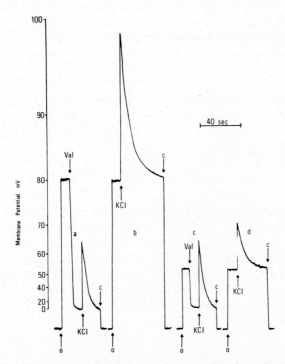

Fig. 3. The effect of 50 mM KCl injections on the intensity of steady-state, one millisecond-delayed light emission from osmotically broken chloroplasts which had or had not been treated with 0·5 μM valinomycin. The cuvette contained 3 ml of chloroplast suspension which was illuminated 2 min before opening the shutter across the photomultiplier. The suspending medium consisted of 0·33 M sucrose, 5 mM N-tris (hydroxymethyl)-methyl-2-aminoethane-sulphonic acid (TES) and 2 mM K⁺ added as KOH to bring the pH to 7·0. The various additions are indicated by arrows and were made by injecting 100 μl of the appropriate stock using a syringe inserted through a light-tight rubber diaphragm. Signals (a) and (b) were obtained with the same preparation while (c) and (d) were obtained on another occasion using a different batch of chloroplasts. The chlorophyll concentration in each case was 25 μg/ml. The opening and closing of the photomultiplier shutter is indicated by o and c respectively. The membrane potential scale was calculated by using the procedure outlined in the text.

before and after valinomycin treatment varied for the two preparations used and was found to be dependent on the size of the valinomycin-sensitive component of the emission.

These results can be explained by consulting the membrane potential scale. To construct this scale it was assumed that the intensity or rate of millisecond emission (L) could be equated with the exponential of the membrane potential ($\Delta\psi$) as indicated in eqn (1)

i.e. $$L\alpha \exp(\Delta\psi) \tag{2}$$

In the millisecond region of the decay the electrical gradient ($\Delta\psi$) could consist of two components, a light induced membrane potential ($\Delta\psi_L$) and a diffusion potential created by a sudden salt addition to the chloroplast suspension ($\Delta\psi_s$).

$$L\alpha \exp(\Delta\psi_L + \Delta\psi_s) \tag{3}$$

The size of the initial diffusion potential created after, for example, a rapid KCl addition can be estimated, assuming K^+ and Cl^- are the main diffusing ions, from the Goldman Voltage Equation (Goldman, 1943; Hodgkin and Katz, 1949).

$$\Delta\psi_s = 58 \log_{10} \frac{[K]_o + \beta[Cl]_i}{[K]_i + \beta[Cl]_o} \, mV \tag{4}$$

where $\beta = P_{Cl}/P_K$ the ratio of the permeability coefficients for Cl^- and K^+ respectively and $[\ ]_i$ and $[\ ]_o$ are the inside and outside concentrations.

In order to apply eqn (4) well washed broken chloroplasts were suspended in a chloride free medium consisting of 0·33 M sucrose, 5 mM *TES and 2 mM K^+ added as KOH to bring the pH to 7·0. Assuming that, after a dark and cold incubation of 30 min in this medium, $[Cl]_i = 0$ and $[K]_i = [K]_o = 2$ mM then the sudden establishment of a KCl gradient would develop a membrane potential ($\Delta\psi_s$) given by eqn (5).

$$\Delta\psi_s = 58 \log \frac{[K]_o}{2 + \beta[Cl]_o} \, mV \tag{5}$$

substituting (5) into (2) gives

$$c'L_s = \frac{[K]_o}{2 + \beta[Cl]_o} \tag{6}$$

where c' is a proportionality constant and L_s is the initial intensity of the salt induced emission.

In order to use eqn (6) to calibrate the intensity of delayed light in electrical units it is necessary to reduce $\Delta\psi_L$ to zero. I have assumed that this occurs,

* TES: N-Tris (hydroxymethyl)-methyl-2-aminoethane-sulphonic acid.

in the absence of light-induced ion gradients, when valinomycin is added to the suspension. As Fig. 3 shows, 5×10^{-7} M valinomycin reduced the emission to a low level and no further inhibition was obtained by adding higher concentrations of this antibiotic. Thus by measuring L_s for at least two different KCl gradients using valinomycin-treated chloroplasts, eqn (6) was solved for c' and β and the relationship between L and $\Delta\psi$ obtained.

From this analysis it can be seen from traces (a) and (b) of Fig. 3 that $\Delta\psi_L = 80$ mV (inside positive) while for traces (c) and (d) $\Delta\psi_L = +52$ mV. The polarity of these electrical gradients is indicated by the finding that valinomycin increases the size of the potential.

Over the course of the work values between $+50$ mV and $+110$ mV have been obtained for $\Delta\psi_L$ although the reason for the variation is not at present known.

Bearing in mind that $\Delta\psi_L$ was different for the two preparations used in the experiments shown in Fig. 3 and that there is an exponential relationship between L and $\Delta\psi$, then the reason for the change in relative magnitudes of the KCl signals, before and after valinomycin treatment, can readily be understood. With valinomycin-treated chloroplasts a 50 mM KCl gradient gave signal peaks corresponding to 66 mV for both preparations (traces a and c). With the same KCl gradient but in the absence of valinomycin (traces b and d), the signals appear quite different although they correspond to the same diffusion potential of 18 mV.

A change of the membrane potential developed by a 50 mM KCl pulse from 18 to 66 mV after valinomycin treatment, corresponds to a change in the P_{Cl}/P_K ratio from about 0·47 to 0·04.

3. Effect of Alkali Metal Cations Before and After Treatment with Valinomycin

The traces shown in Fig. 4 indicate that for chloroplasts not treated with valinomycin there was only a slight difference in the effectiveness of various alkali metal chlorides in bringing about transients in millisecond emission. However, after treatment with valinomycin clear signals could only be detected on injecting chlorides of K^+, Rb^+ and Cs^+. With Li^+ and Na^+ chlorides the signals were barely observable. These results are consistent with the concepts presented above accepting that valinomycin is specific for K^+, Rb^+ and Cs^+ in its ability to increase membrane permeability (Pressman et al., 1967). Before valinomycin treatment the Li^+ and Na^+ chloride additions generated potentials of 14 and 17 mV, respectively, which were capable of significantly increasing L when $\Delta\psi_L = 77$ mV. However, when $\Delta\psi_L$ is reduced to zero then potentials of these magnitudes would have very little effect on L as observed.

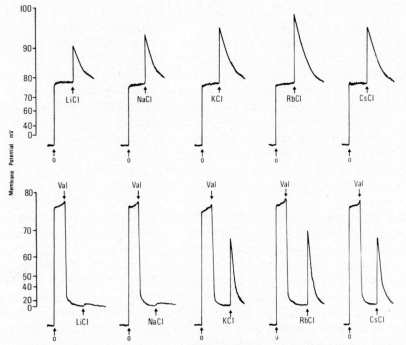

FIG. 4. The ability of various alkali metal chlorides to stimulate millisecond emission from chloroplasts before and after treatment with valinomycin. The salt additions corresponded to 50 mM in the cuvette. Other conditions are the same as given in the legend of Fig. 3.

C. Kinetics of Salt-induced Signals and Ion Fluxes

If the salt-induced millisecond-delayed light transients represent the establishment and fall of a membrane potential then their decay kinetics give a measure of the rate of salt entry into the thylakoid interiors. Since valinomycin increases the size of the KCl-induced potentials then it seems that the thylakoids are more permeable to K^+ than Cl^-. Thus for a KCl gradient the decay of the signals would be expected to be controlled by the rate of chloride penetration. The method of calculating the time course of Cl^- entry is shown in Fig. 5. It can be seen from Fig. 5(a) that the electrical potential fell from 93 mV to 81 mV in 5 s. This corresponds, assuming only net K^+ and Cl^- movement, to an increase of internal KCl from zero to 15 mM. In this way the time course of Cl^- entry can be calculated, as shown in Fig. 5(b). The values obtained were found to fit the first order law shown. The rate constant of $0.075 \, s^{-1}$ corresponds to an initial Cl^- influx of 33 μM $s^{-1} \mu g^{-1}$ chl. In order to express this rate in more meaningful units it is necessary to assume surface and volume parameters of the thylakoids

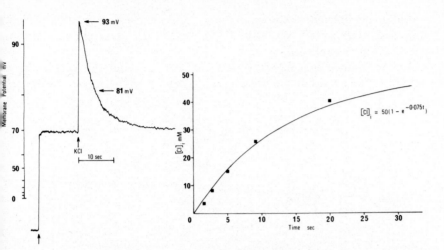

FIG. 5. (a) The kinetics of a 50 mM KCl induced signal. Chlorophyll conc. 36 μg ml^{-1}. Other conditions are the same as given in the legend of Fig. 3. (b) The influx of chloride estimated from the decay of the signal shown in (a). The closed squares are experimental points and the curve has been drawn according to the first order law shown where t = time in seconds and $[Cl]_i$ is the inside chloride concentration in mequiv l^{-1}.

I have taken a single thylakoid to be a flattened disc of 4000 Å diameter and 200 Å thick. These dimensions would mean that a single thylakoid would have a surface area of 27·6 × 10^{-10} cm^2 and a volume of 25 × 10^{-16} cm^3. Junge and Witt (1968) have also taken similar dimensions for a single thylakoid and have further assumed that 10^5 chlorophyll molecules occupy 25 × 10^{-10} cm^2 of thylakoid membrane. This would mean that 1 μg chl is associated with 16·7 cm^2 of membrane surface and a total thylakoid volume of 15·2 × 10^{-6} cm^3. Using these values the above rate of 33 μM s^{-1} μg^{-1} chl would correspond to 3 × 10^{-14} equiv Cl s^{-1} cm^{-2}.

Of course the size of the chloride influx (ϕ_{Cl}) is dependent both on the concentration and electrical gradients, and accepting Goldman's linear field assumption (1943) would be given by:

$$\phi_{Cl} = \frac{F \, \Delta\psi}{RT} P_{Cl} \frac{[Cl]_o - [Cl]_i \exp(-F \, \Delta\psi/RT)}{1 - \exp(-F \, \Delta\psi/RT)} \tag{7}$$

where the symbols have the usual meanings.

For the initial Cl$^-$ influx induced by establishing a KCl gradient eqn (7) can be simplified since the initial internal chloride concentration is zero.

$$\phi_{Cl} = \frac{F \, \Delta\psi}{RT} P_{Cl} \frac{[Cl]_o}{1 - \exp(-F \, \Delta\psi/RT)} \tag{8}$$

In most cases, the size of the membrane potential corresponding to the initial height of the KCl-induced signals before and after treatment with valinomycin differed, as shown in Fig. 3. In these cases, according to eqn (8) the passive influx of Cl⁻ should be different even when the same concentration gradients are established.

Figure 6 shows the analysis of some experimental data for which $\Delta\psi_L = 65.5$ mV. The curves drawn correspond to net Cl⁻ influxes given by the first

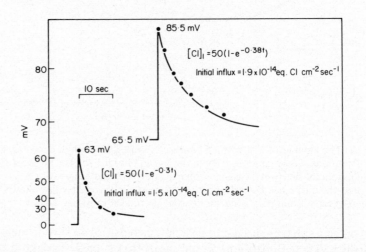

FIG. 6. Kinetic analyses of the signals induced by 50 mM KCl additions to chloroplasts which had (signal a) or had not (signal b) been treated with 0·5 μM valinomycin. The closed circles are points taken from the experimental signals while the curves have been theoretically constructed using the two first order laws shown for chloride influx Before valinomycin treatment $P_{Cl}/P_K = 0.44$ and after $P_{Cl}/P_K = 0.03$. Chlorophyll concentration 30 μg/ml. Other details of the experiment are the same as Figs 3 and 5.

order equations shown and are the best fit to the experimental data. The chloride influx (ϕ_{Cl}) corresponds to 1.9×10^{-14} equiv Cl s⁻¹ cm⁻² before valinomycin treatment and to a slightly lower influx (ϕ_{Cl}^{Val}) of 1.5×10^{-14} equiv Cl s⁻¹ cm⁻¹ after treatment with valinomycin. Using eqn (8) it is possible to check whether the differences in the initial membrane potentials account for the differences in the influxes. Assuming the permeability to Cl⁻ is not altered by valinomycin then the flux ratio for this experiment should have been 1·29 which compares well with the observed ratio of 1·27.

It is worth noting that although the Cl⁻ influx after valinomycin treatment had been reduced, the luminescence signals showed faster decay kinetics. This feature was always observed experimentally and occurs because of the change in the relative permeabilities between K⁺ and Cl⁻ after valinomycin treatment such that the entry of K⁺ has a more significant effect on the rate of fall of the diffusion potential.

D. Pre-treatment with Various KCl Levels

In the experiments presented above the chloroplasts had been incubated in 2 mM K$^+$. If, however, KCl is added before giving a 50 mM KCl pulse, the signals are inhibited. The inhibition is dependent on the quantity of KCl initially added (see Fig. 7). Qualitatively this is what would be expected from

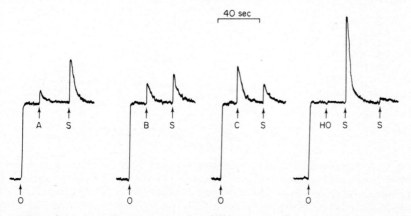

FIG. 7. The effect of pre-injections of various KCl concentrations on the intensity of the 50 mM KCl induced signals. The pre-injections were: (A) 3 mM KCl, (B) 6 mM KCl and at (C) 12 mM KCl. Injections S correspond to 50 mM KCl additions. The chloroplasts had been treated with 0.5 μM valinomycin and the chlorophyll concentration was 26 μg ml^{-1}. Other details are the same as Fig. 3.

the diffusion potential model. Whether the magnitude of the inhibition is consistent with the theoretical approach presented above requires the applications of eqn (9).

$$c'L = \frac{[K]_o + \beta[Cl]_i}{[K]_i + \beta[Cl]_o} \tag{9}$$

For the experiment shown in Fig. 8 it was found that $\beta = 0.03$ and it can be seen that there was good agreement between the inhibition observed and that predicted by the above equation.

IV. Discussion

There seems to be good correlation between the experimental observations and theoretical concepts presented in this and earlier papers (Barber, 1972a; Barber and Varley, 1971a, 1972). Under various conditions it is possible to account for both the magnitude and kinetics of the KCl signals in terms of a diffusion potential model. From this it seems that millisecond delayed light, like the 515 nm shift (Junge and Witt, 1968), is an indicator of electrical

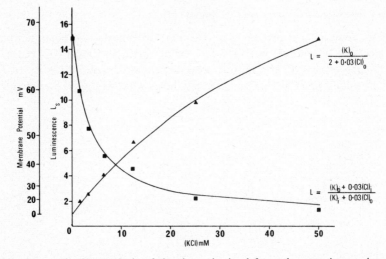

FIG. 8. A quantitative analysis of the data obtained from the experiment shown in Fig. 7. The closed squares are the initial heights of the 50 mM KCl-induced signals after the various pre-injections of KCl at the concentrations indicated. The closed triangles are the initial height of the pre-injection KCl signals. The curves through the experimental points have been drawn according to the equations shown and the membrane potential scale calculated in the way outlined in the text.

gradients across the thylakoid membranes and can be used to obtain a measure of their size and polarity. From the analyses presented it seems that rapidly raising the external KCl level and also illumination of chloroplast suspensions causes the thylakoid interiors to become electrically more positive with respect to the suspending medium. The results also indicate that although the thylakoids are more permeable to monovalent cations than anions, there is only a slight selectivity between the various alkali metals.

Chlorides of calcium and magnesium cause an inhibition rather than a stimulation of delayed light suggesting that the membrane has a relatively low permeability to these divalent cations.

Of course, the assumption that $\Delta\psi_L$ is reduced to zero by valinomycin can only be valid in the absence of ion gradients and may only occur if K^+ is out of equilibrium. If in fact light-induced ion gradients exist across the thylakoid surface then there could be a diffusional, as well as an electrogenic, component of the membrane potential. Without having knowledge of these ion gradients and permeability properties of the membrane it would be impossible to say how valinomycin would effect such a potential. In the above experiments, however, no cofactors were added to the chloroplasts to induce electron flow and consequently there was no evidence that any light-promoted ion gradients existed.

By analysing the decay characteristics of the KCl-induced transients it has been possible to estimate the Cl^- influx under known driving conditions. For a 50 mM Cl^- gradient and potentials in the region of 50–100 mV (inside +ve) the influx of this anion corresponded approximately to 10^{-14} equiv $cm^{-2} s^{-1}$. Accepting that this value is based on rather crude estimates of the thylakoid dimensions, the rate is very low. It is normally assumed that the thylakoids are leaky to Cl^- (Walker and Crofts, 1970). From the above analyses, however, the relatively rapid decay of the KCl-induced signals is not due to a leaky membrane but almost certainly reflects the very large surface area to volume ratio which, as estimated above, could be in the region of 10^6. Using eqn (8) it is possible to obtain a quantitative estimate of the permeability of the thylakoids to Cl^-. Taking the data of Fig. 5 it was found that for a Cl^- gradient of 50 mM and a membrane potential of +93 mV the influx of this anion was $4\cdot0 \times 10^{-14}$ equiv $Cl^- s^{-1} cm^{-2}$. Substituting these values in eqn (8) gives $P_{Cl} = 2\cdot1 \times 10^{-10} cm s^{-1}$.

Although this analysis seems reasonable it gives a P_{Cl} value which is at present difficult to reconcile with some of the experimental findings. It has been estimated that P_K was about 30 times greater than P_{Cl} after valinomycin treatment. This would give a P_K value in the region of 10^{-8} to $10^{-9} cm s^{-1}$. Such values would seem unacceptably low bearing in mind the reported effect of valinomycin on artificial membranes (Andreoli et al., 1967). For the moment I have no clear explanation for this. It could be that the estimates of the P_{Cl}/P_K ratio after valinomycin treatment are incorrect although earlier analyses of KCl-induced 10 s delayed light using a stop–flow system gave very similar values (Barber and Varley, 1971a, 1972). Alternatively a deviation from a Nernst potential relationship when valinomycin is present (i.e. P_{Cl}/P_K very small) could be due to the net transfer of K^+ across the membrane in the absence of diffusing coions. Such a breakdown in the laws of electroneutrality for systems having very large surface area to volume ratios has already been discussed in some detail by Mitchell (1968). Another possibility is that at the valinomycin concentrations employed there may be some loss of cation selectivity resulting in an increase of membrane conductance to protons as well as K^+.

Acknowledgements

The author wishes to thank the Science Research Council and the Royal Society for financial support. Thanks also go to Mr A. Butler and Mr L. Hullis for technical assistance.

References

ANDREOLI, T. E., TIEFFENBERG, M. and TOSTESON, D. C. (1967). J. Gen. Physiol. 50, 2527–2545.
BARBER, J. (1972a). FEBS Letters 20, 251–254.

BARBER, J. (1972b). *Biochim. biophys. Acta* **275**, 105–116.

BARBER, J. and KRAAN, G. P. B. (1970). *Biochim. biophys. Acta* **197**, 49–59.

BARBER, J. and VARLEY, W. J. (1971a). Proc. 2nd Internation Congress of Photosynthesis Research, Stresa, Italy. pp. 963–975. W. Junk, Hague.

BARBER, J. and VARLEY, W. J. (1971b). *Nature, Lond.* **49**, 188–189.

BARBER, J. and VARLEY, W. J. (1972). *J. exp. Bot.* **23**, 216–228.

BULYCHEV, A. A., ANDRIANOV, V. K., KURELLA, G. A. and LITVIN, F. F. (1972). *Nature* **236**, 175–176.

CROFTS, A. R., WRAIGHT, C. A. and FLEISCHMANN, D. E. (1971). *FEBS Letters* **15**, 89–100.

DEAMER, D. W. and PACKER, L. (1969). *Biochim. biophys. Acta* **172**, 539–545.

DILLEY, R. A. and VERNON, L. (1965). *Archs Biochem. Biophys.* **111**, 365–375.

GOLDMAN, D. E. (1943). *J. gen. Physiol.* **27**, 37–60.

HEBER, U. and KRAUSE, G. H. (1971). In "Photosynthesis and Photorespiration (M. D. Hatch, C. B. Osmond and R. O. Slatyer, eds) pp. 218–225, Wiley-Interscience, Sydney.

HODGKIN, A. L. and KATZ, G. (1949). *J. Physiol.* **108**, 37–77.

JAGENDORF, A. T. and URIBE, E. G. (1966). *Brookhaven Symp. Biol.* **19**, 215–245.

JUNGE, W. and WITT, H. T. (1968). *Z. Naturforsch.* **23**, 244–254.

MAYNE, B. C. (1967). *Photochem. Photobiol.* **6**, 189–197.

MITCHELL, P. (1966). *Biol. Rev.* **41**, 445–502.

MITCHELL, P. (1968). Chemiosmotic Coupling and Energy Transduction. Pub. Glynn Research Ltd.

MURATA, N., TASHIRO, H. and TAKAMIYA, A. (1970). *Biochim. biophys. Acta* **150**, 32–40.

PRESSMAN, B. C., HARRIS, E. J., JAGGER, W. S. and JOHNSON, J. H. (1967). *Proc. Natn. Acad. Sci. U.S.* **58**, 1949–1956.

ROTTENBERG, H., GUNWALD, T. and AVRON, M. (1972). *Eur. J. Biochem.* **25**, 54–63.

SALTMAN, P., FORTE, J. G. and FORTE, G. M. (1963). *Exp. Cell Res.* **29**, 504–514.

STREHLER, B. and ARNOLD, W. (1951). *J. gen. Physiol.* **34**, 809–820.

SCHWARTZ, M. (1971). *A. Rev. Pl. Physiol.* **22**, 469–484.

WALKER, D. A. and CROFTS, A. R. (1970). *A. Rev. Biochem.* **39**, 389–428.

WINOCUR, B. A., MACEY, R. I. and TOLBERG, A. B. (1968). *Biochim. biophys. Acta* **150**, 32–40.

IV.2

Proton and Chloride Uptake in Relation to the Development of Photosynthetic Capacity in Greening Etiolated Barley Leaves

Ulrich Lüttge

Fachbereich Biologie (Botanik) der Technischen Hochschule, Darmstadt, Germany

I. Introduction

A major argument against the Lundegårdh hypothesis of a direct coupling of ion uptake to respiratory electron transport has arisen from spatial cytological considerations. Electron flow is confined to the inner membranes of mitochondria. How could it drive ion transport across distant cellular membranes? Similarly the hypothesis of cellular ion uptake depending on photosynthetic electron flow in the thylakoid membranes of chloroplasts has met with considerable scepticism. By contrast, the postulate of a ready availability of ATP in the entire cell has made acceptance of phosphorylation-powered ion uptake mechanisms possible from the spatial cytological point of view.

We now know that both phosphorylation-driven ion uptake mechanisms and mechanisms depending more closely on electron flow or on redox processes do exist (cf. survey of references in the introduction of Lüttge *et al.*,

1971b). The type of energy utilized may vary with the plant material and also with ionic species and with membranes within a given cell type. We also know that phosphorylation-independent energetic linkage of processes outside the organelles to electron transport (i.e. to redox reactions) within the organelles is by no means a cytological impossibility (Heber and Krause, 1971), and conversely, that ready distribution of ATP within the cell is not evident. In chloroplasts metabolic shuttles are required for the transport of $\sim$ P across their envelope, and in addition redox shuttles are important for correlation of cytoplasmic and organelle activities (transport metabolites: Heber and Santarius, 1970; Heber and Krause, 1971; Krause, 1971; Heldt, 1969; Heldt and Rapeley, 1970a, b). Thus the energetic linkage of both $\sim$ P-driven ion uptake mechanisms and uptake mechanisms depending on redox reactions may be mediated by metabolic intermediates. I have called this the biochemical mode of linkage or "coupling" (Lüttge, 1971a, b) In order to investigate this proposal of biochemical linkage it is necessary to design experiments correlating ion uptake with synthesis of specific metabolites, as has been attempted in an earlier investigation (Lüttge et al. 1971a, b).

Phosphorylation is always a biochemical event. By contrast, for electron flow-dependent ion uptake mechanisms an alternative possibility of energetic linkage exists. It is well established that photosynthetic electron flow in the thylakoid membranes of chloroplasts is accompanied by changes of ionic gradients across these membranes (e.g. H^+ fluxes and charge separation according to the Mitchell hypothesis; K^+ fluxes). A photosynthesis-dependent change of gradients at the thylakoid membranes will affect gradients at the chloroplast envelope (i.e. the cytoplasm–chloroplast interphase) and it is possible that it reflects on gradients at the plasmalemma and tonoplast (i.e. the outside–cytoplasm and cytoplasm–vacuole interphases) Hence, photosynthetic electron flow may lead to photosynthesis dependent ion fluxes at these membranes without involvement of phosphorylation and without involvement of metabolites carrying redox equivalents. I call this the biophysical mode of linkage or "coupling" (Lüttge, 1971a, b). To demonstrate this it is necessary to design experiments where ion uptake proceeds in conjunction with photosynthetic electron flow while production of photosynthetic metabolites is drastically reduced or absent.

The greening of dark-grown leaves upon illumination is accompanied by a gradual development of various functions of primary and secondary reactions of photosynthesis (cf. Park and Sane, 1971; Wolf, 1971). This yields ideal material for an evaluation of the relation of light-dependent ion fluxes to particular photosynthetic events. Some experiments with greening barley leaves are described here and discussed in relation to the problems outlined above.

II. Materials and Methods

Barley leaves

Barley plants were grown on an aerated Knop's culture solution at 25°C in the dark. 6 or 7 days after seed germination the etiolated plants were transferred to a controlled environment room at 25°C, 65% r.h. and with 8000–8500 lx light intensity from a Xenon lamp. The time of transfer to the light defines zero time on the greening time scale. These growth conditions and the greening conditions were chosen to resemble as closely as possible those used by Tamas *et al.* (1970). After various periods of greening leaves were harvested, rapidly sliced into small pieces about 1–2 mm wide and transferred to the appropriate experimental solution. Chlorophyll content was determined using the method described by Arnon (1949).

O_2 exchange and $\Delta[H^+]$

A Rank oxygen electrode assembly (Rank Bros., Bottisham, U.K.) was used together with an Ingold combination, flat-ended pH electrode (Dr W. Ingold K.G., Frankfurt a.M., Germany) to measure O_2 exchange and $\Delta[H^+]$ simultaneously, as depicted in Fig. 1 of Hope *et al.* (1972).

Photosynthetic O_2 evolution and light-dependent $\Delta[H^+]$ were calculated from the change in slope of respective recordings for light–dark and dark–light transitions as described by Hope *et al.* (1972). This procedure includes a correction for respiration-dependent O_2 consumption and $\Delta[H^+]$.

CO_2 fixation

CO_2 fixation into formic acid stable products was measured using $H^{14}CO_3^-$. In some experiments CO_2 fixation was followed simultaneously with O_2 evolution and $\Delta[H^+]$ in the Rank O_2 electrode assembly. Counting and the appropriate controls for self-absorption were performed as discussed in detail by Hope *et al.* (1972). Products of CO_2 fixation were determined by separation of ethanol extracts using the thin layer chromatography method developed by Feige *et al.* (1969) as described earlier (Lüttge *et al.*, 1972).

Cl^- uptake

Chloride uptake was followed using $^{36}Cl^-$. Samples were taken 20, 40 and 60 min after incubation of the tissue in the labelled uptake solution and rates of Cl^- uptake were calculated following the absorption period. At the end of the uptake periods leaf slices were washed 2 × 15 min in ice cold unlabelled uptake solution. Slices were counted in a gas flow counter as previously described (Lüttge *et al.*, 1970, 1971a, b, 1972).

Light sources

The light source in the experiments on chloride uptake and in part of the $^{14}CO_2$ fixation experiments was a Philips HQL-lamp of $\sim 20,000$ lx intensity. In the O_2 exchange, $\Delta[H^+]$, and the concomitant $^{14}CO_2$ fixation measurements light was obtained from a low voltage microscope lamp focused on the O_2 electrode vessel. Light intensities were varied using neutral density filters (Jenaer Glaswerk, Schott and Gen. Mainz). Light source and intensity are given for each of the experiments described here in the legends of figures and tables. All intensities are given in lux, as during the course of these investigations a lux-meter was the only equipment available to measure intensities. For the comparative nature of the experiments discussed here, this appears to be sufficient.

Experimental solution

Experimental solution in all experiments was 5 mM KCl, 0·1 mM $CaSO_4$ and 0·25 mM $NaHCO_3$ kept at 25°C. This solution was buffered with 5 mM phosphate buffer (pH 6·0) or 1 mM HEPES (pH 6·7) as indicated in the legends. $H^{14}CO_3^-$ and $^{36}Cl^-$ respectively were added as appropriate for CO_2 fixation and Cl^- uptake measurements.

Potential measurements

Membrane potential was measured using glass microelectrodes filled with 3 M KCl attached to a Keithley electrometer (model 604) via Ag/AgCl contacts. Electrodes were inserted into the cells using Leitz micromanipulators.

Errors

Errors given in figures and tables are standard errors.

Abbreviations

DCMU: dichlorophenyl-dimethyl-urea,
FCCP: *p*-trifluoro-methoxy-carbonyl-cyanide-phenyl-hydrazone,
pBQ: *p*-benzo-quinone,
HEPES: *n*-hydroximethylpiperazine-n'-2-ethanesulphonic acid,
PS I, PS II: photosystems I and II,
NAD: nicotinamide-adenine-di-nucleotide,
NADP: nicotinamide-adenine-di-nucleotide-phosphate,
Ru-5-P: ribulose-5-phosphate,
RuDP: ribulose-1,5-diphosphate,
3-PGA: 3-phosphoglyceric acid,
OAA: oxaloacetate.

III. Results and Discussion

A. Photosynthesis

The kinetics of the development of photosynthetic capacity in greening etiolated plant leaves have been investigated in many laboratories (for literature surveys see: Park and Sane, 1971; Wolf, 1971). This literature indicates the sequence in which the various photosynthetic functions are established. However, information on the exact time after the beginning of illumination at which these various processes begin is extremely variable. This is due to the influence of variations in growth and greening conditions. For comparison with ion uptake results it was therefore necessary to measure some photosynthetic parameters under the conditions of our experiments.

. Photosynthetic electron flow as measured by detection of O_2 evolution

The rate of O_2 evolution on a fresh weight basis increases steadily to a maximum value during the first 8 h after the onset of illumination. The rate of O_2 evolution then decreases somewhat with extended illumination. The chlorophyll content continues to rise slightly with increasing greening period. O_2 evolution is DCMU-sensitive at all greening times tested, but the sensitivity is noticeably reduced after longer greening periods (Fig. 1).

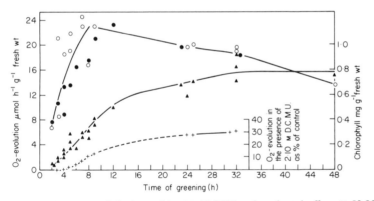

FIG. 1. O_2 evolution at two light intensities ($\bullet$ 10,000 lx, phosphate buffer; $\bigcirc$ 50,000 lx, HEPES), DCMU sensitivity of O_2 evolution at 50,000 lx (+), and chlorophyll content of the material ($\blacktriangle$) as a function of time of greening. Light source: microscope lamp. Rates of O_2 evolution obtained at the two light intensities do not differ significantly. Separate experiments show that O_2 evolution saturates at light intensities above 5000–10,000 lx. Cl uptake described in section II/3 already saturates at about 2000–5000 lx.

The literature (see above) suggests that the various activities of the primary reactions of photosynthesis are established in the following sequence after

the onset of illumination of etiolated plants:
  (i) PS I,
  (ii) PS II, partitions between individual thylakoids, DCMU sensitive
       O_2 evolution
  (iii) Enzyme systems for transfer of electrons from PS I to NADP.
Hence the results shown in Fig. 1 can be taken to demonstrate that under the
conditions of our experiments non-cyclic electron flow is functioning even
at the shorter greening times (2–4 h).

2. *Secondary reactions of photosynthesis as measured by detection of $^{14}CO_2$*
   fixation

In contrast to O_2 evolution, photosynthetic $^{14}CO_2$ fixation into acid
stable products is clearly delayed after the beginning of illumination. Appreciable light-dependent CO_2 fixation is not observed earlier than 3 h after the
onset of illumination (when O_2 evolution is already close to its half maximum
rate). Light dependent CO_2 fixation may take as long as 6 h to become
apparent (Figs 2 and 3). The onset of light-dependent CO_2 fixation for various
tissues has been reported in the literature to occur 2–6 h after beginning of
illumination (cf. Park and Sane, 1971).

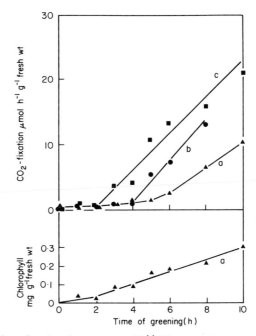

FIG. 2. CO_2 fixation after 5 s of exposure to H $^{14}CO_3^-$ at varied time of greening. (a), (b)
and (c) represent 3 separate experiments. Chlorophyll content was measured in experiment (a) only. HQL lamp: 20,000 lx. Phosphate buffer.

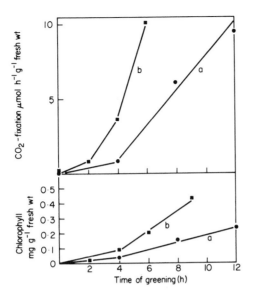

FIG. 3. As Fig. 2 but 15 min exposure to H $^{14}CO_3^-$. 2 separate experiments (a and b).

The author is unaware of literature where simultaneous measurements of the appearance of capacities of primary and secondary photosynthetic reactions in greening etiolates are reported. However, it seems possible that the light-dependent acquirement of activities of some enzymes in the reductive CO_2 fixation pathway and in the Calvin cycle limits CO_2 fixation during the early stages of greening. RuDP-carboxylase is active after 3–6 h (maize: Bogorad, 1967; Chen et al., 1967), RuDP-kinase after 6 h (maize: Chen et al., 1967).

3. Light-dependent synthesis of metabolites

It has been argued that in barley leaves in the initial phase of greening, when CO_2 fixation by RuDP-carboxylase and through the Calvin cycle is limited, $NADPH_2$ generated by noncyclic electron flow serves to enhance CO_2 fixation via β-carboxylation of PEP into malate, i.e. by the fixation mechanism operative in the dark (Tamas et al., 1970). Labelled metabolites found by thin layer chromatography of extracts obtained after 5 s and 15 min, respectively, of $^{14}CO_2$ fixation in the light are shown in Figs 4 and 5. The data show that in our experiments there is no unusual light-driven malate formation at any greening time. Labelling of malate up to 2–4 h after commencement of illumination corresponds to that of dark fixation. Increase of label in malate is accompanied by a more drastic increase of label in 3-PGA

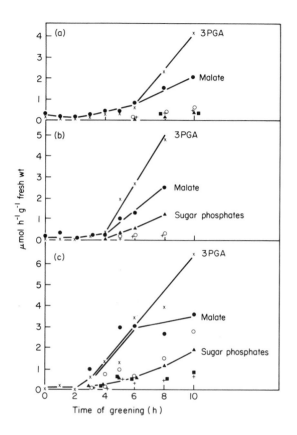

FIG. 4. Labelled metabolites found after 5 s of CO_2 fixation in H $^{14}CO_3^-$. Same experiments as those depicted in Fig. 2. (a), (b) and (c) refer to lines (a), (b) and (c) in Fig. 2. ×—3-PGA; ● malate; ▲ sugar phosphates; + glycine plus serine; ○ alanine; ■ aspartate.

(Figs 4 and 5) and in sugars (Fig. 5) consistent with the C_3 pathway of photosynthesis in barley.

Hence, it was not possible to produce material of a given species with varied major products of light-driven CO_2 fixation. Since malate–OAA is an important shuttle system for the transport of reducing equivalents across the chloroplast envelope (Heber and Krause, 1971), and in view of earlier speculations on a malate-dependent light-driven Cl^- uptake in leaves of C_4 plants (Lüttge et al., 1971b), this would have been of particular interest. However, the tissue obtained in the first 2–4h of the greening time, having vigorous photosynthetic electron flow but very little light-dependent CO_2 fixation still proved to be valuable in our ion transport studies.

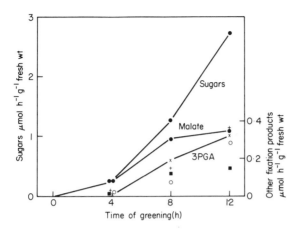

Fig. 5. Labelled metabolites found after 15 min of CO_2 fixation in H $^{14}CO_3^-$. Same experiment as that depicted by lines (a) in Fig. 3. ● sugars, malate (note the different ordinates for sugars and the other fixation products); other symbols as in Fig. 4.

B. Ion Fluxes and Membrane Potential

1. Apparent proton fluxes

Figure 6 shows that over a wide range of greening times and light intensities and in the presence of 2×10^{-6} M DCMU the pH change in the external solution (apparent proton flux: $\Delta\Phi H^+$) is closely correlated with O_2-evolution. The ratio H^+/O_2 is between 1:1 and 1.5:1.

Figure 7 depicts similar data obtained during the first 120 min of greening, where from the results described in sections A.2 and A.3 it was expected that photosynthetic CO_2 fixation is very small. It can be seen that H^+/O_2 again is approximately 1.5:1 (average of the 20 points shown in the graph: $H^+/O_2 = 1.47 \pm 0.06$), while the ratio CO_2/O_2 is clearly smaller by about an order of magnitude (0.15 ± 0.02).

There are two possible explanations of light-dependent *apparent* H^+ fluxes at the outer membrane of intact green cells:

(i) they might simply reflect CO_2 fixation resulting in a shift in the HCO_3^-/CO_2 equilibrium in the solution (Atkins and Graham, 1971; Neumann and Levine, 1971), or

(ii) they may be more directly linked to electron flow and H^+ fluxes at the thylakoid membranes of the chloroplasts (Schuldiner and Ohad, 1969; Ben-Amotz and Ginzburg, 1969; Brinckmann and Lüttge, 1972).

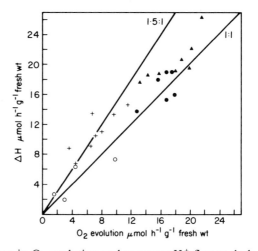

FIG. 6. The change in O_2 evolution and apparent H^+-flux at dark–light transitions. Solution buffered with HEPES. Microscope lamp. ● greening time varied between 3 and 24 h, light intensity 44,000 lx. ○ greening time varied between 6 and 31 h, 2×10^{-6} M DCMU present in the incubation vessel, light intensity 44,000 lx. + greening time 2 h, light intensity varied between 5000 and 50,000 lx. ▲ greening time 24 h, light intensity varied between 5000 and 50,000 lx.

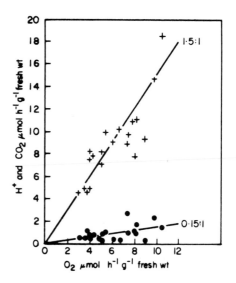

FIG. 7. The change in O_2 evolution and apparent H^+-flux upon dark–light transitions. Buffered with HEPES. Microscope lamp. Light intensities varied between 625 and 10,000 lx. Time of greening varied between 30 and 120 min. + H^+–O_2, ● CO_2–O_2. $\Delta[H^+]$, O_2 and CO_2 were measured simultaneously.

Evidence favouring the first alternative is the finding that intact cells of a *Chlamydomonas* mutant unable to fix CO_2 in the light do not show apparent H^+ fluxes, while the light reactions of photosynthesis are normal and the usual H^+-fluxes are observed with isolated thylakoids from the same cells (Neumann and Levine, 1971). By contrast, working with intact *Elodea* leaves, Hope *et al.* (1972) were able to find experimental conditions where CO_2 fixation was reduced to almost nil, while $\Delta\Phi H^+$ and O_2 evolution remained at high levels (Table I). These results seem to indicate that a fraction

TABLE I. Photosynthesis and apparent proton fluxes in *Elodea canadensis* (from Hope *et al.*, 1972)

		+FCCP	
	Controls	5 μM	10 μM
H^+/O_2	1·03 ± 0·046 (28)	0·71 ± 0·16 (5)	0·63 ± 0·054 (5)
CO_2/O_2	0·61 ± 0·07 (24)	0·033 ± 0·004 (5)	0·023 ± 0·010 (5)

		+pBQ	
	Controls	0·3 mM	0·6 mM
H^+/O_2	1·03 ± 0·046 (28)	0·36 ± 0·11 (6)	0·27 ± 0·06 (10)
CO_2/O_2	0·61 ± 0·07 (24)	0·071 ± 0·019 (6)	0·0074 ± 0·0014 (10)

of the apparent H^+-flux observed between the external solution and the intact cells is independent of CO_2 fixation and more closely linked to photosynthetic electron flow ($H^+/e^- \simeq 0\cdot1$). However, the uncoupler (FCCP) and the artificial electron acceptor (pBQ) used in these experiments to cause differential effects on CO_2 fixation and electron flow are likely to affect permeabilities of the outer cell membrane and the chloroplast envelope, leaving some difficulties for interpretation (cf. Hope *et al.*, 1972; Gimmler and Avron, 1971).

From the experiments described here, the picture emerges still more clearly. No inhibitors had to be used to separate CO_2 fixation and electron flow, and it is evident that $\Delta\Phi H^+$ is virtually independent of the rate of CO_2 fixation but closely linked to O_2 evolution, i.e. e-flow (Fig. 7, compare Fig. 6 with Figs 1, 2, and 3). In the first 120 min of greening time, when $CO_2:O_2 \simeq$ 0·15 and $\Delta[H^+]:O_2 \simeq 1\cdot5$, $H^+:e^-$ is about 0·3. This ratio is higher than that reported by Hope *et al.* (1972) but still much lower than the value of $\simeq 2$ obtained for chloroplasts lacking their outer membrane (Schwartz, 1968). This may be due to the chloroplast envelope being a significant barrier for H^+ movement (Heber and Krause, 1971).

2. Transients of membrane potential

We speculated earlier that the well known photosynthesis-dependent transients of membrane potentials in green cells observed after dark–light and light–dark transitions (e.g. literature reviewed by Pallaghy and Lüttge, 1970; Throm, 1970, 1971a, b) are due to a change from largely respiratory to largely photosynthetic energy conversions and hence directly dependent on electron flow and ion movements (possibly H^+) at the thylakoid membranes (Pallaghy and Lüttge, 1969, 1970). Brinckmann in our laboratory has set out to investigate the light-on and the light-off transients of membrane potential over the greening period of etiolated barley leaves. A typical light-off transient is shown in Fig. 8 (light from a microscope lamp). The parameters checked in relation to greening periods were:

    (i) the duration of the first peak of the transient (d),

    (ii) the amplitude of the transient (a),

    (iii) the time elapsing until half the amplitude is attained ($t_\frac{1}{2}$), and

    (iv) the initial rate of change of potential (tg α).

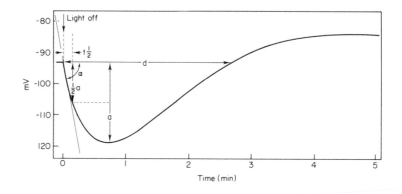

FIG. 8. Typical transient of membrane potential after a light–dark transition. d = duration of peak; a = amplitude; $t_\frac{1}{2}$ = half time for attainment of amplitude; tg α = initial rate of potential change. Greening time 18 h. (From Brinckmann, unpublished.)

Among these parameters only the latter two appear to show clear changes during the greening period. In Fig. 9 it can be seen that the initial rate of change of potential (tg α) increases considerably from very low values at zero greening time (effect of brief illumination during preparation?) and reaches a constant value after about 4 h of greening. Therefore, this phenomenon appears to be independent of CO_2 fixation and of formation of metabolites.

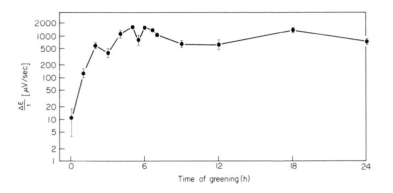

FIG. 9. Semi-log plot of initial rates of potential change after a light dark transition vs. greening times. (From Brinckmann, unpublished results.)

3. $^{36}Cl^-$ uptake

(a) $^{36}Cl^-$ *uptake in the dark and light with air.* Figure 10 shows that the rate of ^{36}Cl$^-$ uptake in air is similar in the light and in the dark. The small increase during the first 4 h of greening is unlikely to be due to the photosynthetic formation of substrates during the greening period, since very little CO_2 fixation occurs in the first 4 h of greening (see A/2 and 3). This increase may depend on metabolic changes in other ways, and hormonal interactions may play a role here (Feierabend 1969, 1970a, b).

(b) $^{36}Cl^-$ *uptake in the dark + N_2.* In the dark + N_2 there is a "basic rate" of uptake of non-exchangeable ^{36}Cl$^-$ of $\sim 1\ \mu$mol h^{-1} g^{-1} which is unaffected by greening (Fig. 10). This uptake is not sensitive to FCCP concentrations, usually causing specific effects (i.e. up to 2×10^{-6} M FCCP, Table II). It may be argued that this uptake reflects binding of Cl$^-$ to exchange sites within or at the surface of the cytoplasm, which is not of further interest in the context of this investigation.

(c) $^{36}Cl^-$ *uptake in the light + N_2.* The rate of ^{36}Cl$^-$ uptake in the light + N_2 rises linearly within the first 4 h of greening from the dark + N_2 rate to the rate obtained in air in either dark or light (Fig. 10). ^{36}Cl$^-$ uptake in the light + N_2 is sensitive to both DCMU (Table III) and FCCP (Fig. 11).

These results on light-dependent Cl$^-$ uptake in nitrogen are still ambiguous, and various explanations can be offered. The sensitivity to the inhibitor of photosynthetic electron flow DCMU and to the uncoupler FCCP indicates that both electron flow and phosphorylation are required. It is not clear whether phosphorylation is photosynthetic or respiratory. The latter might be driven by small amounts of photosynthetically evolved O_2 remaining in the tissue despite the N_2 bubbling.

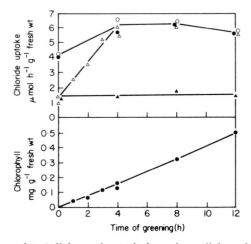

FIG. 10. Chloride uptake. ○ light + air; ● dark + air; △ light + N$_2$; ▲ dark + N$_2$. HQL lamp, 20,000 lx. Phosphate buffer (i.e. conditions identical to those of experiments, depicted in Figs 2, 3, 4, and 5).

TABLE II. FCCP sensitivity of "basic Cl$^-$ uptake rate" (i.e. rate obtained at 0 h greening in D + N$_2$ and L + N$_2$, and at all greening times in D + N$_2$)

FCCP concentration M	Cl$^-$ uptake % of control
2 × 10^{-7}	105
5 × 10^{-7}	111
10^{-6}	96
2 × 10^{-6}	98
5 × 10^{-6}	76

TABLE III. DCMU sensitivity of Cl$^-$ uptake in L + N$_2$ after subtraction of the "basic uptake rate" (average of two experiments)

DCMU concentration M	Cl$^-$ uptake % of control 2 h	5 h
5 × 10^{-7}	21	22
10^{-6}	4	9
2 × 10^{-6}	3	10

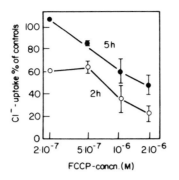

FIG. 11. Uncoupler (FCCP)-sensitivity of Cl⁻ uptake in light + N₂ at 2 and 5 h of greening time. HQL lamp : 20,000 lx. Phosphate buffer.

Since photosynthetic CO_2 fixation is very limited during the first 4 h of greening, i.e. when light-dependent Cl^- uptake in N_2 reaches its maximum rate, it is an interesting question how DCMU-sensitive electron flow is maintained (Jeschke and Simonis, 1967; 1969). Processes other than CO_2 reduction may serve as electron acceptors (Ullrich, 1971, 1972) and perhaps Cl^- uptake itself is involved.

Another interesting question is posed by the finding that greening times required

(i) for attainment of the maximum rate of light-dependent Cl^- uptake in N_2,

(ii) for attainment of the maximum rate of membrane potential change after a light–dark transition, and

(iii) for the onset of light-dependent CO_2 fixation are very similar. Further work is required to determine the meaning of this coincidence.

In summary, the simplest explanation seems to be that Cl^- uptake in light + N_2 depends on photophosphorylation. CO_2 fixation may compete for ATP from photophosphorylation, as there is no further rise in the rate of Cl^- uptake after the onset of CO_2 fixation. Electron transport before the onset of CO_2 fixation may be largely driven by electron acceptors other than CO_2.

IV. Speculative Outlook

Biophysical mode of linkage

In the sense outlined in the introduction the light-driven H^+ fluxes described in section B.1 may be interpreted as an example of the biophysical mode of linkage of ion fluxes at the outer cell barrier to events within the chloroplasts. This is supported by the ratio $H^+ : O_2$ being insensitive to

DCMU and far less sensitive to FCCP than the ratio $CO_2:O_2$. It remains to be seen, whether there are similar findings for fluxes of other ionic species. Transients of membrane potentials (section B.2) may also reflect electron flow and ion fluxes at the thylakoid membranes.

Biochemical mode of linkage

Light-dependent Cl^- uptake in N_2 appears to depend on photophosphorylation. According to Heber and Santarius (1970) a 3-PGA-triosephosphate shuttle is required to transport $\sim P$ out of the chloroplasts. This shuttle carries stoichiometrically equal amounts of $\sim P$ and reducing equivalents (Krause, 1971; Smith, 1971; Fig. 12). Since Cl^- uptake is occurring in the

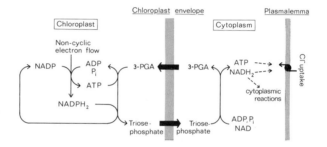

FIG. 12. 3-PGA-triosephosphate shuttle of chloroplasts after Heber and Santarius (1970) and Krause (1971) and its possible relationship to light-dependent Cl^- uptake.

absence of CO_2 fixation, it is most unlikely that concurrent photosynthetic production of transport metabolites is responsible for energetic coupling. Thus the shuttle must be independent of CO_2 fixation, depending only on energetic relations between the cytoplasm and the interior of the chloroplasts. This also offers an explanation of how processes other than CO_2 reduction could provide an electron acceptor for non-cyclic electron flow in the absence of CO_2 fixation (Fig. 12). An example might be ion uptake itself, as indicated by the inhibitory effects of both DCMU and FCCP on Cl^- uptake in light + N_2.

Acknowledgements

These investigations have been made possible by a generous grant from the Deutsche Forschungsgemeinschaft. Herr Brinckmann provided the data on membrane potential transients. These data will be published in detail elsewhere. The skilled technical assistance of Fräulein Erika Ball, Frau G. Kinze and Frau G. Zirke is gratefully acknowledged. Dr Chris Johansen kindly read the manuscript.

References

ARNON, D. I. (1949). *Pl. Physiol.* **24**, 1–15.
ATKINS, C. A. and GRAHAM, D. (1971). *Biochim. biophys. Acta* **226**, 481–485.
BEN-AMOTZ, A. and GINZBURG, B. Z. (1969). *Biochim. biophys. Acta* **183**, 144–154.
BOGORAD, L. (1967). *In* "Harvesting the Sun" (H. San Pietro, F. A. Greer and T. J. Army, eds) pp. 191–210, Academic Press, New York.
BRINCKMANN, E. and LÜTTGE, U. (1972). *Z. Naturforsch.* **27b**, 277–284.
CHEN, S., MACMAHON, D. and BOGORAD, L. (1967). *Pl. Physiol.* **42**, 1–5.
FEIERABEND, J. (1969). *Planta* **84**, 11–29.
FEIERABEND, J. (1970a). *Planta* **94**, 1–15.
FEIERABEND, J. (1970b). *Z. Pflanzenphysiol.* **62**, 70–82.
FEIGE, B., GIMMLER, H., JESCHKE, W. D. and SIMONIS, W. (1969). *J. Chrom.* **41**, 80–90.
GIMMLER, H. and AVRON, M. (1971). *Z. Naturforsch.* **26b**, 585–588.
HEBER, U. and KRAUSE, G. H. (1971). *In* "Photosynthesis and Photorespiration" (M. D. Hatch, C. B. Osmond and R. O. Slatyer, eds) pp. 218–225. Wiley-Interscience, New York, London, Sydney and Toronto.
HEBER, U. and SANTARIUS, K. A. (1970). *Z. Naturforsch.* **25b**, 718–728.
HELDT, H. W. (1969). *FEBS Letters* **5**, 11–14.
HELDT, H. W. and RAPELEY, L. (1970a). *FEBS Letters* **7**, 139–142.
HELDT, H. W. and RAPELEY, L. (1970b). *FEBS Letters* **10**, 143–148.
HOPE, A. B., LÜTTGE, U. and BALL, E. (1972). *Z. Pflanzenphysiol.* **68**, 73–81.
JESCHKE, W. D. and SIMONIS, W. (1967). *Z. Naturforsch.* **22b**, 873–876.
JESCHKE, W. D. and SIMONIS, W. (1969). *Planta* **88**, 157–171.
KRAUSE, G. H. (1971). *Z. Pflanzenphysiol.* **65**, 13–23.
LÜTTGE, U. (1971a). *In* Proceedings of First European Biophysics Congress, Baden (E. Broda, A. Locker and H. Springer-Leder, eds) Vol. III, Membranes, Transport, pp. 119–123. Verlag der Wiener Medizinischen Akademie, Wien.
LÜTTGE, U. (1971b). *A. Rev. Pl. Physiol.* **22**, 23–44.
LÜTTGE, U., PALLAGHY, C. K. and OSMOND, C. B. (1970). *J. Membrane Biol.* **2**, 1–17.
LÜTTGE, U., BALL, E. and VON WILLERT, K. (1971a). *Z. Pflanzenphysiol.* **65**, 326–335.
LÜTTGE, U., BALL, E. and VON WILLERT, K. (1971b). *Z. Pflanzenphysiol.* **65**, 336–350.
LÜTTGE, U., OSMOND, C. B., BALL, E., BRINCKMANN, E. and KINZE, G. (1972). *Pl. Cell. Physiol.* **13**, 505–514.
NEUMANN, J. and LEVINE, R. P. (1971). *Pl. Physiol.* **47**, 700–704.
PALLAGHY, C. K. and LÜTTGE, U. (1969). *Z. Pflanzenphysiol.* **61**, 58–67.
PALLAGHY, C. K. and LÜTTGE, U. (1970). *Z. Pflanzenphysiol.* **62**, 417–425.
PARK, R. B. and SANE, P. V. (1971). *A. Rev. Pl. Physiol.* **22**, 395–430.
SCHULDINER, S. and OHAD, I. (1969). *Biochim. biophys. Acta* **180**, 165–177.
SCHWARTZ, M. (1968). *Nature, Lond.* **219**, 915–919.
SMITH, F. A. (1971). *In* "Photosynthesis and Photorespiration" (M. D. Hatch, C. B. Osmond and R. O. Slatyer, eds) pp. 302–306. Wiley-Interscience, New York, London, Sydney and Toronto.
TAMAS, I. A., YEMM, E. W. and BIDWELL, R. G. S. (1970). *Can. J. Bot.* **48**, 2313–2317.
THROM, G. (1970). *Z. Pflanzenphysiol.* **63**, 162–180.
THROM, G. (1971a). *Z. Pflanzenphysiol.* **64**, 281–296.
THROM, G. (1971b). *Z. Pflanzenphysiol.* **65**, 389–403.
ULLRICH, W. R. (1971). *Planta* **100**, 18–30.
ULLRICH, W. R. (1972). *Planta* **102**, 37–54.
WOLF, F. T. (1971). *Z. Pflanzenphysiol.* **64**, 124–129.

IV.3

The Role of H⁺ and OH⁻ Fluxes in the Ionic Relations of Characean Cells

F. A. Smith and W. J. Lucas

*Botany Department, University of Adelaide,
Australia*

I. Evidence for H⁺ Extrusion from Characean Cells

A. Electrical Effects of pH Changes

There is good evidence that the electrical characteristics of Characean cells cannot be explained satisfactorily in terms of movements of the major ions K⁺, Na⁺ and Cl⁻ (Spanswick *et al.*, 1967). Kitasato (1968) investigated the electrical role of H⁺ in *Nitella clavata* and suggested that there is rapid inward diffusion of H⁺. This must be balanced by active extrusion in order to maintain neutrality of cytoplasmic pH. Large effects of pH changes on the membrane potential of *N. translucens* were subsequently reported by Spanswick (1970). He showed that the hyperpolarization produced by HCO₃⁻ (see also Hope, 1965) was a pH effect.

In contrast to the above results, Lannoye *et al.* (1970) found that increasing the external pH above 6·0 had little effect on the membrane potential of *Chara corallina*, although the potential depolarized as the pH was lowered below 6·0. They explained these findings in terms of changes of the permeability of the plasmalemma to Na⁺ and K⁺. It should be noted that the fluxes of Na⁺ and Cl⁻ measured by Lannoye *et al.* were very different from those normally found in *C. corallina* (see for example, Findlay *et al.*, 1969). They also used solutions of very low buffering power. With buffered solutions, effects of pH changes on the membrane potential of *C. corallina* even larger than those in *N. translucens* have been found (Smith, unpublished results). The technique used was adapted from that described by Walker (1960), and the results are summarized in Fig. 1. As in *N. translucens*, NaHCO₃

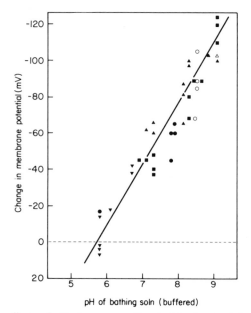

pH of bathing soln (buffered)

FIG. 1. Maximum effects of pH changes on the electrical potential of *Chara corallina*, measured with external electrodes. Cells were held in greased perspex stocks, with one half (the control side) bathed in unbuffered solution. On the other side, unbuffered solutions were replaced by solutions containing sodium dihydrogen phosphate (●), sodium bicarbonate (○), sodium borate (△), imidazole (▲), tris (hydroxymethyl)-methylamine (■) and 2-hydroxyethylpiperazine-2-ethane-sulphonic acid (▼); all at 1–4 mM. Values (measured with an EIL electrometer) show the transmembrane potential in buffered solution relative to that on the control side.

gave the same effect as other buffers at the same pH. With solutions buffered at pH 6·0–7·5 the potential changes (hyperpolarizations) were stable over at least 1 h. However, at higher pH they were of shorter duration: at pH 9 the hyperpolarization decayed after 10–20 min (cf. Spanswick, 1970). Electrical effects of pH changes are described in detail elsewhere in this volume.

B. Effects of Characean Cells on External pH

Direct evidence for H⁺ extrusion in *N. clavata* was provided by Spear *et al.* (1969), who showed that in the light the cell surface is partitioned into areas of acid secretion separated by areas of base secretion. Similar areas occur on *C. corallina* (Smith, 1970). Spear *et al.* suggested that the external alkalinity was due to localized passive H^+ influx, but observations with *C. corallina* Smith, 1970) suggested that alkalinity resulted from localized HCO_3^- nflux, followed by CO_2 fixation and OH^- efflux. This is more in accord with he traditional view of alkalinization of the bulk medium and precipitation of $CaCO_3$ on the cell surface (Arens, 1939). Correlations between rates of HCO_3^- uptake, $CaCO_3$ precipitation and external alkalinization by several Characean species have been discussed previously (Smith, 1968a).

The magnitude of the external pH changes, and their spatial organization have now been measured accurately, using specially constructed pH electrodes (Lucas, 1971; Lucas and Smith, 1973). These had tip diameters of approximately 1·5 mm, making possible measurements of pH values at the cell surface, and at distances 1 mm apart along the cell surface. To prevent dissipation of pH gradients by movement of the bathing solutions the cells were mounted in blocks of dilute agar, with small channels along the cell surface.

It was found that the alkaline areas (pH 8·5–9·5) have sharp peaks, while the acid areas (pH about 5·5) are more uniform. The external pH changes occur rapidly following illumination, and when cells are placed in the dark the pH of the cell surface becomes uniform (about 6·0) after approximately 50 min. These effects are summarized in Figs 2 and 3. With cells free of visible deposits of $CaCO_3$, alkaline regions may appear only temporarily, and sometimes migrate along the cell surface. The presence of $CaCO_3$ stabilizes the alkaline regions. When HCO_3^- and CO_3^{--} were removed from the cell

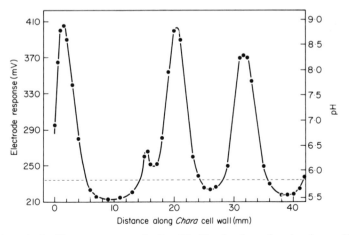

FIG. 2. A typical pH trace, measured after 2 h illumination, showing how pH varies along the cell wall of *Chara corallina*. The broken line indicates the pH of the bathing solution.

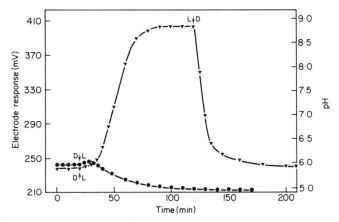

FIG. 3. Changes in pH at an acid (●) and an alkaline (▼) site following transfer from dark to light (D ↑ L), and vice versa (L ↓ D) in the case of an alkaline site.

by prolonged washing the alkaline sites (but not the acid regions) disappeared. They appeared again when low concentrations of $NaHCO_3$ were added to the bathing solution. These results suggested that disappearance or apparent migration occurs when the buffering capacity within the cell wall, provided by HCO_3^- or CO_3^{--} (and subsequent OH^- efflux) is not maintained. In terms of ion transport, it appears that the plasmalemma has numerous sites or carriers capable of exchanging HCO_3^- for OH^-, but that few are operative at any one time.

These results confirm the link between alkalinization and HCO_3^- uptake and agree with the evidence that whereas different Characean species have very different abilities to take up HCO_3^- (Smith, 1968a), their electrical characteristics, which Kitasato (1968) has related to H^+ extrusion, are very similar. If this view is correct, then HCO_3^-/OH^- exchange must be electrically neutral, and independent from the primary H^+ extrusion mechanism. Using results such as those in Figs 2 and 3 it is possible to calculate that the approximate rate of HCO_3^- influx or OH^- efflux is 10–25 pmol cm^{-2} s^{-1} in terms of the alkaline surface area, or about 5 pmol cm^{-2} s^{-1} in terms of the total surface area. This is consistent with measured HCO_3^- uptake values in C. corallina (5–6 pmol cm^{-2} s^{-1}; see Smith, 1968a). Rates of pH change at the acid sites were not studied in sufficient detail to give reliable values for H^+ efflux. Values of 5–20 pmol cm^{-2} s^{-1} for N. clavata were suggested by Spear et al. (1969), while Kitasato (1968) calculated passive H^+ influx values of about 40 pmol cm^{-2} s^{-1} (based on his electrical studies). Rates of H^+ efflux from C. corallina appear to be much smaller than those suggested by Kitasato.

Alkalinization was inhibited both by the uncoupling agent carbonyl cyanide-m-chlorophenylhydrazone (CCCP) and by 3'-(p-chlorophenyl)-1',1'-

dimethylurea (CMU), an inhibitor of non-cyclic photosynthetic electron flow. This shows that it can proceed only in the presence of both photo-synthetic reducing power and a supply of adenosine triphosphate (ATP). However, as alkalinization depends on CO_2 fixation, which also requires both reducing power and ATP, it is not possible to use the inhibitor effects as evidence concerning energy-linked HCO_3^- transport processes at the plasmalemma. CMU also inhibited acidification. Effects of CCCP on H^+ efflux were not studied.

In conclusion, it should be stressed that the experimental system used in these measurements of external pH was artificial in that the bathing solu-tion was stabilized with agar to prevent convective movements. However, in the normal situation, with unbuffered solutions, large pH gradients will still be set up in the cell wall. (This is evident from the fact that $CaCO_3$ becomes deposited on many Characean species under natural conditions.) Measurements of the pH of the bulk solution will only give the balance between net H^+ extrusion, HCO_3^-/OH^- exchange, and changes due to CO_2 levels (i.e. fluctuations resulting from photosynthesis and respiration). This situation must apply to all aquatic plants, with the relative importance of the processes varying in different species and under different conditions.

II. Effects of H⁺ and OH⁻ on Movements of Other Ions

There are a number of ways in which H^+ and OH^- fluxes at the plasma-lemma might interact with other ion movements, whether active or passive. Ionization of molecular components of the plasmalemma could affect rates of diffusion of ions through the membrane, and also the transporting capacity of specific carriers (the conventional ion pumps). More important than this, a primary role for H^+/OH^- separation has been suggested by Spear et al. (1969) and Smith (1970). Spear et al. showed that Cl^- influx into N. clavata was largely confined to the acid regions at the cell surface, and proposed that Cl^- is taken up as HCl, moving into the cell down a pH gradient produced by H^+ extrusion. They pointed out that such a scheme was difficult to reconcile with the hypothesis that Cl^- transport in giant algal cells is closely coupled to influx of K^+ and Na^+ (Raven, 1968; Smith 1967a, 1968b; see also MacRobbie, 1970). However, the considerable difficulties in interpreting the experimental evidence in terms of either chemical or electrical coupling led Smith (1970) to reject the hypothesis and propose instead a scheme similar in some ways to that of Spear et al. It was suggested that Cl^- enters in exchange for OH^-, with H^+ extrusion providing the energy (i.e. the pH gradient). Thus, with H^+ extrusion balanced by influx of K^+ and Na^+, salt accumulation was considered as involving two ion-exchange systems (or antiporters: see Mitchell, 1970) both located in the acid areas of the cell. This scheme is essentially an old one (see references in

Smith, 1970) brought up to date with the suggestion that the H^+ extrusion mechanism might be a Mitchell-type ATPase. Alternatively, evidence from some algal species (not including C. *corallina*) that Cl^- transport is independent of ATP (see MacRobbie, 1970) would implicate vectorially orientated redox reactions in the plasmalemma.

The important feature of this scheme, like that of Spear *et al.*, is that metabolic effects on Cl^- transport would be mediated by changes in the pH gradient across the plasmalemma. These could be brought about not only by changes in the rate of H^+ extrusion, but also by any of the many other metabolic processes which control cytoplasmic pH. These latter are discussed by Raven and Smith (p. 271). This type of scheme provides a likely explanation of the large increases in light-promoted Cl^- transport in Characean cells when low concentrations of $(NH_4)_2SO_4$ or the amine imidazole are added to the bathing solution. It was initially suggested (MacRobbie, 1965, 1966) that these compounds uncoupled photophosphorylation and increased photosynthetic electron flow, which was regarded as the energy source for Cl^- transport. However, the evidence that photosynthesis *in vivo* was not uncoupled by $(NH_4)_2SO_4$ or imidazole (Smith, 1967b; Smith and West, 1969) left the effects with no obvious biochemical explanation in terms of any hypothesis which includes the usual assumption that light-promoted Cl^- transport is limited by the supply of a chemical intermediate from photosynthesis. The suggestion (Smith, 1970) that the effect of $(NH_4)_2SO_4$, imidazole and other amines (such as benzimidazole and tris) is to increase the pH gradient across the plasmalemma is in accord with the known properties of amines, and their effects on ion transport in mitochondria (Chappell and Haarhoff, 1967). Experimental evidence suggests that uptake of the neutral amine $(NH_3, RNH_2, etc.)$ is followed by protonization in the cytoplasm.

Transport of Cl^- into C. *corallina* was shown to be reduced when the external pH was buffered above 7·0 (Smith, 1970). Though this is consistent with a dependence of Cl^- transport on the pH gradient across the plasmalemma, it can be explained in other ways (see above). MacRobbie (1970) has criticized the scheme on the grounds that Cl^- transport should be even more pH-dependent than is the case, but it should be remembered that an artificial increase in external pH will be countered by H^+ extrusion, so that the pH at the cell membrane would be lower than that in the bulk medium. A second criticism by MacRobbie (1970) is that anion transport linked to H^+ or OH^- fluxes would leave the pH of the cell at the mercy of the external anion concentration. This criticism is not valid if H^+ extrusion is considered as a mechanism for regulating cytoplasmic pH, and hence itself affected by changes in cytoplasmic pH (Raven and Smith, this volume: IV.6).

Pre-treatment of C. *corallina* in the dark at high pH (9·0–9·3) followed by transfer to low pH (about 5·3), still in the dark, greatly increases Cl^- influx (Smith, 1971, 1972). This is shown in Table I. The effect is increased by the

TABLE I. Effects of pH on Cl$^-$ influx into *C. corallina* in darkness, following pre-treatment in darkness

Pretreatment conditions (3 experiments; all at 24°C)	Experimental conditions (Cl$^-$ influx in pmol cm^{-2} s^{-1})	
	High pH (9·0 or 9·3)	Low pH (5·3)
pH 9·0 (borate)	0·14 ± 0·02	1·33 ± 0·13
Unbuffered + Na$_2$SO$_4$	—	0·23 ± 0·02
pH 9·0 (borate) + (NH$_4$)$_2$SO$_4$	0·04 ± 0·01	3·35 ± 0·34
pH 5·3 (HEPES) + Na$_2$SO$_4$ + (NH$_4$)$_2$SO$_4$	—	0·30 ± 0·05
pH 9·0 (borate)	0·09 ± 0·02	0·91 ± 0·19
pH 9·3 (tris)	0·03	0·80 ± 0·12
pH 9·3 (tris) + (NH$_4$)$_2$SO$_4$	0·02	3·27 ± 0·39
pH 5·3 (TES) + (NH$_4$)$_2$SO$_4$	—	0·11 ± 0·02

Influx into cells maintained at high pH is compared with influx into cells transferred to solutions at low pH. Effects of pre-treatment in unbuffered solutions or solutions of low pH are also shown. Buffers were: sodium borate at 5 mM; tris(hydroxymethyl)methylamine (tris), tris(hydroxymethyl)methyl-2-amino-ethane-sulphonic acid (TES), and 2-hydroxyethylpiperazine-2-ethanesulphonic acid (HEPES), all at 10 mM. Na$_2$SO$_4$ was 5 mM, and (NH$_4$)$_2$SO$_4$ 0·15 mM. Cl$^-$ influx was measured over a 30 min period, following pre-treatment for 2 h. Data from Smith (1972).

addition of (NH$_4$)$_2$SO$_4$, but is not obtained when cells are pre-treated in the dark at low pH. A large increase is also produced when cells are pre-treated in light at high pH and then transferred to the dark at low pH, thus reversing the normal decrease in Cl$^-$ influx when cells are transferred from light to dark. The simplest explanation of these results is that the pre-treatment at high pH increases cytoplasmic pH and that this is aided by uptake of NH$_3$ where present). During the subsequent period at low pH the artificially large pH gradient across the plasmalemma would allow rapid Cl$^-$ influx. At present the duration of the stimulation is unknown, and there is no direct evidence that the buffering capacity of the cytoplasm is sufficient to maintain a high internal pH after transfer to low external pH. However, the results are extremely difficult to explain with the conventional assumptions that the limiting factor for Cl$^-$ transport is a chemical metabolic intermediate, the supply of which is usually much greater in light than in darkness.

It would be expected that the effects of pH on cation uptake would be the reverse of those on anion uptake, but so far little work has been done with giant algal cells. The addition of HCO$_3^-$ greatly increased influx of Na$^+$

and K^+ into *N. translucens* (Smith, 1965). This can be interpreted as a response to the hyperpolarization caused by the external pH change (see Spanswick, 1970).

If salt accumulation is linked to fluxes of H^+ and OH^- (at the acid sites) then changes in the ionic composition of the bathing solution should affect the pH at the cell surface. Removal of Cl^- in fact decreases the pH of the acid regions (Lucas, unpublished results). This would be expected if the pH gradient produced by H^+ extrusion is not dissipated by Cl^-/OH^- exchange or (according to Spear *et al.*, 1969) HCl uptake. Removal of Cl^- leaves the alkaline regions apparently unaffected. Further interactions between external ions and external pH values are at present under investigation.

III. Conclusions

The conventional approach to ion transport systems in giant algal cells has been to think of them as "pump and leak" systems. Although the possible links between the pumps and their metabolic energy supply have been investigated, little attention has been paid to possible feedback from changes in internal ion levels, or changes in membrane potentials. Passive (downhill) ion fluxes have been considered as completely independent from the active (uphill) fluxes, or at the most linked indirectly via electrical effects. In contrast, according to the concepts put forward by Mitchell (1966, 1968, 1970) and adopted in this paper, ion fluxes can be closely coupled, with the flux of one ion species establishing an energy gradient for the movement of another.

The question as to how much of the energy gradient set up by H^+ extrusion is dissipated by passive H^+ influx (i.e. not linked to any other ion flux) remains unresolved. It may be noted that Kitasato's proposal that the plasmalemma is extremely permeable to H^+ has been questioned by Walker and Hope (1969).

This paper has been limited to discussion of ion transport across the plasmalemma. In Characean cells the rates of transport across the tonoplast are vastly greater than those across the plasmalemma, and MacRobbie's hypothesis of ion transport via vesiculation in the cytoplasm (MacRobbie, 1969, 1970) is an attractive one. Nevertheless, there must be control of pH gradients at the tonoplast, and it is possible that systems similar to those described above (though presumably vectorially the reverse) might also be located at this membrane. The role of H^+ and OH^+ fluxes in ion transport is discussed in a much wider context elsewhere in this volume (IV.6).

IV. Acknowledgements

The experimental work with *C. corallina* was partly financed by the Australian Research Grants Committee. The award of a travel grant from the University of Adelaide to F. A. Smith is gratefully acknowledged.

V. References

ARENS, K. (1939). *Protoplasma* **33**, 295–300.

CHAPPELL, J. B. and HAARHOFF, K. N. (1967). *In* "Biochemistry of Mitochondria" (E. C. Slater, Z. Kaniuga and L. Wojtczak, eds) pp. 75–91. Academic Press, New York.

FINDLAY, G. P., HOPE, A. B., PITMAN, M. G., SMITH, F. A. and WALKER, N. A. (1969). *Biochim. biophys. Acta* **183**, 565–576.

HOPE, A. B. (1965). *Aust. J. biol. Sci.* **18**, 789–801.

KITASATO, H. (1968). *J. gen. Physiol.* **52**, 60–87.

LANNOYE, R. J., TARR, S. E. and DAINTY, J. (1970). *J. exp. Bot.* **21**, 543–551.

LUCAS, W. J. (1971). pH Banding along the cell wall of *Chara corallina*. B.Sc. Honours Thesis, University of Adelaide.

LUCAS, W. J. and SMITH, F. A. (1973). *J. exp. Bot.*, in press.

MACROBBIE, E. A. C. (1965). *Biochim. biophys. Acta* **94**, 64–73.

MACROBBIE, E. A. C. (1966). *Aust. J. biol. Sci.* **19**, 363–370.

MACROBBIE, E. A. C. (1969). *J. exp. Bot.* **20**, 236–256.

MACROBBIE, E. A. C. (1970). *Q. Rev. Biophys.* **3**, 251–294.

MITCHELL, P. (1966). *Biol. Rev.* **41**, 445–602.

MITCHELL, P. (1968). "Chemiosmotic Coupling and Energy Transduction". Glynn Research, Bodmin.

MITCHELL, P. (1970). *Symp. Soc. Gen. Microbiol.* **20**, 121–166.

RAVEN, J. A. (1968). *J. exp. Bot.* **19**, 233–253.

SMITH, F. A. (1965). Ph.D. Thesis, University of Cambridge, England.

SMITH, F. A. (1967a). *J. exp. Bot.* **18**, 716–731.

SMITH, F. A. (1967b). *J. exp. Bot.* **18**, 509–517.

SMITH, F. A. (1968a). *J. exp. Bot.* **19**, 207–217.

SMITH, F. A. (1968b). *J. exp. Bot.* **19**, 442–451.

SMITH, F. A. (1970). *New Phytol.* **69**, 903–917.

SMITH, F. A. (1971). Proceedings of the First European Biophysics Congress (E. Broda, A. Locker and H. Springer-Lederer, eds) Vol. 3, pp. 429–433. Verlag der Wiener Medizinischen Akademie, Vienna.

SMITH, F. A. (1972). *New Phytol.* **71**, 595–601.

SMITH, F. A. and WEST, K. R. (1969). *Aust. J. biol. Sci.* **22**, 351–363.

SPANSWICK, R. M. (1970). *J. Membrane Biol.* **2**, 59–70.

SPANSWICK, R. M., STOLAREK, J. and WILLIAMS, E. J. (1967). *J. exp. Bot.* **18**, 1–16.

SPEAR, D. G., BARR, J. K. and BARR, C. E. (1969). *J. gen. Physiol.* **54**, 397–414.

WALKER, N. A. (1960). *Aust. J. biol. Sci.* **13**, 468–478.

WALKER, N. A. and HOPE, A. B. (1969). *Aust. J. biol. Sci.* **22**, 1179–1195.

Discussion

Lilley asked Barber if he had used his technique on cells. It also seemed important to test the effects of metabolic inhibitors on the giant algal cells by using them on chloroplast preparations. Barber replied that if Lilley meant measurement of membrane potential, he though it was impossible. Millisecond-delayed light emission and the effects of KCl pulses are found only in broken chloroplast preparations; in intact cells the KCl pulse effect is not seen. On the other hand, millisecond light emission is a valuable technique for examining the effects of compounds which in intact systems are reported to affect the high energy state of the system. West then asked Barber if his assumption, that in the presence of valinomycin the potential is zero in the light, might not have been better tested in the presence of CCCP. Barber agreed that if the origin of the potential were an electrogenic H^+ pump and if K^+ and all other ions were in equilibrium, then valinomycin was the wrong compound. One ought to use something which increased H^+ permeability and CCCP or gramicidin would be appropriate. In retrospect Barber felt that at the levels of valinomycin used, which incidentally do affect phosphorylation, H^+ permeability is also increased and this may be the reason why the potential is at such a low level. On the other hand it is also possible that the K^+ equilibrium is affected and that a putative K^+–H^+ exchange system has been disturbed.

Walker then asked if he had tried the effect of nigericin which is known to affect K^+–H^+ exchange. Barber replied that he tends to the idea that membrane potential can reduce the activation energy for the emission process and therefore he has tried to avoid situations where pH gradients develop, because pH is also known to affect the emission process. Nigericin in fact strongly inhibits millisecond-delayed light emission, possibly because there is a component which is pH dependent; it also inhibits the salt-induced signals. Walker then re-phrased the question to ask Barber whether he was sure that the valinomycin state was indeed a zero potential state. Barber then agreed that he had to make two arbitrary choices in (i) the zero potential point and (ii) the peak potential point. He added that gramicidin much reduced delayed light emission; this may be due to an endogenous component which is affected by pH. The zero potential point can be only slightly reduced

232

by gramicidin rather than valinomycin; in the presence of electron acceptors, on the other hand, there is a much more complex situation.

Dainty then asked Barber in what manner he thought he was measuring membrane potential with this system. Furthermore is it a well-known physical effect that a charge-transfer molecule would be so oriented in an electric field of the type that exists here, that diffusion across the membrane would be accelerated as is suggested? Barber said that his interpretation is based on experimental evidence; when the chloroplast preparation is subjected to a salt concentration that would create an electric field of the correct polarity, light emission is greatly increased. Dainty then asked how those who do not agree with the Mitchell hypothesis would explain the data. "Are you not taking the Mitchell hypothesis on trust?" Barber vigorously denied this; his work is quite independent of Mitchell's scheme. His interpretation does not at all arise from the Mitchell scheme; his observations simply happen to give some support to the idea that light emission is affected by membrane potential. Neumann then commented, in regard to charge separation, that there is a close similarity between delayed light emission and fluorescence which is known to be due to the primary charge separation in photosynthesis. The action spectra of delayed light emission and fluorescence are similar and one can therefore conclude that a similar charge separation is operating. Barber said further that others held that delayed light emission is dependent on two parameters: membrane potential and pH gradient. There is a close correlation, Barber concluded, between the build-up of millisecond-delayed light emission and the ability of the system to conduct phorphorylation.

In discussion of Lüttge's paper, Raven asked for how long the H^+ influxes were measured; if nitrate were the electron acceptor, one would expect OH^- production as nitrate is reduced to ammonia. This may then account for some of the observed H^+ uptake which is not connected with CO_2 fixation. Lüttge said that they did not use nitrate as the electron acceptor. The light to dark periods lasted 5 min each; the external solution was buffered and the pH change was negligible. Cram then asked Lüttge which Cl^- was measured, the plasmalemma flux or the tonoplast flux, or a combination of both. Lüttge said that they had no experiments to show at which membrane the Cl^- flux occurred. Cram asserted, "You imply that you are measuring a plasmalemma flux on the diagram at the end of the paper". Lüttge denied this; he said that the electrons can be used equally well if the pumps are at the tonoplast. Cram then asked if section thickness made any difference—Lüttge had presented evidence that 1 and 2 mm thicknesses made no difference—"Do other thicknesses affect the uptake?", Cram asked, "and is there an effect of infiltrating the leaf slices?". Lüttge replied that there were no measurable differences. Smith then asked about the excess reducing power; he said that after 4 h there is an excess equivalent to 25 mM reduced NADPH so that the equivalent of 25 equiv H^+ g^{-1} must be removed from the tissue. Lüttge replied that there was nothing he could say other than that he too was worried about the apparent excess of reducing power. The CO_2 experiments had been repeated many times in an attempt to establish CO_2 fixation which would take up reducing power. Smith then asked which ion balanced the net H^+ uptake. Lüttge said that he did not yet know which ion accompanied the H^+.

Neumann then offered information on the relationship between electron transport coupled H^+ movements in chloroplasts and similar phenomena in intact cells. He gave the example of the greening of a mutant *Chlamy-domonas* where various parameters can be followed during the greening process. He concluded that the evidence showed the pH gradient in the light to be caused by a H^+ pump in the chloroplasts; CO_2 fixation was excluded on the basis of various gassing experiments. He then stressed that pH changes are non-specific; not only are they involved in CO_2 fixation but also in ferricyanide reduction and in ferredoxin reduction. He showed an example where the whole photosynthetic apparatus of *Chlamydomonas* was function-ing, as measured by O_2 evolution, but there was no external pH change. He then suggested a test for Lüttge's system; if the observed pH gradient in the dark is caused by respiration, the addition of *p*-benzoquinone which will inhibit respiration should remove the gradient. Lüttge replied that he was well aware of the work described by Neumann and had in fact discussed it in his paper. It seemed to him that there may be different systems in higher plants and algae. Lilley then commented that he had similar results with *Griffithsia* in the presence of CCCP to those given by Lüttge. He said that he had used an artificial sea water with zero nitrate so that Raven's earlier comment did not apply. However there is sulphate in the medium and he wondered if sulphate might be the terminal electron acceptor. Lüttge said finally that if the shuttle from the chloroplast to the cytoplasm, in his final diagram, were in fact shuttling in the dark, then the H^+ so released in the cyto-plasm might well be taken up and utilized as reducing power in the Krebs cycle.

Turning to Smith's paper, Findlay asked why it was that Smith had revived this old-fashioned method of measuring membrane potential. Smith replied that his method gives similar results to the more modern methods of inserting electrodes into cells and has the great advantage that the cells are left un-damaged. Walker then asked if Smith found similar results using both old-fashioned buffers, i.e. buffers which penetrate, and more modern non-penetrating buffers. Smith said that the stable hyperpolization up to pH 7·5 is found whatever buffer is used. The peak hyperpolarization and subsequent decline are also found irrespective of buffer. Barr then asked if it was correct that the dark–light stimulated Cl^- influx decayed in about 1 h. Smith said that this was so, but the stimulation was in fact from high to low pH; the effect of the pre-treatment disappears after about 1 h. Barr then asked if Smith had made a calculation of the total reserve Cl^- capacity in the cell. Smith said that he had not, but intended to do so.

MacRobbie then commented that if Smith was correct that the stimulated Cl^- flux was due to a high pH in the cytoplasm, then one should detect a depolarization of the membrane potential which would decay with the same time course as the stimulated Cl^- influx; a similar effect ought to be found after pre-treatment with ammonia as well as with high pH. Smith replied that he did not accept that these pre-treatments should depolarize the membrane. MacRobbie retorted that the argument was that pre-treatment increased internal pH and that membrane potential was pH sensitive; therefore the pre-treatment should depolarize the membrane. Smith said that he had only measured cytoplasmic pH immediately after changing the external pH, and there is little change presumably because of natural buffering by organic anions; he had no information on the time course. Walker then commented that the time course of Cl^- flux decay after the pH change is remarkably similar to the decay in Cl^- influx after a period in the light. Raven then said that he had done similar experiments using rather higher reference pH values and found rather different effects. Smith replied that he used pH 5·7 as his reference because this is the value the cell would naturally attain in the light at its acid sites.

Vredenberg then asked whether in these cells with acid and base bands there will be a spatial distribution of types of electrogenic pump with membrane potential profiles along the length of the cell. Smith replied that one should remember that the alkaline bands are not permanent. A young cell with no $CaCO_3$ deposits will develop bands if kept in the light. These bands may stabilize, they may move along the cell or they may disappear. The alkaline regions possibly stabilize when the cell wall pH becomes sufficiently high to give a stable bicarbonate concentration there; this then becomes a bicarbonate pumping site. The pumping stations must be spatially localized; the acid bands are pumping Cl^- and the alkaline bands are pumping bicarbonate. Exactly what this means in terms of voltages and currents is unclear; e.g. the bicarbonate flux is about 25 pmol $cm^{-2} s^{-1}$ while the Cl^- flux is 120 pmol $cm^{-2} s^{-1}$.

Spanswick then commented on the method of changing pH used by Smith. He said that in *Nitella* a change from reference pH 5 to a higher value gives a peak potential pulse. If the same overall pH change is made step-wise the curve is flattened out. Smith then said he would like to comment on a point which had been too often forgotten. H^+ extrusion is partly compensated by CO_2 fixation in green tissue in the light, and it is difficult to estimate H^+ fluxes because of this effect. However he was sure from his own measurements that H^+ efflux is not as high as 100 pmol $cm^2 s^{-1}$.

Cram then asked Smith how the acid and alkali bands originate and he replied that when the light is switched on two processes occur; first there is CO_2 uptake and second there is primary H^+ extrusion. If the H^+ extrusion is in excess in a region, that region will become acid. If the CO_2 uptake in a region is vigorous, the local pH will rise to a level where bicarbonate is present in the cell wall phase. This bicarbonate will then take part in bicarbonate–OH^- exchange and when the local pH exceeds pH 9, $CaCO_3$ will be deposited. Again Smith stressed that the situation is not stable and the bands do move; the more bicarbonate in the external solution, the more the alkali bands predominate on the cell.

Bentrup then asked if Smith could measure which ion is being transported longitudinally along the cell by circuiting a current around. Walker said that that experiment has been done by Barr, by Thain and by Hope and Coster. They did not look for ions other than H^+ because it was thought that the other fluxes had been properly located by Hope and Walker with voltage clamping. They had been examining whether H^+ carried that fraction (50–75%) of the current which cannot be located. In all three laboratories the experiment failed to show any net movement of H^+. Smith replied that one should bear in mind that a cell reacts in such a way as to maintain the status quo. That is why the bathing medium of illuminated cells does not continuously increase in acidity; in the experiments just described by Walker the pH was not measured close to the cell walls. MacRobbie added that measurements of Na^+ and K^+ fluxes in half cells showed that they did not behave in the manner expected of the current-carrying ions. It is important that we should try to measure Cl^- fluxes in a current-carrying system.

IV.4

Proton Movements Coupled to the Transport of β-Galactosides into *Escherichia coli*

I. C. West

*Glynn Research Laboratories,
Bodmin, Cornwall, England*

I. Introduction

The transport of β-galactosides into *Escherichia coli* is genetically determined by the *y* gene of the lactose operon. The protein specified by this gene has been partially isolated, and is estimated to have a molecular weight of around 31,000. (For recent reviews see Kepes, 1971; Kaback, 1970; Kennedy, 1970.) In metabolically active cells it mediates the accumulation of β-galactosides (provided they are not metabolized), but in the presence of azide (an uncoupler of oxidative phosphorylation) and iodoacetate (an inhibitor of glycolysis) the same system appears to catalyse facilitated diffusion, resulting in an equal concentration of galactoside inside and outside the cell (Winkler and Wilson, 1966). Winkler and Wilson also showed that azide and iodoacetate greatly lowered the apparent Michaelis constant for exit of β-galactoside until it was the same as that for entry of the sugar, suggesting that the metabolically-induced accumulation is due, at least in part, to a

237

reduced affinity of the outgoing carrier for internal sugar. However, there was no indication from those experiments as to the chemical nature of the linkage between metabolism and the accumulation of β-galactosides.

The possible involvement of ATP in a primary active transport system was carefully examined. Scarborough et al. (1968) showed that specially washed cells (presumed to be depleted of ATP) would hydrolyse o-nitrophenyl-galactoside faster if treated with ATP, but this was not an effect on the affinity constant, and therefore probably not related to the phenomenon analysed by Winkler and Wilson. Evidence against the involvement of ATP in the accumulation of β-galactosides was provided by Pavlasova and Harold (1969), who showed that uncouplers of oxidative phosphorylation could almost completely abolish the accumulation of β-galactosides while causing little or no reduction in the cellular concentration of ATP. This important result forced one to seek some chemical compound or group other than ATP as the link between metabolism and the accumulation of β-galactosides, and to examine the effects of uncouplers other than that of inhibiting oxidative phosphorylation.

The role of Na^+ ion gradients in the concentrative transport of sugars and amino acids in animal tissues was becoming increasingly well documented (Stein, 1967). However, Stein concluded (Kolber and Stein, 1966; Stein, 1967) that Na^+ and K^+ ions were not involved in the uptake of β-galactosides into E. coli by a coupled transport system, because replacing KCl in the suspension medium with sucrose or NaCl had too small an effect on the kinetic constants of o-nitrophenylgalactoside transport to account for the high concentration gradient which can be established. Mitchell (1967) then pointed out that his earlier suggestion (Mitchell, 1963) that H^+ ions were carried inwards with the β-galactoside by a symport (or co-transport) system, i.e. that H^+ ions constituted the link between metabolism and the accumulation of β-galactosides, had still not been tested.

This symport hypothesis is shown diagrammatically in Fig. 1 (taken from Mitchell, 1963). Mitchell had already postulated (Mitchell, 1961a) the H^+ ion-translocating function of the respiratory chain shown in the lower part of Fig. 1, and had shown that uncouplers of oxidative phosphorylation (e.g. dinitrophenol) catalyse the net movement of H^+ ions across lipid membranes (Mitchell, 1961b). Therefore, when Kepes (1960) and Horecker et al. (1961) showed that dinitrophenol caused the rapid efflux of accumulated β-galactosides from respiring bacteria, it was logical to postulate the β-galactoside–H^+ symport shown in the top part of Fig. 1; the equilibration of the electrochemical potential of H^+ ions via dinitrophenol would lead to the equilibration of the electrochemical potential of β-galactoside if it were the previous disequilibration of H^+ ions that led to the accumulation of β-galactoside. A direct coupling between the inflow of H^+ ions and β-galactoside molecules was one possible mechanism (the simplest) by

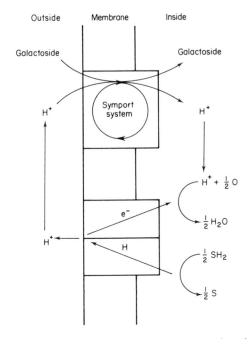

FIG. 1. Diagram of cyclic coupling between the electron-translocating system thought to be present in the plasma-membrane of *E. coli*, and a β-galactoside–H$^+$ symport system which, it is suggested, may translocate β-galactosides into the cell. (From Mitchell, 1963.)

which the electrochemical potential gradient of H$^+$ ions could lead to β-galactoside accumulation.

The experiments described in this paper, in part already published (West, 1970; West and Mitchell, 1972a), were designed to test the suggestion of Mitchell, that the net transport of β-galactosides into *E. coli* is coupled to the net transport of H$^+$ ions in the same direction; i.e. that the β-galactoside transport system is a β-galactoside–H$^+$ symporter. However, as with the phosphate transport system of the mammalian mitochondrion, it cannot be decided on the present evidence whether the system is a β-galactoside–H$^+$ symporter, or a β-galactoside–OH$^-$ antiporter, because the great permeability of biological membranes to H$_2$O and the rapid equilibration of the H$_2$O $\rightleftharpoons$ H$^+$ + OH$^-$ system makes it very difficult to distinguish between the movement of H$^+$ in one direction and the movement of OH$^-$ in the opposite direction. In this paper the movement of H$^+$, and the movement of OH$^-$ in the opposite direction, from which it cannot here be distinguished, will both be described in terms of the "effective" movement of H$^+$.

II. Experimental Results

A. Acid–Base Changes During Galactoside Transport

1. Extracellular

Cells of *E. coli* strain ML 308-225 (constitutive for β-galactoside transport but lacking β-galactosidase) were suspended anaerobically at pH 7·2 in a medium containing: 1 mM tris buffer, 1 mM sodium iodoacetate, 150 mM NaCl. The presence of iodoacetate and the absence of oxygen prevented glycolytic and respiratory reactions from causing outward H^+ ion movements. The addition of an anaerobic solution of lactose caused an alkalification of the extracellular medium measured with an H^+ ion-sensitive glass electrode (Fig. 2). This experiment was repeated with a strain (ML 35) lacking a

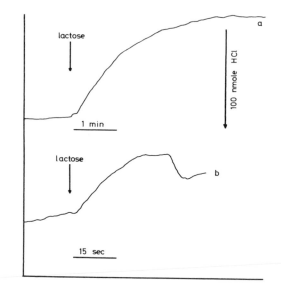

Fig. 2. Strip-chart recordings of pH electrode responses showing decrease in H^+ ion activity in the outer medium on adding lactose to anaerobic suspensions of *E. coli* ML 308–225 in (a) 150 mM NaCl, or (b) 50 mM NaSCN, 150 mM NaCl. (From West, 1970.)

functional *y* gene product, and with suspensions of ML 308-225 treated with formaldehyde or with N-ethylmaleimide (both of which prevent lactose transport), and in each case there was no alkalification of the outer medium on the addition of lactose. It was concluded that the transport of lactose into the cell caused the disappearance of H^+ ions from the outer medium.

2. Intracellular

It was important to show that the alkalification of the outer medium was not due to a general sequestering of H^+ ions inside the cells, but was due to their effective translocation inwards across the membrane, by showing that the intracellular medium became acidic when the outer medium went alkaline. The technique for the measurement of the internal pH was based on that of Kashket and Wong (1969) using the weak acid 5,5-dimethyl-oxazolidine-2,4-dione (DMO*), which penetrates the cells much more rapidly in the protonated than in the deprotonated form. In calculating the internal pH it has been assumed: (1) that the acid dissociation constant of DMO is 4.79×10^{-7} ($pK_a = 6.32$), (2) that the protonated form penetrates the cell membrane sufficiently rapidly to be always at equilibrium, (3) that the deprotonated form does not penetrate at all, (4) that the volume of cell water is $2.7 \ \mu l \ mg^{-1}$ cell dry weight (West, 1970).

Cells (20 mg dry weight) were washed in 10 ml of a solution containing 100 mM NaCl, 10 mM K_2HPO_4, 1 mM iodoacetic acid and HCl to bring the pH to 7.0, and were resuspended in 1.5 ml of the same solution. Nitrogen gas was bubbled gently through the suspension for 20 min, and then 25 μl of [^{14}C]DMO solution was added. Samples (50 μl) were taken at intervals,

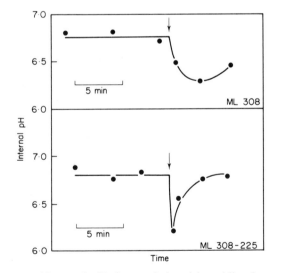

FIG. 3. Time-course of internal pH changes induced by adding lactose to anaerobic suspensions of *E. coli*, estimated from the distribution of DMO. (See text for details.) Lactose added at arrows. Upper figure using strain ML 308 (β-galactosidase-positive), lower figure using strain ML 308-225 (β-galactosidase-negative).

* Abbreviations: DMO, 5,5-dimethyloxazolidine-2,4-dione; FCCP, carbonylcyanide *p*-tri-fluoromethoxyphenylhydrazone.

filtered on Millipore filters, and washed quickly with 5 ml of solution having the same composition and temperature (26°) as the initial suspension medium. The filters were dried, and the ^{14}C estimated in a liquid scintillation counter. After 15 min incubation, 2·5 μl of anaerobic 1·0 M lactose solution was added and sampling continued. The external pH was checked with a glass electrode, and samples of the whole suspension were dried and their ^{14}C content estimated as before.

It can be seen in Fig. 3 that lactose addition did cause an acidification of the cell interior. With the strain possessing β-galactosidase activity (ML 308) the fall of internal pH was of longer duration than with the strain lacking β-galactosidase activity (ML 308-225), the difference presumably being due to the continued net inflow of lactose which is possible when lactose is constantly being removed from the inside of the cell by hydrolysis. Similarly, the alkalification of the external medium was larger and of longer duration in the β-galactosidase-positive than in the β-galactosidase-negative strain (West, 1970).

From these experiments it is concluded that the inflow of β-galactoside into E. coli is coupled to the inflow of acid equivalents across the cell membrane.

B. Electrogenicity of Galactoside Transport

It has been suggested (West, 1970) that the inflow of β-galactoside is an electrogenic process, when it was shown that the osmotic swelling caused by lactose penetration was accelerated by replacing Cl$^-$ with the more rapidly penetrating SCN$^-$ ion. It was argued that the build-up of positive charge in the bacteria caused by the effective inflow of H$^+$ ions could be more rapidly discharged by the inflow of SCN$^-$ than by the slower inflow of Cl$^-$. This suggestion has been reinforced by the following more direct measurements (West and Mitchell, 1972a, b).

Following Pavlasova and Harold (1969), valinomycin was used to make cells permeable to K$^+$ ions after first treating the cells with tris and EDTA to increase their sensitivity to this K$^+$-conducting antibiotic. These K$^+$-permeable cells were suspended in 200 mM sucrose with 30 mM choline chloride (to provide ionic strength) and KCl to give a final K$^+$ concentration of 0·25 mg ions K$^+$/litre. Glass electrodes were used to follow changes of both H$^+$ ion and K$^+$ ion activities, in experiments similar to those of Fig. 2. It can be seen in Fig. 4 that the addition of β-galactoside caused an outward movement of K$^+$ ions synchronous with the effective inward movement of H$^+$ ions, and by correction for baseline drift and the difference in scale, it can be seen (Fig. 5) that the quantity of K$^+$ ion translocated outwards almost exactly equalled the quantity of H$^+$ ion effectively translocated inwards. If the K$^+$ is taken to be at electrochemical equilibrium, the movement of that

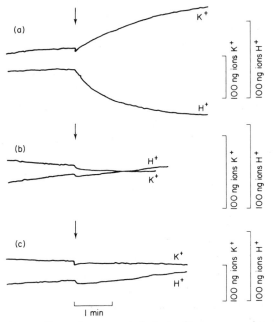

FIG. 4. Strip-chart recordings of uncorrected electrode responses showing increase in K^+ ion and decrease in H^+ ion activities in the outer medium on adding thiomethyl-galactoside to anaerobic suspensions of K^+-permeable *E. coli* ML 308-225. The temperature-controlled experimental vessel, volume 4·0 ml, contained: 24·3 mg cell dry wt, 200 mM sucrose, 30 mM choline chloride, 0·25 mM KCl, 0·1 mg carbonic anhydrase and further additions as indicated. After anaerobic incubation at 25° and at pH 7·0–7·1 for 35 min, 50 μl of anaerobic 0·4 M thiomethylgalactoside solution was injected (at arrows). (a), No further additions; (b), N-ethylmaleimide, 0·5 mM final concentration; (c), FCCP, 2·5 μM final concentration. (From West and Mitchell, 1972b.)

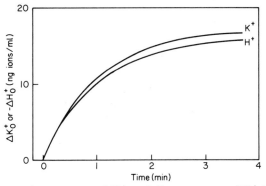

FIG. 5. Time-course of appearance of K^+ and disappearance of H^+ ions in the outer medium on adding thiomethylgalactoside to an anaerobic suspension of *E. coli*. The continuous recordings of Fig. 4 have been corrected for baseline drift and brought to the same scale. (From West and Mitchell, 1972b.)

ion demonstrates that the translocation of acid equivalents caused by the presumed β-galactoside inflow is not electrically neutral, but corresponds to the translocation of singly charged ions; either H^+ ions inwards or OH^- ions outwards.

The movement both of H^+ ions and of K^+ ions caused by β-galactoside addition is inhibited by the uncoupler FCCP (Fig. 4 curves (c)), and by inhibiting β-galactoside entry with N-ethylmaleimide (Fig. 4 curves (b)). The uncoupler presumably acts by allowing the H^+ ions effectively transported inwards with the β-galactoside to return immediately to the outer medium, while the effect of N-ethylmaleimide implies that it is the inward transport of β-galactoside that causes the observed ion movements, and not merely the addition of β-galactoside to the outer medium.

C. Respiratory H^+ Ion Movements

It is now widely accepted that the reduction of oxygen by the respiratory chain of intact mitochondria and aerobic bacteria causes the electrogenic ejection of H^+ ions. By employing techniques similar to those used by Scholes and Mitchell (1970) on *Micrococcus denitrificans*, the ejection of H^+ ions during respiratory pulses has also been observed from *E. coli*. The acidification of the medium caused by injecting a pulse of air-saturated KCl solution into an anaerobic suspension of *E. coli* is shown in Fig. 6. The outward

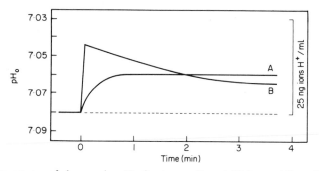

FIG. 6. Time-course of changes in pH of outer medium (pH_o) on adding air-saturated KCl solution to anaerobic suspensions of normal *E. coli* ML 308-225. The experimental vessel contained, in 4·0 ml: 9·8 mg cell dry wt, 1·5 mM glycylglycine, 0·1 mg carbonic anhydrase and either (A) 150 mM KCl, or (B) 100 mM KCl, 50 mM KSCN. After anaerobic incubation at 25° and at pH 7·0–7·1 for 40 min, 50 μl of KCl solution (air-saturated, CO_2-free) was injected at zero time. (From West and Mitchell, 1972b.)

movement of H^+ appears to be impeded by the negative charge which builds up on the cells, for the presence of the freely-permeating SCN^- ion greatly increased the rate of H^+ ejection (Fig. 6). Under the conditions of Fig. 6,

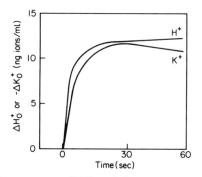

IG. 7. Time-course of appearance of H^+ and disappearance of K^+ ions in the outer 1edium on adding air-saturated KCl solution to an anaerobic suspension of K^+-ermeable *E. coli* ML 308-225. Suspension and preincubation as in Fig. 4. At zero time 0 μl of air-saturated, CO_2-free, solution containing 200 mM sucrose and 30 mM choline hloride was injected. (From West and Mitchell, 1972b.)

1e small quantity of added oxygen was very rapidly consumed (in less than s; confirmed with an oxygen electrode) and the external pH therefore eturned gradually, with simple first-order kinetics, back to the original alue. The half-time of this exponential decay of the pH gradient is a charac-:ristic of the rate of diffusion through the membrane of H^+ and OH^- ions, nd any other basic or acidic group which can combine reversibly with H^+ r OH^- ions.

The electrogenicity of the respiratory ejection of H^+ ions was confirmed y repeating the oxygen-pulse experiments with K^+-permeable cells sus-ended in a sucrose/choline chloride medium as in the experiment of Fig. 4. can be seen (Fig. 7) that the outflow of H^+ ions was almost exactly equalled y an inflow of K^+ ions.

. Effect of pH and Electrical Gradients

pH gradient

Respiratory pulses in the presence of SCN^- were used to establish a radient of pH across the membrane in the virtual absence of a membrane otential. Plotted semi-logarithmically, the time-course of the exponential :cay of the pH gradient was a straight line in the absence of lactose (Fig. 8 irve A). In the presence of 5 mM lactose (Fig. 8 curve C), the rate of decay as initially higher but declined towards the rate in the absence of lactose. /hen lactose transport was inhibited by N-ethylmaleimide, the accelerating lect of lactose on the decay of the pH gradient was completely removed, ; can be seen by comparing curve B (N-ethylmaleimide alone) and curve D J-ethylmaleimide with lactose).

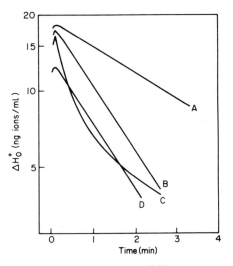

FIG. 8. Semi-logarithmic plots of the time-course of disappearance of H^+ ions from the outer medium following a respiratory pulse. Suspension, preincubation and injection of air-saturated KCl as in Fig. 6B, but with the following additions: A, No addition; B, 0·2 mM N-ethylmaleimide; C, 5 mM lactose; D, 0·2 mM N-ethylmaleimide and 5 mM lactose. (From West and Mitchell, 1972b.)

This experiment indicates, as does the experiment of Fig. 2, that the transport of lactose is coupled to the movement of acid equivalents, but the driving force was in the one case the lactose gradient and in the other case the pH gradient.

2. Electrical gradient

A gradient of electrical potential across the membrane was established by rapidly injecting an anaerobic solution of KSCN into an anaerobic suspension of bacteria in 150 mM KCl. Because the SCN^- ion is considerably more mobile across the membrane than is the K^+ ion, there is a transient diffusion potential established with the cell interior negative. The effective movement of H^+ ions into the cell due to this electrical membrane potential was followed with a glass electrode.

Curve (a) in Fig. 9 shows that in the presence of uncoupler (FCCP) there was a considerable effective inflow of H^+ caused by the injection of KSCN. In the absence of FCCP (curve (b)) there was much less effective H^+ movement, presumably because a greater proportion of the diffusing SCN^- was accompanied by ions other than H^+. The effective inflow of H^+ was increased by the presence of β-galactoside. Once again, N-ethylmaleimide completely abolished the extra H^+ movement associated with the presence of β-galactoside (Compare curves (c) and (d).)

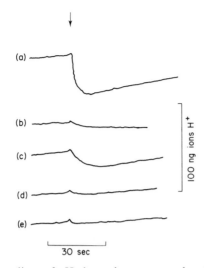

FIG. 9. Strip-chart recordings of pH electrode responses showing decrease in H^+ ion activity in the outer medium on adding KSCN to anaerobic suspensions of normal *E. coli* ML 308-225. The 4 ml contained: 20·2 mg cell dry wt, 150 mM KCl, 1·5 mM glycylglycine, 0·1 mg carbonic anhydrase. At arrow, 25 μl of anaerobic 0·6 M KSCN solution was injected. The following were added to give the final concentrations indicated: (a), 2 μM FCCP; (b), No addition; (c), 10 mM thiomethylgalactoside; (d), 0·5 mM N-ethylmaleimide; (e), 0·5 mM N-ethylmaleimide and 10 mM thiomethylgalactoside. (From West and Mitchell, 1972b.)

Thus it seems that not only does the transport of β-galactoside cause a movement of charge (as shown in Fig. 4) but an electrical field will cause the movement of β-galactoside, and it is concluded that the process of β-galactoside transport is in fact equivalent to the movement across the membrane of β-galactoside with H^+ ions.

III. Discussion

A. The Symport Hypothesis and the Redox Hypothesis

The hypothesis which stimulated the experiments described in this paper is that the β-galactoside transport system of *E. coli* catalyses the coupled movement of β-galactoside and H^+ ions (or OH^- ions) such that the net flow of β-galactoside on the porter cannot occur without a stoichiometric net flow of H^+ ions in the same direction (or OH^- ions in the opposite direction), and that a net flow of H^+ ions (or OH^- ions) on the porter cannot occur without a net flow of β-galactoside. The coupling may not in practice be 100% tight and the stoichiometry need not be 1 H^+ ion : 1 β-galactoside molecule.

It is pointed out that such a coupled system would produce a concentration gradient of β-galactoside whenever H^+ ions are not at electrochemical equilibrium across the membrane. Furthermore, it has been shown that H^+ ions are not at equilibrium when the cells are respiring.

Reeves (1971) showed that the respiratory oxidation of D-lactate by membrane vesicles, prepared from *E. coli* in the manner of Kaback, led to the formation of a pH gradient in experiments analogous with those described in this paper using whole cells. The present hypothesis can therefore be extended to cover the accumulation by membrane vesicles of β-galactosides, and indeed of any compound or group which can traverse the membrane with a positive charge or with H^+ ions bound reversibly (either to the carrier, as is presumably the case with β-galactosides, or to the transported substance itself). Kaback and co-workers (Barnes and Kaback, 1970; Kaback and Milner, 1970; Kerwar *et al.*, 1972) have reported the accumulation of a number of sugars, sugar phosphates, and amino acids by respiring vesicles, while Bhattacharyya (1970) demonstrated the accumulation of Mn^{++} ions, and Bhattacharyya *et al.* (1971) the accumulation of K^+ ions by membrane vesicles.

Kaback and Barnes (1971) postulated that the transport carriers themselves were redox carriers which mediated the flow of reducing equivalents between the various substrates and oxygen and in the process underwent a cyclic conversion between a form having high affinity and one having low affinity for the transported substances. The suggestion probably arose from the observation that electron transport between D-lactate and oxygen is inhibited by N-ethylmaleimide (Kaback and Barnes, 1971), which is known to combine with the β-galactoside carrier and inhibit β-galactoside transport (Fox and Kennedy, 1965), though Kaback and Barnes point out that there could be more than one site of N-ethylmaleimide action. However, the hypothesis of Kaback and Barnes does not provide an adequate explanation of the action of uncouplers in inhibiting the accumulation of β-galactoside and causing the rapid efflux of already accumulated sugar. Furthermore, the accumulation of K^+ in the presence of valinomycin cannot be accommodated by the hypothesis without postulating a hidden redox carrier which can transport K^+ only in the presence of valinomycin.

By contrast, Mitchell's chemiosmotic hypothesis would explain K^+ uptake in the presence of valinomycin in terms of the single, well-documented, property of valinomycin—that of increasing the permeability of lipid membranes to K^+ ions. In Fig. 7 of the present paper, the uptake of K^+ ions during a respiratory pulse was quantitatively accounted for by the electric field established by the outward flow of H^+ ions. Similarly, the action of uncouplers on β-galactoside accumulation can be entirely explained in terms of their property of increasing the H^+ ion conductance of lipid membranes.

3. Cycling Intermediates

The data presented in the experimental section show that there is an ffective flow of H^+ ions caused by the flow of β-galactoside molecules, ut there is nothing to indicate which group reversibly combines with the H^+ on (or OH^- ion). The H^+ ions (or OH^- ions) need not even be transported y the y gene carrier-protein, but could be connected to the flow of β-galactoside indirectly through an unknown intermediate. Thus, an inflow f Na^+ ions could be coupled to the inflow of β-galactoside, the Na^+ subsequently exchanging against H^+ on an entirely separate porter (the Na^+/H^+ ntiporter postulated by Mitchell, 1961a). The combination of β-galactoside–Na$^+$ symport and Na^+/H^+ antiport would give the same overall reaction s the directly coupled β-galactoside–H^+ symport. As there is no observable ag between the addition of β-galactoside and the beginning of effective H^+ nflow, only an extremely rapid Na^+/H^+ exchange would be compatible ith the existing results. However, the suggestion of Stock and Roseman 1971) that the accumulation of galactosides by *Salmonella typhimurium* is by a Na$^+$-coupled transport system led to a closer examination of the possible ole of Na^+ as a cycling intermediate in the transport of β-galactosides by . *coli*. It was found that the transport of β-galactoside (as opposed to its ccumulation) was not greatly affected by the Na^+ ion concentration over the ange 50 μM–50 mM, and in any case Na^+ did not enter the cell during β-galactoside inflow (West and Mitchell, 1972b). Thus, β-galactoside transport s not coupled to Na^+ transport, and Na^+ is not an intermediate between the ow of galactoside and the effective flow of H^+, at least in the case of . *coli*. There remains the possibility that phosphate may act as such an ntermediate, as discussed by Mitchell (1973).

Acknowledgements

The author is indebted to Dr Wilfred Stein, in whose laboratory this work was nitiated, to Dr Peter Mitchell and Dr Jennifer Moyle for collaboration and helpful iscussion, and to Mr Alan Jeal, Mr Robert Harper and Miss Stephanie Phillips for ssistance. This work was supported in part by a research fellowship from the Science .esearch Council, in part by a special research grant from the Science Research Council, nd in part by Glynn Research Ltd.

References

ARNES, E. M. and KABACK, H. R. (1970). *Proc. natn. Acad. Sci. U.S.* **66**, 1190–1198.
HATTACHARYYA, P. (1970). *J. Bacteriol.* **104**, 1307–1311.
HATTACHARYYA, P., EPSTEIN, W. and SILVER, S. (1971). *Proc. natn. Acad. Sci. U.S.* **68**, 1488–1492.
OX, C. F. and KENNEDY, E. P. (1965). *Proc. natn. Acad. Sci. U.S.* **54**, 891–899.
ORECKER, B. L., OSBORNE, M. J., McLELLAN, W. L., AVIGAD, G. and ASENSIO, C. (1961). *In* "Membrane Transport and Metabolism" (A Kleinzeller and A. Kotyk, eds) pp. 378–387. Academic Press, New York.

KABACK, H. R. (1970). *A. Rev. Biochem.* **39**, 561–598.

KABACK, H. R. and BARNES, E. M. (1971). *J. biol. Chem.* **246**, 5523–5531.

KABACK, H. R. and MILNER, L. S. (1970). *Proc. natn. Acad. Sci. U.S.* **66**, 1008–1015.

KASHKET, E. R. and WONG, P. T. S. (1969). *Biochim. biophys. Acta* **193**, 212–214.

KENNEDY, E. P. (1970). *In* "The Lactose Operon" (J. R. Beckwith and D. Zipser, eds) pp. 49–92. Cold Spring Harbor Laboratory, New York.

KEPES, A. (1960). *Biochim. biophys. Acta* **40**, 70–84.

KEPES, A. (1971). *J. Membrane Biol.* **4**, 87–112.

KERWAR, G. K., GORDON, A. S. and KABACK, H. R. (1972). *J. biol. Chem.* **247**, 291–297.

KOLBER, J. and STEIN, W. D. (1966). *Biochem. J.* **98**, 8P.

MITCHELL, P. (1961a). *Nature, Lond.* **191**, 144–148.

MITCHELL, P. (1961b). *Biochem. J.* **81**, 24P.

MITCHELL, P. (1963). *Biochem. Soc. Symp.* **22**, 142–169.

MITCHELL, P. (1967). *Adv. Enzymol.* **29**, 33–87.

MITCHELL, P. (1973). *J. Bioenergetics* **4**, 63–91.

PAVLASOVA, E. and HAROLD, F. M. (1969). *J. Bacteriol.* **98**, 198–204.

REEVES, J. P. (1971). *Biochem. biophys. Res. Commun.* **45**, 931–936.

SCARBOROUGH, G. A., RUMLEY, M. K. and KENNEDY, E. P. (1968). *Proc. natn. Acad Sci. U.S.* **60**, 951–958.

SCHOLES, P. and MITCHELL, P. (1970). *J. Bioenergetics* **1**, 309–323.

STEIN, W. D. (1967). "The Movement of Molecules across Cell Membranes". Academi Press, New York.

STOCK, J. and ROSEMAN, S. (1971). *Biochem. biophys. Res. Commun.* **44**, 132–138.

WEST, I. C. (1970). *Biochem. biophys. Res. Commun.* **41**, 655–661.

WEST, I. C. and MITCHELL, P. (1972a). *Biochem. J.* **127**, 56P.

WEST, I. and MITCHELL, P. (1972b). *J. Bioenergetics* **3**, 445–462.

WINKLER, H. H. and WILSON, T. H. (1966). *J. biol. Chem.* **241**, 2200–2211.

IV.5

Effects of Tris-buffer on Ion Uptake and Cellular Ultrastructure

R. F. M. Van Steveninck, C. J. Mittelheuser and M. E. Van Steveninck

Botany Department, University of Queensland
Australia

I. Introduction

Tris-hydroxymethylamino methane (Tris) which was introduced by Gomori (1946) as an inert buffer has since been shown to have various side effects through participation in reactions involving enzymes (Mahler, 1961; Dahlquist, 1961; Fleming and Pegler, 1963; Jørgensen and Jørgensen, 1967; Betts and Evans, 1968) including those of the glycolytic pathway (Buse *et al.*, 1964) and choline esterases (Pavlič, 1967). It has also been shown to affect the isolation of mitochondria (Stinson and Spencer, 1968), sucrose storage (Humphreys and Garrard, 1969), amino acid incorporation (Bewley and Marcus, 1970) and cell division in disks of storage tissue (Kahl *et al.*, 1969), but two of the many effects recorded are of special interest here, viz. its effect on ion transport (Van Steveninck, 1961, 1966a, b, 1972; Kholdebarin and Oertli, 1970), and on electron transport in the Hill reaction (Jacobi, 1961; Yamashita and Butler, 1969; Yamashita et al., 1971) when pre-incubation in the light proves necessary (Ikehara and Sugahara, 1969).

II. Effects on Storage Root Tissues

The induction of a salt accumulatory system in thinly sliced (1 mm) disks of storage tissues generally takes place during aerobic washing for a period of 1–2 days (Van Steveninck, 1961), the duration of this induction period being dependent on the species, variety and condition of the tissue at the time of slicing. Beetroot slices exposed to 10^{-2} M Tris buffer, however, are capable of an immediate net accumulation of K^+ at a rapid rate until the external concentration of K^+ is depleted to less than 10^{-7} M (Fig. 1a, b and c). 10^{-2} M Tris ($pK_a = 8 \cdot 14$) at pH 8 is much more effective than at pH 6·3, and the latter is less effective than 10^{-3} M Tris at pH 8. Considering $pK_a = pH + \log(RN^+H_3)/(RNH_2)$, the actual concentrations of the non-protonated form of Tris are $1 \cdot 4 \times 10^{-4}$ M, $7 \cdot 2 \times 10^{-3}$ M and $7 \cdot 2 \times 10^{-4}$ M respectively, a ratio of 1:50:5. This suggests that the non-protonated form of Tris is effective in causing immediate cation uptake. This effect of Tris can be reversed by two rinses of 15 min duration each in deionized water 2 h after the application of Tris buffer (Fig. 1b). Although the raising of the pH of the external medium from 6·3 to 8·0 might in itself stimulate cation uptake, a similar change of pH using phosphate buffers has an insignificant effect on net K^+ uptake (Van Steveninck, 1964). Furthermore it was found that Tris shows little or no effect on the Cl^- accumulatory system during or after its development, but in certain instances an inhibitory effect on net Cl^- uptake has been recorded (Van Steveninck, 1964).

A. Tissue Specificity

The Tris effect on development of cation uptake is specific for certain tissues, e.g. beetroot, parsnip and carrot accumulate cations at a rapid and sustained rate immediately the disks are exposed to 10^{-2} M Tris at pH 8, while cation uptake in swede, artichoke, potato and turnip disks is relatively unaffected by the same treatment (Fig. 2).

B. Molecular Structure

The presence of OH-groups of Tris is essential for its effect on ion uptake. Tris (= 2 amino-2 hydroxymethyl-1:3 propanediol), 2 amino-2 methyl-1:3 propanediol, 2 amino-2 methyl-1-propanol and 2 amino-2 methyl propane (= tertiary butylamine) at pH 8 form a series which shows decreasing activity, the latter having practically no effect (Fig. 3a). However, tri-ethanolamine

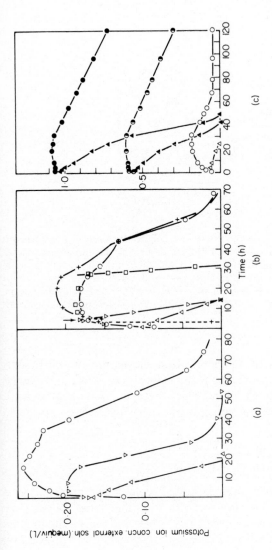

FIG. 1. The effect of Tris buffer on the potassium balance between disks of beetroot tissue and the external solution. (a) 4·5 g freshly sliced tissue (25 disks) in 150 ml solution; ○, distilled water pH 6·3–6·4; △, 10^{-2} M Tris–HCl pH 7·9–8·0; ▽, 10^{-2} M Tris–HCl pH 6·0–6·3. (b) tissue and solution as in (a); ○, distilled water pH 6·2–6·5, △, 10^{-1} M solution pH 7·85–8·0; ▽, 10^{-2} M Tris, 4 h from start of experiment added as 7·5 ml 2×10^{-1} M solution pH 7·85–8·0; □, 10^{-2} M Tris–HCl, 27 h from start of experiment, pH 8·0; +, 10^{-2} M Tris–HCl, pH 8·0 for 2 h, followed by two rinses of 15 min duration each, 1 ml 0·019 M KCl added to make up for K^+ lost during rinsing of the tissue. (c) tissue and solution as in (a); ○, distilled water, pH 5·55–6·15; △, 10^{-2} M Tris–HCl, pH 7·80–7·95; ●, $0·5 \times 10^{-3}$ M KCl solution pH 5·50–5·95; ▲, $0·5 \times 10^{-3}$ M KCl; +, 10^{-2} M Tris–HCl, pH 7·80–7·90; ◉, 10^{-3} M KCl, pH 5·50–6·00; ▲, 10^{-3} M KCl + 10^{-2} M Tris–HCl pH 7·50–7·80.

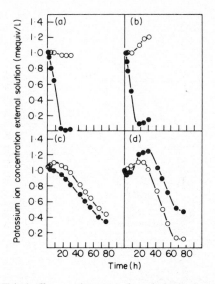

FIG. 2. The effect of Tris buffer on the potassium balance between disks of various tissues and the external solution. (a), carrot phloem tissue; (b), parsnip phloem tissue, (c), artichoke tissue (*Helianthus* sp) and (d), swede tissue. All 25 disks per 150 ml solution. ○, disks in 10^{-3} M KCl solution, pH 6·05–6·85; 6·00–6·70; 6·10–7·20 and 6·00–7·10 for (a), (b), (c) and (d) respectively; ●, disks in 10^{-3} M KCl + 10^{-2} M Tris–HCl solution, pH 7·95 → 7·60; 7·95 → 7·50; 7·95 → 8·00, 7·95 → 8·00 for (a), (b), (c) and (d) respectively.

and Tris are equally effective, indicating that the nature of substituents rather than the degree of substitution at the N level is of prime importance, e.g., tri-ethanolamine is much more effective than ethanolamine while N,N-di-methylamino-ethanol is practically ineffective (Fig. 3b; Van Steveninck, 1961).

C. Interactions with Glucosamine and Glucose

The cationic hydroxamine, D-glucosamine, completely prevents Tris-induced K^+ uptake at pH 7·8 (Fig. 4a). On the other hand, N-acetyl-D-glucosamine and D-glucose are without any effect, but the D-glucosamine inhibition of Tris-stimulated cation uptake is completely reversed on addition of an equimolar concentration of D-glucose (Fig. 4B). The reversal becomes less effective after 24 h, but a further addition of glucose re-establishes the reversal effect (Fig. 4b). The reversal is glucose specific and it was suggested that stimulation of a specific hexokinase (ATP-D-glucose-6-phospho-transferase) might be operative in causing the Tris effect (for further particulars see Van Steveninck, 1966b).

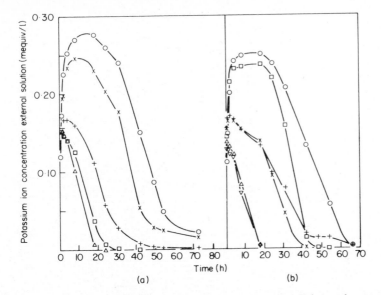

FIG. 3. The effect of a number of amines, structurally related to Tris, on the potassium balance between disks of beetroot tissue and the external solution (25 disks per 150 ml of solution). (a) ○, distilled water, pH 6·35–6·70; △, 10^{-2} M 2-amino-2-methyl 1:3 propanediol, pH 8·05–8·30; +, 10^{-2} M 2-amino-2-methyl 1-propanol, pH 8·00–8·10, adjusted with small amounts of base during the course of the experiment; ×, 10^{-2} M 2-amino-2-methyl propane (tertiary butylamine), pH 7·80–7·95, adjusted with base. (b) ○, distilled water, pH 6·25–6·55; △, 10^{-2} M Tris, pH 8·1–8·2; ▽, 10^{-2} M triethanolamine, pH 7·8–8·2, adjusted with base; □, 10^{-2} M N,N-di-methylaminoethanol pH 7·8–8·0 (adjusted); ×, 10^{-2} M methylaminoethanol, pH 7·7–8·2 (adjusted); +, 10^{-2} M ethanolamine, pH 7·4–8·1 (adjusted).

D. Ion Fluxes

An analysis of fluxes of K^+ and Cl^- in beet disks using ^{42}K and ^{36}Cl during and after the development of ion uptake capabilities shows a relatively small increase of K^+ influx and marked decrease of apparent efflux thus causing net uptake of K^+ (Van Steveninck, 1962). Over the same period apparent influx of Cl^- increases markedly with a concurrent small reduction of apparent efflux.

The increased rates of net accumulation of K^+ obtained with Tris are due to a large increase in influx and a decrease of efflux (Fig. 5). In fresh disks Tris causes a greater net loss of Cl^- by increasing the efflux of Cl^- and decreasing the influx with respect to the control treatment. Therefore, a large excess of net cation over anion uptake appears to occur. If after some time of unequal uptake, K^+ is depleted from the external solution then net Cl^- uptake still proceeds at a rapid rate. Hence, it appears possible for cation

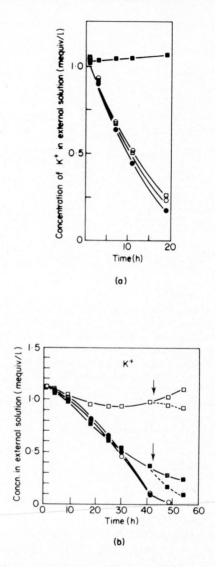

FIG. 4. The effects of D-glucosamine, N-acetyl-D-glucosamine, and D-glucose on the potassium balance between disks of beetroot tissue and the external solution consisting of 10^{-2} M Tris–HCl at pH 7·8. (a) 25 disks (4·7 g tissue) per 125 ml solution; $\bigcirc$, 10^{-2} M Tris + 1·03 mM KCl; $\bullet$, 10^{-2} M Tris + 10^{-2} M D-glucose + 1·02 mM KCl; $\square$, 10^{-2} M Tris + 10^{-2} M N-acetyl-D-glucosamine + 1·01 mM KCl; $\blacksquare$, 10^{-2} M Tris + 10^{-2} M D-glucosamine + 1·01 mM KCl. (b) 50 disks (9·98 g tissue) per 250 ml solution; $\bigcirc$, 10^{-2} M Tris + 1 mM KCl; $\bullet$, 10^{-2} M Tris + 5 × 10^{-3} M D-glucose + 1 mM KCl; $\square$, 10^{-2} M Tris + 5 × 10^{-3} M D-glucosamine + 1 mM KCl; $\blacksquare$, 10^{-2} M Tris + 5 × 10^{-3} M D-glucosamine + 5 × 10^{-3} M D-glucose + 1 mM KCl; $\downarrow$, D-glucose equivalent to a concentration of 5 × 10^{-3} M added at $42\frac{1}{4}$ h.

apparent while in the presence of 1 mM KCl a proton output of 35·2 mequiv kg^{-1} of tissue is measured. The difference in proton output of Tris treatment and control accounts for about 90% of the excess of K$^+$ absorbed by the tissue over the same period of time. When disks of storage tissue are placed in 10^{-3} M KCl solution a progressive rise in external pH occurs, but this is followed by a distinct fall after the onset of net K$^+$ uptake (Table II). This characteristic fall in pH occurs only in those tissues which are capable of an immediate response to Tris treatment, e.g. carrot, parsnip and beetroot. In the presence of Tris a distinct drop of pH occurs only in those tissues which show an immediate stimulation of cation uptake (Table II).

Table II. Effect of 10^{-2} M Tris buffer on changes in pH of solutions in which disks of various tissues were immersed

Tissue	Treatment	pH of external solution	
		After 30 min	After 76½ h
Artichoke	25 disks of each	6·10 →[a]	7·20
Swede	tissue placed in	6·00 →	7·10
Carrot	160 ml 10^{-3} M	6·05 → 6·85 (42)[b] → 6·60	
Parsnip	KCl	6·00 → 6·70 (42)[b] → 6·45	
Beetroot		5·95 → 6·55 (30)[b] → 5·40	
Artichoke	25 disks of each	7·95 →	8·00[c]
Swede	tissue placed in	7·90 →	8·00[c]
Carrot	160 ml 10^{-3} M	7·95 →	7·55[d]
Parsnip	KCl plus	7·95 →	7·50[d]
Beetroot	10^{-2} M Tris	7·90 →	7·70[d]

[a] Arrows indicate gradual fall or rise towards minimum or maximum pH respectively.
[b] Time (h) of pH measurement when net uptake commenced.
[c] No Tris effect: K$^+$ absorption not stimulated.
[d] Tris effect: K$^+$ absorption greatly stimulated.

F. Organic Acids

The induced excess cation uptake causes a distinct change in organic acid contents of the beetroot disks. Normally, during the ageing of the disks total organic acid content decreases. This may be due partly to utilization in metabolism or leakage of organic acids into the external solution, especially in fresh disks (Dale and Sutcliffe, 1959). Yet, differences in organic acid contents between control and Tris-treated disks correspond with differences in ion movement (Table III), e.g. at 48 h Tris-treated disks show an excess of K$^+$ over Cl$^-$ uptake of 27·2 mequiv kg^{-1} fresh wt while controls lost 5·0 mequiv kg^{-1} fresh wt more K$^+$ than Cl$^-$, the total difference of 32·2 mequiv

TABLE III. Effect of Tris buffer (10^{-2} M, pH 7·9) and of added potassium chloride (10^{-3} M) on the organic acid content of beetroot disks[a] and on the movements of K$^+$ and Cl$^-$ ions

| Treatments | Period of incubation (h) | Organic acid (mequiv kg^{-1} fresh wt) | | | | | Total organic acid (mequiv kg^{-1} fresh wt) | Net movement[b] of | |
		Malate	Citrate	Lactate	Pyrrolidone carboxylate	Unknown		K$^+$ (mequiv kg^{-1} fresh wt)	Cl$^-$ (mequiv kg^{-1} fresh wt)
Control	0	16·8	22·2	1·7	2·1	0·5	43·3	—	—
KCl	48	9·2	16·5	1·3	0·9	0·3	28·1	−7·4	−2·4
KCl + Tris	48	23·4	31·2	3·6	1·6	0·5	60·3	+21·4	−5·8
KCl	117	4·2	10·1	0·5	1·2	0·4	16·4	+20·2	+22·0
KCl + Tris	117	6·0	17·2	0·5	1·6	0·4	25·7	+21·4	+15·8

[a] 212 disks (38·9 g) per 900 ml solution.
[b] + = net uptake; − = net loss from the disks to the external solution.

kg^{-1} fresh wt being in agreement with the difference in organic acid contents of $32\cdot2$ mequiv kg^{-1} fresh wt. By 117 h a considerable proportion of malate and citrate have been metabolized. The difference between amounts of organic acids in Tris-treated disks and controls is diminished, possibly as a result of K^+ depletion of the external solution by the Tris-treated disks, which continue to accumulate Cl^- in the absence of K^+.

3. Ultrastructural Changes

The plasmalemma is consistently absent in beetroot disks which have been exposed to Tris buffer and then fixed directly in OsO_4 (Figs 6 and 7). In addition to the very poor structural preservation of the outer region of cytoplasm near the cell walls, the pieces of tissue fail to turn black as routinely observed after exposure to OsO_4. The lack of preservation of the outer membrane may be regarded as a fixation artifact. This assumption is reinforced by a recent report (Litman and Barrnett, 1972) that urea has a similar effect on OsO_4 fixation. Litman and Barrnett suggested that OsO_4 fixation

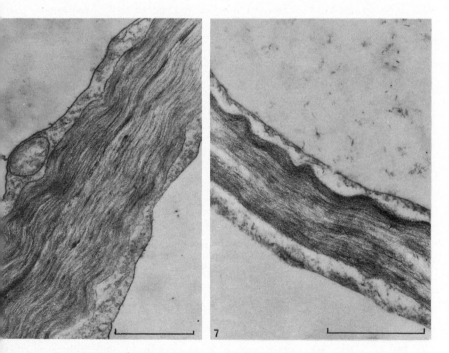

FIGS 6 and 7. The effect of Tris on structural preservation by OsO_4. Fresh disks of beetroot were aerated in water (Fig. 6) or in 10^{-2} M Tris pH 8 (Fig. 7) for 4 h before fixation with 2 % buffered OsO_4. Compare the preservation of the tonoplast, plasmalemma and cytoplasm. In all electron micrographs the marker represents 1 μm.

of proteins occurs through a process of H-bonding and that urea and OsO_4 compete in this process.

III. Effects on Leaf Tissues

A. Ion Transport

Kholdebarin and Oertli (1970) studied the effect of Tris buffer on salt uptake and organic acid synthesis in barley leaf tissue in light and dark and recorded that all the Tris–anion combinations tried caused a reduction in Cl^- uptake and a stimulation of organic acid synthesis to compensate for excess cation uptake. As in storage tissue only the non-protonated form of Tris proved to be effective. It was suggested that the Tris inhibition occurs through a block in the photosynthetic non-cyclic electron flow. Tris inhibition of net Cl^- uptake within 3 h both in light and dark was confirmed in wheat leaf segments by Mittelheuser and Van Steveninck (1972) (Table IV)

TABLE IV. The effect of Tris buffer[a] and light on the Cl^- contents of 2 mm segments of primary wheat leaves[b]

Time (h)	Treatment	Cl^- (mequiv kg^{-1} fresh wt)
0	—	42
3	Control light	48
3	Control dark	47
3	Tris light	32
3	Tris dark	33
6	Control light	74
6	Control dark	70
6	Tris light	32
6	Tris dark	40

[a] 200 mM Tris–SO_4 pH 8·0 + 10 mM NaCl + 0·5 mM $CaSO_4$ Control solutions 10 mM NaCl + 0·5 mM $CaSO_4$.
[b] Segments washed for 1 h in 0·5 mM $CaSO_4$, 130–150 mg fresh wt incubated in 50 ml solution in 100 ml Erlenmeyer flasks at 26°C on shaker in dark or under 2600 lm m^{-2} supplied by a 400 Watt Hg-vapour lamp.

B. Ultrastructural Changes

Tris (2×10^{-2} M) at pH 8 in the light, but not in the dark, induces marked changes in chloroplast structure in segments of wheat and barley leaves (Mittelheuser and Van Steveninck, 1972). The organization of both grana

FIGS 8–13. The effect of Tris on leaf ultrastructure. Wheat and barley leaves were treated and fixed for electron microscopy (using glutaraldehyde followed by osmium tetroxide) as described by Mittelheuser and Van Steveninck (1972).

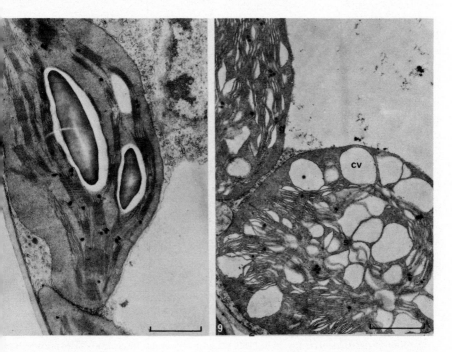

FIG. 8. Chloroplast of wheat leaf tissue after incubation in the control medium in the light for 6 h.

FIG. 9. Chloroplast of wheat leaf tissue after light incubation in 0·2 M Tris for 3 h. Note chloroplast vacuoles (cv), distorted grana lamellae (g) and stroma lamellae (s).

and stroma lamellae is altered after treatment for 3 h, with the formation of small vacuoles within the chloroplast (Figs 8 and 9). After more prolonged treatment (6 h or longer), or alternatively treatment with high concentrations of Tris (0·2 M), the chloroplasts are transformed into a mass of membrane coils (Fig. 10). Deposits may be seen in the chloroplast vacuoles (Fig. 11) and in the cell vacuoles which also contain membrane coils probably arising from the tonoplast (Fig. 12). Even though Tris treatment may result in extreme distortion of the chloroplast membranes and the tonoplast, there was no apparent effect on other organelles (Figs 12 and 13). Furthermore, when Tris is replaced by 0·02 M Tricine at pH 8 or 0·02 M sodium phosphate buffer at pH 8, these effects are not evident.

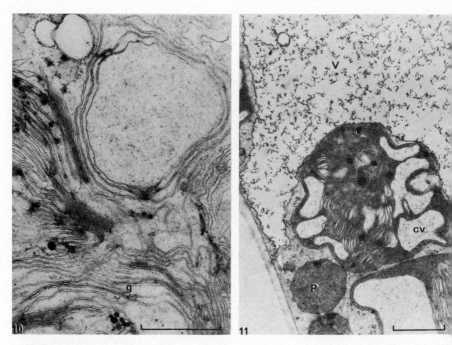

FIG. 10. Chloroplast of wheat leaf tissue after light incubation in 0·2 M Tris for 6 h. Note remnants of grana lamellae (g), and the absence of outer chloroplast membrane.

FIG. 11. Deposit-containing vacuoles in chloroplasts of barley leaf tissue after light incubation in 0·02 M Tris for 24 h. Note structurally intact microbody (P) and deposits in the cell vacuole (V).

IV. Mechanism of Action

Tris, a relatively simple hydroxamine, is unlikely to have a highly specific mechanism of action. This conclusion is substantiated by the numerous effects recorded including those effects on a wide range of enzymes (see I).

Apparently, the hydroxyl groups are an essential feature for activity while the degree of substitution at the N-level is of minor importance. Reversible binding of hydroxamines to protein molecules was recently recorded by Rawitch and Gleason (1971).

A. Unequal Cation–Anion Uptake

Figure 14 gives a diagrammatic representation of two mechanisms that could explain excess cation uptake in freshly sliced storage tissue. One (upper half) is based on exchange of metabolically produced H^+ for K^+ at the plasmalemma followed by secretion of K^+ and organic acids into the vacuole.

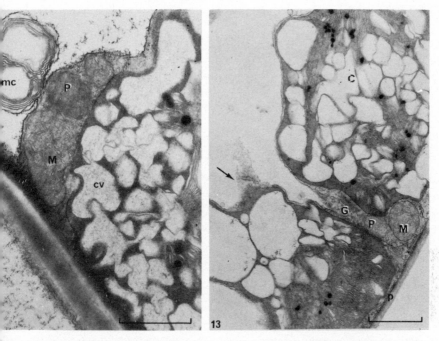

FIG. 12. Barley leaf tissue after light incubation in 0·2 M Tris for 24 h showing vacuolar membrane coil (mc), part of which appears to have arisen from the tonoplast. Note structurally intact mitochondrion (M) and microbody (P) and also chloroplast vacuoles (cv) containing deposits.

FIG. 13. Barley leaf tissue after light incubation in 0·02 M Tris for 24 h. Note intact mitochondrion (M), microbody (P), Golgi body (G), and plasmalemma (p). The tonoplast is discontinuous (arrow) and the lamellar system of the chloroplasts (C) is grossly distorted.

This scheme is represented by a large mitochondrion because of their implied role in ion transport processes (Robertson, 1968), and because, in beetroot parenchyma tissue, the thickness of the cytoplasmic layer is often appreciably less than 0·5 μm so that the tonoplast is distended by the mitochondria.

The other mechanism (lower half) depends on the presence of HCO_3^- in the external solution. The concentration of HCO_3^- is determined by the pH of the external solution, and by a possible conversion of respiratory CO_2 to bicarbonate at the outer surface of the cell. Bicarbonate ions would then accompany the K^+ ions across the plasmalemma and in the relatively acid environment of the cytoplasm, pyruvate would convert the resulting CO_2 to form malate which in turn would be secreted together with K^+ into the vacuole. Both mechanisms account for the observed acidification of the external solution, but in the first mechanism the metabolically produced

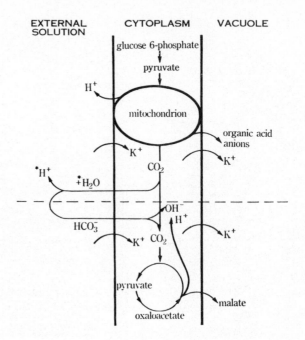

FIG. 14. Diagrammatic representation of two alternative mechanisms which coul
explain excess cation uptake in beetroot tissue.

protons must negotiate the plasmalemma, while in the second mechanisr
the protons (*) find their origin in the external solution and remain there.

B. Tris Carbamate vs HCO_3^-

Hurd (1959) stressed the importance of HCO_3^- in maintaining electrica
neutrality, and Poole (1968) showed that the presence of 0.2 mM HCO_3
causes the membrane potential in beet to become 50 mV more negative
However, HCO_3^- is not present as such in an alkaline Tris solution becaus
Tris-base combines with CO_2 to form Tris carbamate ($RNHCOO^-$) muc
more quickly than the hydration of CO_2 to form bicarbonate (Edsall an
Wyman, 1958). This factor especially seems to preclude direct involvemen
of HCO_3^- in maintaining excess cation uptake.

In the absence of Tris, HCO_3^- may still play an important role. Hope (196！
implied the presence of a light driven electrogenic pump in Chara australi
Raven (1968) suggested the same for Hydrodictyon, and Smith (1968) fo
Characean cells in general. The contention that in the presence of Tris, th
carbamate may serve as the accompanying anion was dispelled by Va
Steveninck (1966a) on the basis of nitrogen determinations of the tissu
Furthermore triethanolamine, which is just as effective as Tris, is considere

not to form a carbamate (Edsall and Wyman, 1958). The possible activity of a carbonic anhydrase system operating at the cell surface was investigated. However, 2-acetyl-1, 3,4-thiadiazole-5-sulphonamide (Diamox), a specific inhibitor of carbonic anhydrase, at concentrations ranging from 10^{-5} M to 5×10^{-4} M did not affect K^+ uptake, in the presence or absence of Tris. Samiy et al. (1961) recorded that Tris did not affect carbonic anhydrase activity in vitro.

C. Metabolic Role of Tris

The reasons for the prevention of the Tris effect by D-glucosamine and its reversal with D-glucose are not understood at present. These observations suggest the involvement of a glucose specific hexokinase. Beetroot is known to store large quantities of reducing sugars, up to an average concentration of 0·1 M (MacDonald and de Kock, 1958), and it is also known that Tris induces conditions of hypoglycaemia in a number of animal tissues (Bennett and Tarail, 1961; Buse et al., 1964). Therefore it could be suggested that a Tris-induced accelerated rate of glucose catabolism will provide an additional source of protons for exchange with K^+ or Na^+, or alternatively that glucose phosphate might act as a carrier for K^+ or Na^+ transport.

D. Proton Movement and Proton Acceptors

The establishment and maintenance of proton gradients across membrane-bound structures is at present a much discussed aspect of energy transduction. It has been established that, as a result of respiratory processes, mitochondria develop a proton gradient with a relatively higher activity of protons externally (Mitchell and Moyle, 1965), while light-induced proton gradients in chloroplasts are established in the opposite direction, i.e. the activity of protons is higher in the internal phase (Jagendorf and Uribe, 1966; Schwartz, 1968). Rottenberg et al. (1972) measured a light-induced Δ pH of more than 2 units using ^{14}C-methylamine, and assuming that $(H^+)_{in}/(H^+)_{out} = (R - NH_3^+)_{in}/(R - NH_3^+)_{out}$ applies when the non-protonated amine is in equilibrium.

Any compound which might interfere with the movement of protons across membranes or modify the permeability of membranes with respect to protons would be expected to influence energy transfer and ion transport. It could be postulated that the non-protonated Tris base (RNH_2) would penetrate through membrane structures much more readily than its protonated form (RN^+H_3). It may then act in two ways simultaneously:

a) As a proton acceptor causing osmotic phenomena such as vacuolation because the protonated form cannot readily "escape". The presence of deposits in these spaces after glutaraldehyde–OsO_4 fixation provides some evidence as Tris forms reaction products with a range of aldehydes (Hauptmann and Gabler, 1968).

With respect to proton gradients, it is significant that in leaf tissue Tris causes a severe disruption of chloroplast lamellar structure while the mitochondrial cristae remained undisturbed.

(b) As a chaotropic agent causing unfolding and swelling of protein molecules resulting from interactions between hydroxyl groups and H-bonds and other side chains containing $-COO^-$ and $\equiv N^+H$ groups (Klotz and Ayers, 1953). This is particularly evident in beetroot tissue where H-bonding of Tris molecules at the outer boundary of the cell appears to interfere with the OsO_4-fixation process. At these sites and possibly in the cytoplasm, Tris may again act as a proton acceptor, thus facilitating the transfer of cations and organic acids across the tonoplast. Such a proposition seems feasible for beetroot tissue as the tonoplast appears to remain intact. Tris-induced excess cation accumulation of the order of 30 mequiv kg^{-1} fresh wt in storage tissues is unlikely to be located in the cytoplasm which constitutes only about 2% of the tissue volume in beetroot (Briggs et al., 1958).

In conclusion, all recorded phenomena involving Tris seem to point towards its role as a proton acceptor and emphasize the major importance of proton gradients in ion movements between organelles and cytoplasm and between cytoplasm and the external solution.

Acknowledgement

This work was supported by the Australian Research Grants Committee.

References

BENNETT, T. E. and TARAIL, R. (1961). Ann. N.Y. Acad. Sci. 92, 651–661.
BETTS, G. F. and EVANS, H. J. (1968). Biochim. biophys. Acta 167, 193–196.
BEWLEY, J. D. and MARCUS, A. (1970). Phytochemistry 9, 1031–1033.
BRIGGS, G. E., HOPE, A. B. and PITMAN, M. G. (1958). J. exp. Bot. 9, 128–141.
BUSE, M. G., BUSE, J., MCMASTER, J. and KRECH, L. H. (1964). Metabolism 13, 339–353.
DAHLQUIST, A. (1961). Biochem. J. 80, 547–551.
DALE, J. E. and SUTCLIFFE, J. F. (1959). Ann. Bot. N.S. 23, 1–21.
EDSALL, J. T. and WYMAN, J. (1958). "Biophysical Chemistry", Vol. 1. Academic Press, New York.
FLEMING, I. D. and PEGLER, H. F. (1963). Analyst 88, 967–968.
GOMORI, G. (1946). Proc. Soc. exp. Biol. 62, 33–34.
HAUPTMANN, S. and GABLER, W. (1968). Z. Naturforsch. 23b, 111–112.
HOPE, A. B. (1965). Aust. J. biol. Sci. 18, 789–801.
HUMPHREYS, T. E. and GARRARD, L. A. (1969). Phytochemistry 8, 1055–1064.
HURD, R. G. (1959). J. exp. Bot. 10, 345–358.
IKEHARA, N. and SAGAHARA, K. (1969). Bot. Mag. Tokyo 82, 271–277.
JACOBI, G. (1961). Naturwiss. 48, 577.
JAGENDORF, A. T. and URIBE, E. (1966). Proc. natn. Acad. Sci., U.S.A. 55, 170–177.
JØRGENSEN, B. B. and JØRGENSEN, O. B. (1967). Biochim. biophys. Acta 146, 167–172.
KAHL, G., ROSENSTOCK, G. and LANGE, H. (1969). Planta 87, 365–371.

KHOLDEBARIN, B. and OERTLI, J. J. (1970). Z. Pflanzenphysiol. 62, 231–236.
KLOTZ, I. M. and AYERS, J. (1953). Disc. Faraday Soc. 13, 189–196.
LITMAN, R. B. and BARRNETT, R. J. (1972). J. Ultrastr. Res. 38, 63–86.
MACDONALD, I. R. and DE KOCK, P. C. (1958). Ann. Bot. N.S. 22, 429–448.
MAHLER, H. R. (1961). Ann. N.Y. Acad. Sci. 92, 426–439.
MITCHELL, P. and MOYLE, J. (1965). Nature, Lond. 208, 147–151.
MITTELHEUSER, C. J. and VAN STEVENINCK, R. F. M. (1972). Aust. J. biol. Sci. 25, 517–530.
PAVLIČ, M. (1967). Biochim. biophys. Acta 139, 133–137.
POOLE, R. J. (1968). Pl. Physiol. 43, S–50.
RAVEN, J. A. (1968). J. exp. Bot. 19, 193–206.
RAWITCH, A. B. and GLEASON, M. (1971). Biochem. biophys. Res. Comm. 45, 590–597.
ROBERTSON, R. N. (1968). "Protons, Electrons, Phosphorylation and Active Transport". Cambridge University Press, London and New York.
ROTTENBERG, H., GRUNWALD, T. and AVRON, M. (1972). Eur. J. Biochem. 25, 54–63.
SAMIY, A. H., OKEN, D. E., REES, S. B., ROBIN, E. D. and MERRILL, J. P. (1961). Ann. N.Y. Acad. Sci. 92, 570–578.
SCHWARTZ, M. (1968). Nature, Lond. 219, 915–919.
SMITH, F. A. (1968). J. exp. Bot. 19, 207–217.
STINSON, R. A. and SPENCER, M. (1968). Can. J. Biochem. 46, 43–50.
VAN STEVENINCK, R. F. M. (1961). Nature, Lond. 190, 1072–1075.
VAN STEVENINCK, R. F. M. (1962). Physiologia Pl. 15, 211–215.
VAN STEVENINCK, R. F. M. (1964). Physiologia Pl. 17, 757–770.
VAN STEVENINCK, R. F. M. (1966a). Aust. J. biol. Sci. 19, 271–281.
VAN STEVENINCK, R. F. M. (1966b). Aust. J. biol. Sci. 19, 283–290.
VAN STEVENINCK, R. F. M. (1972). Physiologia Pl. 27, 43–47.
YAMASHITA, T. and BUTLER, W. L. (1969). Pl. Physiol. 44, 435–438.
YAMASHITA, T., TSUJI, J. and TOMITA, G. (1971). Pl. Cell Physiol. 12, 117–126.

IV.6

The Regulation of Intracellular pH as a Fundamental Biological Process

J. A. Raven

Department of Biological Sciences, The University of Dundee
Scotland

F. A. Smith

Department of Botany, University of Adelaide
Australia

I. Introduction

A large amount of data has been accumulated on various aspects of H^+ uptake or loss in relation to the metabolism in cells and isolated organelles. This paper sets out to summarize the main findings, and the hypotheses to which they have given rise; and to attempt to produce a coherent view of H^+ and OH^- fluxes measured in many different systems.

II. Effects of Metabolism on the pH of Cells and their Environment

A. General Considerations: the Value and Constancy of Intracellular pH

Many metabolic reactions lead to the production or consumption of H^+ inside cells (see IID). Nevertheless, the internal pH does not show large fluctuations, as shown by the pH of expressed sap (e.g. Hiatt, 1967a; Hiatt and Hendricks, 1967; Kirkby and Mengel, 1967; Raven, 1970). Considerable pH changes occur in the external solution (e.g. Street and Sheat, 1958; Dijkshoorn, 1962; Hiatt, 1967a; Raven, 1970), suggesting that a major mechanism for maintaining a relatively constant internal pH is the loss of excess H^+ or OH^- to the medium.

The cytoplasm is only a small fraction of the total cell volume; on the grounds of pH optima of many enzymes, a relatively constant pH in the range 6·0–7·5 would be expected for the cytoplasm (Lehninger, 1970). However, it is likely that the pH of the cytoplasm can vary by 0·5 units or so depending on metabolic circumstances (cf. III). The pH of the expressed sap of higher plants probably reflects the pH of the vacuole. It is generally in the range 5·0–6·0, but values nearer 2·0 have been recorded (see III).

B. External Production and Consumption of H^+

A convenient way of measuring H^+ fluxes between cells and their aqueous environment is to determine the pH of the external solution. However, some of the H^+ changes measured in this way reflect H^+ production or consumption in the medium due to uptake or loss of the un-ionized form of weak electrolytes (e.g. H_2CO_3, NH_4OH). Re-equilibration of the ionized and un-ionized forms in the medium can greatly change external pH without the occurrence of H^+ or OH^- fluxes across the cell membrane. The pH can increase (e.g. during photosynthesis with CO_2: Raven, 1970) or decrease (e.g. during respiration: Pitman, 1970; or assimilation of NH_4OH: MacMillan, 1956; Budd and Harley, 1962a, b).

Changes in pH due to these processes can mask pH change due to H^+ movements between the medium and the cell. Thus for a cell growing photosynthetically on CO_2 and NH_4^+, there is a net H^+ loss from the cells during growth (see IIC*1*); yet the measured pH change in the medium may be an increase due to CO_2 depletion in photosynthesis.

C. Internal Production and Consumption of H^+

1. Cell growth on CO_2 or hexose as carbon source and NH_4^+ as nitrogen source

During cell growth when the carbon species entering the cell is neutral CO_2 (photosynthetic growth: see IIB) or hexose (heterotrophic growth) and the nitrogen species is NH_4^+ (see Lycklama, 1963), the substrates for synthesis of compounds containing CHON carry a net positive charge. However, at physiological pH the products formed carry a net negative charge (ignoring the negative charge due to phosphate, which is taken up as an anion). This is because the cell constituents have an excess of free carboxyl groups over free amino groups (Lehninger, 1970; Kirkby, 1969). Maintenance of internal pH involves removal of H^+ ions and their replacement by exogenous cations, e.g. K^+. This is a general case of excess cation absorption associated with carboxylate synthesis (Kirkby, 1969), which is generally studied (by plant physiologists) in the absence of a nitrogen source (Leggett, 1968).

2. Cell growth with HCO_3^- as carbon source, or NO_3^- as nitrogen source

In certain aquatic plants, photosynthesis at high pH proceeds with HCO_3^- as the carbon species entering the cell (Raven, 1970). CO_2 is fixed and the excess OH^-, left when the processes considered above have occurred, is lost from the cell. The H^+ production in the cell at very high external pH is less than that discussed in the previous paragraph since ammonia would probably enter as NH_4OH (references in IIB).

Nitric acid has such a low pK relative to extracellular pH values that it must always enter cells as NO_3^-. Reduction of NO_3^- to NH_4OH produces one OH^- (Dijkshoorn, 1962); more OH^- is produced as NH_4OH is converted to NH_4^+. The H^+ generated in the processes considered in IIC1 are inadequate to neutralize all this OH^-; some of the excess OH^- is neutralized by additional synthesis of organic acid, while the rest is lost from the cell (Dijkshoorn et al., 1968; Kirkby, 1969). Similar considerations apply to $SO_4^=$ reduction, although it is quantitatively small compared with NO_3^- reduction (Dijkshoorn and van Wijk, 1967).

D. H^+ Exchanges Related to Energy Coupling in Electron Transport Phosphorylation

These H^+ exchanges occur across "coupling membranes", i.e. the outer membranes of prokaryotes and the cristae and thylakoid membranes of the mitochondria and chloroplasts of eukaryotes (Mitchell, 1966, 1968, 1970; Greville, 1969). In these membranes, downhill electron transport, or ATP breakdown, bring about active transport of H^+ (see IIE). Two hypotheses have been proposed to account for the relationship of this H^+ transport to energy coupling.

In the chemiosmotic hypothesis the H^+ free energy gradient is the means of coupling electron transport to phosphorylation. Electron transport creates an H^+ free energy gradient sufficient to drive the ATPase–H^+ pump backwards, i.e. H^+ flux downhill drives ATP synthesis. H^+ recirculation is an essential feature of this hypothesis.

The chemical intermediate hypothesis involves a (hypothetical) chemical "high energy intermediate", which couples electron transport to phosphorylation, and also powers the H^+ pump. Here H^+ transport occurs as an alternative to ATP synthesis, and no H^+ circulation occurs if electron flow and ATP synthesis are completely coupled. In both hypotheses a low P_{H^+} is required for the coupling membrane, since in either case H^+ leakage uncouples electron flow from phosphorylation.

E. Active and Passive Transport of H^+ and OH^-

From the Nernst equation, the values of the potential difference across the plasmalemma of plant cells (the cytoplasm being 40–180 mV negative with respect to the outer solution: Scott, 1967; Gutknecht and Dainty, 1968) imply a cytoplasmic pH 0·5–3·5 units more acid than that in the medium if H^+ is at passive flux equilibrium. Only in a medium of appropriate and constant pH (e.g. the sea) could passive H^+ distribution be consistent with the cytoplasmic pH values quoted in IIA. Thus transport of H^+ or OH^- against a free energy gradient is, in general, required.

In media of low pH an H^+ extrusion pump occurs (Pitman, 1970; Saddler, 1970; Slayman, 1970). This certainly applies to situations in which there is a net internal production of H^+ in metabolism (see IIIC1). It also applies to the case of excess internal OH^- production in NO_3^- assimilation if P_{H^+} is so high as to allow more H^+ to enter the cell than is required to neutralize the metabolically produced OH^-. When a net OH^- loss to a very alkaline medium occurs, OH^- loss is probably active (Raven, 1970).

There is considerable evidence, largely from electrical studies, for a substantial passive permeation of H^+ in a number of eukaryotic cells (e.g. Kitasato, 1967; Slayman, 1970; Rent et al., 1972). P_{H^+} may be as high as $10^{-3}\ cm\ s^{-1}$; this adds to the H^+ disposal problem of a cell with a net metabolic production of H^+ and an inwardly directed H^+ diffusion gradient; active H^+ effluxes of more than 10 pmol $cm^{-2}\ s^{-1}$ have been suggested. A high P_{H^+} would aid in the disposal of OH^- in cells with a net OH^- production in metabolism, again assuming an external pH low enough to give an inwardly-directed passive driving force on H^+.

In coupling membranes direct estimates of the energy gradient are difficult to make because direct measurements of the electrical component of the driving force are technically almost impossible (Lassen et al., 1971). Indirect estimates suggest that the H^+ fluxes supported by downhill electron transport and by ATPase activity are active, and that P_{H^+} is low (Mitchell, 1966, 1968, 1970; Greville, 1969).

F. H^+ Transport in Relation to other Solute Transport Processes at the Outer Membrane of Eukaryotes

Robertson (1968) reviews the evolution of the hypotheses which have been put forward to account for ion transport at the outer membrane of eukaryotes. A recent hypothesis (Smith, 1970) relates an H^+ gradient established by an H^+ pump powered by ATP or redox reactions to cation accumulation, and (via OH^- efflux down a free energy gradient) to active accumulation of Cl^- (see also Smith and Lucas, this volume: IV.3).

III. Reappraisal

Our proposal is that H^+ (and OH^-) pumps arose as mechanisms for regulating the internal pH of cells, and that this is still a major function for them. As has been argued in IIE, the H^+ extrusion pump is probably the basic one. This pump exchanges H^+ for K^+ either electrogenically (Slayman, 1970) or by chemical coupling. The coupling to K^+ rather than Na^+ may be related to a requirement for volume regulation in wall-less cells by a K^+–Na^+ pump (Gutknecht, 1970), with subsequent enzyme adaptation to a K^+-rich medium (Evans and Sorger, 1966). Thus the walled cells of plants, fungi and prokaryotes can have both K^+–H^+ and K^+–Na^+ exchange mechanisms, although the osmoregulatory requirement for the K^+–Na^+ pump disappeared with the origin of the cell wall. In a growing cell K^+–H^+ exchange (i.e. net cation accumulation) will predominate over the K^+–Na^+ exchange characterized in steady-state systems.

The energy supply to the H^+ pump is not well characterized in most eukaryote outer membranes. The H^+ pump in coupling membranes can be driven by either ATP or redox reactions (see IID); the same two energy sources are probably involved at the outer membrane of eukaryotes (Slayman, 1970; Luttge et al., 1970; Siekevitz, 1965).

The H^+ pump can be involved in turgor generation in walled cells. Synthesis of organic acids, coupled to K^+–H^+ exchange, leads to accumulation of K^+ organate$^-$ in excess of purely biochemical requirements for the anion. This increases the internal osmotic pressure and hence turgor. In vacuolate cells much of the extra organic anions and K^+ are located in the vacuole (Oaks and Bidwell, 1970).

The most plausible explanation for this accumulation of "excess" K^+ organate appears to be that of Johnson et al. (1963), and of Hiatt (1967a, b) and Hiatt and Hendricks (1967). Here the H^+ extrusion pump may be regarded as being stimulated in some way by the requirements of turgor generation. This raises the internal pH (within the permitted range: see IIA), and stimulates the synthesis of organic acids, which stabilizes the internal pH and provides anionic partners for the K^+ taken up in exchange for H^+.

One specific aspect of turgor regulation with which H^+ extrusion can be associated is stomatal opening; the K^+ influx which is involved in opening in many species is not associated with the influx of a specific anion (e.g. Thomas, 1970).

In bacteria and fungi K^+ organate$^-$ accumulation is the predominant ionic mechanism for maintaining the cellular osmotic pressure (Christian and Waltho, 1962; Shere and Jacobsen, 1970; Slayman, 1970). Active influx of inorganic anions occurs only in the case of the metabolized anions ($H_2PO_4^-$, $SO_4^=$, NO_3^-) and not, apparently, for Cl^- (Christian and Waltho, 1962; Kotyk and Janacek, 1969; Shere and Jacobsen, 1970).

An alternative method of turgor generation is to accumulate both cation and anions from the medium. This is the main method in most vacuolate algae (Gutknecht and Dainty, 1968). Higher plant cells (including pauci vacuolate cells: Gerson and Poole, 1972) and pauci-vacuolate algae (Barber 1968; Schaedle and Hernandez, 1969) have an intermediate mechanism in which organic anions are synthesized *and* inorganic anions (e.g. Cl^- are accumulated from the medium. The change in capacity for salt absorption by slices of storage tissues aged in water or dilute $CaSO_4$ can be regarded as a changeover from internal (organic) anion movement to external (inorganic anion uptake. Cation uptake remains relatively constant during ageing (van Steveninck, 1966). In those plants with an extremely acid sap (see IIA) there must be movement of organic acids into the vacuole, which thus acts as an internal sink for H^+ (see Ranson, 1965).

As regards the role of the H^+ extrusion pump in inorganic anion accumulation, it is possible that K^+ enters in exchange for H^+, while Cl^- exchanges for OH^-. Thus the loss of OH^- in exchange for Cl^- may be construed, in terms of both pH regulation and turgor generation, as equivalent to internal OH^- neutralization by organic acid synthesis. Smith (1970, 1971) has presented evidence that Cl^- influx in *Chara* is increased when the ratio of internal to external OH^- is increased. Present evidence does not allow a definite decision to be made as to whether the Cl^- influx is energized by the OH^- gradient, or whether some other energy source is required. It is likely that both cation and anion pumps (and passive movements) are controlled in some way by the requirements of regulation of intracellular pH, even if the involvement of H^+ and OH^- transport is not as close as that envisaged by Smith (1970, 1971). The hypothesis of MacRobbie (1969, 1970) involves uptake of "salt" ($KCl + NaCl$) from the medium by pinocytosis, followed by rapid transfer to the vacuole. It is difficult to extend this mechanism to embrace H^+ regulation; such mechanisms, which only apply to "salt" uptake, must be supplementary to ion transport mechanisms which can accommodate H^+ regulation.

The data on changes in the ionic composition of plants in response to different nitrogen sources (Dijkshoorn *et al.*, 1968; Kirkby, 1969) may also be reasonably accounted for in terms of the operation of an H^+ pump. NH_4^+ nutrition leads to a lower total ionic content which may be related to competition between NH_4^+ and other ions during uptake; and to the decreased cytoplasmic pH which suppresses organic acid synthesis and (to a lesser extent) inorganic anion uptake. The production of OH^- during NO_3^- assimilation can explain the extra organic anions present in plants grown on NO_3^-. An explanation in terms of H^+ pumping has greater generality than the hypotheses proposed to account solely for pH regulation during nitrogen assimilation (Dijkshoorn, 1962).

Thus the concept of an active H^+ extrusion pump serves to unify a number of phenomena in plant ion relations. It also serves as a logical starting point for the evolution of chemiosmotic coupling of electron transport to phos- phorylation. Mitchell (1970) proposes that this coupling arose from the presence, in the same cell outer membrane, of a redox-powered and an ATP- powered H^+ extrusion pump. The redox pump could produce a larger H^+ free energy gradient than could the ATP-driven pump (using ATP from fermentation), thus coupling of electron flow to ATP synthesis arose via H^+ circulation. Mitchell (1970) proposed that the H^+ pumps arose as a means of setting up an H^+ gradient which could drive other transport processes, coupled to the back diffusion of H^+. We propose that both the coupling of other solute transport processes, and the coupling of electron transport to phosphorylation, arose from H^+ pumps which had evolved as a means of regulating intracellular pH.

In eukaryotes, the coupling membrane has been removed from the cell outer membrane, by division of labour or endosymbiosis, into the energy- transducing organelles. The organelles carry out coupled electron transport, and also H^+-coupled solute transport (Mitchell, 1970). In the outer mem- brane, H^+ transport occurs which regulates intracellular pH and also intra- cellular ion content.

Ion transport mechanisms at the tonoplast have received little attention in this paper. Rungie and Wiskich (1972a) have suggested that part of the anion-stimulated ATPase activity of turnip microsomes represents a tono- plast ATPase vectorially the reverse of the plasmalemma ATPase proposed by Smith (1970). However, it is possible that ion accumulation in the vacuole involves vesicles or "minivacuoles" (MacRobbie, 1969; 1970). ATPase activity is also shown by microsomal fractions derived from the endoplasmic reticulum; these also have (nonphosphorylating) electron transport systems (see Rungie and Wiskich, 1972b). Thus it is possible that electron transport, ATPase activity and ion transport can occur at all biological membranes, but that the degree of coupling of any two (or all three) processes varies.

References

BARBER, J. (1968). *Nature, Lond.* **217**, 876–878.
BUDD, K. and HARLEY, J. L. (1962a). *New Phytol.* **61**, 138–149.
BUDD, K. and HARLEY, J. L. (1962b). *New Phytol.* **61**, 244–255.
CHRISTIAN, J. H. B. and WALTHO, J. A. (1962). *Biochim. biophys. Acta* **65**, 508–509.
DIJKSHOORN, W. (1962). *Nature, Lond.* **194**, 165–167.
DIJKSHOORN, W. and VAN WIJK, A. L. (1967). *Pl. Soil* **26**, 129–157.
DIJKSHOORN, W., LATHWELL, D. J. and DE WIT, C. T. (1968). *Pl. Soil.* **29**, 369–390.
EVANS, H. J. and SORGER, G. J. (1966). *A. Rev. Pl. Physiol.* **17**, 47–76.
GERSON, D. F. and POOLE, R. J. (1972). *Pl. Physiol., Lancaster* **50**, 603–607.
GREVILLE, G. D. (1969). *In* "Current Topics in Bioenergetics" (D. R. Sanadi, ed.), Vol. 3, pp. 1–78. Academic Press, New York.
GUTKNECHT, J. (1970). *Am. Zool.* **10**, 347–354.
GUTKNECHT, J. and DAINTY, J. (1968). *A. Rev. Oceanogr. Mar. Biol.* **6**, 163–200.
HIATT, A. J. (1967a). *Pl. Physiol. Lancaster* **42**, 294–298.

HIATT, A. J. (1967b). *Z. Pflanzenphysiol.* **56**, 233–245.
HIATT, A. J. and HENDRICKS, S. B. (1967). *Z. Pflanzenphysiol.* **56**, 220–232.
JOHNSON, R. E., JACKSON, P. C. and ADAMS, H. R. (1963). *Pl. Physiol., Lancaster* **38**, xxv.
KIRKBY, E. A. (1969). *In* "Ecological Aspects of the Mineral Nutrition of Plants" (I. E. Rorison, ed.), pp. 215–235. Blackwells, Oxford.
KIRKBY, E. A. and MENGEL, K. (1967). *Pl. Physiol., Lancaster* **42**, 6–14.
KITASATO, H. (1967). *J. gen. Physiol.* **52**, 60–87.
KOTYK, A. and JANACEK, K. (1970). "Cell Membrane Transport". Plenum Press, New York.
LASSEN, U. V., NIELSEN, A-M., POPE, L. and SIMORSEN, L. O. (1971). *J. Membrane Biol.* **6**, 269–288.
LEGGETT, J. E. (1968). *A. Rev. Pl. Physiol.* **19**, 333–346.
LEHNINGER, A. L. (1970). "Biochemistry". Worth, New York.
LÜTTGE, U., PALLAGHY, C. K. and OSMOND, C. B. (1970). *J. Membrane Biol.* **2**, 17–30.
LYCKLAMA, J. C. (1963). *Acta Bot. Neerl.* **12**, 361–423.
MACMILLAN, A. (1956). *J. exp. Bot.* **7**, 113–126.
MACROBBIE, E. A. C. (1969). *J. exp. Bot.* **20**, 236–256.
MACROBBIE, E. A. C. (1970). *J. exp. Bot.* **21**, 335–344.
MITCHELL, P. (1966). "Chemiosmotic coupling in oxidative and photosynthetic phosphorylation". Glynn Research, Bodmin.
MITCHELL, P. (1968). "Chemiosmotic coupling and energy transduction". Glynn Research, Bodmin.
MITCHELL, P. (1970). *Symp. Soc. gen. Microbiol.* **20**, 121–166.
OAKS, A. and BIDWELL, R. G. S. (1970). *A. Rev. Pl. Physiol.* **45**, 787–790.
PITMAN, M. G. (1970). *Pl. Physiol., Lancaster* **45**, 787–790.
RANSON, S. K. (1965). *In* "Plant Biochemistry" (J. Bonner and J. E. Varner, eds) pp. 493–525. Academic Press, New York.
RAVEN, J. A. (1970). *Biol. Rev.* **45**, 167–221.
RENT, R. K., JOHNSON, R. A. and BARR, C. E. (1972). *J. Membrane Biol.* **7**, 231–244.
ROBERTSON, R. N. (1968). "Protons, Electrons, Phosphorylation and Active Transport". Cambridge University Press.
RUNGIE, J. M. and WISKICH, J. T. (1972a). *Pl. Physiol., Lancaster*, in press.
RUNGIE, J. M. and WISKICH, J. T. (1972b). *Aust. J. biol. Sci.* **25**, 89–102.
SADDLER, H. D. W. (1970). *J. gen. Physiol.* **55**, 802–821.
SCHAEDLE, M. and HERNANDEZ, R. (1969). Abstr. XI Int. Bot. Congr. (Seattle), p. 191.
SCOTT, B. I. H. (1967). *A. Rev. Pl. Physiol.* **18**, 409–418.
SHERE, S. M. and JACOBSEN, L. (1970). *Physiologia Pl.* **23**, 51–62.
SIEKEVITZ, P. (1965). *Fed. Proc. Fedn Am. Socs biol. Sci.* **24**, 1153–1155.
SLAYMAN, C. L. (1970). *Am. Zool.* **10**, 377–392.
SMITH, F. A. (1970). *New Phytol.* **69**, 903–917.
SMITH, F. A. 1(971). Proceedings of the First European Biophysics Congress (E. Broda, A. Locker and H. Springer-Lederer, eds) pp. 429–433. Verlag der Weiner Medizinschen Akademie, Vienna.
STREET, H. E. and SHEAT, D. E. G. (1958). *In* "Encyclopaedia of Plant Physiology", (W. Ruhland, ed.), Vol. 8, pp. 150–165. Springer, Berlin.
THOMAS, D. A. (1970). *Aust. J. biol. Sci.* **23**, 981–989.
VAN STEVENINCK, R. F. M. (1966). *Aust. J. biol. Sci.* **19**, 271–281.

Discussion

Spanswick asked West if he had tried to measure the potential using highly mobile ions as some Russian workers had done. West said that they had not but that they did not measure the potential as such. What they did was to assume K^+ in equilibrium in the presence of valinomycin, and calculate the potential from the internal and external K^+ concentrations. Cram commented that West's system was very similar to the Na^+-dependent sugar and amino acid system in various animal cells. West replied that it does have striking similarities, but that they are sure that it is not dependent on Na^+ influx, but on H^+ influx. Raven then asked West why the work of Kaback has been rejected; it is based on the specificity of various electron donors to the electron transport chain. West said that the basis for his rejection of Kaback's model is that it depends on K^+ transport, and that he is certain that K^+ is passively equilibrated across the membrane in the presence of valinomycin. He adduced the evidence from the artificial membrane studies to show that valinomycin is a very effective ionophore for K^+. Higinbotham then asked on the mode of action of N-ethyl maleimide on blocking galactose transport. West said that N-ethyl maleimide binds specifically and tightly to the galactose carrier, probably to an SH site near the active centre. Cram then asked if there is any evidence that Kennedy's protein (this is the N-ethyl maleimide-bound protein) is a carrier for galactose. West replied it is a product of the Y gene, it is membrane-bound, it has the specificities one would expect of the carrier, it does react with N-ethyl maleimide and this reaction is blocked by thiodigalactoside. These factors together constitute good evidence that this is the carrier.

Walker then commented on Van Steveninck's paper; the effects of tris on chloroplast preparations in the light are similar to those of methionine in the light, where chloroplast distortion is produced with characteristic membrane contortions. Van Steveninck said that is correct; these effects only occur in the light and may result from a specific aldehyde–amine interaction during fixation. The tris enters the chloroplasts as a neutral molecule and is then ionized. There may be a disruption of the H bonds between the protein–lipid complexes, so producing the structural alteration in the chloroplast. Neumann then said that the effects mentioned by Walker were observed within minutes. Van Steveninck added that the distortion in wheat and barley leaf slices take place within 3 h; in chloroplast preparations, the effect is more rapid. Van Steveninck then agreed with Neumann that tris had a slight uncoupling effect; it blocks on the pathway leading to oxygen evolution, but the blockage is quite reversible. Barber said that photosystem II inhibition only occurs at 0·8 M tris—a very high level.

Smith then commented on the technique; in *Chara* tris stimulates Cl^- influx, but not K^+ and in the tris solutions HCl is always present, whereas in the controls HCl is not present. To what extent are the effects observed due to Cl^- rather than to tris? Van Steveninck said the effects were also found with tris sulphate. Shone then asked if there was any information on the concentrations of tris in the organelles. Van Steveninck said he had no information; they had attempted nitrogen assays, but the amounts of endogenous nitrogen in the tissue mean that the small amounts from the tris are neglible. Pitman then asked if the swelling observed in the electron micrographs takes place on the addition of tris rather than during E.M. fixation Had Van Steveninck observed the chloroplasts during preparation by light microscopy? Van Steveninck said he had not, but in dark-treated chloroplasts the effect is not found and both preparations are fixed identically.

Commenting on Raven and Smith's paper, Barr said if there is H^+ extrusion into the vacuole (which is likely because the electrochemical situation favours a low pH in the cytoplasm) and if Spanswick is correct that conductance is primarily a flux through an electrogenic pump, then what does resting potential mean for a microelectrode in the vacuole? Raven said, "Yes I think the tonoplast is the plasmalemma running in the opposite direction The pH in the vacuole is confused by the presence of organic anions there." He then cited the brown alga *Desmarestia* which has 1N H_2SO_4 in the vacuole; there must presumably be H^+ secretion from the cytoplasm into the vacuole. Barber then asked how Raven and Smith's present publication related to Raven's past work showing cation movements in connection with different energy sources. Raven said that he thought the transport now under discussion is a method of regulating pH which may then be tied into a cation–anion coupled system of the kind described in his earlier publications. Thus, the K^+–H^+ ion exchange he is discussing here is rather different from the K^+–Na^+ exchange discussed in his earlier papers. He speculated that the original K^+–H^+ exchange in primitive cells has evolved partly to a K^+–Na^+ exchange. During this evolution, the Na^+–K^+ exchange has in some cases picked up ouabain sensitivity.

Cram then suggested that H^+ control in algae can be maintained by excreting H^+; but in a higher plant cell—in a leaf for instance—extruding

H$^+$ into the free space does not get one very far. Raven agreed that the situation in higher plants is rather more confusing. In those plants which reduce nitrate in the leaves, it is possible that the OH$^-$ produced is circulated downward in the phloem. Excess H$^+$ production in the leaves is more difficult to deal with since the phloem pH is 8, and therefore one imagines it to be a rather poor transport system for H$^+$. Smith then commented that it is interesting that ion uptake into the leaf slices of higher plants is much less affected by light than is ion uptake into giant algal cells. This may be due to the fact that light effects do involve pH alterations, and algal cells can have large external pH variations, whereas leaf cells in higher plants are in a much more controlled pH environment. Further, in higher plant cells K$^+$ uptake with H$^+$ efflux and involvement of organic anions is perhaps the natural thing, and age-induced Cl$^-$ uptake is somewhat an artefact. Raven also commented on the distinction Smith had made between higher plants and algal cells. He said that higher plant leaf cells, for example, are not normally supplied with ions in the light, and therefore most of the ion accumulation occurs in the dark, whereas algal cells have ions surrounding them at all times and in the dark they may conserve energy by shutting off ion accumulation.

Cram said that he did not agree with their ideas in ageing tissue storage discs. In carrot, bicarbonate fluxes increase 10-fold after 1–4 days of washing exactly paralleling the increase in Cl$^-$ flux. Furthermore, the net increase in Cl$^-$ uptake is not due to an increase in Cl$^-$ influx across the plasmalemma. The influx remains more or less the same, but Cl$^-$ efflux decreases with ageing. Thus, one cannot simply imagine that the switch is from a K$^+$–H$^+$ exchange to K$^+$ uptake linked with Cl$^-$. Lüttge then made the point that the work on leaf discs differs from work on algae in that the tissue of leaf discs is already damaged. He said that, in his work on *Atriplex*, he found a difference in light stimulation on ion uptake depending upon the physical size of the leaf discs he was using. Lüttge then gave details of K$^+$ uptake in freshly cut and aged maize leaf slices. There is no light stimulation with fresh slices, but there is light stimulation with aged slices. In the dark, aged and fresh slices behave similarly for K$^+$ uptake. Raven said that some giant algal cells which do not show light-stimulated Cl$^-$ influx can be brought to this state by giving them 12 h in the dark and then repeating the experiment.

MacRobbie then asked Raven and Smith to speculate further on the tonoplast situation. She said, "If it is the plasmalemma working in reverse, would you consider the pump to be electrogenic, but not very much so, because the tonoplast potential is small? Are there also the other fluxes, K$^+$ going into the cytoplasm and Cl$^-$ into the vacuole?" Raven agreed that this is a difficult problem because it does seem that Na$^+$ goes into the vacuole, and that K$^+$ is near equilibrium across the tonoplast. Raven said he would not like to speculate further until he knew what the very high conductance at the tonoplast really meant. Spanswick then commented that the tonoplast situation is very different from the plasmalemma in that there are high K^+ concentrations on either side, so that even if an electrogenic H^+ pump were present it may not show up because of the high K^+ permeability at the tonoplast.

Jeschke then commented on work in Würzburg on the pH dependence of Na^+ and K^+ uptake by a unicellular alga *Ankistrodesmus*. Na^+ uptake has an optimum at pH 7–8 while K^+ shows a steady increase as pH increases. Raven asked whether these experiments were in the presence of Cl^- or PO_4^-. In *Hydrodictyon* K^+ and Cl^- seem to go in together and so do Na^+ and PO_4^-. Walker commented that in 1969 he, Hope and Simpson gave a paper which showed Cl^- efflux to be low in the light and high in the dark. There is a speculation in the paper that this Cl^- may efflux on a pump when the energy supply is removed. He now realized that it is a Cl^-–OH^- coupled pump. This proposal by Raven and Smith may be a small revolution in scientific thinking. When the pH gradient set up in the light is removed then Cl^-–OH^- exchange will run backward and out will come Cl^- in the dark. This may be a buffering system in the dark to eject OH^- from the cell.

Finally, Pallaghy supported Lüttge's earlier assertion that leaf slices are disturbed systems; he said that, using similar dye techniques in *Elodea* leaves to those used for the giant algae, large H^+ fluxes can be detected; in stomata, K^+ values rise from 80 to 300 mM during the turgor alterations necessary for opening the stoma. Raven commented that it is difficult to know where the H^+ ions go when they leave the guard cells in this case. Pallaghy agreed but pointed out that there is a large dilution effect into the very much larger epidermal cells, so that the pH change in these need not be very large.

Section V

Na$^+$–K$^+$ Transport and Halophytes

Chairman: M. Pitman

V.1

K+-Stimulated Na+ Efflux and Selective Transport in Barley Roots

Wolf Dietrich Jeschke

Botanisches Institut I der Universität Würzburg
Germany

I. Introduction

Measurement of membrane potentials (Scott *et al.*, 1968, Pierce and Higinbotham, 1970) and of fluxes by compartmental analysis (Pitman and Saddler, 1967; Etherton, 1967) has provided evidence for an active extrusion of Na+ at the plasmalemma as the mechanism of K–Na-selectivity in higher plants. However, the sodium efflux pump has not been demonstrated directly, for obvious reasons. A Na+ efflux pump operates from a rather limited source, the Na+ content of the cytoplasm. If the pump were to work efficiently it would soon remove most of the cytoplasmic sodium content. Thus a K+-stimulated Na+ efflux should be transient. Another reason is that a sodium-rich cytoplasmic phase is required from which the pump may draw its sodium. Unfortunately but significantly, one of the properties

285

of root cell cytoplasm also seems to be a high K^+ and a low Na^+ content (Pitman and Saddler, 1967; Scott et al., 1968).

Nevertheless the attempt was made to obtain a K^+-stimulated Na^+ efflux in barley roots which had been grown with Na^+ and without K^+. When K^+ was added at the start of efflux experiments the Na^+ efflux appeared—by indirect evidence—to have increased in comparison with the control without K^+ (Jeschke, 1970).

However, these measurements using compartmental analysis are limited by the polarity of efflux in excised roots: transport through the cut ends of the xylem may contribute considerably to the apparent total efflux J'_{co} (Weigl, 1969; Pitman, 1971).

In this paper the modified compartmental analysis introduced by Pitman (1971) is applied to the *transient* fluxes of Na^+ that occur on the addition of K^+ to barley roots, although this method originally applied only to steady state fluxes and conditions.

The measurements confirm earlier conclusions (Jeschke, 1970) and permit quantitative estimates and a localization of the K^+-dependent Na^+ efflux in barley roots.

II. Methods and Estimation of Fluxes

A. Abbreviations (compare Fig. 3)

J_{oc}, J_{co}	influx and efflux at the plasmalemma of the cortical cells, [μmol g^{-1} fresh wt h^{-1}], [as for the following fluxes]
J_{cv}, J_{vc}	influx and efflux at the tonoplast
R'	transport from the cytoplasm (of the xylem parenchyma cells) through the xylem
J'_{co}	apparent efflux from the symplasm of the entire excised roots: $J'_{co} = J_{co} + R'$
M_c, M_v	cytoplasmic or vacuolar content [μmol g^{-1} fresh wt]
$J_{out\,steady}$	quasi-steady efflux of ^{22}Na at the end of the elution [cpm g^{-1} fresh wt h^{-1}]
$J_{out\,init.}$	initial efflux of ^{22}Na (cytoplasmic component), cf. Fig. 1, [cpm g^{-1} fresh wt h^{-1}]
$\sigma_o, \sigma_c, \sigma_v$	specific activity of ^{22}Na [cpm μmol^{-1}] in the outside solution, cytoplasmic phase (symplasm) or vacuole
k_v, k_c	rate constants of exchange of ^{22}Na between vacuole or cytoplasmic phase and outside solution [h^{-1}]
CCCP	carbonyl-cyanide-m-chlorophenylhydrazone

B. Material

Barley plants (*Hordeum distichum* L., var. "Müllers Sommergerste") were cultivated for 7 days in the greenhouse in liquid culture under "low salt" conditions in a 0·5 mM $CaSO_4$ solution. The solution was not aerated. The roots were about 5 cm long when harvested.

C. Measurement of Na$^+$ Fluxes and Transport

Steady state fluxes of Na$^+$ at the plasmalemma and tonoplast and Na$^+$ transport through the xylem were determined by methods similar to the modified compartmental analysis given by Pitman (1971), which have been described in detail elsewhere (Jeschke, 1972).

Roots were excised and mounted in vessels with two chambers in such a way that the major part (2·5–3 cm) with the tips was in chamber S, while the cut ends (2–4 mm) were in chamber X. The roots were sealed between the chambers with silicone grease and "Parafilm".

To start an influx of Na$^+$ and ^{22}Na a labelled solution was added to S (20·5 ml) and X (9·5 ml); it contained, in mequiv l^{-1}: NaH_2PO_4/Na_2HPO_4 (PH 5·8): 1; $Ca(NO_3)_2$: 6; $MgSO_4$: 1, but no K$^+$. The solution in both chambers was aerated. Influx and transport of ^{22}Na and Na$^+$ was measured by counting the radioactivity in S and X. To ensure constant σ_o and $[Na^+]_o$, ^{22}Na and Na$^+$ were added as often as required. At the end of the influx period (after 21–22 h) the roots continued to take up ^{22}Na and Na$^+$ in S but transported the same amount to X. Hence accumulation *in* the roots had ceased and influx was equal to the total efflux:

$$J_{oc} = J'_{co} \quad \text{or} \quad J_{oc} = J_{co} + R', \quad \text{and} \quad J_{cv} = J_{vc}. \tag{1}$$

All Na$^+$ in the roots (except the small endogenous content) was taken up from the labelled solution, so that

$$\sigma_o = \sigma_c = \sigma_v. \tag{2}$$

After equilibration an unlabelled solution of the same composition was added, and the efflux of ^{22}Na in S and X was measured in consecutive samples for several hours (Fig. 1). The samples were dried, 2,5-diphenyl-oxazole in toluene was added, and the β-emission of ^{22}Na was measured in a liquid scintillation counter (Beckman Instruments CPM 100). After removal of the scintillator the Na$^+$ and K$^+$ content was determined by flame photometry. After elution the roots were sampled, weighed, dried, and analysed for ^{22}Na, Na$^+$ and K$^+$.

D. Evaluation of Efflux Measurements

The efflux in S can be considered entirely as an efflux through the root cortex. The efflux in X consists of xylem efflux and a small contribution of the 2–4 mm root cortex. The latter could not be estimated, since efflux through the older cortex in X does not necessarily correspond to the cortical efflux in S. Thus the efflux in X was taken to be an efflux through the xylem. It is expressed relative to the total fresh weight of the roots. The loss of ^{22}Na in S is related to the fresh weight in S since only this part of the root contributes to this efflux.

If the loss of ^{22}Na in S and X is plotted logarithmically against time (Fig. 1), three phases of cortical efflux are clearly seen: a rapidly decreasing loss from the free space, an intermediate (cytoplasmic) component and a steady, almost constant loss, $J_{\text{out steady(S)}}$. The latter can be attributed to the loss of tracer from the vacuoles (Pitman, 1963, 1971; Cram, 1968), which decreases slowly with the rate constant k_v. The loss of ^{22}Na in X consists mainly of two parts, an efflux from a cytoplasmic phase and a quasi-steady efflux from a vacuolar phase; there is less free space efflux than in S (Fig. 1) consistent with the small amount of cortical efflux in X.

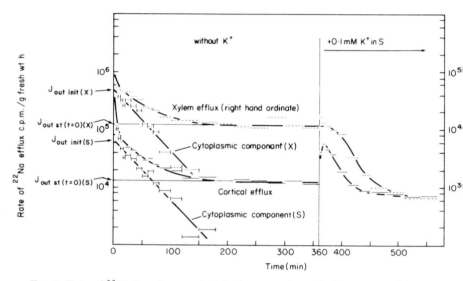

FIG. 1. Rate of ^{22}Na loss from excised barley roots through the cortex and the xylem. Experimental conditions: 21 h influx, 1 mM Na$^+$ + 21 μCi ^{22}Na/mmol, 0 K$^+$; 0–360 min: elution in inactive solution 1 mM Na$^+$, 0 K$^+$; 24–26°C; 360–570 min: re-elution in the same solution but + 0·1 mM KCl in S. The quasi-steady efflux in S and X was extrapolated by the factor 1·071/360 min (calculated from k_v).

As shown in Fig. 1, the quasi-steady efflux in S and X can be extrapolated to $t = 0$, yielding $J_{\text{out steady}(t=0)}$. The sum of the steady losses in S and X is related to J'_{co} and the steady state value of σ_c. If the *symplasm* of the roots behaves as *one rapidly mixed cytoplasmic phase*, which seems to be true (cf. Läuchli et al., 1971), its specific activity σ_c will be given by σ_v, by the flux into it J_{vc}, and by the fluxes out J_{cv} and J'_{co}. With (2) we may therefore write (cf. Pitman, 1971)

$$J_{\text{out steady}(t=0)(S+X)} = J'_{co}\sigma_v \frac{J_{vc}}{J'_{co} + J_{cv}} = J'_{co}\sigma_o \frac{J_{vc}}{J'_{co} + J_{cv}}. \qquad (3)$$

In S and X, $J_{\text{out steady}}$ may be subtracted from the measured ^{22}Na efflux.

In this way an efflux component is obtained which decreases linearly with time (Fig. 1). This is the cytoplasmic component of efflux in S or X which decreases—parallel with σ_c—at a rate corresponding to k_c. In S or X linear extrapolation to $t = 0$ yields the cytoplasmic efflux component at the start of elution $J_{\text{out init.}}$ (Fig. 1). It can be shown that

$$J_{\text{out init.(S+X)}} = J'_{co}\sigma_o\frac{J'_{co}}{J'_{co} + J_{cv}}.\tag{4}$$

As shown for the conventional compartmental analysis by Cram (1968), $J'_{co}/(J'_{co} + J_{cv})$ describes that part of cytoplasmic activity that is lost to the external solution.

With (2) the ^{22}Na efflux at $t = 0$ may be converted to Na$^+$ efflux. Further, at $t = 0$, the total loss of Na$^+$, i.e. $J_{\text{out steady(S+X)}} + J_{\text{out init.(S+X)}}$, is J'_{co} or with (1) J_{oc}. Similarly $J_{cv} = J_{vc}$ may be calculated from (3). As in the conventional efflux analysis (Pitman, 1963; Cram, 1968) $M_c = k_c/(J'_{co} + J_{cv})$. M_v can be obtained by chemical means or from

$$k_vM_v = \frac{J'_{co} \cdot J_{vc}}{J'_{co} + J_{cv}}.\tag{5}$$

As has been shown by Pitman (1971), the separate determination of efflux in S and X further permits an estimation of the Na$^+$ efflux through the plasmalemma of the cortical cells J_{co} and of the transport through the xylem R'. J_{co} is obtained as $(J_{\text{out steady}(t=0)} + J_{\text{out init.}})_S \cdot \sigma_o^{-1}$ and R' similarly as $(J_{\text{out steady}(t=0)} + J_{\text{out init.}})_X \cdot \sigma_o^{-1}$.

The evaluation of transient Na$^+$ fluxes will be described together with the results.

III. Results

A. Steady State Fluxes of Sodium

The efflux of ^{22}Na from barley roots in a solution containing 1 mM Na$^+$ but no K$^+$ under steady state conditions is shown on the left hand side of Fig. 1 (0–360 min). The roots had been loaded to flux equilibrium with Na$^+$ and ^{22}Na for 21 h. As outlined in the methods, the fluxes of Na$^+$ at the plasmalemma J_{oc} and J_{co}, at the tonoplast J_{cv} and J_{vc}, and the transport R' through the xylem may be estimated from the efflux values at $t = 0$: $J_{\text{out steady}(t=0)}$ and $J_{\text{out init.}}$. The estimated fluxes are given in Fig. 3A: considerable steady state fluxes of Na$^+$ are seen to occur in barley roots in the absence of K$^+$; the transport of Na$^+$ through the xylem occurs at a similar rate as that of Cl$^-$ in the same species (Pitman, 1971).

B. Effect of Potassium on Sodium Efflux

If low salt barley roots are loaded with Na$^+$ and ^{22}Na as in Fig. 1, i.e. without K$^+$, and if these roots are then washed out in an unlabelled solution containing K^+, a different efflux pattern is observed (Fig. 2). The loss of $^{22}Na^+$ through the root cortex (S) immediately increases due to the cytoplasmic efflux component (compare Fig. 1 and inset A in Fig. 2). Possibly after a short lag the rate of efflux decreases rapidly until a steady state efflux is reached. On the other hand, the initial efflux through the xylem scarcely alters in comparison with the control (see $J_{out\,init.(X)}$ in Figs 1 and 2); after a longer lag the efflux in S is decreased to a low steady state efflux at a similarly high rate as in S.

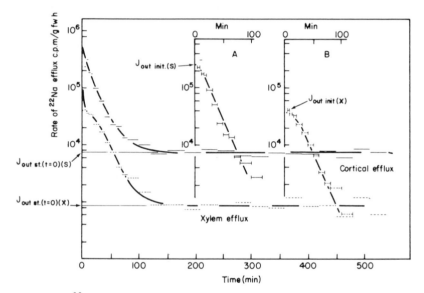

FIG. 2. Rate of ^{22}Na loss from barley roots in presence of K^+. Experimental conditions: influx as in Fig. 1, without K^+; elution in inactive solution containing 0·2 mM KCl in addition to the K^+-free solution (1 mM Na^+); 22–23°C. Factor for extrapolation of $J_{out\,steady}$ in S and X: 1·045/400 min. Inset A and B: Cytoplasmic components of efflux. The extrapolated values $J_{out\,init.}$ and $J_{out\,steady(t=0)}$ may be compared directly to those of Fig. 1 (equal σ_o).

Obviously K^+ has a strong influence on the Na^+ fluxes in barley roots and therefore the fluxes are no longer in a steady state, as is prerequisite for an evaluation of fluxes by compartmental analysis. However, if on the addition of K^+ the Na^+ fluxes are changed almost instantaneously, then σ_c and σ_v can be equated with σ_o (2) and J_{co} as well as R' can be calculated from the initial loss of $^{22}Na^+$ from the tissue, in the same way as under steady state conditions. Hence the initial fluxes $J_{co(+K)(t=0)} = J_{out\,init.(S)} + J_{out\,steady(t=0)(S)}$ and $R'_{(t=0)} = J_{out\,init.(X)} + J_{out\,steady(t=0)(X)}$ may be obtained and are given in Fig. 3B. As is seen, the addition of 0·2 mM K^+ and presumably

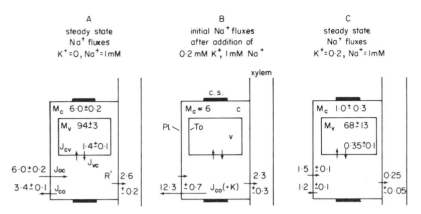

FIG. 3. Estimated values of individual fluxes and transport [μmol/g fresh wt h] and content [μmol/g fresh wt] of Na⁺ in barley root cells. The "cell" is intended to represent a series of cells between the root cortex and the xylem; Pl = plasmalemma, To = tonoplast, V = vacuole, C = symplasm (cytoplasmic phase) of the roots, c.s. = Casparian strip. (A) Steady state Na⁺ fluxes in the absence of K⁺. Values shown as mean $\pm$ S.E.M. (17 determinations). (B) Initial Na⁺ fluxes after addition of 0·2 mM K⁺. Fluxes from experiments as in Fig. 2 or re-elution experiments similar to Fig. 1; (6 determinations). (C) Steady state Na⁺ fluxes in presence of 0·2 mM K⁺ and 1 mM Na⁺ (2 determinations).

the influx of K⁺ results in a K⁺-stimulated Na⁺ efflux through the plasmalemma of the cortical cells amounting to 12·3 μmol g⁻¹ fresh wt h⁻¹.

On the other hand, R' is almost unchanged initially by the addition of K⁺ (Fig. 3B), but then it drops as indicated by the decrease in the xylem efflux of ²²Na⁺ (Fig. 2, see also below under F.).

C. K⁺-stimulated Na⁺ Efflux in Re-elution Experiments

In the flux analysis described above a contribution by the free space efflux to the K⁺-stimulated ²²Na efflux from the cytoplasm cannot be entirely excluded. This factor can be minimized if K⁺ (+Na⁺) is added when the roots have been first eluted without K⁺ to a steady state for 6 h. Under these conditions there is practically no further efflux of ²²Na from the free space, since its ²²Na content has already been exchanged and since the K⁺- exchangeable Na⁺ in the free space is rather low in the presence of an excess of Ca⁺⁺ and Mg⁺⁺ (together 7 mN).

The result of such a "re-elution" of roots in the presence of 0·1 mM K⁺ is shown in Fig. 1. Similarly to Fig. 2 the ²²Na efflux through the root cortex is increased transiently and drastically. The xylem efflux, however, is decreased after a time lag.

From these transient ²²Na efflux rates the initial value of $J_{co(+K)}$ can be estimated in a similar way. In the steady state of elution before addition of

K^+, σ_c is given by

$$\sigma_{c(t)} = \sigma_v \cdot \frac{J_{vc}}{J'_{co} + J_{cv}} = \sigma_o\, e^{(-k_v \cdot t)} \frac{J_{vc}}{J'_{co} + J_{cv}}. \qquad (6)$$

The fluxes and k_v can be obtained from the efflux analysis in the absence of K^+ during the first 6 h (Figs 1 and 3A). It can be assumed that K^+ produces an increased efflux of Na^+ *and* $^{22}Na^+$ without a sudden change of σ_c. Thus $J_{co(+K)}$ can be calculated from $\sigma_{c(t=6\,h)}$ and the K^+-stimulated efflux of ^{22}Na. The values of $J_{co(+K)}$ obtained in this way correspond to those obtained by the method described in the preceding paragraph.

D. K^+-dependent Na^+ Efflux as a Consequence of Potassium Influx

Obviously this re-elution method opens the way to the study of K^+-stimulated Na^+ efflux under varied experimental conditions. Figure 4 shows the re-elution of barley roots with 0·05 mM K together with 1 mM Na^+.

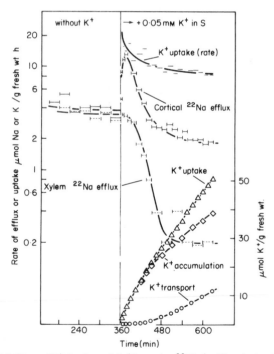

FIG. 4. Effect of 0·05 mM K^+ (+1 mM Na^+) on the $^{22}Na^+$ efflux in barley roots. 0–360 min: efflux measurement in 1 mM Na^+-solution as in Fig. 1. Beginning at 360 min: K^+ was added to the wash-out solutions in S, but not in X. The ^{22}Na efflux was converted to μmoles Na^+/g fresh wt h by the use of $\sigma_{c(360min)}$, cf. (6). K^+ uptake in the same roots was determined from the disappearance of K^+ in S, K^+ transport was measured as the increase of K^+ in X. K^+ accumulation is the difference between uptake and transport.

Even this low $[K^+]_o$ produces a strong stimulation of the Na$^+$ efflux through the cortex and a consecutive inhibition of the xylem efflux.

Figure 5 presents the K$^+$-dependent portion of J_{co}, i.e. the K$^+$-stimulated efflux minus the efflux without K$^+$, at different concentrations of K$^+$. The stimulation reaches its maximum value above 0·1 mM K$^+$, and at 0·015 mM K$^+$ the K$^+$-dependent Na$^+$ efflux is already about one-third of its maximum value.

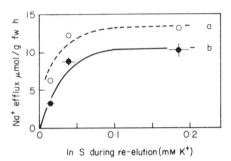

FIG. 5. Effect of K$^+$ concentration in chamber S (containing the root tips) on the K$^+$-stimulated (a) and K$^+$-dependent (b) Na$^+$ efflux at the plasmalemma of cortical cells of barley root. Results of re-elution experiments similar to Figs 1 and 4, means of two determinations. The K$^+$-dependent Na$^+$ efflux is the difference between K$^+$-stimulated and steady state Na$^+$ efflux (without K$^+$).

It has been tacitly assumed so far that the stimulation of the Na$^+$ efflux occurs as a consequence of an influx of K$^+$ ions. The K$^+$ influx should then be comparable to or larger than the K$^+$-dependent Na$^+$ efflux. In Fig. 4 the uptake, accumulation and transport of K$^+$ has been measured simultaneously with the efflux of Na$^+$ in the same roots. In these low-K roots the rate of K$^+$ uptake, which approximates during the initial period to the K$^+$ influx, is well above the value of the K$^+$-dependent Na$^+$ efflux (Fig. 4).

It may be significant that the secondary drop in the Na$^+$ efflux after the transient increase is accompanied by a corresponding decrease in the uptake of K$^+$ (Fig. 4).

It follows from similarities in their dependence on $[K^+]_o$ that the K$^+$ influx promotes the Na$^+$ efflux. According to Epstein et al. (1963) the K$^+$ influx at 0·015 mM K$^+$ amounts to about 38 % of the maximal rate. This corresponds to about one-third saturation of the K$^+$-dependent Na$^+$ efflux at the same concentration (Fig. 5).

E. Metabolic Linkage of the K$^+$-stimulated Na$^+$ Efflux

The effect of CCCP, the uncoupler of oxidative phosphorylation, was studied. Complications arise from the fact that CCCP affects the fluxes of

Na^+ in barley roots. Thus 1 μM CCCP increases the cortical ^{22}Na efflux though more slowly and to a lesser degree than does an addition of K^+. If K^+ is added 1 h *after* CCCP further increase in the Na^+ efflux can hardly be detected.

On the other hand, CCCP has only a limited effect on the steady state efflux of ^{22}Na from the xylem. But when K^+ is added, the K^+-dependent drop in Na^+ transport through the xylem is strongly inhibited, as is the transport of K^+.

From this it may be concluded that CCCP inhibits the K^+-stimulated Na^+ efflux and its consequences, which points to a linkage with respiratory metabolism.

First experiments with ouabain showed little if any effect by this inhibitor.

F. Steady State Sodium Fluxes in the Presence of K^+

Since the K^+-stimulated Na^+ efflux greatly exceeds the sodium influx, it should produce a decrease in the cytoplasmic sodium content and hence a decrease in sodium transport R', provided that this occurs symplasmatically (Jeschke, 1972).

The secondary K^+-dependent changes can be seen from Fig. 3C which shows the steady state sodium fluxes in the presence of 0·2 mM K^+. As had been anticipated, M_c and R' have been decreased strongly. Also J_{oc} is inhibited (Fig. 3C), and this may be attributed to competition between K^+ and Na^+ influxes (Epstein, 1966). Yet it seems significant that the drop in J_{oc} is less than the drop in M_c or R'. Hence selective (low) transport of sodium cannot be the consequence of selective influx alone; it must be due rather to the K^+-stimulated Na^+ efflux.

Of particular interest is the relatively small drop in the steady value of J_{co} (Fig. 3C), which will be referred to in the discussion.

IV. Discussion

Potassium and sodium uptake into the roots and shoots of several higher plant species is highly selective, with a preference for potassium. At first sight paradoxically, their shoots contain less sodium and more potassium than the roots from which they derive their ion supplies. Selectivity has been attributed to preferential uptake of potassium by the sites of system I in the roots at low concentrations (Epstein, 1966; Rains and Epstein, 1967). The sharper discrimination between K and Na in the shoot has been explained in addition by Na^+ retention in the roots and by selective withdrawal of Na^+ from the xylem sap in the roots and stems (Wallace et al., 1965).

Alternatively Pitman and Saddler (1967) attributed the K–Na-selectivity in barley to a Na^+ efflux pump at the plasmalemma of barley root cells providing an explanation for the selectivity of the roots as well as for the higher K/Na ratio found in the shoots. A sodium efflux pump at the plasmalemma produces a low Na^+ and high K^+ content within the root *cytoplasm* and the K/Na ratio in the shoots reflects the ratio in the root cytoplasm if ions are transported symplasmatically to the xylem and hence to the shoot.

In the present paper a K^+-stimulated Na^+ efflux at the plasmalemma of the cortical cells of barley roots has been measured directly. As a K^+-stimulated Na^+ efflux would be expected from a sodium efflux pump, the results lend strong support to the view of Pitman and Saddler (1967).

As has been shown in the results, an influx of K^+ and not the K^+ content in the roots stimulates the Na^+ efflux. The influence of the K^+ concentration points to a linkage with system I of K^+ uptake (Fig. 5); even at 2 mM K^+ the K^+-dependent Na^+ efflux has the same, maximal value as at 0·2 mM K^+. Consistent with a dependence on K^+ influx, K^+ uptake in the roots exceeds the Na^+ efflux and drops as the transient K^+-dependent Na^+ efflux gets smaller (Fig. 4). The results in Fig. 4 suggest that only part of the K^+ uptake is linked to the Na^+ efflux while the K^+ uptake remaining after about 2 h (Fig. 4) is presumably linked mainly to an uptake of anions.

Nevertheless, a K^+-dependent Na^+ efflux would also be necessary to maintain the low steady state Na^+ content in the cytoplasm in the presence of K^+ (Fig. 3C). Evidence for the continued operation of a K^+-dependent Na^+ efflux may be obtained from the ratio J_{co}/M_c. In absence of K^+ (Fig. 3A) this ratio is 0·57 while in its presence a ratio J_{co}/M_c of 1·2 is found (Fig. 3C). This points to a relatively higher sodium efflux in presence of K^+ compared with conditions without K^+.

As suggested by Pitman and Saddler (1967) selective transport of K^+ and Na^+ results from the sodium efflux pump. This is supported by the finding that the K^+-dependent drop in the xylem transport of Na^+ (Figs 1 and 4) is preceded by K^+-stimulated Na^+ efflux through the root cortex (Fig. 4). By an evaluation of the kinetics of cortical Na^+ efflux and xylem transport, it is found that sodium transport decreases with a lag of about 14 min behind the decrease in the cytoplasmic Na^+ content of the roots. From this it may be concluded that selective K^+ and Na^+ transport is a consequence of the K^+-stimulated Na^+ efflux.

An additional withdrawal of sodium from the xylem sap (Wallace *et al.*, 1965) does not seem probable in barley roots. In our experiments the roots were at a steady state with 1 mM Na^+ and probably saturated with Na^+ before the addition of K^+. On the other hand, competition and selectivity of K^+ and Na^+ *influxes* (Epstein, 1966) is certainly present in barley roots, as indicated by the depression of $J_{oc(Na^+)}$ by K^+ (cf. Figs 3A and C). However, selectivity of influx is not sufficient for an explanation of selective uptake and transport through the root or in whole barley plants (see paragraph F).

From these results a synthesis between influx selectivity (Rains and Epstein 1967) and K–Na discrimination by the operation of the sodium efflux pump may be achieved.

K–Na-selectivity by means of a sodium efflux pump would be limited when sodium influx exceeded the efflux rate of the pump. K–Na-selectivity in barley roots ceases when the outside Na^+ concentration reaches about 10 mM (Rains and Epstein, 1967). At this concentration the uptake of sodium rises to a value of about 12 μmol g^{-1} fresh wt h^{-1}. This rate is similar to the maximal K^+-stimulated Na^+ efflux in barley roots (Figs 5 or 3B). It is suggested therefore that sodium influx by system II and the capacity of the efflux pump determine the limits of K–Na-selectivity in barley.

Acknowledgement

This investigation was supported by the *Deutsche Forschungsgemeinschaft*. Thanks are extended to Miss Hedwig Eschenbacher for skilful and untiring technical assistance.

References

CRAM, W. J. (1968). *Biochim. biophys. Acta* **163**, 339–353.
EPSTEIN, E. (1966). *Nature, Lond.* **212**, 1324–1327.
EPSTEIN, E., RAINS, D. W. and ELZAM, O. E. (1963). *Proc. natn. Acad. Sci. U.S.A.* **49**, 684–692.
ETHERTON, B. (1967). *Pl. Physiol.* **42**, 685–690.
JESCHKE, W. D. (1970). *Planta* **94**, 240–245.
JESCHKE, W. D. (1972). *Planta*, **106**, 73–90.
LÄUCHLI, A., SPURR, A. R. and EPSBEIN, E. (1971). *Pl. Physiol.* **48**, 118–124.
PIERCE, W. S. and HIGINBOTHAM, N. (1970). *Pl. Physiol.* **46**, 666–673.
PITMAN, M. G. (1963). *Aust. J. biol. Sci.* **16**, 647–668.
PITMAN, M. G. (1971). *Aust. J. biol. Sci.* **24**, 407–421.
PITMAN, M. G. and SADDLER, H. D. W. (1967). *Proc. natn. Acad. Sci U.S.A.* **57**, 44–49.
RAINS, D. W. and EPSTEIN, E. (1967). *Pl. Physiol.* **42**, 319–323.
SCOTT, B. I. H., GULLINE, H. and PALLAGHY, C. K. (1968). *Aust. J. biol. Sci.* **21**, 185–200.
WALLACE, A., HEMAIDAN, N. and SUFI, S. M. (1965). *Soil Sci.* **100**, 331–334.
WEIGL, J. (1969). *Planta* **84**, 311–323.

V.2

The Ionic Relations of Seedlings of the Halophyte *Triglochin maritima* L.

R. L. Jefferies

School of Biological Sciences, University of East Anglia Norwich, England

I. Introduction

There have been a number of studies of the salt tolerance of higher plants (Collander, 1941; Arisz *et al.*, 1955; Black, 1956; Scholander *et al.*, 1962, 1966; Greenway and Rogers, 1963; Atkinson *et al.*, 1967; Rains and Epstein, 1967; Hill, 1967a, b, 1970a, b; Elzam and Epstein, 1969a, b; Epstein, 1969; Greenway and Osmond, 1970; Pallaghy, 1970; Lüttge, 1971). The results of these studies taken together draw attention to the selective absorption by halophytes of ions such as potassium when the plants are placed in a saline medium and the data also indicate that there is an active extrusion of sodium from the tissues of such plants. These results are essentially similar to those obtained from studies of marine algae in which all cells apparently pump out sodium and most cells actively absorb potassium and chloride under saline conditions (Gutknecht and Dainty, 1968). In addition the cytoplasm of all cells and the vacuole of most cells is electronegative with respect to external sea water.

In this study the movements of sodium and potassium between the roots of the halophyte *Triglochin maritima* and the external environment have been investigated in order to determine whether the general conclusions outlined above are applicable. Concentrations of sodium and potassium in the cyto-plasm and vacuole of the cortical cells of the root have been estimated and the results compared with values for the internal concentrations of these ions in halophytic bacteria and marine algae.

II. Materials and Methods

A. Material

Seeds of *Triglochin maritima* were collected from plants growing in the upper levels of Stiffkey salt-marshes, Norfolk, England (Grid Reference TG964441). The seeds were sown on the surface of sand which was kept moist by additions of tap water. Under glasshouse conditions germination occurred over a period of 21 days. Subsequently the trays of sand containing the seed-lings were placed in the laboratory and the seedlings which were used in the different experiments possessed root lengths of approximately 5 cm and shoot lengths of between 2·5 and 4·0 cm.

B. Flux Measurements

Before each experiment seedlings were washed free of sand and each inserted into a hollow glass tube which was placed immediately above the surface of the appropriate solution so that only the roots, which were free of the tube, were suspended in the solution. One of the following procedures took place before the start of each experiment. When pseudo-steady-state conditions were required, plants were pretreated in a non-radioactive solu-tion of composition identical with the experimental solution for a period of between 72 and 96 h. Non-steady-state conditions prevailed in some experi-ments when plants were placed directly in experimental solutions after they had been removed from the sand culture.

During the period of pretreatment and the actual experimental period plants were subjected to a temperature of 25°C and continuous illumination, supplied from mercury and tungsten lamps (500 W m^{-2}). Unless otherwise stated, in all experiments the solution bathing the roots was artificial sea water which consisted of the chloride salts of sodium, potassium and calcium at 0·5 M, 0·01 M and 0·01 M respectively. This solution was diluted with distilled water in order to produce solutions of different salinities.

1. Influx measurements

At the start of the experiments either ^{42}K or ^{24}Na (as KCl and NaCl obtained from Radiochemical Centre, Amersham) was introduced into the solutions to give a final activity of approximately 0·004 μCi ml^{-1}. The addition of isotope did not significantly alter the ionic concentrations of the solutions. At appropriate intervals of time after the introduction of the isotope six plants were harvested from a treatment, washed in distilled water for 10 s, blotted dry and the roots and shoots weighed separately. Thereafter the roots and shoots were placed in separate scintillation phials which contained 1 M HCl (10 ml) and the radioactivity counted in a Packard Tricarb Scintillation Counter using Cherenkov counting techniques (Ballance and Johnson, 1970). The specific activities of potassium and sodium in the experimental solutions were measured in a similar manner and the results calculated as moles of ion absorbed g^{-1} fresh wt of root min^{-1}.

2. Efflux measurements

As the analysis of the results of the efflux of ions from the roots is based on the assumption that pseudo-steady-state conditions exist in each experiment, the ionic compositions of the pretreatment solution, the radioactive solution and the experimental solution were identical. After the period of pretreatment the roots of plants were submerged in the radioactive solution for a period not exceeding 20 h and thereafter each plant was removed from its glass tube and washed in distilled water for 10 s. Subsequently it was placed in another glass tube which fitted into the cap of a scintillation phial containing 10 ml of non-radioactive solution of identical ionic composition. The glass tube was positioned so that only the roots were submerged in the solution. The scintillation phials were changed frequently during the efflux of the isotope and the radioactivity present in each sample counted. At the termination of an experiment a seedling was separated into root and shoot, each organ weighed, and the radioactivity present in the tissues measured as described above. The calculation and analysis of the results are discussed later.

C. Chemical Analyses

Chemical analyses were made of shoot and root material. Tissue was soaked in 1·0 M HNO$_3$ for at least 24 h and this solution filtered and diluted as appropriate.

Potassium, sodium and calcium contents of the filtrate were analysed by the use of a Unicam S.P.90 Atomic Absorption Spectrophotometer. Chloride was estimated from the results of electropotentiometric titrations (Ramsay *et al.*, 1955).

D. Electrical Potential Measurements

Potential differences between the cortical cells of the roots and the external solution were measured with an internal 3 M KCl glass micropipette electrode referred to an external 3 M KCl agar bridge. This electrode and the reference electrode, which was placed in the bathing solution that flowed continuously past the roots, were connected via calomel half-cells to a high impedance voltmeter. Measurements of electrical potential were made on cells in a 2 cm segment of root approximately 1·0 cm from the root tip.

III. Results

The relationships between the concentrations of ions present in the external solutions at different salinities and the corresponding concentrations in the roots and shoots of seedlings of *T. maritima* grown at these salinities are shown in Fig. 1. At external calcium and potassium levels above 1·0 mM the internal concentrations of these two ions in the root and shoot are invariant. In contrast, the mean internal concentrations of sodium and chloride ions in both root and shoot are related to the prevailing external concentrations of these two ions. At high salinities, high concentrations of sodium and chloride ions occur in the shoot, indicating that the shoot system does not constitute an effective barrier to the movement of these ions. In addition, the ratio of the mean internal concentration of sodium and chloride ions is close to unity.

In order to examine the driving forces acting on the movement of ions into the roots, resting potentials have been measured between the cortical cells and the external solution under pseudo-steady-state conditions. Although the cells are small similar values of potential were obtained wherever the tip of the microelectrode rested inside a cell. It is assumed therefore that the potential difference between the vacuole and the cytoplasm is less than 3 mV and values of resting potentials are shown in Fig. 1. This information is used later in the analysis of the data of the effluxes of ions. Because the potassium concentrations in the roots of *Triglochin maritima* appear to be similar at the different external salinities and because there is evidence that in some marine algae potassium is pumped into the plants (Gutknecht and Dainty, 1968) a series of experiments was undertaken to establish the characteristics of the influx of potassium in the roots of this halophyte.

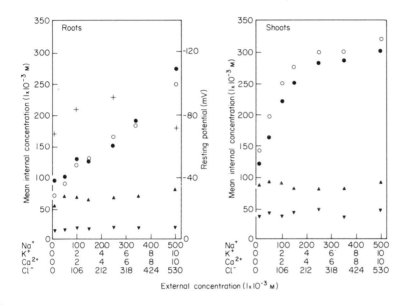

FIG. 1. The mean internal concentration of ions in the roots and shoots of plants of *Triglochin maritima* grown at different salinities together with the resting potentials between the cortical cells of the root and the external solutions.
Key:
○, chloride; ●, sodium; ▲, potassium; ▼, calcium; +, resting potential.

A. Influx of Potassium

Unless otherwise stated, the plants were not pretreated with artificial sea water but were taken from the sand cultures, washed in 0·5 mM $CaSO_4$ and used directly in each experiment. In Fig. 2 the influx of potassium into the roots of intact plants under non-steady-state conditions is shown in relation to different concentrations of sodium in the external solution. The concentrations of potassium and calcium in all solutions were 0·1 mM and 3 mM respectively and the data are based on frequent measurements of rate over a period of 1 h. The presence of sodium chloride in the external solution at certain concentrations results in an elevated influx of potassium compared with that in the absence of sodium chloride as observed initially by Parham (1971). These high rates of uptake into the tissues appear to be specifically linked with the sodium ion as the presence of other cations and anions in the experimental solutions does not produce a similar response (Fig. 3). The influx of potassium is sensitive to CCCP at a concentration of 2×10^{-6} M (Table IB). The presence of the uncoupler at all concentrations of sodium chloride reduces the influx of potassium by approximately 85% compared with the rate in the absence of CCCP. The conclusion from these

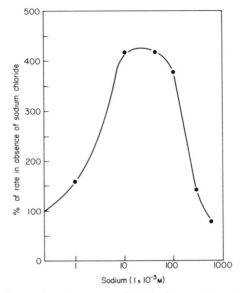

FIG. 2. The influx of potassium into roots of *Triglochin maritima* at different concentrations of sodium chloride in the external solutions. The concentration of potassium chloride in all solutions was 1×10^{-4} M and the concentration of calcium chloride 3×10^{-3} M. The results are expressed as a percentage of the influx of potassium in the absence of sodium chloride in the experimental solution. (The influx values are given in Table II.)

TABLE IA. Effect of CCCP on the influx of K^+ in roots of *Triglochin maritima*

Artificial sea water	No CCCP	CCCP (1×10^{-6} M)
1×10^{-3} M	K^+ influx (mol $\times 10^{-8}$ g fresh wt^{-1} min^{-1})	
Na$^+$ 500 K$^+$ 10 Ca^{++} 10	19·6	19·9

Plants pre-treated in artificial sea water for 72 h, and influx based on measurements of rate over a period of 60 min.

results is that under these non-steady-state conditions, where the external concentration of potassium is 1×10^{-4} M and the plants have received no pretreatment, there is an inwardly directed potassium pump operating in the roots of the seedlings.

At high concentrations of potassium in the external solution most of the movement into the tissue appears to be passive, as the influx is effectively

TABLE IB. Effect of CCCP on the influx of potassium in roots of *Triglochin maritima* at different levels of sodium

External solution			Influx of K^+ (mol $\times 10^{-8}$ g fresh wt^{-1} min^{-1})		Ratio of fluxes $\times 100$
Na^+ (1×10^{-3} M)	K^+ (1×10^{-3} M)	Ca^{++} (1×10^{-3} M)	No CCCP	CCCP (1×10^{-6} M)	
0	0·1	3·0	0·151	0·0127	14·1
1	0·1	3·0	0·147	0·0273	18·4
10	0·1	3·0	0·288	0·0388	13·5
100	0·1	3·0	0·275	0·0313	11·4
500	0·1	3·0	0·081	0·0132	18·2

Influx based on rate measurements over a period of 60 min.
Plants not pre-treated in appropriate experimental solution before experiment.

R. L. JEFFERIES

TABLE II. Effect of different external concentrations of sodium chloride on the influx of potassium in roots of *Triglochin maritima* at three external concentrations of potassium

External concentration of Na$^+$ 1×10^{-3} M	External concentration of K$^+$ 0.1×10^{-3} M	Influx of K$^+$ mol $\times 10^{-8}$ g fresh wt^{-1} min^{-1}	As % of rate in absence of Na$^+$ (K$^+$ influx)	External concentration of K$^+$ 1×10^{-3} M	Influx of K$^+$ mol $\times 10^{-8}$ g fresh wt^{-1} min^{-1}	As % of rate in absence of Na$^+$ (K$^+$ influx)	External concentration of K$^+$ 1×10^{-3} M	Influx of K$^+$ mol $\times 10^{-8}$ g fresh wt^{-1} min^{-1}	As % of rate in absence of Na$^+$ (K$^+$ influx)
0	0.1	0.50	100	1.0	3.40	100	10	51.92	100
1	0.1	0.72	143	1.0	3.70	108	10	48.12	98
10	0.1	1.68	336	1.0	5.30	159	10	57.20	111
50	0.1	1.79	358	1.0	4.90	145	10	43.12	83
100	0.1	1.59	318	1.0	4.30	128	10	47.36	91
200	0.1	0.64	128	1.0	4.40	129	10	40.78	79
500	0.1	0.43	86	1.0	3.90	114	10	41.44	80

Calcium at a concentration of 3 × 10⁻³ M in all solutions.

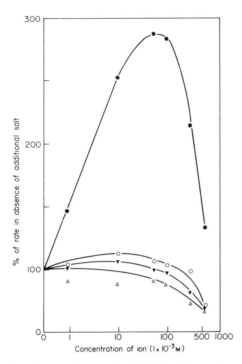

Fɪɢ. 3. The influx of potassium into roots of *Triglochin maritima* in the presence of different concentrations of the salts in the external solution. The concentration of potassium chloride in all solutions was 1×10^{-4} M and the concentration of calcium chloride 3×10^{-3} M. The results are expressed as a percentage of the influx of potassium in the absence of the appropriate salt in the experimental solution.

Key:

O—O K^+ influx in the presence of sodium sulphate
O—O K^+ influx in the presence of lithium chloride
▼—▼ K^+ influx in the presence of magnesium chloride
△—△ K^+ influx in the presence of ammonium chloride

insensitive to the presence of CCCP when the external concentration of potassium is 10 mм (Table IA). At this high external concentration of potassium, unlike the situation at a low external concentration of potassium, the influx of potassium does not increase when sodium chloride is present in the culture medium (Table II). These results suggest that at high salinities similar to that of sea water much of the potassium is entering the tissue by passive diffusion and that the contribution made by the pump to the total influx at these external concentrations of potassium is small.

B. Efflux of Sodium and Potassium

In order to determine whether sodium and potassium are moving passively within the cortical cells of the root the Ussing–Teorell flux equation has been used as a test for the independent passive movement of an ion. This equation, which was deduced by Ussing (1949) and Teorell (1949), is as follows:

$$\frac{J_{in}}{J_{out}} = \frac{C_{Jo}}{C_{Ji}(\exp z_j FE/RT)} = \frac{\bar{\mu}_j^o}{\bar{\mu}_j^i} \tag{1}$$

where J^i is the flux of the ion j from outside to inside the membrane and J^o is the flux from inside to outside the membrane; $\bar{\mu}_j^o$ is the electrochemical activity of ion j outside and $\bar{\mu}_j^i$ is the corresponding activity inside the membrane. In order to test whether an ion is moving in accord with this equation in the root cells it is necessary to have information on the fluxes, concentration gradients and electropotential gradients. Much of the primary data can be obtained from measurements of efflux of these ions.

C. Estimates of Compartments and Fluxes

Efflux of labelled potassium or sodium from the roots was measured over a period of about 8 h. From counts of the activity remaining in the roots at the end of the elution together with the activity in each of the washings an efflux curve was constructed. One of the assumptions made in the analysis of the data is that the movement of isotopes of potassium and sodium during the course of the wash-out period between the root and shoot is small relative to movement of these isotopes from the root to the outside solution. Support for this is based on the amount of isotope which appears at the cut end of a root during the wash-out period when a microcap is attached to the root. This is less than 4 % of that which passes from the root into the bathing solution. Seedlings of this species are very slow-growing and there is no appreciable water flux from the cut end of the root into the glass capillary tube. The activity present in the shoot systems at the end of an experiment was not included in the analysis of the efflux of ions. If the efflux is measured over a period exceeding 600 min instead of 500 min the form of the efflux curve at the end of an experimental period is complex and is marked by an increase in the efflux of tracer compared with the rate of loss between 200 and 500 min. This increase in the rate of loss of isotope is assumed to represent the movement of tracer from the shoot to the root. Hence, provided the efflux of isotope is terminated after 500 min and the activity present in the shoots at this time not included in the analysis, it is assumed that the efflux of isotope represents movement from the root to the external solution.

The interpretation which is placed on such data depends, amongst other things, on the spatial relationships which are considered to exist between different cell compartments. The cell wall, cytoplasm, and vacuole are generally considered to be arranged in series so that ions moving between the outside solution and the vacuole traverse the cytoplasm. Although this model was used initially in studies of giant algal cells (Diamond and Solomon, 1959; MacRobbie and Dainty, 1958; MacRobbie, 1964) its applicability to studies of higher plant cells also has been successfully demonstrated (Pitman, 1963; Cram, 1968; Macklon and Higinbotham, 1970; Pierce and Higinbotham, 1970; Poole, 1971) and therefore in this investigation the serial model has been used as the basis for the interpretation of the present data.

The efflux curve in each case is made up of first-order rate losses from each of the compartments within the tissue. The slowest compartment is assumed to be the vacuole and its contribution is indicated by the final linear part of the curve. If this component is subtracted from the total activity at each time interval and the data re-plotted, the final linear part of the curve which is obtained is considered to represent the cytoplasmic compartment. On subtraction of this component from the remaining activity at each time interval an additional phase results which is assumed to represent the free space and Donnan free space. From this graphical analysis the efflux rate constant (k) and the half-time $(t_{\frac{1}{2}})$ of exchange can be estimated for each phase. For the purpose of calculating the amounts of an ion in each phase it is assumed that the specific activities of the ion in the cytoplasm and free space are the same as that of the external solution and hence the apparent amount of an ion in each of these compartments is taken as equal to the counts per minute at the intercept of the efflux curve divided by the specific activity of the external solution. The amount of an ion in the vacuole (Q_v) is given by the difference between the total amount of the ion estimated chemically for the tissue and the sum of the ion in the other compartments estimated by specific radioactivity. The apparent influx across the tonoplast is estimated from the amount of tracer accumulated in the vacuole during the loading period.

Table III contains the results of the analysis of efflux and the apparent influx to vacuole for both sodium and potassium at different external concentrations of these ions. The rate constants for the efflux of the two ions from each compartment are of similar magnitude although the constants for the respective compartments differ markedly from one another. Of particular interest is the relative short half-time for the efflux of sodium and potassium from the vacuole compared with corresponding values for other plants (Macklon and Higinbotham, 1970; Poole, 1971).

Although the graphical analysis gives values of radioactivity of compartments and their efflux constants it does not reveal the amounts of stable ions present within each compartment.

The values of content given in Table III are apparent values which are not corrected for concurrent opposing fluxes, and they represent in particular a considerable underestimate of the cytoplasmic content as the data do not

TABLE III. Analysis of efflux from immersed roots of *Triglochin maritima* and apparent influx to vacuole

External solution (artificial sea water) concn of Na$^+$ and K$^+$ only	Compartment	k min^{-1}	$t_{\frac{1}{2}}$ min	Apparent content μmol/g fresh wt^{-1}	Isotopic equilibration %	Apparent influx μmol g^{-1} min^{-1}
				VALUES FOR SODIUM		
Na$^+$ 1 × 10^{-3} M						
1	Vacuole	1·00 × 10^{-3}	693	41·0	14	0·48
	Cytoplasm	1·07 × 10^{-2}	65	5·0	100	?
	Free space	0·139	5	23·0	100	?
10	Vacuole	0·62 × 10^{-3}	1117	72·0	85	6·00
	Cytoplasm	1·61 × 10^{-2}	43	1·2	100	?
	Free space	0·173	4	6·5	100	?
100	Vacuole	0·64 × 10^{-3}	1076	72·0	92	6·00
	Cytoplasm	0·58 × 10^{-2}	120	11·2	100	?
	Free space	0·173	4	90·7	100	?
100	Vacuole	0·90 × 10^{-3}	797	47·0	91	10·13
	Cytoplasm	0·86 × 10^{-2}	80	12·0	100	?
	Free space	0·139	5	70·0	100	?
500	Vacuole	0·85 × 10^{-3}	820	130·0	79	10·8
	Cytoplasm	0·87 × 10^{-2}	80	11·4	100	?
	Free space	0·139	5	134·0	100	?

TABLE III (cont.)

VALUES FOR POTASSIUM

K^+ 1×10^{-3} M						
0.2	Vacuole	0.70×10^{-3}	1069	54.0	2.7	0.12
	Cytoplasm	1.10×10^{-2}	65	0.12	100	~
	Free space	0.173	4	0.86	100	~
2.0	Vacuole	0.56×10^{-3}	1242	45.0	51	5.54
	Cytoplasm	0.95×10^{-2}	73	3.0	100	~
	Free space	0.139	5	21.0	100	~
2.0	Vacuole	0.84×10^{-3}	822	60.0	16.8	0.84
	Cytoplasm	0.99×10^{-2}	70	3.7	100	~
	Free space	0.139	5	6.5	100	~
4.0	Vacuole	1.39×10^{-3}	500	60.0	21.0	1.05
	Cytoplasm	0.63×10^{-2}	110	2.6	100	~
	Free space	0.139	5	7.2	100	~
8.0	Vacuole	0.83×10^{-3}	850	15.0	81.0	5.03
	Cytoplasm	1.24×10^{-2}	56	13.0	100	~
	Free space	0.173	4	52.0	100	~

FIG. 4. Notations used in compartmental analysis

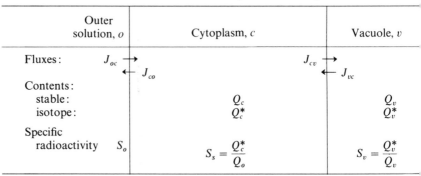

take into account the concomitant transfer of ions from the vacuole to the cytoplasm during the wash-out procedure.

A number of authors have given procedures for arriving at estimates of the fluxes and the concentrations of ions within compartments (MacRobbie and Dainty, 1958; Diamond and Solomon, 1959; Pitman, 1963; MacRobbie, 1964; Etherton, 1968; Cram, 1968; Macklon and Higinbotham, 1970; Pierce and Higinbotham, 1970). Figure 4 illustrates the components of the system. J is the flux of an ion (direction indicated by subscripts and arrow and has the units of moles per g fresh wt of tissue per minute in this case. Q is the chemical amount in a compartment where the units are moles per g or tissue, and Q^* is the amount of isotope in a compartment and is in counts per fresh wt per min. S_o is the specific radioactivity of the external solution at cpm mol^{-1}. The essential equations are

$$\mathrm{d}\frac{Q_c^*}{\mathrm{d}t} = J_{oc}S_o - (J_{cv} + J_{co})\frac{Q_c^*}{Q_c} + J_{vc}\frac{Q_v^*}{Q_v} \tag{2}$$

and

$$\mathrm{d}\frac{Q_v^*}{\mathrm{d}t} = J_{cv}\frac{Q_c^*}{Q_c} - J_{vc}\frac{Q_v^*}{Q_v} \tag{3}$$

where t is time and the remainder of the notations are those given above. Provided Q_c and Q_v are constant during the experiment the equations may be re-written as

$$Q_c\,\mathrm{d}\frac{S_c}{\mathrm{d}t} = J_{oc} - (J_{cv} + J_{co})S_c + J_{vc}S_v \tag{4}$$

and

$$Q_v\frac{\mathrm{d}S_v}{\mathrm{d}t} = J_{cv} - J_{vc}S_v \tag{5}$$

Calculations of the compartmental amounts and fluxes can be made using the following relationships

$$J_{oc} = k_c I_c + \frac{I_v}{tup} \tag{6}$$

where I_v/tup is the apparent influx to the vacuole (Table III); k_c is the rate constant for the second slowest phase believed to be cytoplasmic and I_c is the apparent content of this phase

$$J_{vc} = J_{oc} \cdot \frac{k_v Q_v}{k_c I_c} \tag{7}$$

where Q_v is the amount of ion in the vacuole

$$J_{co} = k_c I_c + k_v Q_v \tag{8}$$

It is assumed that as the ionic content of the compartments is essentially constant

$$J_{oc} - J_{co} = J_{cv} - J_{vc} \tag{9}$$

Thus

$$J_{vc} = J_{oc} - J_{co} + J_{cv} \tag{10}$$

Finally

$$Q_c = I_c \cdot \frac{J_{co} + J_{cv}}{S_c \cdot J_{co}} \tag{11}$$

S_c, the specific activity of the cytoplasm at the start of the efflux is given by

$$\frac{J_{oc}}{J_{cv} + J_{co}} \tag{12}$$

The estimates of fluxes and sodium and potassium contents in the different compartments are given in Table IV.

D. Estimates of Compartment Volumes

Although the results provide information on the amounts of sodium and potassium in the cytoplasm and vacuole, the concentrations of these ions in the compartments cannot be determined unless the volumes of these compartments are known. Measurements were made of cell size and the cytoplasmic and vacuolar volumes calculated in order to provide some estimate of concentrations of the two ions in the compartments (Table IV). The

TABLE IV. Unidirectional fluxes of K$^+$ and Na$^+$ and estimated amounts of these ions in cytoplasm and vacuole together with estimated cytoplasmic volumes

External solution (artificial sea water)	Cytoplasm						Vacuole		^{a}Estimated cytoplasmic volume as a percentage of total cell volume
	J_{oc}	J_{co}	Q_c		J_{cv}	J_{vc}	Q_v		
	mol × 10^{-8} g^{-1} min^{-1}		mol × 10^{-6} g^{-1}	1 × 10^{-3} M	mol × 10^{-8} g^{-1} min^{-1}		mol × 10^{-6} g^{-1}	1 × 10^{-3} M	volume
Na$^+$ (1 × 10^{-3} M)									
1	1·05	4·69	8·23	74·8	4·00	7·60	59·43	75·2	11%
10	8·00	6·26	16·53	97·0	19·58	17·84	55·80	76·4	17%
100	6·48	10·07	38·00	110·0	1·05	2·66	45·00	81·1	34·5%
100	20·98	14·91	34·14	98·7	14·58	8·57	25·49	45·9	34·5%
500	20·68	15·16	46·00	148·0	10·79	5·28	84·01	142·4	31%
K$^+$ (1 × 10^{-3} M)									
0·2	0·25	3·63	6·54	38·5	3·52	6·90	47·60	65·2	17%
2·0	4·50	8·70	10·84	31·4	2·00	6·20	52·64	94·8	34·5%
2·0	5·54	5·17	11·14	32·3	5·20	4·83	26·89	48·5	34·5%
4·0	2·69	10·02	26·04	75·5	6·41	13·75	36·79	66·3	34·5%
8·0	21·11	28·31	21·95	70·8	8·85	16·05	6·85	11·6	31%

^{a} Free space and cell wall volume taken as 10% of the total cell volume.

measurements were based on an examination of hand-cut sections of the root tissue using a light microscope.

E. Application of Ussing–Teorell Flux Ratio Equation

It is possible to apply the Ussing–Teorell flux ratio equation with the information on fluxes and amounts of ions in the cytoplasm and estimates of compartment volumes, together with data of electrical potentials (Table V). The conclusions which may be made from the data assuming the applicability of the serial model and a reasonable accuracy in measurements of electropotentials, compartment concentrations and fluxes are as follows.

The ratios of fluxes of sodium ions across the tonoplast are in accord with those predicted using the flux ratio equation at all the different external concentrations of sodium (1–500 mM) and therefore this ion appears to be at electrochemical equilibrium between the cytoplasm and vacuole. At low external concentrations of sodium (1–10 mM) the movement of this ion across the plasmalemma also appears to be passive as conditions of flux equilibrium prevail. However, at higher external concentrations of sodium (100–500 mM) the ratios of electro-chemical activity between the outer solution and the cytoplasm are much larger than the corresponding flux ratios.

TABLE V. Application of Ussing–Teorell flux ratio equation to uni-directional fluxes

External solution (artificial sea water)	Cytoplasm		Vacuole	
	$\dfrac{J_{oc}}{J_{co}}$	$\dfrac{\bar{\mu}_{out}}{\bar{\mu}_{in}}$	$\dfrac{J_{cv}}{J_{vc}}$	$\dfrac{\bar{\mu}_{out}}{\bar{\mu}_{in}}$
$Na^+ 1 \times 10^{-3}$ M				
1	0·22	0·18	0·52	0·99
10	1·28	1·70	1·10	1·34
100	0·64	27·20	0·40	1·35
100	1·41	30·00	1·64	2·15
500	1·36	55·51	2·04	1·04
$K^+ 1 \times 10^{-3}$ M				
0·2	0·07	0·09	0·51	0·59
2·0	1·07	1·78	1·07	0·69
2·0	0·52	1·92	0·32	0·34
4·0	0·27	1·72	0·47	1·18
8·0	0·75	2·27	0·55	6·37

The efflux of sodium from the cytoplasm to the external solution is against the electro-chemical gradient which suggests that when these external concentrations of sodium prevail there is an active outward movement of this ion at the plasmalemma. Hence, under conditions where the external salinity is similar to that of sea water the maintenance of conditions of flux equilibrium at the tonoplast appears to be linked with existence of an efflux pump at the plasmalemma.

In contrast under pseudo-steady-state conditions the movement of potassium across the plasmalemma appears to be passive at all external concentrations of this ion, although close agreement of the ratios of electro-chemical activity and the flux ratios does not occur at high external concentrations of potassium which suggests that, under such conditions, potassium may be exported from the tissues. The movement of this ion across the tonoplast also appears to be passive when the external potassium levels are between 0·2 mM and 2·0 mM, although at higher external concentrations of potassium the ratio of electrochemical activity between the cytoplasm and vacuole is apparently larger than the corresponding flux ratio.

F. Electrical Potential Measurements during Non-Steady-State Conditions

The evidence presented above indicated that sodium at low external concentrations moves passively across the plasmalemma and tonoplast. In an attempt to discover if the electrical potential could be regarded as a sodium diffusion potential at these external concentrations plants were pretreated in a solution of artificial sea water or diluted artificial sea water for 72 h and measurements made of the resting potential under pseudo-steady-state conditions. Thereafter the plants were bathed in a series of solutions which contained different concentrations of sodium and which represented "step-up" or "step-down" conditions and the transient changes in potential followed. The results shown in Table VI clearly indicate that under some conditions large changes in potential occur when the sodium concentration in the external solution is altered. The largest changes in potential occur in those plants which have been pretreated in a solution containing a low concentration of sodium (10 mM). If sodium benzene sulphonate is substituted for sodium chloride the changes in potential which occur as the external sodium concentration is altered are very similar to those obtained when sodium chloride is present. Alterations of the potassium or calcium concentration (10 ×) only result in a small change in potential (± 5 mV). Therefore in plants pretreated in solutions which contain low concentrations of sodium (10 mM) the potential adjustments appear to be closely linked with changes in the external sodium concentration and the

TABLE VI. Resting potentials and transient potentials in roots of *Triglochin maritima* associated with changes in the external concentration of sodium

External Na$^+$ (mM)	Electrical potential (mV)	External Na$^+$ (mM)	Electrical potential (mV)	External Na$^+$ (mM)	Electrical potential (mV)
100	-83^a	10	-65^a	500	-74^a
10	-121	100	-27	50	-78
1·0	-164	250	-14	1	-115
100	-85	500	-6	500	-80
250	-77	1000	-2	1000	-80
500	-76	1	-57	100	-78
1000	-66	10	-63		
100	-79	100	-29		

Note: left, middle, and right blocks are each labelled "Sequence of concentration changes" (with a downward arrow).

^{a} Resting potential under pseudo-steady-state conditions.

potential can be regarded largely as a sodium diffusion potential. However, where plants have been grown in solutions which contain high concentrations of sodium (100–500 mM) the change in potential associated with an alteration in the external sodium concentration is small, except where the sodium level is 1 mM or 10 mM. As discussed, under pseudo-steady-state conditions there appears to be an outwardly directed sodium efflux pump at the plasmalemma which is operating at high external concentrations of sodium and consequently passive diffusion cannot account entirely for the movement of sodium across the plasmalemma. In addition, the lack of response of the electrical potential to changes in concentration of sodium in the external solution is associated with alterations in the permeability coefficient for the entry of sodium across the plasmalemma. Under conditions where the plants are pretreated in solutions containing low concentrations of sodium (1–10 mM) the permeability coefficient for the movement of sodium from the outside solution to the cytoplasm is higher than that of potassium but where the plants are pretreated in solutions of high salinity the permeability coefficients for the two ions are similar to one another. These differences appear to be linked with the adjustments which the cortical cells undergo when subject to changing levels of salinity (Table VII).

TABLE VII. Permeability coefficients of roots of *Triglochin maritima* for the influx of sodium and potassium

External solution (1×10^{-3} M)		Na^+	K^+
		10^{-10} cm s^{-1}	10^{-10} cm s^{-1}
Na^+	1·0		
K^+	0·1	7·15	—
Ca^{++}	0·1		
Na^+	10		
K^+	0·2	5·61	0·89
Ca^{++}	0·2		
Na^+	100		
K^+	2	1·25	1·63
Ca^{++}	2		
Na^+	400		
K^+	8	—	1·85
Ca^{++}	8		
Na^+	500		
K^+	10	0·24	—
Ca^{++}	10		

Calculated from unidirectional fluxes across the plasmalemma according to the Goldman Equation

$$J_{oc} = -P_J \frac{Z_J FE}{RT} \frac{C_J^0}{1 - \exp Z_J FE/RT}$$

$$(1 \text{ g fresh wt of tissue} \equiv 800 \text{ cm}^2)$$

IV. Discussion

The results given in this paper which are based on the use of the Ussing–Teorell flux ratio equation strongly support the conclusion of Pierce and Higinbotham (1970), that ion-selectivity resides largely in the plasmalemma of higher plant cells. Assuming the validity of the serial model and the assumption that the rates of movement of isotope between the shoot and the root are insignificant during the period of efflux, the chief electro-potential barrier appears to lie between the outside solution and the cytoplasm, as the difference in electrochemical activity of sodium and potassium between the cytoplasm and the vacuole is small. The results of compartmental concentrations, half-times and permeability coefficients are similar in general to those values obtained from a large number of higher plant cells which are summarized by Pierce and Higinbotham (1970). One of the major difficulties, however, is that the estimates of ions in the cytoplasm include ions that may be associated with fixed sites or ions which accumulate in cell organelles and hence the activity of ions at the boundary of the plasmalemma may be different from the estimates. It does not appear that in these experiments this has led to serious difficulties, although a knowledge of activities within the cytoplasm may have resulted in a closer agreement between the flux ratios and the ratios of electrochemical activity in instances where the movement of ions appears to be passive. Another reason that may account for some error in the cytoplasmic values is where the amount of label in the wall space is about 10-fold or more of that in the cytoplasm (Pitman, 1963). In a number of treatments the amount of labelled sodium in the cell wall compartment was appreciably greater than that in the cytoplasm, which may have resulted in poor estimates of sodium concentrations in the cytoplasm. However, the reasonably close agreement between the flux ratios and the ratios of electrochemical activities suggests that the errors in the estimates are not appreciable.

The conclusions of this study may be summarized as follows. At all salinities the sodium ion appears to be in electrochemical equilibrium between the cytoplasm and vacuole and therefore it is assumed that there is passive movement of this ion across the tonoplast boundary. At low salinities (1–10 mM) there is also passive diffusion across the plasmalemma but at high salinities the evidence indicates that sodium is pumped from the cytoplasm to the outside solution and the ion ceases to be in electrochemical equilibrium under these conditions. Unlike the situation at low salinities the potential cannot be regarded as a sodium diffusion potential and this change in the response of the potential to sodium ions at high salinities is associated with the functioning of an efflux pump at the plasmalemma and changes in the permeability of the plasmalemma to sodium. Where the plants have been grown under pseudo-steady-state conditions, at most external concentrations of potassium, this ion appears to be in electrochemical equilibrium between the cytoplasm and outside solution, and flux equilibrium also prevails between the cytoplasm and vacuole. However at high external concentrations of potassium there are indications that this ion may no longer be in electrochemical equilibrium between the outside solution, the cytoplasm and the vacuole. Recent results, which will be published elsewhere, show that at very high salinities potassium as well as sodium is pumped from the cytoplasm to the outside solution. Whether potassium is pumped from the vacuole to the cytoplasm under these circumstances is still not resolved. It is evident from data presented in this paper that where the external concentration of potassium is low a weak inwardly directed pump which is activated by sodium ions operates at the plasmalemma.

Although the complexity of the tissues of the root prevents accurate measurements, in general the results appear to be similar to those conclusions of Gutknecht and Dainty (1968) which are based on a review of a large number of marine and brackish water algae and which were mentioned in the introduction.

Sodium appears to be an important ion involved in osmotic adjustment in the cells of this halophyte. At low salinities the cell membranes appear to offer relatively little resistance to the exchange of sodium between the root and the external solution. At different external salinities, the internal sodium and chloride contents of the roots change although these changes appear to be associated largely with the cell wall and free space compartment. Unfortunately data of chloride fluxes are not yet complete but it appears that chloride is actively transported into the tissues at certain salinities in *Triglochin maritima*. It is of interest that over a wide range of salinities the concentration of sodium and the concentration of potassium in the cytoplasm and vacuole do not change appreciably and the ratio of these ions approaches unity in both compartments. Furthermore with the exception of the sodium pump at the plasmalemma these concentrations are achieved as a result of the passive movement of sodium and potassium across the cell membranes.

Of interest is the possible existence of a potassium efflux pump located at the tonoplast in the cells of the roots of *Triglochin maritima*. Blinks and Jacques (1929) and Blinks (1930) suggested that this ion is pumped from the vacuole into the cytoplasm in *Halicystis osterhoutii* and Rigler (cited in Gutknecht and Dainty, 1968) suggests that a similar situation exists in *Bryopsis plumosa*. There is also evidence that *Codium* cells have a potassium efflux pump located at the tonoplast (Kesseler, 1965; Gutknecht, cited in Gutknecht and Dainty, 1968). Clearly there is a need to establish whether this potassium efflux pump at the tonoplast in *Triglochin maritima* operates at high external salinities, and whether the movement of potassium from the vacuole to the cytoplasm is linked with chloride movement. In situations where conditions of flux equilibrium between the outside solution and the cytoplasm do not exist, such as when changes in external salinity occur, a potassium efflux pump may operate at the tonoplast in the root cells of *Triglochin maritima*, thereby regulating potassium levels in the cytoplasm. There is a considerable amount of evidence that potassium is actively transported into the cells of certain marine algae from the external solution (cf. Gutknecht and Dainty, 1968), however the functioning of the potassium pump is not linked directly with the presence of sodium in the external medium as appears to occur in the roots of *Triglochin maritima*. At present the exact role of sodium in facilitating the transport of potassium is not known. In marine bacteria one of the functions of sodium is to permit the transport of substrates into the cells and this requirement for sodium appears to be specific (Drapeau and MacLeod, 1963). Rains and Epstein (1967) have shown that in the mangrove, *Avicennia marina*, sodium chloride appears to increase the uptake of potassium. Some marine algae such as *Porphyra perforata* respond to an increase in external salinity by increasing the potassium concentration in the cytoplasm (Eppley and Cyrus, 1960).

A possible link between increased salinity and the maintenance of generally high potassium levels in the cytoplasm is of interest in relation to the known sensitivity of protein synthesis in bacteria to the presence of sodium. As discussed in the reviews of Larsen (1967) and Jefferies (1972) there is evidence that the tolerance of halophytic bacteria to high salinities is linked with the persistence of high levels of intracellular potassium in cells. However halophytic higher plants such as *Triglochin maritima* do not appear to adjust to saline conditions in a manner similar to halophytic bacteria which grow in extremely saline habitats. As discussed elsewhere in this volume these halophytic bacteria appear to represent "open systems" in which ions move relatively easily between the cells and the ambient solution. Under the imposition of saline conditions there is a marked increase in the internal ion content of the bacteria, particularly the potassium content. Consistent with this finding are the results of *in vitro* enzyme studies which show that maximum activation of enzymes extracted from bacteria grown in highly saline conditions only occurs when concentration of ions in the assay medium is high (Larsen, 1967). In contrast enzymes extracted from the cells of halophytic higher plants show maximum activation at relatively low concentrations of ions in the assay medium (Flowers, 1972a, b, Greenway and Osmond, 1970) and furthermore the cells appear to be less permeable to ions than in the case of species of halophytic bacteria. It is of significance that the estimated concentrations of ions in the cytoplasm and vacuole of the cells of roots in *Triglochin maritima* are similar to those concentrations of ions which are associated with maximum activation of enzymes from halophytic higher plants in studies *in vitro*. These results indicate that different groups of organisms do not necessarily employ the same strategies in overcoming the deleterious effects of saline conditions.

Acknowledgements

I wish to thank Mr N. Perkins for competent technical assistance. Dr H. Greenway, Mr E. Tarr, and Dr J. Thain read the draft of the manuscript and made a number of valuable criticisms. Mrs J. Crook and Mrs J. Stocks kindly typed the manuscript. The author gratefully acknowledges the receipt of an NERC grant for some of the work reported in this paper.

References

ARISZ, W. H., CAMPHIUS, I. J., HEIKENS, H. and TOOREN, A. J. VAN. (1955). *Acta. Bot. neerl.* **4**, 322–338.
ATKINSON, M. R., FINDLAY, G. P., HOPE, A. B., PITMAN, M. G., SADDLER, H. D. W. and WEST, K. R. (1967). *Aust. J. biol. Sci.* **20**, 589–599.
BALLANCE, P. E. and JOHNSON, S. (1970). *Planta* **91**, 364–368.
BLACK, R. F. (1956). *Aust. J. biol. Sci.* **9**, 67–80.
BLINKS, L. R. (1930). *J. gen. Physiol.* **13**, 223–229.
BLINKS, L. R. and JACQUES, A. G. (1929). *J. gen. Physiol.* **13**, 733–737.
COLLANDER, R. (1941). *Pl. Physiol.* **16**, 691–720.
CRAM, W. J. (1968). *Biochim. biophys. Acta* **163**, 339–353.
DIAMOND, J. M. and SOLOMON, A. K. (1959). *J. gen. Physiol.* **42**, 1105–1121.
DRAPEAU, G. R. and MACLEOD, R. A. (1963). *Biochem. biophys. Res. Commun.* **12**, 111–115.
ELZAM, O. E. and EPSTEIN, E. (1969a). *Agrochimica* **13**, 187–195.
ELZAM, O. E. and EPSTEIN, E. (1969b). *Agrochimica* **13**, 196–206.
EPPLEY, R. W. and CYRUS, C. C. (1960). *Biol. Bull. mar. biol. Lab., Woods Hole* **118**, 55–65.
EPSTEIN, E. (1969). *In* "Ecological Aspects of the Mineral Nutrition of Plants". (I. H. Rorison, ed.) pp. 345–355. Blackwell Scientific Publications.
ETHERTON, B. (1968). *Pl. Physiol., Lancaster* **43**, 838–840.
FLOWERS, T. J. (1972a). *J. exp. Bot.* **23**, 310–321.
FLOWERS, T. J. (1972b). *Phytochem.* **11**, 1881–1886.
GREENWAY, H. and ROGERS, A. (1963). *Pl. Soil.* **18**, 21–30.
GREENWAY, H. and OSMOND, C. B. (1970). *In* "The Biology of *Atriplex*". (R. Jones, ed.) pp. 49–56. C.S.I.R.O. Canberra.
GREENWAY, H. and OSMOND, C. B. (1972). *Pl. Physiol., Lancaster* **49**, 256–259.
GUTKNECHT, J. and DAINTY, J. (1968). *A. Rev. Oceanogr. mar. Biol.* **6**, 163–200.
HILL, A. E. (1967a). *Biochim. biophys. Acta* **135**, 454–460.
HILL, A. E. (1967b). *Biochim. biophys. Acta* **135**, 461–465.
HILL, A. E. (1970a). *Biochim. biophys. Acta* **196**, 66–72.
HILL, A. E. (1970b). *Biochim. biophys. Acta* **196**, 73–79.
JEFFERIES, R. L. (1972). *In* "The Estuarine Environment". (R. S. K. Barnes and J. Green, eds) pp. 61–85. Elsevier, Amsterdam.
KESSELER, H. (1965). *In* Proc. Fifth Mar. Biol. Symp., Botanica Gothoburgensia **3**, 103–111.
LARSEN, H. (1967). *Adv. microb. Physiol.* **1**, 97–132.
LÜTTGE, U. (1971). *A. Rev. Pl. Physiol.* **22**, 23–44.
MACROBBIE, E. A. C. (1964). *J. gen. Physiol.* **47**, 859–877.
MACROBBIE, E. A. C. and DAINTY, J. (1958). *Physiol. Pl.* **11**, 782–801.
MACKLON, A. E. S. and HIGINBOTHAM, N. (1970). *Pl. Physiol., Lancaster* **45**, 113–138.
PALLAGHY, C. K. (1970). *In* "The Biology of *Atriplex*". (R. Jones, ed.) pp. 57–62. C.S.I.R.O. Canberra.
PARHAM, M. (1971). Ph.D. Thesis, University of East Anglia.
PIERCE, W. S. and HIGINBOTHAM, N. (1970). *Pl. Physiol.* **46**, 666–673.
PITMAN, M. G. (1963). *Aust. J. Biol. Sci.* **16**, 647–668.
POOLE, R. J. (1971). *Pl. Physiol., Lancaster* **47**, 731–734.
RAINS, D. W. and EPSTEIN, E. (1967). *Aust. J. biol. Sci.* **20**, 847–857.
RAMSAY, J. A., BROWN, R. H. J. and CROGHAN, P. C. (1955). *J. exp. Biol.* **32**, 822–829.
SCHOLANDER, P. F., HAMMEL, H. T., HEMMINGSEN, E. and GAREY, W. (1962). *Pl. Physiol., Lancaster* **37**, 722–729.

SCHOLANDER, P. F., BRADSTREET, E. D., HAMMEL, H. T. and HEMMINGSEN, E. A. (1966).
   Pl. Physiol. **41**, 529–532.
TEORELL, T. (1949). *Arch. Sci. Physiol.* **3**, 205–219.
USSING, H. H. (1949). *Acta physiol. Scand.* **19**, 43–56.

V.3

Cations and Filamentous Fungi: Invasion of the Sea and Hyphal Functioning

D. H. Jennings

Department of Botany, University of Liverpool
England

I. Introduction

In the sea, growing on solid organic matter, there is a well defined fungal flora of Phycomycetes, Ascomycetes and Fungi Imperfecti (but only two Basidiomycetes). The evidence we now have (Jones *et al.*, 1971) indicates that this flora is associated with its marine habitat because the individual species require sea water for sporulation and germination. Vegetative growth can occur equally well in fresh water, given that the fungus is receiving the appropriate carbon and nitrogen sources (Jones and Jennings, 1964). Many terrestrial species can grow equally well in sea water media. However, they cannot reproduce in the marine habitat (Jones *et al.*, 1971).

Other work has shown that both marine and terrestrial members of the Phycomycetes, Ascomycetes and Fungi Imperfecti are able to tolerate relatively high concentrations of sodium chloride (Jones, 1963; Jones and Jennings, 1965; Slayman and Tatum, 1964; Tresner and Hayes, 1971). Tresner and Hayes examined the tolerance to sodium chloride of 770 species of fungi and showed that 50% (total 666) of species in the above classes were capable of tolerating 2·0 M sodium chloride. Basidiomycetes uniformly had a much greater sensitivity to salt. Although this work provides no clear data about rates of vegetative growth, the accumulated data from all sources show that, apart from Basidiomycetes, the growth of filamentous fungi is remarkably independent of the sodium chloride concentration in the medium. This is unlike higher plants (Jennings, 1968).

We now have enough information to show why such fungi are so tolerant of sodium chloride. This information is of considerable relevance to studies on ion transport in plants and also to our understanding about how fungal hyphae function.

Since we know (Jones and Jennings, 1965) that the ability of a fungus to grow in the presence of sodium chloride depends on whether or not sufficient potassium can be absorbed, particular attention must be given to the transport and retention of this ion within fungal hyphae.

II. Potassium Transport System

The transport of potassium has been most extensively studied in *Neurospora crassa* (Lester and Hechter, 1958, 1959; Slayman, C. L., 1965a, b; Slayman, C. W., 1970; Slayman and Slayman, 1968, 1970; Slayman and Tatum, 1964, 1965a, b). There is also information for *Dendryphiella salina* (Jones and Jennings, 1965; Allaway and Jennings, 1970a; Jennings and Aynsley, 1971), *Fusarium oxysporium* (Shere and Jacobson, 1970a, b) and *Neocosmospora vasinfecta* (Budd, 1969).

In *N. crassa* potassium transport is metabolically driven (Lester and Hechter, 1958; Slayman and Tatum, 1965). Evidence for this comes from the metabolic inhibitors sodium azide and DNP. The former at 10^{-4} M gives 85% inhibition of respiration and almost complete inhibition of potassium uptake. Loss of potassium of any magnitude from the mycelium only occurs at higher concentrations of the inhibitor.

The transport system has a high degree of specificity for potassium. The system also transports rubidium. In *N. crassa*, the affinity for rubidium is 40% of that for potassium at 5 mM concentration of each ion (Slayman and Tatum, 1964). In *N. vasinfecta* a 5-fold excess of rubidium gives 50% inhibition of potassium uptake from 1 mM concentration (Budd, 1969).

Little sodium is transported. Thus is *N. vasinfecta*, there is no significant effect on the potassium level in the mycelium up to 100 equivalent excess of sodium over potassium in the external medium (Budd, 1969). The most striking case is provided by the marine fungus *Asteromyces cruciatus*. When grown in a glucose–tryptone medium (potassium concentration 0.11 mM) there is no difference in the mycelial potassium content, when sodium is present in the medium at a concentration as high as 600 mM (Jones, 1963).

Looking at transport in *N. crassa* in more detail, the system is probably also responsible for the rapid change of potassium between the mycelium and the external medium (Slayman and Tatum, 1965a). Such exchange has a V_{max} of 20 mM per kg of cell water per min, which is the same as exchange of potassium for sodium and hydrogen ions. Further, a single gene change can produce mycelium in which K–K exchange and net potassium uptake are equally affected (Slayman and Tatum, 1965; Slayman, C. W., 1970). The Michaelis constants for external potassium for K–K exchange and K–Na–H exchange are however quite different: 1·0 mM in the former instance and 11·8 mM in the latter. A similar situation is observed with red blood cells (Glynn *et al.*, 1970) but all the evidence from biochemical studies indicates that only a single transport system is involved (Glynn *et al.*, 1971).

Slayman and Slayman (1970) have examined the kinetics of net potassium flux into the cytoplasm as a function of the potassium concentration and pH of the medium. At low pH (4·0–6·0) net flux is a simple exponential function of time which obeys Michaelis kinetics as a function of potassium concentration. At high pH, potassium uptake is more complex, obeying sigmoid kinetics. The data have been fitted satisfactorily by two different two-site models. In one, the transport system is thought to contain both a carrier site responsible for potassium uptake and a modifier site: for a hydrogen ion at low pH and a potassium ion at high pH. Armstrong and Rothstein (1964, 1967) have made a similar suggestion for potassium transport in yeast, where at high pH, net potassium uptake has a low K_m but high V_{max}, whereas at low pH the V_{max} decreases (which is said to be an effect of the hydrogen ion at the modifier site) and the K_m increases (competition between hydrogen ions and potassium at the modifier site). The second model of Slayman and Slayman postulates a transport system consisting of multiple subunits, each with an active site for potassium, hydrogen ions being allosteric activators.

The potassium–sodium–hydrogen ion pump is electrogenic. Metabolic inhibitors (cyanide, carbon monoxide, dinitrophenol, azide, anoxia and low temperature) all cause rapid changes in the membrane potential (Slayman, C. L., 1965b). Thus, within 1 min of adding 1 mM sodium azide to the external medium, already containing 10 mM KCl + 1 mM $CaCl_2$ + 2% sucrose, the potential (inside negative) shifts from -227 mV to -19 mV. No significant change in membrane resistance can be detected. Slayman *et al.* (1970) have also shown that the decay in potential brought about by a metabolic inhibitor (in this case 1 mM cyanide) is paralleled by a drop in the mycelial concentration of ATP such that the voltage/time curve is superimposable upon the ATP/time curve, with rate constants for both corresponding to a half time of 3·7 s.

The current-carrying ion in the system is hydrogen. This ion remains the only candidate after all other possibilities have been removed in experiments where it was shown that the potential could not be changed by changing the anion composition of the medium nor by changing the sodium content of the mycelium (Slayman, C. W., 1970). It is interesting to find that the latter ion has no effect, in view of its role in the membrane potential of many animal tissues (Baker, 1966).

III. Passive Fluxes

Estimates of the passive loss of potassium from *N. crassa* indicate that the greater portion of potassium uptake is via an active mechanism. Passive loss has been estimated in three ways (Slayman and Slayman, 1968): from measurements of loss into distilled water and into buffer from azide-poisoned cells and from flux measurements at the minimum extra-cellular concentration allowing non-growing cells to remain in a steady state. These measurements respectively give values of 0.5, 0.3 and 0.7 mM kg cell water^{-1} min^{-1} which is less than 3% of the maximum net flux. When calculated in more standard units the passive flux would therefore be about 0.4 pmol cm^{-2} s^{-1}

D. salina behaves like *N. crassa* in showing only a very small loss of potassium into distilled water (Jones and Jennings, 1965). Jennings and Aynsley (1971) have examined potassium exchange between the mycelium and the external medium at 2°C. The fungus was grown at 21°C, in two media identical except that one contained potassium labelled with ^{42}K. The cultures were transferred to 2°C for a period of time, the fungus separated from the medium, the radioactive mycelium placed in the non-radioactive medium and the efflux of ^{42}K measured. At the lower temperature, growth is reduced by at least 75% of that at 21°C, with no growth on occasion. The fungus was treated as being in potassium flux equilibrium.

It was of considerable interest to find that the mycelium in all instances behaves as a multi-compartment system with regard to potassium, with about 75% of the ion being in the slowest exchanging compartment. Efflux from this compartment into a medium containing 2.5 mM potassium was calculated to be of the order of 0.14 pmol cm^{-2} s, assuming a uniform concentration of potassium within the mycelium. This value is comparable with the passive flux values for *N. crassa* given above. Technical difficulties prevented the determination of fluxes from the faster exchanging compartments.

Jones and Jennings (1965) postulated that the major cause of reduced growth of fungi, brought about by raising the sodium concentration of the growth medium, was an increased passive efflux of potassium from the hyphae. Their evidence for this was the ability of bivalent cations, particularly calcium, to prevent the toxic effect of sodium ions so that potassium loss is reduced to those levels in the absence of sodium ions. Further potassium loss in the presence of high concentrations of sodium was shown to be relatively temperature-insensitive.

Jennings and Aynsley (1971) showed in the same experiments as described above that when mycelium is grown in medium containing also 50 mM sodium the rate of potassium efflux at 2°C is doubled. There is a need to measure the passive flux of potassium in the presence of higher concentrations of sodium but the data do in part confirm what Jones and Jennings have suggested.

It seems that this effect of raising the concentrations of sodium ions in the medium is a relatively unspecific effect on the ability of the mycelium to retain solutes and that raising the potassium concentration of the medium has a similar effect. Thus Allaway and Jennings (1970b) showed that 200 mM sodium brings about large losses of mannitol and arabitol and probably other carbon compounds from *D. salina*. Calcium ions prevent this effect of sodium. Jennings and Austin (1973) showed that 200 mM sodium and potassium both reduce the uptake of the non-metabolizable sugar 3-O-methyl glucose by bringing about an increased efflux of the sugar. Calcium ions nullified the effect of the two monovalent cations.

Studies on the effect of cations in the external medium on the intercellular potential of *N. crassa* are in keeping with these observations (Slayman, C. L., 1965a). With extra-cellular solutions containing only sucrose as the other solute, increasing the concentration of potassium chloride (or sulphate) moves the internal potential towards zero, becoming less negative, at 45 mV per log unit over the range 0·1–10 mM. Above 10 mM, the change of potential is greater: 87 mV per log unit. Sodium has a very similar effect. Calcium at 1 mM diminishes the influence of an increasing concentration of potassium on sodium such that the potential changes are only 17 and 9 mV per log unit respectively. Calcium exerts its effect when the potassium concentration is 20 mM.

The model of Slayman, C. L. (1965b) (Fig. 1) can be used to explain the above observations. It consists of two distinct electromotive forces in parallel —a diffusion potential (E_d) and an electrogenic potential (E_e). They are linked through resistances R_d and R_e. The model requires only that R_d behave like an integral resistance across a potassium-specific membrane (relative to sodium) in order that the internal potential should vary with $\log [K]_0$ along a slope that is significantly less than the Nernst slope at low $[K]_0$ and significantly greater at high $[K]_0$. The decline in potential difference with increasing potassium concentration would be made up of a decrease of E_d and a decrease in the fraction of E_e which could be measured across the resistance network. The effect of calcium would be accounted for if calcium simply increased R_d. Similar proposals have been put forward to explain the membrane potentials of *Nitella translucens* in the presence and absence of calcium (Spanswick *et al.*, 1967).

Calculated R_d/R_e values (Slayman, C. L., 1965b) show that raising calcium from 0 to 1 mM appears to increase the ratio by 7-fold, and raising potassium

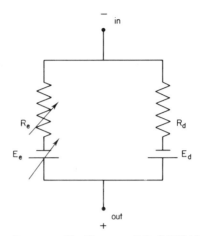

FIG. 1. The electrical circuit proposed by Slayman, C. L. (1965b) for the internal potential of *Neurospora crassa*. E_d, diffusion potential; E_e electrogenic potential; R_d and R_e internal or intrinsic resistances of E_d and E_e respectively.

from 10 to 100 mM decreases the ratio 3-fold. If R_e remains constant this 3-fold decrease means a 40% drop in R_d.

On the basis of the findings with *D. salina*, one can also suggest that, with increasing concentration of potassium or sodium in the external medium, the membrane becomes not only more permeable to potassium ions but it becomes much less specific for that cation. This suggestion is put forward using the leakage of organic solutes as the criteria for changed specificity of the membrane to passive monovalent cation movement. On this basis, increasing concentrations of hydrogen ions have a similar effect to increasing concentrations of potassium or sodium ions, since at pH values less than 4·0 considerable quantities of mannitol and arabitol are lost to the medium (Holligan and Jennings, 1972b). Support for this comes from observations on *N. crassa* (Slayman, C. L., 1965a). With a change of pH in the external medium from 9·0 to 8·0, the internal potential shifts (more positive) about 12 mV per pH units but when the change is 4·0 to 3·0, the shift is about 33 mV per pH unit. These facts by themselves could be due to the character-istics of the hydrogen diffusion potential *per se* or a more general control by pH of membrane permeability. The latter possibility is more preferable, especially since cells maintained at pH 3·0 leak more potassium.

This is an appropriate point at which to refer in more detail to the effect of pH on the membrane potential in *N. crassa*. Since hydrogen ions, as well as potassium and sodium ions, at increasing concentration in the external medium shift the membrane potential towards zero, all three ions must be involved in the diffusion potential. The influence of anions can be excluded (Slayman, C. L., 1965a). However, if this is so, the membrane must be 100–

000 times more permeable to hydrogen than potassium and the pH immediately inside the membrane near to 3·0. This need not be an impossibility in view of the results for yeast which indicate a space close to the cell surface responsible for acid production (Conway and Downey, 1950; Rothstein, 1956). High membrane permeabilities for hydrogen ions have been suggested before. Kitasato (1968) has estimated the permeability coefficient of the alga *Nitella clavata* for hydrogen ions to be three orders of magnitude larger than that for potassium. There are, however, difficulties in testing this hypothesis because of the inherent difficulties in measuring the partial fluxes of hydrogen ions and the cytoplasmic pH next to the plasma-lemma of the hypha.

IV. Ion Balance within the Hypha

If hydrogen ions are lost from a hypha in exchange for potassium ions, the cytoplasmic pH should rise. Experiments using DNP for measuring the internal pH (Slayman and Slayman, 1968) show that there is a rise of 0·3 pH units over an interval of 3 min when cells lost 50% of the total K-induced hydrogen loss.

However, sodium and hydrogen loss need not be the only mechanism for balancing potassium uptake. In *F. oxysporium*, the balance is brought about by loss of magnesium ions as well as hydrogen and a gain in organic acids (Shere and Jacobson, 1970a). On an absolute basis, the largest increase was in succinic acid. This is like yeast. Conway and Brady (1950) showed that when yeast is fermenting glucose, addition of potassium chloride to the medium results in excretion of hydrogen ions and accumulation of potassium succinate in the cells. The exchange of magnesium for potassium ions in *F. oxysporium* is insensitive to DNP and may be non-metabolic in nature. It is worth noting that in yeast magnesium transport into the cell is essentially irreversible, but since the process is coupled to glycolysis, it is also unaffected by DNP (Rothstein *et al.*, 1958 and Jennings *et al.*, 1958).

Phosphate groups must often make a significant contribution to the negative charges within mycelia. In *N. crassa* phosphorus levels of 300 mM/kg cell water have been measured (Slayman and Slayman, 1968).

V. Uptake of Chloride and Phosphate

Slayman and Slayman (1968) measured the influx of chloride, sulphate and phosphate into *N. crassa*. The influx is independent of whether or not potassium or sodium is present. The initial influx of the anions is as follows: chloride 2·0, sulphate 0·2 and phosphate 1·0 mM/kg cell water/min. That of chloride is no more than an eighth of potassium influx under the same conditions.

Budd (1969) has found that potassium uptake by mycelium of *N. vasinfectc* originally grown in a low potassium medium is accompanied by only smal amounts of chloride. Miller and Budd (1971) have pointed out that thi chloride uptake, though small, is probably active if the internal potential i like that for *N. crassa*, namely negative. The presence of glucose in the external medium inhibits net chloride uptake. Both influx and efflux of the ion appear to be affected.

Low rates of anion uptake are not necessarily the rule. Shere and Jacobsor (1970b) have shown that when *F. oxysporium* is grown on a low phosphoru medium, the rate of phosphate uptake is equal to the rate of potassiun uptake. However, when the fungus is grown in a medium of high phosphat content, only a small fluctuating uptake is observed. Potassium uptake i much retarded.

It would be reasonable to assume that part of the potassium uptake i driven by the transport of phosphate. In keeping with this, phosphate uptak from calcium was shown to take place with only a small uptake of the cation charge balance inside the cell being maintained, in part, by a drop in organi acid content.

VI. Comment

I think it is clear why those fungi we find in the sea are able to grow there and at the same time show us no specific requirement for sea water fo vegetative growth. It would seem that fungi growing in the sea are secondaril marine, since the properties of their outer membranes appear to be very like those of their terrestrial counterparts. These properties allow the latte fungi to grow almost equally well in saline media.

The salient properties are:

   (i) the presence of a pump which is highly specific for potassium in the external medium and which is responsible for removing sodium ion from the cell and the absence of any pump capable of taking i sodium.

   (ii) low passive permeability to sodium ions and anions, with these latter ions being pumped into the mycelium for the most part a low rates.

This situation is very different from halophytic higher plants where botl sodium and chloride ions are absorbed in large quantity (Jennings, 1968) Marine fungi effectively exclude both these ions from their mycelia. This i possible because, unlike higher plants, the appropriate osmotic pressur within the hyphae is not so greatly dependent upon absorption of ions fron the external medium. A large component of the osmotic pressure is due t soluble carbohydrates, particularly mannitol and other sugar alcohols In *D. salina* mycelium growing in a standard glucose–mineral mediun

TABLE I. The contribution made by mannitol and arabitol to the excess osmocity (Δ osm) within the hyphae of *Dendryphiella salina* in response to varying dilutions of sea water for 6 h. The medium contained glucose at 50 g ml^{-1}. Mycelium grown as described by Allaway and Jennings (1970a) and the ratio of intracellular water/dry weight taken as 2·54 (Slayman and Tatum, 1964). Osmocity of mannitol and sea water obtained from data of West and Selby (1967). That of arabitol taken as being the same as mannitol

% Sea water in medium	Osmocity mol l^{-1}	ARABITOL			MANNITOL			Total osmocity mol l^{-1}	Δ osm mol l^{-1}
		mg 100 mg dry weight^{-1}	mol l^{-1} intracellular water	Osmocity mol l^{-1}	mg 100 mg dry weight^{-1}	mol l^{-1} intracellular water	Osmocity mol l^{-1}		
100	0·575	10·1	0·395	0·245	9·0	0·195	0·111	0·356	0·245
50	0·265	8·9	0·23	0·133	9·0	0·195	0·111	0·244	0·133
25	0·13	5·9	0·15	0·084	6·8	0·15	0·084	0·168	0·057
12·5	0·07	5·8	0·15	0·084	7·6	0·65	0·093	0·177	0·066
6·25	0·03	3·8	0·09	0·048	5·7	0·125	0·063	0·111	0
0	0	3·8	0·09	0·048	5·8	0·125	0·063	0·111	—

contains mannitol at around 0·25 M and arabitol at 0·15 M. Further increase in the level of these sugar alcohols can contribute to osmotic adjustment (Table I). These sugar alcohols are readily synthesized from sugars, such as glucose (Holligan and Jennings, 1972a, b), which are readily available as breakdown products of the timber substrates on which the fungi are usually found.

It would seem that Basidiomycetes have permeability properties different from those of Phycomycetes and Ascomycetes. As far as I am aware, hardly any work has been done on ion relations of Basidiomycetes. One paper by Harley and Wilson (1959) does, however, help us. They showed that excised beech mycorrhizal roots in which the fungal partner is a Basidiomycete, lose considerable amounts of potassium into the medium at temperatures above 20°C, when the oxygen tension is below 3% and when low concentrations of metabolic inhibitors are present. Thus the hyphae appear to have different properties, with respect to the uptake and retention of potassium, from those of the fungi referred to above. Some work on ion relations of selected Basidiomycetes is clearly required.

Exponential growth of a filamentous fungus is very closely dependent upon the availability of combined carbon in the medium, especially when other nutrients are not limiting (Borrow et al., 1961). It is therefore not surprising to find that the balance of charge when potassium enters hyphae can be met by anions whose production is closely related to carbohydrate metabolism—condensed phosphates and organic acids. It is unlikely that the latter compounds will be produced directly from pyruvate via the Krebs cycle owing to the drain on organic acids from this source as a consequence of anabolic processes such as amino acid and protein synthesis. Anaplerotic pathways must be involved (Kornberg, 1970) accounting for R.Q.s of less than unity always observed with growing fungi (Altman and Ditmer, 1971). There is a need to find out whether or not there is any relationship between potassium and bicarbonate uptake, the anion being taken in and metabolized (as carbon dioxide) by one or more anaplerotic pathways. In N. crassa, bicarbonate ions seem to play no role in potassium uptake but this may be because of the high phosphate content of the cells (see Shere and Jacobson, 1970b).

The data that we now have indicate that the potassium–sodium–hydrogen pump is likely to be an ATPase. The evidence in favour of this is the very close relationship between decay in the membrane potential and mycelial ATP level shown by C. L. Slayman. But as well as this, the rate of respiration and glucose utilization of mycelium of D. salina responds to cations in a similar manner to red blood cells for which the data are interpreted in terms of rate of functioning of the sodium potassium ATPase (Whittam et al., 1964; Allaway and Jennings, 1971; Jennings and Austin, 1973).

It is now imperative that an attempt should be made to isolate the fungal ATPase and compare its properties with those from other tissues, to attempt to provide an explanation, in molecular terms, as to why the fungal potassium–sodium–hydrogen pump can only transport sodium in one direction (though the possibility of Na–Na exchange needs examination) and why hydrogen not sodium ions carry the electric current.

There are two pieces of evidence that a hypha does not behave uniformly along its length with respect to potassium. First, germinating spores of *N. crassa* (Slayman and Tatum, 1964) and regenerating hyphal fragments of *D. salina* (Jennings and Aynsley, 1971) have a considerably lower potassium/sodium ratio than mycelium which has been growing for some time (Table II). Second, Slayman and Slayman (1962) have shown that the membrane potential decreases as one moves from the mature parts of a hypha to its tip.

TABLE II. Potassium/sodium ratios in mycelium of *Neurospora crassa* (calculated from data in Fig. 1 of Slayman and Tatum, 1964) and *Dendryphiella salina* (Jennings and Aynsley, 1971) at different times during growth in media containing 36·8 mM potassium and 8·4 mM sodium and 0·11 mM potassium and 7·6 mM sodium respectively

N. crassa[a]		D. salina[b]	
Time (hours)	K/Na	Time (hours)	K/Na
0	8·5	10	0·0049
2	6·5	24	0·11
4	4·5	30	0·42
6	15·0	48	0·25
8	12·0	54	0·085[c]
10	12·5		
12	14·0		
14	16·5		

[a] End of lag period at $3\frac{1}{2}$ h.
[b] End of lag period at $7\frac{1}{2}$ h.
[c] All the potassium in the medium depleted by this time.

It is worthwhile exploring the significance of these two observations. Both suggest that there are fewer potassium pumps at the hyphal tip since both potassium/sodium sensitivity and membrane potential are for the most part under the control of these pumps. As well as this, the observations might mean that the hyphal tip has a higher passive permeability to cations, since fewer potassium pumps need not lead to a higher level of sodium in the hyphae. If this were so, it would explain the presence of a compartment in *D. salina* exchanging potassium with the medium more rapidly than the bulk of the mycelium (Jennings and Aynsley, 1971).

If these conclusions are correct, and if we accept that hyphae along a good proportion of their length from the tip are essentially continuous tubes (because of the incomplete nature of the septa), we can see a similarity between a growing hypha and a developing xylem vessel in the stele of a growing root. Therefore, from the point of view of potassium, a hypha can be considered as a tube open at one end with the cation moving through the sides of the tube in a virtually irreversible manner. If this is so, then as for the xylem vessel, there will be a standing-gradient osmotic flow (Anderson *et al.*, 1970). There will be a bulk flow of fluid from the older part of the hypha to the tip.

Some support for this comes from the observations of Cowan *et al.* (1972) on potassium movement in mycelium of *Phycomyces blakesleeanus*. They found that the rate of translocation ^{42}K was dependent upon the direction of growth of the mycelium: essentially three times faster when the movement was from older to younger parts than vice versa. Cowan *et al.* explain their findings on the basis of the presence of indigenous potential gradient along the hyphae, since the conditions of their experiment did not allow a bulk flow of fluid through the hyphae driven by evaporation. They quote in support of their hypothesis the variation in membrane potential along the hyphae demonstrated by Slayman and Slayman (1962).

The presence of a standing-gradient osmotic flow is a more attractive alternative, especially since the fluid flow would have a directional effect on the movement of vesicles carrying cell wall material from the Golgi apparatus to the extreme tip (Grove *et al.*, 1970). Further, Cowan *et al.* (1972) appear to be misinterpreting the observations of Slayman and Slayman (1962). Since the membrane potential contains a large electrogenic component, information about potentials along a hypha cannot tell us about the intra-cellular potentials. This is only possible if potential across the membrane is a diffusion potential. Of course, a standing-gradient osmotic flow will only occur if the concentration of other solutes along the hypha does not act against it, e.g. an excess of soluble organic compounds at the tip. This seems unlikely since there appears to be very little detectable variation in osmotic pressure along a hypha (Robertson and Rizvi, 1968) and there is evidence that there are mechanisms within a hypha for maintaining a constant concentration of soluble carbohydrates (Jennings and Austin, 1973).

References

ALLAWAY, A. E. and JENNINGS, D. H. (1970a). *New Phytol.* **69**, 567–579.
ALLAWAY, A. E. and JENNINGS, D. H. (1970b). *New Phytol.* **69**, 581–593.
ALLAWAY, A. E. and JENNINGS, D. H. (1971). *New Phytol.* **70**, 511–518.
ALTMAN, P. L. and DITMER, D. S. (1971). "Respiration and Circulation". Federation of American Societies for Experimental Biology, Bethesda.
ANDERSON, W. P., AIKMAN, D. P. and MEIRI, A. (1970). *Proc. R. Soc. Lond.* B **174**, 445–458.

ARMSTRONG, W. McD. and ROTHSTEIN, A. (1964). *J. gen. Physiol.* **48**, 61–71.
ARMSTRONG, W. McD. and ROTHSTEIN, A. (1967). *J. gen. Physiol.* **50**, 967–988.
BAKER, P. F. (1966). *Endeavour* **25**, 166–172.
BORROW, A., JEFFREYS, E. G., KESSEL, R. H. J., LLOYD, E. G., LLOYD, P. B. and NIXON, I. (1961). *Can. J. Microbiol.* **7**, 227–276.
BUDD, K. (1969). *J. gen. Microbiol.* **59**, 229–238.
CONWAY, E. J. and BRADY, T. G. (1950). *Biochem. J.* **47**, 360–369.
CONWAY, E. J. and DOWNEY, M. (1950). *Biochem. J.* **47**, 347–355.
COWAN, M. C., LEWIS, B. G. and THAIN, J. F. (1972). *Trans. Br. mycol. Soc.* **58**, 103–112.
GLYNN, I. M., HOFFMAN, J. F. and LEW, V. L. (1971). *Phil. Trans. R. Soc.* B **262**, 91–102.
GLYNN, I. M., LEW, V. L. and LÜTHI, U. (1970). *J. Physiol.* **207**, 371–391.
GROVE, S. N., BRACKER, C. E. and MORRÉ, D. J. (1970). *Am. J. Bot.* **57**, 245–266.
HARLEY, J. L. and WILSON, J. M. (1959). *New Phytol.* **58**, 281–298.
HOLLIGAN, P. M. and JENNINGS, D. H. (1972a). *New Phytol.*, **71**, 569–582.
HOLLIGAN, P. M. and JENNINGS, D. H. (1972b). *New Phytol.*, **71**, 583–594.
JENNINGS, D. H. (1968). *New Phytol.* **67**, 899–911.
JENNINGS, D. H. and AUSTIN, S. (1973). *J. gen. Microbiol.*, in press.
JENNINGS, D. H. and AYNSLEY, J. S. (1971). *New Phytol.* **71**, 713–723.
JENNINGS, D. H., HOOPER, D. C. and ROTHSTEIN, A. (1958). *J. gen. Physiol.* **41**, 1019–1026.
JONES, E. B. G. (1963). Ph.D. thesis, Leeds University.
JONES, E. B. G. and JENNINGS, D. H. (1964). *Trans. Br. mycol. Soc.* **47**, 619–625.
JONES, E. B. G. and JENNINGS, D. H. (1965). *New Phytol.* **64**, 86–100.
JONES, E. B. G., BYRNE, P. and ALDERMAN, D. J. (1971). *Vie Milieu.* Suppl. No. 22, 265–280.
KITASATO, H. (1968). *J. gen. Physiol.* **52**, 60–87.
KORNBERG, H. L. (1970). *Biochem. Soc. Symp.* **30**, 155–171.
LESTER, G. and HECHTER, O. (1958). *Proc. natn. Acad. Sci. U.S.A.* **44**, 1141–1149.
LESTER, G. and HECHTER, O. (1959). *Proc. natn. Acad. Sci. U.S.A.* **45**, 1792–1801.
MILLER, A. G. and BUDD, K. (1971). *J. gen. Microbiol.* **66**, 243–245.
ROBERTSON, R. N. and RIZVI, S. R. H. (1968). *Ann. Bot.* **32**, 279–291.
ROTHSTEIN, A. (1956). *Disc. Faraday Soc.* No. 21, 229–238.
ROTHSTEIN, A., HAYES, A. D., JENNINGS, D. H. and HOOPER, D. C. (1958). *J. gen. Physiol.* **41**, 585–594.
SHERE, S. M. and JACOBSON, L. (1970a). *Physiol. Pl.* **23**, 51–62.
SHERE, S. M. and JACOBSON, L. (1970b). *Physiol. Pl.* **23**, 294–303.
SLAYMAN, C. L. (1965a). *J. gen. Physiol.* **59**, 62–92.
SLAYMAN, C. L. (1965b). *J. gen. Physiol.* **49**, 93–116.
SLAYMAN, C. L., LU, C.Y-H. and SHANE, L. (1970). *Nature, Lond.* **226**, 274–276.
SLAYMAN, C. L. and SLAYMAN, C. W. (1962). *Science* **136**, 876–877.
SLAYMAN, C. L. and SLAYMAN, C. W. (1968). *J. gen. Physiol.* **52**, 424–443.
SLAYMAN, C. W. (1970). *Biochim. biophys. Acta* **211**, 502–512.
SLAYMAN, C. W. and SLAYMAN, C. L. (1970). *J. gen. Physiol.* **55**, 758–786.
SLAYMAN, C. W. and TATUM, E. L. (1964). *Biochim. biophys. Acta* **88**, 578–592.
SLAYMAN, C. W. and TATUM, E. L. (1965a). *Biochim. biophys. Acta* **102**, 149–169.
SLAYMAN, C. W. and TATUM, E. L. (1965b). *Biochim. biophys. Acta* **109**, 184–193.
SPANSWICK, R. M., STOLAREK, J. and WILLIAMS, E. J. (1967). *J. exp. Bot.* **18**, 1–16.
TRESNER, H. D. and HAYES, J. A. (1971). *Appl. Microbiol.* **22**, 210–213.
WEST, R. C. and SELBY, S. M. (1967). "Handbook of Chemistry and Physics", Cleveland, Ohio.
WHITTAM, R., AGAR, M. E. and WILEY, J. S. (1964). *Nature, Lond.* **202**, 1111–1112.

V.4

Study of the Efflux and the Influx of Potassium in Cell Suspensions of *Acer pseudoplatanus* and Leaf Fragments of *Hedera canariensis*

R. Heller*, C. Grignon* and D. Scheidecker†

I. Effluxes of K^+ and the Other Ions in an Exsorption System

(cell suspensions of *Acer pseudoplatanus*)

A. Demonstration and Extent of the Potassium Efflux

When the cell suspensions of *A. pseudoplatanus*, obtained and cultured according to the usual techniques (Grignon, 1969) are allowed to absorb ^{42}KCl (all the solutions contain sucrose at 5×10^{-2} M) and then to exsorb in ^{39}KCl, the fraction of radioactivity absorbed that remains in the cells depends not only on the conditions of absorption but also on exsorption, and furthermore on times of exsorption and volume of solution used. Hence the so-called "non-exsorbable" fraction of the mass absorbed would not seem to be a definite amount controlled only by the conditions of absorption.

* Laboratoire de Physiologie végétale (nutrition minérale), Université de Paris 7, 2 Place Jussieu 75221, Paris 5ᵉ, France.
† Laboratoire du Phytotron, CNRS, 91-Gif-sur-Yvette, France.

Furthermore, if the exsorption medium is a very large volume or if it is periodically renewed, it is practically impossible to stabilize this fraction in similar periods of time to those used for absorption; hence there is a permanent efflux liable to affect almost the whole of the amount absorbed and the expressions "non-exsorbable fraction" or "irreversible absorption" only reflects a certain difficulty in the exit of the ions but not an impossibility.

This efflux is not produced by an artefact since the exsorption medium is the same as the absorption medium (simply with ^{39}KCl replacing ^{42}KCl); there is certainly an efflux accompanying influx. Figure 1 shows an experiment

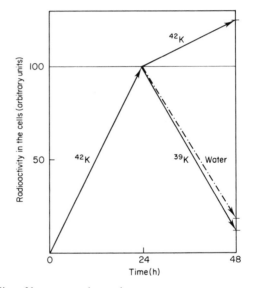

FIG. 1. Reversibility of long-term absorption.

indicating the respective importance of the two processes: cells are placed for absorption in ^{42}KCl (10 mM l^{-1}) for 24 h; the incorporated radioactivity R (100 %) is measured in a sample, then one part is maintained in absorption in ^{42}KCl while another part is allowed to exsorb in water and a third in a solution of ^{39}KCl (the exsorption medium was periodically renewed). The efflux (unidirectional, since renewal of the medium practically suppresses reabsorption) is 81 % of R in water, and 87 % in KCl, whereas the net influx is 24 %.

It is even possible to observe positive *net effluxes* under conditions of absorption when the external concentration is sufficiently low, less than 0·8 µequiv ml^{-1} (Fig. 2): the measurement should be made on stable isotope ^{39}K (otherwise the unidirectional efflux would be measured) in the medium (the variation is too small to be measured in the cells).

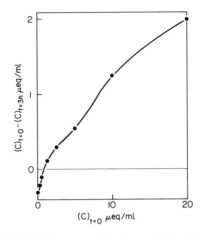

FIG. 2. K^+ absorption (3 h, 21°C) measured by the variation of concentration occurring in the medium.

B. Study of Unidirectional Efflux in Water

The appearance of ^{39}K in the water enables one to trace the gradual emptying of ^{39}K from the cells.

After 24 h absorption at 21°C in a ^{39}KCl (10 mm l^{-1}) solution, a 20 ml sample of suspension is drained and the cells plunged in 20 ml of exsorption solution (sucrose alone). After time Δt, sufficiently short (from 30 s at the beginning of the experiment to 2 h at the end) for efflux $\phi = \Delta n/\Delta t$ to be considered constant and for the external concentration to be considered negligible, the medium is removed and immediately replaced by 20 ml of fresh solution. This periodic return to a concentration of zero enables one to consider ϕ as a unidirectional flux. The exsorbed amount (Figs 3 and 4) at time t is:

$$y = f(t) = \sum_0^t \phi \Delta t = \sum_0^t \Delta n$$

(ϕ in μequiv mn^{-1} per milliatom gram of cellular nitrogen; in principle this should be related to a surface area, but it is difficult to estimate the latter and the samples taken had values that were sufficiently close for the fluctuations of mass and area to be of the same magnitude; moreover, the cytoplasmic mass is mainly localized at the surface).

Analysis of the curve obtained (Fig. 3) was undertaken by a classical serial analysis, which postulates that efflux is the sum of simultaneous partial effluxes each following kinetics of the first order; the amount exsorbed at time t is then:

$$y = \sum_i y_{mi}(1 - e^{-k_it})$$

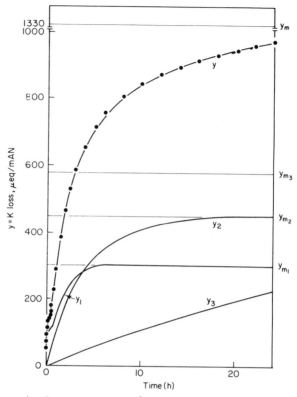

FIG. 3. K$^+$ exsorption in sucrose 5×10^{-2} M. y is the experimental curve; y_1, y_2 and y_3 are the theoretical components of y. (y in μequiv per milliatom gram of cellular N).

each individual kinetic being of the form:

$$\ln \left(1 - \frac{y_i}{y_{mi}} \right) = -k_i t = \frac{t}{T_i} \ln 0{\cdot}5$$

(k_i: rate constant; T_i: half-time period; y_{mi}: maximum exsorbable amount or content of system i.)

The asymptote of $\ln (1 - y/y_m) = f(t)$ represents the slowest component, enabling one to determine the parameters and hence the value of each point. By the difference between these and the experimental figures, one obtains the resultant of the other components which are extracted successively until the final curve obtained resembles a straight line.

The initial postulate (simultaneous flux with first order kinetics) is admissible for "surface-limited"; but not for "volume-limited" fluxes. For example, if the limitation depends on the displacement of ions within the cell volume considered as a sphere, then the diffusion equations are as follows:

$$\ln\left(1 - \frac{y_i}{y_{mi}}\right) = A - Bt$$

for fairly long periods of time (joining up to the point of origin reveals the establishment of the steady state and not an additional component). An analytical method makes it possible to eliminate this ambiguity (Ling et al., 1967): the shape of the curves $y_i/y_{mi} = f(\sqrt{t})$ is in fact characteristic of the model chosen.

The application of previous methods has shown that the kinetics of efflux may be represented (Fig. 3 and Table I) by the sum of three components, the first of which, y_1, is diphasic and the two others obey first order kinetics. The component y_1 by its shape as function of time, $f(\sqrt{t})$ (Fig. 6), seems to reflect a diffusion within the cell volume, followed in the second hour by a desorption or other form of exsorption of the first order.

TABLE I. Components of the efflux of potassium into water (capacities in μequiv per milliatom gram of cellular N)

Components	Characteristics	Order	Half-time	Content
y_1	diphasic:			
	y_1': volume-limited	—	30 s	$\left.\begin{matrix}114\\201\end{matrix}\right\}$ 315
	y_1'': surface-limited	1	1 h	
	(point of inflexion at $t = 45$ mn)			
y_2	surface-limited	1	2 h	450
y_3	surface-limited	1	34 h	580

These data will be discussed later, together with information obtained from the influence of temperature on efflux (Fig. 4). It should be noted, however, that the part played by the cell volume is confirmed by the fact that if exsorption is interrupted, for example from the 8th to the 40th minute, by keeping the drained cells out of the medium before replunging them in the exsorption solution, the exsorption starts up again at a greater speed than previously. It seems as if the ions come together before being exsorbed and that this movement is a limiting factor for subsequent exsorption.

C. Unidirectional Efflux in Water

Effluxes into water have been measured for Cl^-, Na^+, Ca^{++} and Mg^{++} at the same time as the efflux of K^+; but whereas K^+ and Cl^- were loaded

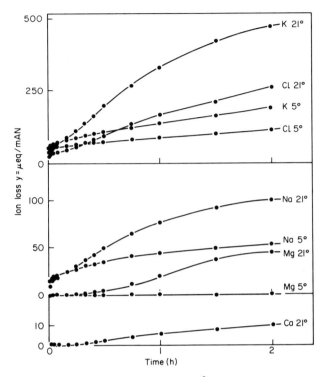

FIG. 4. Effluxes of some ions in sucrose 5×10^{-2} M. In all cases the y_1' component alone persists in 5°C. (y in μequiv. per milliatom gram of cellular N).

TABLE II. Components of the effluxes of different ions in water
(same units of content as Table 1 VL = volume-limited; SL = surface-limited)

Ions	Components	Characteristics	Order	Half-time	Content
Cl⁻	y_1	diphasic $\begin{cases} y_1' \text{ VL} \\ y_1'' \text{ SL} \end{cases}$	— 1		
	y_2	SL	1	1·8 h	227
	y_3	SL	1	30 h	364
Na⁺	y_1	diphasic $\begin{cases} y_1' \text{ VL} \\ y_1'' \text{ SL} \end{cases}$	— 1		
	y_2	SL	1	3·4 h	51
	y_3	SL	1	106 h	65
Ca⁺⁺	y_1	sigmoid, SL	—	—	—
Mg⁺⁺	y_1	sigmoid, SL	—	—	—

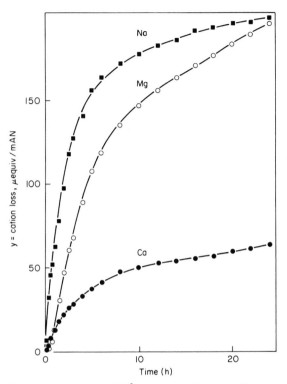

FIG. 5. Cation effluxes in sucrose 5×10^{-2} M. (y in μequiv. per milliatom gram of cellular N).

during the 24 h prior to placing in exsorption, the other types of ions were absorbed before loading in KCl.

The results obtained (Table II) show that exsorption of Cl^- and Na^+ appears to obey the same rules as that of potassium, but with much longer half-times in the case of Na^+, which is in good agreement with the low permeability of cell membranes to this ion; in contrast, it was not possible during the time of the experiment, to extract y_2 or y_3 type components for Ca^{++} and Mg^{++}.

In the case of the latter two ions, the sigmoid type curve (Figs 4 and 5) does not enable one to attribute simple first order kinetics to them. However, analysis of a profile as $f(\sqrt{t})$ (Fig. 6) led to its being considered as surface-limited. Hence, it is still possible that it is first order kinetics, but is preceded by a latent period or induction period. This phase y'_1 would appear to be avoided and essentially the efflux of Mg^{++} and Ca^{++} into water reduces to a component of the type y''_1.

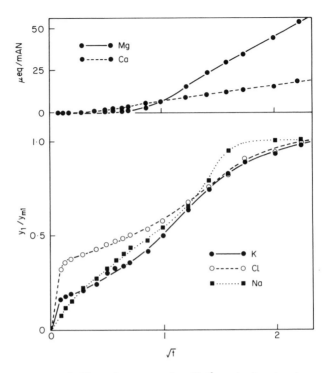

Fig. 6. y_1 component of effluxes in sucrose 5×10^{-2} M: the fractional approach of y_1 to its maximum (K, Cl, Na) or the total loss, y (Ca, Mg), are plotted against $\sqrt{t}$. (y in μequiv. per milliatom gram of cellular N, t in hours)

II. Influx of K^+ and Concomitant Effluxes in KCl

A. Influx of $^{42}K^+$

A 15-day-old suspension (the K^+ content of the medium is 4 mequiv l^{-1}) is washed for one hour in a sucrose suspension, then introduced into the absorption medium KCl = 10 mM l^{-1} labelled with ^{42}K.

After a variable time t, an aliquot of the suspension is drained and immediately treated with hot 0·1 N HCl. The radioactivity of this extract gives the total net absorption at time t. A second aliquot is washed for 30 min in 200 times its volume of unlabelled KCl, then submitted to the same treatment, its radioactivity indicates the non exsorbable fraction in KCl (under the conditions and time limits of the experiment).

The methods of analysis described above show that:

(1) The influx in the exchangeable fraction can be subdivided into two components: y_0 = constant (36 μequiv milliatom g^{-1} of N), which represents an instantaneous and permanent adsorption, which remains apparent in cells

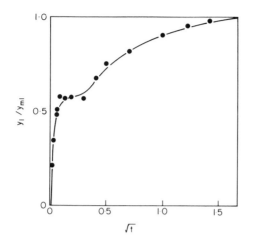

FIG. 7. y_1 component of the ^{42}K incorporation curve: profile against $\sqrt{t}$. (t in hours).

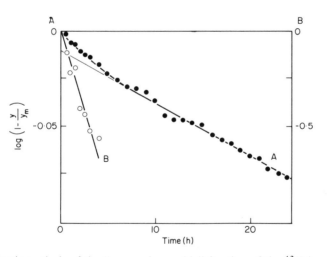

FIG. 8. Kinetic analysis of the "non-exchangeable" fraction of the ^{42}K incorporated. A, experimental data; B, after subtraction of the slowest component.

killed by alcohol; y_1 the profile of which in $f(\sqrt{t})$ (Fig. 7) is characteristic of a diffusion followed by an adsorption or of entry into a subcellular compartment.

The influx $(y_0 + y_1)$ reaches its maximum value after about 2 h and subsequently does not vary.

(2) The influx of the non exchangeable fraction (Fig. 8) can be subdivided into two first order components:

$$y_2 : T = 1 \cdot 8 \text{ h}, \ y_{m2} = 34 \ \mu\text{equiv milliatom g}^{-1} \text{ of N}$$

$$y_3 : T = 109 \text{ h}, \ y_{m3} = 1400 \ \mu\text{equiv milliatom g}^{-1} \text{ of N}$$

(y_{m2} and y_{m3}: apparent contents, since one is dealing with net influx).

The profile in $f(\sqrt{t})$ shows the general aspect of "surface-limited fluxes".

B. Unidirectional Effluxes of K^+, Na^+, C^{++} and Mg^{++}

During absorption of K^+ (as well as Cl^-) there is an efflux of K^+, Na^+, Ca^{++}, and Mg^{++}. The experimental method is exactly the same as that described in section I: a charge of ^{42}KCl during 24 h, then transfer into a solution of ^{39}KCl (and sucrose) with periodic renewals so as to avoid reabsorption.

(1) For potassium, the unidirectional efflux recorded can be subdivided into three components similar to those of the efflux in a pure sucrose solution, with the same half-times (see Table I).

(2) For the other ions, importance is given not so much to the efflux itself as to the part played in the absorption of K^+. This exchange efflux is estimated by the difference between the efflux in a pure sucrose solution and a solution containing KCl as well as sucrose.

The curves of these effluxes by exchange (Fig. 9) showed a marked difference between calcium and the other two ions: all of the exsorbable Ca^{++} is eliminated in 1 h whereas Na^+ and Mg^{++} continue to be exsorbed during at least 20 h.

Even though the kinetic analysis of the efflux of Ca^{++} reveals two independent first order components ($T = 30$ s and $T = 10$ min) the profile in $f(\sqrt{t})$ shows that it is preferable to consider that there is only one component consisting of two successive surface-limited stages (diffusions or desorptions).

The exchange effluxes of Mg^{++} and Na^+ are complex. Their profile in $f(\sqrt{t})$ suggests, as in the case of Ca^{++}, that there are successive but more numerous surface-limited diffusions or desorptions from the subcellular compartments.

The following experiment confirms this difference in localization. If one allows a brief period of absorption (1 min) to be followed by 30 min exchange in ^{39}KCl, all of the ^{42}K is removed as well as a certain quantity of the three other cations; if there is a certain lapse of time between absorption and exsorption, the longer the delay, the smaller is the quantity of ^{42}K removed whereas the contrary is true in the case of the Mg^{++} and Na^+; the quantity of Ca^{++} exchanged does not depend on time factors. Hence the migration of ^{42}K towards less and less exchangeable sites is accompanied by reverse migration of Na^+ and Mg^{++}; Ca^{++} is only exchanged from the surface of the cell.

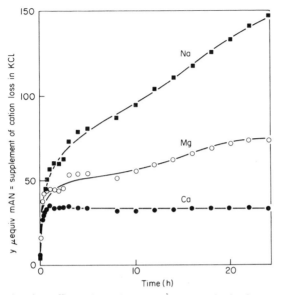

FIG. 9. Additional cation effluxes brought about by KCl (cf. Fig. 5).
(y in μequiv per milliatom gram of cellular N)

III. Complementary Experiments

(Leaf fragments of *Hedera canariensis*)

From the experiments performed for the purpose of studying absorption
and migration of calcium (Scheidecker and Andreopoulos-Renaud, 1971),
we shall only mention some data relative to the present paper.

A. Efflux and Influx

In the two experiments reported here the quantities exsorbed and absorbed
were measured by means of ^{45}Ca and ^{42}K (double labelling); in the measurements of exsorption, reabsorption was avoided by operating at 2°C
(absorption at 25°C).

(1) After one hour of absorption in a labelled complete nutrient medium
($[Ca^{++}] = 8.1$ mequiv l^{-1}) fragments of leaf blade were allowed to exsorb
in water, in complete nutrient medium (non-labelled) or in a solution of
$CaSO_4$ (4 mM l^{-1}) (Fig. 10). Exsorption is very slight in water, whereas it is
quite considerable in the presence of calcium, a fact which fully confirms
the previous observations with cells of *Acer*.

(2) After variable absorption times d in a complete nutrient medium
($[Ca^{++}] = 8.1$ mequiv l^{-1}, $[K^{+}] = 5.5$ mequiv l^{-1}) the leaf blades are
placed for exsorption in an identical medium (but non-labelled) during the
same periods of time. The irreversible fraction (under these conditions)
increases regularly with d in the case of potassium, whereas for calcium, it

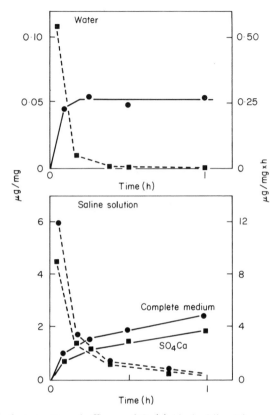

FIG. 10. Exsorbed amounts and effluxes of Ca^{++} (dashed lines) in water or in a saline solution.

rapidly reaches a plateau (Fig. 11). Furthermore, the ratio of the exsorbed amounts to the absorbed amounts in the first few minutes, is nearly 90% for K^+ and only 45% for Ca^{++}. It seems as though the potassium penetrates into the free space from which it can easily be removed or can proceed slowly and regularly towards more interior sites, whereas calcium is retained as soon as it enters on superficial sites from which it can only be eliminated by exchange.

Finally, the data obtained from leaf tissues of *Hedera* are in perfect agreement with the conclusions derived from the kinetic studies performed in cells of *Acer*.

B. Influence of Light on Net Influx

The influence of light on potassium has been mentioned in a number of studies (notably those of Rains, 1967; Nobel, 1969; Lüttge *et al.*, 1970) and

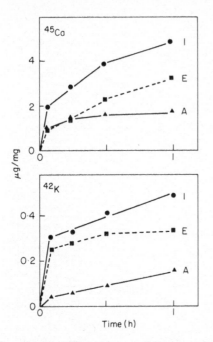

FIG. 11. Absorption (I), exsorption (E) and irreversible absorption (A = I − E) by leaf fragments.

was also observed in the leaf fragments of *Hedera* (Scheidecker *et al.*, 1971). With this material, the main conclusions are:

(1) It is impossible to demonstrate any effect of light on the entry of calcium, even when this ion is present alone and when the experiment lasts 24 h, whether the concentration is low or high ($0.5–20$ mequiv l^{-1}), or the accompanying anion either $SO_4^=$ or Cl^-;

(2) In contrast light clearly favours the entry of K^+. Leaf fragments were placed in absorption (0.5 mM KCl + 0.5 mM CaSO$_4$) in the dark or in the light (white fluorescent and incandescent, 72.000 erg cm^{-2} s^{-1}) then after variable times in exsorption for 30 min, the net influx increases from 0.06 μg min^{-1} per g of dry material to 0.12 μg min^{-1} (Fig. 12). This influence exclusively affects the non-reversible fraction.

(3) The total lack of effect of a pretreatment in dark red light or by lighting with low energy monochromatic light (0.3 mE cm^2 s^{-1}, i.e. approximately 500 erg cm^2 s^{-1} at 728 nm) in a range of wave lengths from 440 to 728 nm, weakens any hypothesis of the purely catalytic effect of light, as produced by the phytochrome. The effect of light in this case may be due to a relationship between the absorption of K^+ and photosynthetic activity.

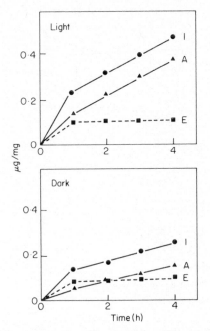

FIG. 12. Effect of light on absorption (I), exsorption (E) and irreversible absorption (A = I − E) by leaf fragments.

IV. Discussion and Conclusions

(1) The observations provided by the experimental results on the existence and importance of a permanent efflux of K^+ underline, in agreement with many other authors, the value of concepts of free space and reversible and irreversible phases of absorption. An almost total exit of K^+, during the washing out of cells, takes place without any visible short-term modifications of physiological behaviour of the cells, which is in good agreement with the biochemical oligodynamic role of K^+ in cell metabolism (Cohen, 1972).

(2) The serial analysis of unidirectional efflux of K^+ in water has made it possible to distinguish three main components. The first, y_1, consists of two phases, one volume-limited resembling efflux outside free space, and the other surface-limited resembling desorption; its existence reconciles opinions as to the roles of diffusion and adsorption (Franklin, 1969; Wacquant and Passama, 1971) in the reversible phase of absorption. The two other components, y_2 and y_3, with long periods, are both suggestive of first order kinetics. Assuming this hypothesis, they reflect that two distinct barriers have been crossed, whether by simple diffusion or by means of transporters according to Epstein's model (with the condition that the

concentrations of the compartments are low with respect to K_M, otherwise the kinetics would be in the order less than one).

(3) The validity of the previous interpretations finds support in the study of the effluxes of other ions. In water Na^+ and Cl^- obey kinetics of the same type as K^+ (only in the case of Na^+ a lower value of the permeabilities is confirmed); whereas the kinetics of the efflux of Ca^{++} and Mg^{++} are apparently reduced to phase y''_1, which would indicate that there is no liberation of these ions by simple diffusion from the more interior compartments. The additional efflux observed when there is a possibility of exchange with absorbed K^+ creates a clear distinction between calcium and the other cations and seems to indicate a very superficial binding, in an exchangeable form.

This latter conclusion is confirmed by the observations obtained from the leaves of *Hedera* and from the roots of various plants (Salsac, 1970); it is in agreement with the fact that the absorption of calcium, with different materials, is largely independent of external doses (except for abnormally low or abnormally high ones); that it is little dependent on temperature and not influenced by oxygen, metabolic inhibitors or light.

(4) Light affects entry into the last compartments representing the space occupied by the so-called "irreversible" absorption (a characteristic which, as we have already seen, is only approximate). Its intervention is not (solely) catalytic but is related to photosynthesis.

In cell suspensions, we observed that the slow, i.e. irreversible, components were first order. In the *Chlorella* the light-dependent influx is first order (Barber, 1968). The results obtained with leaf fragments agree with these observations.

(5) The fluxes of K^+ and Na^+, measured under very different conditions obey fairly similar kinetics (Table II). This analogy justifies the association of each of the kinetic components to a permanent subcellular compartment, representing a cytological localization and/or a particular condition of the ions and/or mode of transport.

In the above discussion we have not taken into account the undulating curves shown by the kinetics of influx. These undulations, observed on several occasions (Grignon and Salsac, 1969) and in various material, are quite real, but we cannot be sure that they are not the consequences of a shock effect. This point of discussion will form the subject of future studies, but the existence of these undulations does not contradict the existence of distinct compartments.

The results of the kinetic analysis of efflux of monovalent ions on coenocytic algae, beetroot or carrot tissues, coleoptiles of oat, epicotyls of pea, etc. all showed the presence of 3–4 components, to which correspond, respectively, the periods of 30 s, a few minutes, 1–2 h and from tens to hundreds of hours (MacRobbie and Dainty, 1958; Pitman, 1963; Cram,

1968; Pierce and Higinbotham, 1970; Macklon and Higinbotham, 1970; Poole, 1971). They were attributed to cytological compartments arranged in series: surface film, walls, cytoplasm and vacuole. A model of this type may be convenient in interpreting the complex structure of the efflux observed in cells of *Acer pseudoplatanus*; the association of y'_1 with the superficial film and the walls, of y_2 and y_3 with the cytoplasm and vacuole constitutes a working hypothesis that none of the facts observed can refute.

The component y''_1 presents a more difficult problem. It is easily mistaken for y'_1 during serial analysis which lacks in sensitivity as far as the number of components is concerned (Sheppard, 1962). The point of inflexion which makes it clearly evident in free cells is perhaps not visible in organs or tissues because of the heterogeneity of the material, which may explain why it has not been mentioned. Its relatively brief period (1 h) attributes an important role to the free space. However, the influence of temperature (Fig. 4) and the particular characteristics of its sensitivity to pH (Grignon, 1969) are in favour of an association of this phase with the plasmatic structures rather than with the walls. A similar component observed in the roots of lupin, during exsorption of bivalent cations, is sensitive to detergents (A. Lamant, personal communication). The compartment associated with y''_1 may thus represent the surface of the plasmalemma (or the interstitial regions of the cytoplasm). The detailed study of exsorption of Rb by barley roots resulted in the same hypothesis of a cytoplasmic free space (Vallée-Shealthiel, 1963).

A picture of the distribution of different ions in cell compartments may thus be inferred from the exsorption curves. The monovalent ions exist in the free form (diffusible in water) in all these regions. This condition seems excluded in the case of bivalent ions of the walls and the cytoplasmic surface which are all bound (exchangeable only). With regard to the distribution of bound ions, the distinction is no longer between the mono- and bivalent ions, but between Ca^{++} and the other ions: the former seems to be confined to the walls and on the surface of the cytoplasm. It might even seem that almost all of the Ca^{++} is in this state, and that its exit into water is in fact an exchange with H^+, or with cations of internal origin (Ca^{++} is not exsorbed in water that is not renewed, as opposed to the other cations). This superficial nature of the localization of Ca^{++} is confirmed in *Acer pseudoplatanus* by the incorporation curves of this ion (Grignon, 1969), and by its exsorption during entry of K^+ (Fig. 9). The same conclusions were reached from a detailed study of the movements of Ca^{++} in the roots of lupin and horse-bean (Salsac, 1970).

Hence, the few experiments mentioned above have provided some indications as to the contribution of the different compartments in absorption and exsorption of ions. It should simply be borne in mind that the phenomena appear to be much more complex than is suggested by the classical schematization into two phases, one reversible, the other irreversible, but the perfect agreement and the mutual support of the data acquired by kinetic analyses with that obtained by more traditional physiological means should be borne in mind.

References

BARBER, J. (1968). *Biochim. biophys. Acta* **163**, 141–149.
COHEN, A. C. (1972). *C.R. Hebd. Acad. Sci., Paris* **274**, 682–685.
CRAM, W. J. (1968). *Biochim. biophys. Acta* **163**, 339–353.
FRANKLIN, R. E. (1969). *Pl. Physiol.* **44**, 697–700.
GRIGNON, C. (1969). *Bull. Soc. Fr. Physiol. végét.* **15**, 193–211.
GRIGNON, C. and SALSAC, L. (1969). *C.R. Hebd. Acad. Sci., Paris* **268**, 73–75.
LING, G. N., OCHSENFELD, M. M. and KARREMAN, G. (1967). *J. gen. Physiol.* **50**, 1807–1820.
LÜTTGE, U., PALLAGHY, C. K. and OSMOND, C. B. (1970). *J. Membrane Biol.* **2**, 17–30.
MACKLON, A. E. S. and HIGINBOTHAM, N. (1970). *Pl. Physiol., Lancaster* **45**, 133–138.
MACROBBIE, E. A. C. and DAINTY, J. (1958). *J. gen. Physiol.* **42**, 335–353.
NOBEL, P. S. (1969). *Pl. Cell Physiol.* **10**, 597–605.
PIERCE, W. S. and HIGINBOTHAM, N. (1970). *Pl. Physiol.* **46**, 666–673.
PITMAN, H. G. (1963). *Aust. J. biol. Sci.* **16**, 647–668.
POOLE, R. J. (1971). *Pl. Physiol., Lancaster* **47**, 731–734.
RAINS, D. W. (1967). *Science* **156**, 1382–1383.
SALSAC, L. (1970). Sci. thesis, Paris.
SCHEIDECKER, D. and ANDREOPOULOS-RENAUD, U. (1971). *C.R. Hebd. Acad. Sci., Paris* **273**, 576–579.
SCHEIDECKER, D., JACQUES, R., ANDREOPOULOS-RENAUD, U. and CONNAN, A. (1971). *C.R. Hebd. Acad. Sci., Paris* **272**, 3142–3145.
SHEPPARD, C. W. (1962). "Basic principles of the tracer method". John Wiley and Sons, N.Y., p. 246.
VALLEÉ-SHEALTHIEL, M. (1963). Sci. thesis, Paris.
WACQUANT, J. P. and PASSAMA, L. (1971). *C.R. Hebd. Acad. Sci., Paris* **272**, 711–714.

Discussion

Pitman suggested that the meeting should first consider how relevant the techniques used on algae are for higher plants. Cram said that there is no evidence in higher plants of a straight-through component to the vacuole as there supposedly is in *Nitella*, but we have evidence that the vacuole and the cytoplasm are connected. He pointed out consequences of having a possible straight-through flux to the vacuole; this would mean that one's estimates of influx and efflux at the plasmalemma are substantially correct, using the traditional flux analysis. The tonoplast fluxes, on the other hand, will be over-estimates.

Shone then questioned some of the conclusions reached by Jeschke. He said first that the efflux is principally from the cortical cells and that their studies at Letcombe on Na^+–K^+ selectively in maize showed that Na^+ is found chiefly in the endodermal layer. This then suggests that any Na^+–K^+ selectivity is inefficient elsewhere in the cortex. Shone suggested that the Na^+ selectivity is a re-absorption of Na^+ from the xylem vessels as the xylem fluid is on its way to the leaves of the plant. Thus there may well be cells in the root which can take up Na^+ in preference to K^+. Jeschke said that this did not apply in his experiment; the roots had been pre-treated with Na^+ and with only a very small amount of external K^+, the xylem fluid contained large preferential amounts of K^+. He thought it improbable that Na^+ could be reabsorbed from the xylem vessels in tissue which was already saturated. Further, he had not been able to Na^+-saturate maize roots; they are very efficient at discriminating against Na^+.

Pallaghy then showed a slide demonstrating points of inflection on efflux curves from the roots of a variety of species in both fresh and aged tissue. The points of inflection point a caution in calculating unidirectional fluxes across plasmalemma and tonoplast. Heller was interested to see Pallaghy's support in these different types of tissue and said that the most important component is his y_1'' which corresponds to the plasmalemma and is sensitive both to temperature and pH. There is a discontinuity in the pH variation of y_1'', which corresponds to the pK value of phospholipids rather than pectic acids; thus it seems to be associated with the plasmalemma, rather than the cell wall. Furthermore, isolated plasmalemma material has bound K^+, Ca^{++} and Na^+; the amounts of Ca^{++} present differ in calcicoles and calcifuges.

Page then commented that a difficulty in multi-compartmental analysis is in assuming a uniform phase in each compartment. To Jeschke "can you therefore give more information on Läuchi's work on mixing in the cytoplasm?" Jeschke replied by adducing the similarity in the half-times of efflux from xylem and cortex as evidence for good mixing. Higinbotham commented that they had information to suggest that there is not an exclusively serial compartmentation. The parallel component he thought might be from the wall through the cytoplasm. Bentrup then said that they had evidence on cell suspensions from cultures which showed that the K^+ and Na^+ permeabilities were exactly the reverse of what is found in the intact organism.

Meares commented on the heterogeneity of the tissue and how a perturbation in the external medium could lead to a series of superimposed transients in the different cell types. When K^+ is introduced into the external medium and starts to be taken up at the plasmalemma, is there not then a diffusion potential which will cause an efflux of Na^+ from those cells? The time course of events in the cytoplasm is too slow to keep pace, so that there are transient peaks and alterations at each plasmalemma. Jeschke replied to Meares that he was not assuming that the cytoplasm-specific activity is constant in time except when he calculated the K^+-stimulated Na^+ efflux. Secondly, he did not know what the coupling is between Na^+ and K^+ fluxes—indeed it may well be an electrical coupling. Ouabain does not effect the Na^+–K^+ selectivity; CCCP inhibits the K^+-stimulated Na^+ efflux. If Ca^{++} is omitted for a short time prior to the addition of K^+ there is a much increased K^+-stimulated Na^+ efflux.

Kylin then spoke in comment on Jennings' paper; he said that he had studied two strains of yeast, one of which is salt resistant; the salt resistant strain has a Na^+–K^+ exchange, the other has a K^+–H^+ exchange. The ratio of K:Na is higher in the salt-resistant yeast in all nutrient media. The salt concentration in the cells is much less than in the medium under all conditions, so that there is no possibility of achieving osmotic balance by salt; the osmo-regulation must be achieved with organic molecules. The salt-resistant yeast regains lost water under exposure to osmotic shock much more quickly than does the normal yeast, which takes up to 4 h before it starts to regain water. The rate at which yeast, fungi etc. can regain water is important in their ability to establish halophytic niches in the environment.

Higinbotham then reported on some NMR studies on Na^+ and K^+ in living higher plant cells. Most of the Na^+ and K^+ is apparently in the free state but there is a small component of about 10–12 μequiv g^{-1} which may be in a bound state. This may be important for Na^+ in Jefferies' case where he has used cytoplasmic Na^+ levels to compute vacuolar fluxes. Notice for example that Jefferies reports that 500 mM external Na^+ produces 11 mM cytoplasmic Na^+; according to the NMR results, a high proportion of this may be bound so that the reported vacuolar fluxes may be in error. Ginzburg said that NMR studies on ions were very difficult, and that many of the previous reports of bound Na^+ and K^+ are in fact incorrect. B. Hill agreed with Ginzburg that it was difficult to detect Na^+ by NMR; she said further that a distinction should be drawn between obligate halophytes and halophytes which can successfully adapt to low salt conditions. A bacterium from the Dead Sea which she has studied can grow in solutions from 50 mM NaCl to 5 mM NaCl. It shows a marked increase in internal Na^+ above 1 M external Na^+ concentration, with concomitant decreases in respiration, amino acid incorporating enzymes etc.

Jefferies then made a point about the levels of salt necessary to bring about full activation of enzymes in plants. He said that techniques usually employ an excess of substrate, but with lower substrate concentrations the sensitivity to Na^+ is raised to a much higher level. Further, these enzymes are often highly pH-sensitive. White then said, "Jeschke postulates a Na^+ efflux pump operating maximally at 10 mM external Na^+, while Jefferies postulates a passive Na^+ efflux in the range 1–2 mM external Na^+ with an active Na^+ efflux in the 100–500 mM region. Is it likely that two such widely different situations pertain in two different species?" Finally White questioned Jefferies' assumption that the rate of isotope movement from the root to the shoot is small in comparison with the efflux rate. Jeschke replied that he had not said that the Na^+ efflux pump operated maximally at 10 mM; it operates maximally at 0·1 mM external K^+. In all his experiments 1 mM Na is present in the external solution. The reference to 10 mM external Na^+ was in confirmation of Epstein's statement that Na^+–K^+ selectivity disappears at 10 mM external Na^+. At this level of external Na^+, the Na^+ influx is equal to the Na^+ efflux on the proposed pump. Jefferies said in reply to White there was no reason in his mind why his Na^+ effluxes should not have the characteristics they do, and Jeschke's have the characteristics they do; proteins are very specific in their affiinities for ions. In regard to the comment of root to shoot isotope movement, Jefferies said that experiments with excised roots had shown that xylem isotope flux was negligible compared to efflux into the external solution.

V.5

Biochemical and Cytochemical Studies of *Suaeda maritima*

T. J. Flowers and J. L. Hall

School of Biological Sciences, University of Sussex
England

I. Introduction

In general plants are not very tolerant of high ionic concentrations and the growth of most crop plants is inhibited by sodium chloride levels above 0·1 M (Nieman, 1962). There are, however, a number of plant species, including many members of the Chenopodiaceae, that are exceptionally tolerant of saline conditions and whose growth may even be stimulated by the high levels of sodium (0·48 N) and chloride (0·56 N) found in sea water. Growth under these conditions must, however, be associated with an adjustment of the cellular water potential at least to less than −30 bars, the osmotic potential of sea water, if water is to be maintained within the symplast. This could be achieved either by the uptake of ions from the environment or, if ion exclusion occurs, by the internal production of an osmotically active solute. In the former case the ions may be either sequestered within the vacuoles or distributed throughout the protoplast. This second alternative is exhibited by halophilic bacteria and the high cytoplasmic ion content is associated with the evolution of enzyme systems which require high salt concentrations for optimal activity (Baxter and Gibbons, 1956; Larsen, 1967).

The question thus arises of whether the plant enzymes are tolerant of high sodium chloride levels, as found in halophilic bacteria, or whether they are protected from these high ionic levels by the sequestering of ions within the vacuoles. The determination of the effect of sodium chloride on enzyme systems from halophytes *in vitro* should therefore indicate whether the enzymes are adapted to these saline conditions and also give an indication of the levels of salt to be found in the cytoplasm.

This report describes a biochemical and cytochemical study of *Suaeda maritima* grown under saline and non-saline conditions. The effect of increasing salt concentration was studied in relation to a number of soluble enzymes and to an amino acid-incorporating microsomal fraction. The sites of ATPase activity and salt accumulation were investigated by cyto-chemical techniques. The results are discussed in relation to the mode of salt accumulation and salt tolerance by halophytes.

II. Materials and Methods

Suaeda maritima (L.) Dum. was grown much as previously described (Flowers, 1972). Briefly, seeds were germinated in sand and transferred to solution culture after about 3 weeks. For plants grown under saline conditions, sodium chloride was added at not more than $2 \, \mathrm{g} \, \mathrm{l}^{-1} \, \mathrm{day}^{-1}$ to give a final concentration of $0.34 \, \mathrm{M}$ sodium chloride and the plants were used approxi-mately five weeks after transfer to the solution culture. Peas (*Pisum sativum* L. cv. Alaska) were grown either in sand irrigated with culture solution or on grids over water.

Microsomal fractions were prepared and amino acid incorporation measured as described elsewhere (Hall and Flowers, 1973). Chlorophyll con-centration was estimated by the method of Arnon (1949). Otherwise the analytical methods and enzyme preparations were as described by Flowers (1972).

For the cytochemical studies, ATPase activity was localized by the method described by Hall (1971). Material for electron microscopy was processed by standard procedures and embedded in an Epikote–Araldite mixture. Sections were post-stained with uranyl acetate and lead citrate unless stated otherwise.

III. Results

A. Growth

Suaeda maritima is a characteristic plant of the lower levels of salt marshes and thus tolerates high levels of salinity in the field. Grown in culture solution, both fresh and dry weight production were maximal with the addition of 0·2–0·35 M sodium chloride (Fig. 1). Even at 0·68 M sodium chloride, growth measured in these terms was 1·5 times as great as in the absence of sodium chloride. The protein content per plant was only about 1·3 times as great under saline conditions (0·34 M) and when expressed on a dry weight basis was half that of plants grown in the absence of salt (Table I). In contrast, peas were very sensitive to sodium chloride (Bernstein and Hayward, 1958; Nieman, 1962) and both fresh and dry weight were much reduced by 0·1 M sodium chloride.

The increase in weight of the *Suaeda* plants grown in the presence of salt was associated with an increase in leaf size. Sections of leaves showed that this was largely associated with an increase in the size of the cells surrounding the vascular tissue (Fig. 2). No distinct morphological differences could be detected between sections of stems and roots of plants grown in the presence and absence of salt.

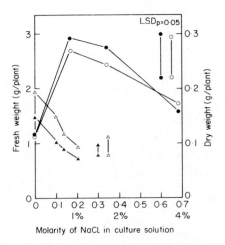

Fig. 1. The effect of sodium chloride added to the culture solution on dry weight (open symbols) and fresh weight (black symbols) of *Suaeda maritima* (circles) and *Pisum sativum* (triangles). Reproduced with permission from Flowers (1972).

As well as the marked increase in fresh and dry weight of *Suaeda* produced by 0·34 M sodium chloride, there was a pronounced difference in the colour of the plants. Those grown in the presence of sodium chloride were lighter in colour than those grown in its absence and had a specific chlorophyll content lower by some 3·5 times (Table I). Salt also reduced the dark respiration rate of the leaves (Table I).

Salinization of the culture solution causes a considerable increase in the sodium and chloride content of the plant tops. Sodium and chloride levels

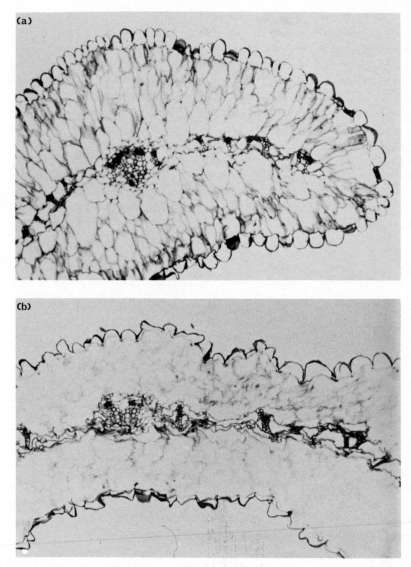

FIG. 2. Paraffin-embedded sections of leaves of *Suaeda maritima* grown in the presence (a) and absence (b) of salt. × 125.

were some thirty times higher in plants grown in the presence of 0·34 M sodium chloride, with concentrations reaching about 0·6 N when based on the overall water content. Potassium levels, on the other hand, were some 5·5 times lower in the salt-grown plants. The pea plants, whose growth was so noticeably reduced by salt showed a similar pattern of ion uptake although

TABLE I. Some characteristics of *Suaeda* grown in the presence and absence of sodium chloride

Added sodium chloride (M)	0	0·34
Protein (mg/g dry weight)	60	30
Chlorophyll (mg/g fresh weight)	1·34	0·38
(mg/g dry weight)	10·3	3·6
Respiration (μl O_2/g fresh weight/h)	349	117
Ion contents		
Sodium (mequiv/g dry weight)	0·21	6·6
(mequiv/g water)	0·02	0·65
Chloride (mequiv/g dry weight)	0·20	6·1
(mequiv/g water)	0·02	0·59
Potassium (mequiv/g dry weight)	1·77	0·32
(mequiv/g water)	0·19	0·03

much lower magnitude. The sodium, potassium, and chloride levels recorded
ter growth in 0·2 M sodium chloride were 0·2, 0·1 and 0·3 N respectively.
The major difference, therefore, between the response of the two species
saline conditions was the stimulation of growth in *Suaeda* associated
ith a high ion content and the inhibition of growth in *Pisum* associated
ith a low ion content.

. Enzyme Activity

To study the effect of salinization on enzyme systems of halophytes and
lt-sensitive plants, the effect of sodium chloride was determined on the
tivity of four soluble enzymes isolated from the tops of *Suaeda* and *Pisum*.
he enzymes investigated were malic dehydrogenase, glucose-6-phosphate
hydrogenase, peroxidase and an acid phosphatase. The effect of 0·17 M
d 0·33 M sodium chloride was determined at the optimal pH and at pH
lues above and below this. For the dehydrogenases, the effect of added
lt was independent of the pH and in both cases the enzymes prepared from
aeda, whether grown in salt or not, were just as sensitive to salt as the
zymes prepared from *Pisum* (Table II).
Unlike the dehydrogenases, the effect of sodium chloride on peroxidase
as very much a function of the pH, the greatest inhibition occurring at pH
lues (4·9) on the acid side of the optimum (pH 5·5) with no significant
fect being observed at pH 7·0. However, as with the dehydrogenases, there
as no difference in the response to added salt between the enzymes prepared
om *Pisum* and *Suaeda* (Table II).
Acid phosphatase was assayed using ATP as substrate and the activity
om both species was maximal in the pH range 4·5–5·5. This acid phos-
atase was relatively insensitive to added sodium chloride, the only

TABLE II. The effect of sodium chloride, added *in vitro*, on the activities of some enzyme of *Pisum* and *Suaeda*

Enzyme		Molarity of added sodium chloride						
		0	0·167		0·33		LSD$_p$ = 0·05	
Malic dehydrogenase	*Pisum*	5·4	3·3	(61)	2·2	(41)	0·5	
(ΔOD/min/mg protein at	*Suaeda*–0^a	3·7	2·2	(59)	1·2	(32)	0·2	
pH 7·1)	*Suaeda*–0·34	4·1	2·5	(64)	1·6	(39)	0·35	
Glucose-6-phosphate								
dehydrogenase	*Pisum*	13·9	10·3	(74)	6·0	(43)	1·7	
(ΔOD × 10^3/min/mg protein)	*Suaeda*–0	30·0	19·2	(64)	13·5	(45)	4·4	
(pH 7·1 for *Pisum* and 7·8 for								
Suaeda)	*Suaeda*–0·34	25·2	18·8	(75)	14·1	(56)	2·2	
Peroxidase								
(ΔOD/min/mg protein at	*Pisum*	9·0	5·3	(59)	3·2	(36)	1·3	
pH 4·9)	*Suaeda*–0	6·6	3·5	(53)	2·8	(42)	0·5	
	Suaeda–0·34	4·3	1·9	(44)	1·3	(30)	0·4	
Acid phosphatase	*Pisum*	106	116	(109)	113	(107)	28·3	
(nmol Pi/min/mg protein at	*Suaeda*–0	30·3	28·8	(95)	27·2	(90)	1·5	
pH 5)	*Suaeda*–0·34	94·4	91·9	(97)	83·1	(88)	4·5	

^{a} *Suaeda* grown in the absence of salt.
The figures in parentheses are the rates relative to those in the absence of added salt.

significant effect observed on the activity from *Pisum* being a 34 % inhibition by 0·33 M salt at pH 7·0. The enzyme activity from *Suaeda* was sensitive to 0·33 M sodium chloride at all three pH values employed (4·0, 5·5, and 6·6 with a mean inhibition of 19 %. It may be inferred from these results that the enzymes of *Suaeda* are not similar to those of the halophilic bacteria that are adapted to high sodium chloride concentrations but are in some way protected from the high salt levels that build up in the aerial parts.

C. Protein Synthesis

A microsomal fraction has been prepared from *Suaeda* shoots grown under both saline and non-saline conditions which incorporated ^{14}C-leucine into protein at a rate of 468×10^{-15} and 796×10^{-15} mol/mg protein 15 min respectively. The effect of increasing sodium chloride concentration on the rate of amino acid incorporation in the presence of 50 mM potassium is shown in Fig. 3. Incorporation by microsomal fractions from both salt and non-salt-grown plants was inhibited by sodium chloride to a similar extent and this was marked even at the lowest concentration (25 mM employed. The inhibition was independent of the time of incubation in the assay medium (from 15 to 40 min) and of both magnesium and potassium

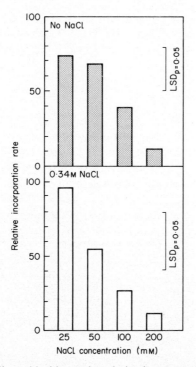

FIG. 3. The effect of sodium chloride on the relative incorporation rate of ^{14}C-leucine into the microsomal fraction of *Suaeda* grown under saline and non-saline conditions. In addition to the sodium added as chloride, all the assays contained 15 mN sodium as the salts of phosphocreatine and ATP.

TABLE III. The effect of sodium and potassium concentrations on the rate of incorporation of leucine into the microsomal fraction (from Hall and Flowers, 1973)

Potassium chloride (mM)	Sodium chloride[a] (mM)	Leucine incorporated (10^{-15} mol/mg protein/ 15 min)
50	0	447
100	0	253
200	0	85
50	100	91
100	100	106
200	100	59
100	50	159
100	200	93

$LSD_{p=0.05} = 158$
[a] See also Fig. 3.

concentrations. In fact at concentrations greater than 50 mM potassium was as inhibitory as sodium (Table III). Similar effects of monovalent cations have been reported for rat-liver microsomal preparations (Samson, personal communication).

D. Cytochemical Observations

The general fine structure of the root cells (Figs 4 and 6) showed no distinct differences from the structure of non-halophytes and no marked changes in fine structure were observed between *Suaeda* roots grown under non-salt and saline conditions. Roots incubated in the ATPase staining medium showed heavy activity at the plasmalemma (Fig. 4) and at the tonoplast

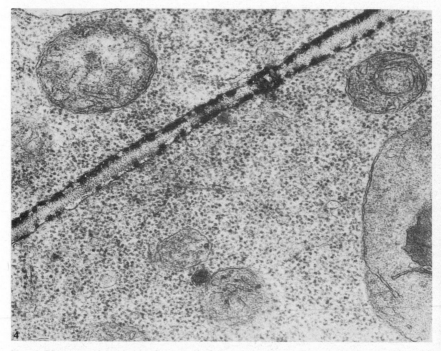

FIG. 4. Electron micrograph of root cell of *Suaeda maritima* grown in the presence of salt and incubated for ATPase activity showing heavy staining in the plasmalemma and plasmodesmata. × 38,000.

(Fig. 5). Frequently cells showing high tonoplast activity did not show staining at the plasmalemma. The fine precipitate visible in the vacuoles of the salt-grown plants (e.g. Fig. 6) was also present in *Suaeda* grown in the absence of salt. Such precipitates are not normally visible in the roots of glycophytic species such as peas and maize.

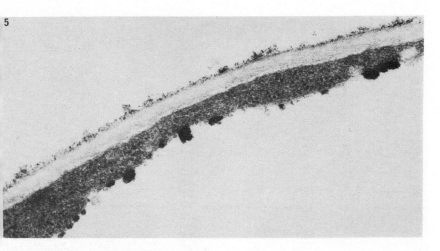

FIG. 5. Electron micrograph of vacuolated root cell grown and treated as in Fig. 4. Showing ATPase activity at the tonoplast. × 42,000.

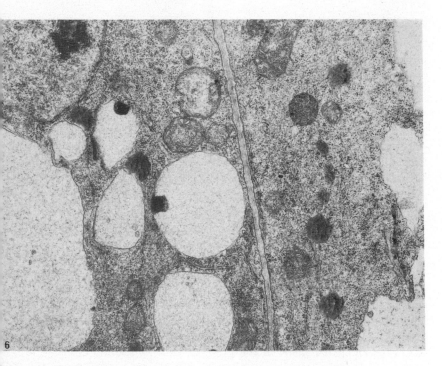

FIG. 6. General appearance of root cells of *Suaeda maritima* grown in the presence of salt and showing the fine precipitates in the vacuoles. × 14,000.

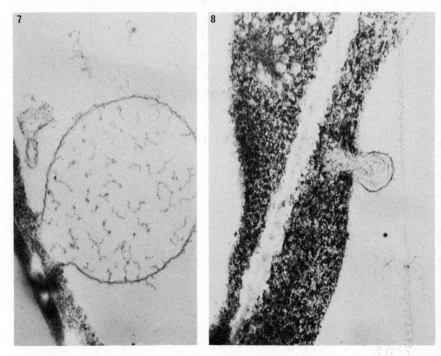

FIGS 7 and 8. Electron micrographs of root cells of *Suaeda maritima* grown in the presence of salt and showing membrane vesiculations from the plasmalemma (Fig. 7) and tonoplast (Fig. 8). Fig. 7. × 34,500. Fig. 8. × 27,500.

In the highly vacuolated cortical cells behind the meristem and in the leaf cells many large vacuoles filled with a fine precipitate were observed within the main vacuole close to the cell surface. These vesicles appeared to arise both from plasmalemma (Fig. 7) and tonoplast (Fig. 8) and frequently exhibited ATPase activity.

IV. Discussion

The increase in growth induced in *Suaeda* by sodium chloride is typical of the more tolerant halophytes such as the *Salicornia* spp. (van Eijk, 1939; Baumeister and Schmidt, 1962; Webb, 1966). In both, fresh and dry weights are promoted particularly by 0·2–0·3 M sodium chloride. Growth in these saline solutions is accompanied by a massive increase in sodium and chloride in the tissues indicating an ionic adjustment of the water potential. Depression of potassium in the plant in the presence of high sodium chloride levels appears to be a fairly common response (Black, 1956; Ashby and Beadle,

1957; Williams, 1960; Greenway, 1962, 1968; Norkrans and Kylin, 1969). It must be realized however, that these data give no indication of the distribution of sodium and potassium throughout the protoplast.

The four soluble enzymes prepared from *Suaeda* and from *Pisum* showed no significant difference in their sensitivity to high sodium chloride concentrations added to the incubation medium. All, except acid phosphatase, showed appreciable inhibition. These results support the observations of Greenway and Osmond (1972) who reported no differences in salt sensitivity when comparing a variety of enzymes extracted from both salt-tolerant and sensitive species and from commercial sources. Thus unlike halophilic bacteria (Larsen, 1967; Hochstein and Dalton, 1968; Aitken *et al.*, 1970), in higher plants there is no evidence of adaptation to the high ionic levels found in the cells of salt-tolerant species. This lack of adaptation is also illustrated by studies of the effect of sodium chloride on amino acid incorporation into protein by a microsomal fraction isolated from *Suaeda*. Although the rate of incorporation was generally higher in the preparations isolated from the plants grown without salt, the addition of sodium chloride had a similar inhibitory effect in both systems. It is interesting that incorporation was even inhibited by low sodium chloride concentrations (40 and 65 mM) since these levels often stimulate isolated enzymes (Evans and Sorger, 1966). Salt has previously been shown to inhibit both amino acid uptake and incorporation into acid precipitates by excised pea root tips and there was some evidence that growth under saline conditions caused damage to incorporation sites (Poljakoff-Mayber and Meiri, 1969). These results contrast with those obtained with halophilic bacteria where high potassium levels (4 M) are necessary for ribosomal stability (Bayley and Kushner, 1964).

This lack of adaptation of various enzymic and protein-synthesizing systems to growth at high salt levels strongly suggests that these systems are not subjected to high ionic concentrations in the living cell. The apparent enigma in *Suaeda* of sodium-inhibited protein synthesis and yet sodium-stimulated growth may be reconciled if the cytoplasm is low in sodium which is sequestered in large amounts in the vacuoles. The consequential osmotic movement of water into the vacuole and increase in turgor is seen as the factor promoting growth.

The vesiculation apparent in both roots and leaves of *Suaeda* may be important with regard to the sequestering of salt within the vacuoles. The possibility cannot, however, be ruled out at this time that the vesicles are artefacts of fixation. Staining for ATPase activity showed a high reaction at the plasmalemma and it is interesting that Jennings (1968) has proposed that a sodium activated ATPase at the plasmalemma could bring about ATP synthesis in the presence of high levels of sodium in the external medium.

References

AITKEN, D. M., WICKEN, A. J. and BROWN, A. D. (1970). *Biochem. J.* **116**, 125–134.
ARNON, D. I. (1949). *Pl. Physiol., Lancaster* **24**, 1–14.
ASHBY, W. C. and BEADLE, N. C. W. (1957). *Ecology* **38**, 344–352.
BAYLEY, S. T. and KUSHNER, D. J. (1964). *J. mol. Biol.* **9**, 654–669.
BAUMEISTER, W. and SCHMIDT, L. (1962). *Flora* **152**, 25–56.
BAXTER, R. M. and GIBBONS, N. E. (1956). *Can. J. Microbiol.* **2**, 599–606.
BERNSTEIN, L. and HAYWARD, H. E. (1958). *A. Rev. Pl. Physiol.* **9**, 25–46.
BLACK, R. F. (1956). *Aust. J. biol. Sci.* **9**, 67–80.
EIJK, M. VAN (1939). *Rec. trav. bot. neerl.* **36**, 559–657.
EVANS, H. J. and SORGER, G. J. (1966). *A. Rev. Pl. Physiol.* **17**, 47–76.
FLOWERS, T. J. (1972). *J. exp. Bot.* **23**, 310–321.
GREENWAY, H. (1962). *Aust. J. biol. Sci.* **15**, 16–38.
GREENWAY, H. (1968). *Israel J. Bot.* **17**, 169–177.
GREENWAY, H. and OSMOND, C. B. (1972). *Pl. Physiol., Lancaster* **49**, 256–259.
HALL, J. L. (1971). *J. Microsc.* **93**, 219–225.
HALL, J. L. and FLOWERS, T. J. (1973). *Planta* **110**, 361–368.
HOCHSTEIN, L. I. and DALTON, B. P. (1968). *J. Bact.* **95**, 37–42.
JENNINGS, D. H. (1968). *New Phytol.* **67**, 899–911.
LARSEN, H. (1967). "Advances in Microbial Physiology" (A. H. Rose and J. F. Wilkinson, eds) Vol. I, pp. 97–132. Academic Press, London and New York.
NIEMAN, P. H. (1962). *Bot. Gaz.* **123**, 279–285.
NORKRANS, B. and KYLIN, A. (1969). *J. Bact.* **100**, 836–845.
POLJAKOFF-MAYBER, A. and MEIRI, A. (1969). *In* "The Response of Plants to Changing Salinity". Technical Report of the Volcani Institute, Jerusalem.
WEBB, K. L. (1966). *Pl. Soil.* **24**, 261–268.
WILLIAMS, M. C. (1960). *Pl. Physiol., Lancaster* **35**, 500–505.

V.6

Adenosine Triphosphatases stimulated by (Sodium + Potassium): Biochemistry and Possible Significance for Salt Resistance

Anders Kylin

Botanical Institute, University of Stockholm
Sweden

I. Introduction

The starting point for our interest was the demonstration of a mechanism for extrusion of sodium in the microalga *Scenedesmus* (Kylin, 1964). The mechanism showed *in vivo* properties (Kylin, 1966), which reminded one of the $(Na^+ + K^+)$-activated ATPases that Skou (1964) and others had connected with active sodium extrusion and potassium-for-sodium exchange in animal tissues. Extrusion of sodium took place not only aerobically but also anaerobically (ATPase!); small amounts of Na markedly increased the retention of Rb (and by inference K) and Cs, thus showing an inside site, where Na was the ion preferred; and small amounts of Na increased the uptake of Rb, without much inhibition by great amounts of Na, thus showing an outside site with Rb (K) as the preferred ion.

At the time the observation was interesting, since the question of sodium extrusion was controversial in plant physiology. Contrary to the workers in human or zoophysiology, most of the groups concerned with ion transport in plants maintained that the rate-limiting and regulating steps were to be

Abbreviations used: DOC = deoxycholate; DTT = dithiothreitol.

found only in the uptake processes. Evidence to the contrary was scarce and found from salt water algae (review by Eppley, 1962), which were regarded as exceptions; or from electrophysiological measurements, which could mean sodium extrusion but which could also be interpreted in the classical sense as reflecting only different permeabilities for different ionic species (review by Dainty, 1962).

When contemplating the next step to be taken, it was decided to initiate a search for $(Na^+ + K^+)$-stimulated ATPases in plants, analogous to those searches connected with sodium extrusion in animals (Skou, 1964). This meant making subcellular fractions. *Scenedesmus* was not a suitable material, due to its hard cell walls, which take such forces to smash that too much general damage is done to the contents. It was also felt that work with a higher plant might be more generally interesting than work with a microalga.

For several reasons, sugar beet was chosen as the first material to work with. It is an agricultural plant, so its relation to mineral nutrients is well known; a remarkable feature is that sodium in the fertilizer will increase the dry weight of the crop, and that lack of sodium will lead to symptoms similar to those of potassium deficiency (El-Sheik *et al.*, 1967). It is tempting to think of this effect as connected with potassium-for-sodium exchange, so that a certain amount of sodium must be available for extrusion in order to get enough potassium in and to keep the potassium inside. Furthermore, we associated the effect of sodium with the origin of sugar beet, which was developed from the halophytic *Beta maritima*. It seemed likely that if sodium extrusion exists in higher plants, it should have a greater ecological importance for halophytes than for glycophytes. Conversely, halophytes and their derivatives might be good starting points in a search for $(Na^+ + K^+)$-activated ATPases.

Another point to consider before the search started was the buffer system for the homogenization. Tris is a common alternative, but the investigations of Van Steveninck (1962) had shown that the lag phase for uptake of potassium to red beets was abolished by tris. In our minds this indicated the danger that tris might interfere in a sodium-for-potassium exchange system, and we looked out for other buffers. The final choice was 0·03 M histidine-HCl at pH 7·2 with 0·20 M sucrose as osmoticum.

II. Preparation and Properties

For the first investigations, the fractions were made broad, so that none of the activity should disappear. The gross particles were spun down by centrifugation for 20 min at 2500 g, and the supernatant treated for 1 h with 0·1 % deoxycholate, after which the particles for study were centrifuged off at 20,000 g for 1 h. As measured by the liberation of P_i the fraction showed two ATPase optima for pH, one at pH 5·75 and one at pH 6·75. The optima for Mg were different at the two pH values. At the higher pH, further increase in the ATPase activity could be obtained by the proper additions of sodium and potassium (Hansson and Kylin, 1969).

The $(Na^+ + K^+)$ dependence of the ATPase activity proved to be complicated and without precedence in the animal systems so far investigated. To clarify the relationship, we worked at a constant addition of (NaCl + KCl) to the basal Mg–ATP–histidine–HCl medium, with the ratios of Na:K varying in all proportions. Some stimulation is obtained with all-sodium, optimum stimulation at Na:K 1:1, then progressively less till Na:K is about 1:8, and finally a second high at all-potassium. The theoretical interpretation is that either there are two differently $(Na^+ + K^+)$-activated ATPases present; or there is an allosteric transformation, so that at the site originally activated by Na, Na is first outcompeted by increasing K proportions, after which transformation occurs so that the site can be activated by K (Hansson and Kylin, 1969).

Investigations on the $(Na^+ + K^+)$-stimulated ATPase from the salt-excreting leaves of the mangrove *Avicennia nitida* further stress the technical precautions that must be taken. In addition to the specific ionic effects, which in this case gave three Na:K ratios (8:2, 5:5, and 2:8) for peak activity of the ATPase, the general ionic and osmotic strength of the medium are also important. At supraoptimal total concentrations of (NaCl + KCl), the ATPase activity was suppressed and no peaks could be seen (Kylin and Gee, 1970).

The method chosen to activate the membrane fragments also has consequences for the specificity of the synergistic action of Na and K. In the above investigations we used the swelling action of deoxycholate (DOC) to get a higher and more reproducible type of ATPase activity than in the untreated ("control") fragments. Even higher specific activities are now being obtained by protection of SH groups with agents like cystein (10 mM) or dithiothreitol (DTT; 25 mM). For a great number of preparations from roots of sugar beet seedlings, the following $(Mg^{2+} + Na^+ + K^+)$-induced specific activities have been obtained at the optimal ratio of Na:K 1:1. The data are given as liberated P_i in nmol (mg protein)$^{-1}$ min^{-1} and rounded off to the nearest 5: Control 50, DOC 70, Cystein 175, DTT 265. About half of this is due to $(Na^+ + K^+)$.

As regards specificity, the activation with DTT is superior to the others. As compared with the control, it does not increase the percent activation due to the monovalents in the low-points all-sodium and Na:K 1:8. The percent $(Na^+ + K^+)$-activation over Mg^{2+} alone at the optimal Na:K ratio of 1:1 increases from 105 in the control to 135 in the DTT preparation. Besides the specific activation, the DOC and cystein treatments lead to an increase in the percent activation that can be induced by any combination of (Na + K), so that an unspecific activation by monovalent cations is

superimposed on the synergistic effect of certain ratios of Na:K (Hansson
et al., 1973).

The search for $(Na^+ + K^+)$-activated ATPases in plants by other groups
has so far led to the detection of activities more or less unspecifically stimulated by monovalent cations (Atkinson and Polya, 1967; Gruener and
Neumann, 1966; Fisher and Hodges, 1969; Horowitz and Waisel, 1970) or
insufficiently substantiated claims about synergistic effects of $(Na^+ + K^+)$
(Brown and Altschul, 1964; Brown et al., 1965). The alleged cytochemical
demonstration of Na- and K-activated ATPases in pine roots (McClurkin
and McClurkin, 1967) is doubtful: the pictures published show ATPase
activity, but the present author finds it difficult to trace the effects of the
monovalents. Since one must now raise the question of how generally the
$(Na^+ + K^+)$ synergism occurs in ATPases from higher plants, these cases
should be reinvestigated with regard to the composition of the medium as
well as using a denser spacing of the Na:K ratios and more than one total
level of $(Na^+ + K^+)$. Bean tissues (Phaseolus vulgaris: Lai and Thompson,
1971) seem to contain $(Na^+ + K^+)$-stimulated ATPase activity, although
effects due to ionic strength or osmotic conditions are not critically excluded.
Sodium and potassium influenced the ATPase activity of plasmalemma
preparations from the highly salt-tolerant microalga Dunalielia tertiolecta,
although the scarcity of the material made it possible to use only a few ratios
of Na:K (Jokela, 1969). Correlation between the transport of Rb and Rb-
stimulated ATPase activity was shown for corn, wheat, and barley (Fisher et
al., 1970).

Ouabain is the classical inhibitor of $(Na^+ + K^+)$-stimulated ATPases
in animal systems. So far, it has not been found to be markedly inhibitory in
the plant systems (Brown and Altschul, 1964, Brown et al., 1965, Hansson,
G. and Karlsson, J., unpublished); also, extrusion of sodium from Scenedesmus was unaffected under the conditions used (Kylin, unpublished). Sodium
efflux in carrot (Cram, 1968), Nitella (MacRobbie, 1962) and Hydrodictyon
(Raven, 1967) was partly inhibited by ouabain.

III. Lipids and Measurements of Surface Charge

A remarkable correlation between salt resistance in rootstocks of grape
and their lipid composition was found by Kuiper (1968a, b). In the five
types of rootstock available, increasing capacity for chloride uptake (salt
sensitivity) was correlated with increasing percentage of monogalactose
diglyceride in the root lipids. In model experiments this lipid highly increased the transport of chloride through a layer of pentanol between two
water phases, whereas other lipid types were less effective. The percentage
of phosphatidyl choline was positively correlated with salt tolerance, and
in the model system phosphatidyl choline allowed exchange of sodium for

potassium. Lignoceric acid (saturated C_{24} in a straight chain) occurred mainly in the phosphatidyl choline and phosphatidyl ethanolamine and ranged from 0·8 to 11·9 % of the total fatty acids, positively correlated with salt tolerance.

There is also evidence from more sophisticated model membranes that the lipid composition can explain most of the properties of passive permeability for water (Graziani and Livne, 1972) and salts (Hopfer et al., 1970; Papahadjopoulos, 1971). For membrane-linked, enzymatic processes, the possibility has been raised that the surfactant lipids direct the orientation and conformation of the proteins (Benson, 1964; Green and Tzagoloff, 1966). In fact, Kuiper (1972) succeeded in reconstituting a defatted ATPase from bean roots by addition of either phosphatidyl choline or sulpholipid. The two lipids gave different temperature dependencies to the reconstituted enzyme, so that the lipids did partly determine the characteristics of the ATPase.

With the above facts and theories in mind, we set out to determine the lipid and fatty acid composition of the sugar beet preparations. First we investigated the sequence (whole roots of sugar beet seedlings)–(control fragments)–(DOC–activated fragments)–(fragments inactivated by prolonged DOC treatment) (Kylin, Kuiper, and Hansson, 1972). Later the sequence (control fragments)–(DTT-activated fragments)–(fragments inactivated by the combination of DTT and DOC) was used (Hansson et al., 1973).

The intact roots contain a high proportion of sulpholipid (37 % of the total lipids) and comparatively little phosphatidyl choline (3 % of the total lipid). Almost all the phosphatidyl choline of the root follows the control fragments, but only about 5 % of the total sulpholipid. In both sequences, activation of the $(Na^+ + K^+)$-stimulated ATPase is accompanied by non-proportionate losses of the zwitterionic phosphatidyl choline and by good retention of the acidic sulpholipid. In the DTT treatment, which gives the sharpest specificity (cf. above), the separation is almost clean, with 90 % loss of phosphatidyl choline and 90 % retention of sulpholipid. It may be noted that sulphatides increased more than any other lipid fraction when the salt gland in duck was activated (K. A. Karlsson et al., 1971).

In our DOC treatment, the lost phosphatidyl choline and the retained sulpholipid were both rich in a saturated fatty acid with 26 C atoms, including a cyclopropanol ring. Due to the clean fractionation between the two lipids, the fatty acid fractionation within them could not be established in the DTT preparations. The C_{26} acid has been observed in relation to a couple of physiological adaptations (Kuiper and Stuiver, 1972—earliness in plants correlated with high C_{26} in the sulpholipid; Bervaes et al., 1972—dehardening of pines correlated with turnover from digalactosyl diglyceride to monogalactosyl diglyceride and their contents of C_{26}).

The two types of inactivating treatments had different effects. Prolonged treatment with DOC led to non-proportionate losses of the (pigments + neutral lipid) fraction and of the part of the sulpholipid that contains the C_{26} acid. A combined treatment with DTT and DOC for the standard time had its main effects in the combined (sterol glycoside + cerebroside + digalactosyl diglyceride) fraction, and again in the sulpholipid; both fractions being almost completely lost.

A loss of zwitterionic lipids and retention of acidic ones may be traceable as increased surface negativity of the particles. The presence of C_{26} fatty acid should increase the charge around the polar groups (Nash, 1965), and with the fractionation observed, make the sulpholipid even more negative. A strong negative charge at a site could be important for the binding of the highly mobile Na and K ions, and conversely, such binding could be necessary to give the lipoprotein the proper conformation to act as an ATPase (Kylin et al., 1972; cf. Hendrickson and Reinertsen, 1971).

Data around these possibilities are being collected by particle electrophoresis (J. Karlsson et al., 1971). Silica particles can be coated with the membrane fragments, and the electrical mobility determined in a Zytopherometer from Zeiss–Oberkochen. In the DOC treatments, negative sites are set free so that the net charge per surface unit is maintained, despite the strong swelling that occurs. The maintenance of the surface negativity is true also in the DTT preparations, but here the swelling is much less. The different biochemistry behind the two denaturing treatments is reflected also in their surface charges: despite the strong swelling in (DTT + DOC), the negativity per unit area is increased, whereas it decreases under a prolonged treatment with DOC alone.

Furthermore, the surface charge varies with the ionic additions (Karlsson, J. and Tribukait, B., unpublished). Magnesium will make the fragments less negative, ATP will increase their negativity. In a combined treatment (MG + ATP), the Mg effect is the dominating one. The monovalents may modify the charges, but it is not yet clear how this influence combines with the ATP and Mg^{++}.

IV. The ATPases of High- and Low-salt Beets

Returning to one of the starting points, namely that sodium extrusion should have a greater ecological significance for halophytes than for glycophytes, this makes a paradox when put side by side with the classical data of Collander (1941), showing that the proportion of Na to K in higher plants increases with increasing salt tolerance. Active extrusion of sodium should be expected to give low ratios of Na:K inside the plants, as was actually found by Atkinson et al. (1967), when they compared the salt-excreting mangrove Aegialitis with the non-excreting Rhizophora. Grown under salt stress, the salt-tolerant yeast *Debaryomyces* seems to exchange sodium for potassium and acquires a lower ratio of Na : K than non-tolerant *Saccharomyces*, which under the same conditions shows more of a proton for potassium exchange (Norkrans and Kylin, 1969).

Higher plants contain large central vacuoles, yeasts do not. The physiologically important role of Na^+ extrusion is to protect the plasma from overhydration due to the hydrated sodium ions (Rothstein, 1964), but higher plants must also have enough solutes in their vacuoles to maintain their turgor pressure. The refined electrochemical and ion flux measurements of later years have given evidence for sodium extrusion from the cytoplasm both to the medium through the plasmalemma and through the tonoplast to the vacuole (Spanswick and Williams, 1964; Spanswick *et al.*, 1967; Pierce and Higinbotham, 1970).

These facts were combined by Kylin and Hansson (1971) into the hypothesis that, as regards sodium, the cytoplasm is the real "inside" of the cell. Optimal salt resistance in a vacuolated cell would then be achieved by sodium extrusion to both the medium and the vacuole, the sodium arriving in the vacuole being turned from a danger to a help in maintaining the turgor pressure. Since the vacuole is a major part of their cells, it can be understood why Collander (1941) found that halophytic higher plants have a high ratio of Na : K. It can also be understood that the reverse is true for non-vacuolated yeast (Norkrans and Kylin, 1969), where the vacuoles are minimal.

Support for the hypothesis was derived from a study on inbred lines of sugar beet (Kylin and Hansson, 1971). Four strains were available through the courtesy of Dr. Olof Bosemark of the Swedish Sugar Company. The one extreme is a low-salt strain with a low ratio of Na : K under field conditions, the other extreme is high in salt and has a high ratio of Na : K. Two strains are intermediate, the intermediate strains being more viable than the extremes. Isolating the $(Na^+ + K^+)$-stimulated ATPases from these strains and testing them at a wide range of Na and K concentrations, the low-salt, low-sodium strain contained one major type of ATPase, with optimum activity at high K concentrations and with only limited influence from Na. The high-salt, high-sodium extreme also had one major type of ATPase, but now with an optimum at low or no potassium and highly affected by sodium. The intermediate strains had two types of ATPases each, similar to the normal field beets used by Hansson and Kylin (1969). It is tempting to think of the extremes as derived by selection, the low-salt one for a transport ATPase localized in the plasmalemma, the high-salt one for a transport ATPase in the tonoplast, each ATPase working in accordance with the electrochemical and flux measurements cited above. The two peak ratios of Na : K found in the intermediate strains as well as in the field type of beets would then represent two different ATPase systems.

Corresponding data were found for ATPases from a similar series that was developed in rye grass (P. Nelson and P. J. C. Kuiper, personal communication).

Acknowledgements

The above investigations have been supported by different grants, the most consistent ones from the Swedish Council for Natural Science Research, the Swedish Council for Agriculture and Forestry, and from C. F. Lundström's Foundation. The manuscript was prepared during a sabbatical stay with Professor Y. Waisel at the Botany Department of the University of Tel-Aviv, despite the energetic attempts of Professor Waisel to lure the author from the dull path of duty and out to the spring meadows and desert halophytes of Israel.

References

ATKINSON, M. R. and POLYA, G. M. (1967). *Aust. J. biol. Sci.* **20**, 1069–1086.
ATKINSON, M. R., FINDLAY, G. P., HOPE, A. B., PITMAN, M. G., SADDLER, H. D. W. and WEST, K. R. (1967). *Aust. J. biol. Sci.* **20**, 589–599.
BENSON, A. A. (1964). *A. Rev. Pl. Physiol.* **15**, 1–16.
BERVAES, J. C. A. M., KUIPER, P. J. C. and KYLIN, A. (1972). *Physiol. Pl.* **27**, 231–235.
BROWN, H. D. and ALTSCHUL, A. M. (1964). *Biochem. biophys. Res. Commun.* **15**, 479–483.
BROWN, H. D., NEUCERE, N. J., ALTSCHUL, A. M. and EVANS, W. J. (1965). *Life Sci.* **4**, 1439–1447.
COLLANDER, R. (1941). *Pl. Physiol.* **16**, 691–720.
CRAM, W. J. (1968). *J. exp. Bot.* **19**, 611–616.
DAINTY, J. (1962). *A. Rev. Pl. Physiol.* **13**, 379–402.
EL-SHEIK, A. M., ULRICH, A. and BROYER, T. C. (1967). *Pl. Physiol.* **42**, 1202–1208.
EPPLEY, R. W. (1962). In "Physiology and Biochemistry of Algae", (R. Lewin, ed.), pp. 255–266. Academic Press, New York.
FISHER, J. and HODGES, T. K. (1969). *Pl. Physiol.* **44**, 385–395.
FISHER, J. D., HANSEN, D. and HODGES, T. K. (1970). *Pl. Physiol.* **46**, 812–814.
GRAZIANI, Y. and LIVNE, A. (1972). *J. Membrane Biol.* **7**, 275–284.
GREEN, D. E. and TZAGOLOFF, A. (1966). *J. Lipid Res.* **7**, 587–602.
GRUENER, N. and NEUMANN, J. (1966). *Physiol. Pl.* **19**, 678–682.
HANSSON, G. and KYLIN, A. (1969). *Z. Pflanzenphysiol.* **60**, 270–275.
HANSSON, G., KUIPER, P. J. C. and KYLIN, A. (1973). *Physiol. Pl.* **28**, 430–435.
HENDRICKSON, H. S. and REINERTSEN, J. L. (1971). *Biochem. biophys. Res. Comm.* **44**, 1258–1264.
HOPFER, U., LEHNINGER, A. L. and LENNARZ, W. J. (1970). *J. Membrane Biol.* **3**, 142–155.
HOROWITZ, C. T. and WAISEL, Y. (1970). *Experientia* **26**, 941–942.
JOKELA, A. C.-C. T. (1969). Ph.D. Dissertation, University of California, San Diego, U.S.A.
KARLSSON, J., TRIBUKAIT, B. and KYLIN, A. (1971). *Proc. 1st Europ. Biophys. Congr.* **III**, 75–79.
KARLSSON, K. A., SAMUELSON, B. E. and STEEN, G. O. (1971). *J. Membrane Biol.* **5**, 169–184.
KUIPER, P. J. C. (1968a). *Pl. Physiol.* **43**, 1367–1371.

UIPER, P. J. C. (1968b). *Pl. Physiol.* **43**, 1372–1374.

UIPER, P. J. C. (1972). *Physiol. Pl.* **26**, 200–205.

UIPER, P. J. C. and STUIVER, C. E. E. (1972). *Pl. Physiol.* **49**, 307–309.

YLIN, A. (1964). *Biochem. biophys. Res. Comm.* **16**, 497–500.

YLIN, A. (1966). *Pl. Physiol.* **41**, 579–584.

YLIN, A. and GEE, R. (1970). *Pl. Physiol.* **45**, 169–172.

YLIN, A. and HANSSON, G. (1971). *In* "Potassium in Biochemistry and Physiology".
   Int. Potash Inst. Colloquium, Skokloster, Sweden.

YLIN, A., KUIPER, P. J. C. and HANSSON, G. (1972). *Physiol. Pl.* **26**, 271–278.

AI, Y. F. and THOMPSON, J. E. (1971). *Biochim. biophys. Acta* **233**, 84–90.

ICCLURKIN, I. T. and MCCLURKIN, D. C. (1967). *Pl. Physiol.* **42**, 1103–1110.

IACROBBIE, E. A. C. (1962). *J. gen. Physiol.* **45**, 861–878.

ASH, T. (1965). *In* "Surface Activity and the Microbial Cell". Soc. Chem. Ind.
   (London) Monographs **19**, 122–135.

IORKRANS, B. and KYLIN, A. (1969). *J. Bact.* **100**, 836–845.

APAHADJOPOULOS, D. (1971). *Biochim. biophys. Acta* **241**, 254–259.

IERCE, W. S. and HIGINBOTHAM, N. (1970). *Pl. Physiol.* **46**, 666–673.

AVEN, J. A. (1967). *J. gen. Physiol.* **50**, 1607–1625.

OTHSTEIN, A. (1964). *In* "The Cellular Functions of Membrane Transport", (J. F.
   Hoffman, ed.), pp. 23–39. Prentice-Hall, Englewood Cliffs, N.J.

KOU, J. C. (1964). *Progr. Biophys. mol. Biol.* **14**, 131–166.

PANSWICK, R. M. and WILLIAMS, E. J. (1964). *J. exp. Bot.* **15**, 193–200.

PANSWICK, R. M., STOLAREK, J. and WILLIAMS, E. J. (1967). *J. exp. Bot.* **18**, 1–16.

AN STEVENINCK, R. F. M. (1962). *Physiol. Pl.* **15**, 211–215.

V.7

Enzymatic Approaches to Chloride Transport in the *Limonium* Salt Gland

B. S. Hill and A. E. Hill

Botany School, University of Cambridge
England

Limonium is an extreme example of a euryhaline salt-marsh plant, in that it grows quite happily on fresh water or double seawater. The salt gland it possesses seems to confer on it the ability to regulate its internal ionic composition in response to changes in the medium, and in view of this adaptability it would be surprising if this could be accomplished with a constitutive transport system. In fact, it does appear that the activation of the gland by a salt load is a true inductive process (Shachar-Hill and Hill, 1970) and that higher loading induces higher pump levels (Hill, 1970). Low salt tissue transferred to a 0·1 M NaCl solution enters a puromycin- and dactinomycin-sensitive stage in which no activity can be detected, after which all the parameters of transport such as secretory potential, short-circuit current, chloride extrusion and volume flow rise to steady state values, as shown in Fig. 1. The whole process takes between 3 and 4 h, and is independent of the diffusive pathway to the gland cells (Hill, 1970). As to the actual ion pumping, it has emerged as a result of experiments where ion fluxes are followed during clamping of the secretory potential, that chloride is pumped but sodium is merely entrained in an internal potential gradient deep within the gland (Hill and Hill, 1972a), and indeed all our other flux data is in agreement with this. Where the coupling of this chloride transfer to metabolism is concerned, ATP seems to be responsible (Hill and Hill, 1972b), and we are quite content to go along with most animal physiologists who seem to assume that ATP is always guilty of driving ion pumping unless proved innocent. Obviously the next step with this system is to aim for a cell-free preparation, and we have looked at the ATPase activity of a microsomal fraction from *Limonium* in much the same way as it has become

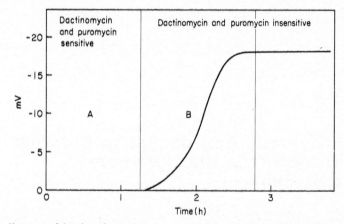

FIG. 1. A diagram of the rise of secretory potential in the *Limonium* salt gland as a function of time after salt-loading. The lag period, A, is sensitive to inhibitors of RNA transcription and protein synthesis, whilst the rise to the steady state, B, is insensitive but highly temperature-dependent.

customary to study the ubiquitous sodium pump. To complement the transport studies, the properties of a microsomal fraction should display the following characteristics: (1) a fraction of the ATPase activity should be Cl-stimulated, (2) the Cl-stimulated fraction should show a substantial increase after salt-loading, (3) the increase should be blocked by puromycin at the same concentration as that used to block the increase in electrical activity in the gland, and (4) the time course of the increase should be of the order of a few hours. That all these conditions should have to be fulfilled might seem a little severe, but we considered that if this were not the case then either the ATPase assay or the transport studies would have been partially based upon artifacts and the system could not be readily understood. The only clear differences between leaf tissue in a short-circuit chamber and leaf tissue salt-loaded in bulk is that the diffusion path of ions to the gland cells in the former is very short, since the lower cuticle has been removed by abrasion, whereas in cut leaf tissue the diffusion path is lateral via the cut edges.

In the experiments 20 g of *Limonium* leaf were finely cut and suspended in solution overnight with relatively bright tungsten illumination (although the induction seems to occur very well in the dark too). After a good wash the tissue was homogenized at high speed in ×7 volumes of medium (Tris 0·025 M, ascorbate 0·025 M, sucrose 0·5 M, brought to pH 7 with NaOH) at 4°C. The homogenizing time was one minute, to ensure that most of the glands were broken open; the gland cells are encased in a tough cuticular envelope. The homogenate was then filtered through muslin and spun at 6000 g for 10 min to sediment cell debris and organelles. The supernatant was centrifuged at 100,000 g for one hour and the microsomal pellet re-suspended in tris sulphate (0·025 M at pH 7). The size of the resulting pellet or its protein content was in fact little altered by spinning at 10,000 g for 20 min, so we assume that the bulk of the mitochondria were removed by the 6000 g spin. The resuspended pellet was assayed immediately. The assay medium (Tris sulphate buffer pH 7, 0·025 M, Mg^{++} 0·004 M, Tris ATP 0·003 M) contained either sodium chloride (0·1 M) or sodium sulphate (0·033 M). These two media have the same ionic strength to eliminate any effects of salt stimulation of the enzymatic activities as opposed to a specific ion stimulation. The microsomal fraction was incubated for an hour at 25°C and liberated phosphate determined by Fiske and Subbarow's method, after stopping the reaction with ice-cold trichloracetic acid.

In the first set of experiments we were content to induce transport over-night in 0·1 M NaCl and measure the relative ATPase activity of the micro-somal fraction in chloride medium as compared to sulphate medium (Fig. 2).

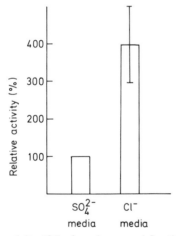

FIG. 2. Relative ATPase activity (%) of a microsomal fraction from induced tissue in sodium sulphate or sodium chloride media of identical ionic strength (0·1 M).

It can be seen that the stimulation is quite appreciable, amounting very conveniently to exactly 300 %. The actual amounts of extractable activity vary considerably, and this may well be due to variation in the average age of the leaf samples chosen for assay, reflected probably in the amount of protein that comes down in every sample. The Cl-stimulated ATPase activity is very labile, however, and although we try to keep to a strict time schedule during extraction, and always to work under constant conditions, it is probable that we lose a fair amount of the activity during centrifugation. Whether or not the system is indeed so labile, or whether if sufficiently cleaned it would ever become as stable as the Na, K–ATPase of animal origin, remains to be seen.

In the second set of experiments we took material from low-salt plants, i.e. plants that have grown on dilute Hoagland solution without any chloride ions, and induced some in 0·1 M NaCl overnight whilst an uninduced control was run in water. As can be seen from the results in Fig. 3, where A is the uninduced control and B is the induced material, induction results in a 350 % increase in activity of the Cl-stimulated ATPase. In some experiments the activity in the uninduced control was zero, although this was not generally the case. This in fact mirrors the case found during transport studies, where sometimes a gland preparation responds immediately to chloride ions, and the system seems to be induced but merely lacks chloride. The uninduced Cl-ATPase levels may therefore be due either to some prior induction of the system or to a similar activity in the mesophyll cells which could be regarded as "constitutive".

In a third set of experiments we preloaded tissue with puromycin (125 mgl^{-1}) overnight with a control in water; with the puromycin still present both batches were then salt-loaded with 0·1 M NaCl for 4–10 h, and assayed. In each experiment some uninduced material was included in the assay, and Fig. 3 in fact represents a lumping of all the data from this and the

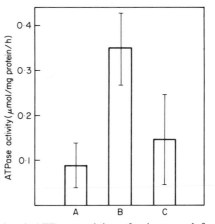

FIG. 3. The Cl-stimulated ATPase activity of microsomal fractions isolated from *Limonium* tissue under the following conditions; A, uninduced material, B, induced material, C, material preloaded with puromycin at 125 mg/l and then induced. All inductions were done with a salt load of 0·1 M NaCl.

second set of experiments, where C is the assay of the puromycin-loaded tissue. The inhibitor substantially reduced the level of induced activity to a value not significantly different from that of the uninduced material. Two other observations need comment. In Fig. 4 the results of two experiments

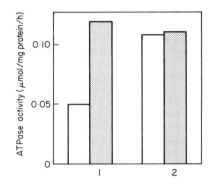

FIG. 4. The Cl-stimulated ATPase activity of a microsomal fraction isolated from tissue induced in 0·1 M NaCl (unhatched), and tissue induced in 0·1 M NaCl and then loaded with puromycin at 125 mg/l (hatched) overnight. 1 and 2 are separate but identical experiments, in which it can be seen that puromycin has no inhibitory action if applied at a later stage.

are shown in which puromycin was added to previously induced tissue and incubated overnight. It seems quite clear that it is not inhibitory. This again accords with the transport studies where this inhibitor has no effect on the short-circuit current after its induction. The other observation concerns the time-course of induction, which we have not studied in any detail. Figure 4 represents two experiments in which tissue was induced for only 4 h prior to assay, and in both of these cases the induced level of Cl-ATPase activity was within the standard error of the mean for the total. This is an indication that Cl-ATPase activity can be induced within this time to a level comparable to the norm.

Although it is only possible to draw parallels between a vectorial transport system and an enzymic process occurring in the same tissue we feel that in *Limonium* the parallels are consistent enough for us to assume that the Cl-stimulated ATPase activity is that of the electrogenic chloride pump described in this tissue. The enzymatic properties of the pump also confirm that ATP is in fact the energy source for transport, and that the pump is being synthesized *de novo* in response to the salt loading. Also, in as much as the transport is a function of the gland cells in this plant we can say with reasonable confidence that these cells are sites of synthesis and pump assembly, though possibly not exclusively. We have no evidence as to the induction of this enzyme in the mesophyll cells, to which of course the gland cells are symplastically connected by plasmodesmata. It is possible that the Cl-ATPase is restricted to the gland cells, but also quite probable that the concentration there is merely very high indeed, and the enzyme is present in the mesophyll but has a low activity. To make any continuing progress with the analysis of this system we must attempt to clean the preparation, and to discover how its active life can be prolonged; only then can one investigate such interesting questions as the Mg^{++} ion and pH dependence in any detail.

To end on an ecological note, we might speculate upon the role of the pump induction in the life-cycle of the species. In the salt marsh *Limonium* is found on the mud flat bordering the waterline, which may be of a composition varying from brackish to sea water. Its ecological niche is therefore a saline gradient, often in two dimensions, and as new leaves break they presumably set their own level of salt extrusion to suit the prevailing salinity in their immediate vicinity. Salt glands are not common to all the species of the flats, however, and must represent only one specialized way of solving the problem.

References

HILL, A. E. (1970). *Biochim. biophys. Acta* **196**, 73.
HILL, A. E. and HILL, B. S. (1972a). *J. Membrane Biol.* (in press).
HILL, B. S. and HILL, A. E. (1972b). *J. Membrane Biol.* (in press).
SHACHAR-HILL, B. and HILL, A. E. (1970). *Biochim. biophys. Acta* **211**, 313.

V.8

Electrical Potentials in the Salt Gland of *Aegiceras*

T. E. Bostrom* and C. D. Field†

*Botany Department (Biophysics), University of Queensland
Brisbane, Australia*

I. Introduction

The mangrove *Aegiceras corniculatum* is one of a number of species of tropical and subtropical mangroves that are able to reduce their internal salt content by the process of secretion of salt solution from glands situated in the leaves. In *Aegiceras* the concentration of the exudate may exceed that of seawater and is usually some 10–20 times greater than that of the xylem sap, which is similar to seawater in ionic composition (Scholander *et al.*, 1962). This work and studies of other plants possessing analogous secretory systems (Arisz *et al.*, 1955; Hill, 1967b) indicate that it is likely that osmotic work is done during the exudation process and that the process is dependent on metabolic processes in the leaf. It has been demonstrated (Hill, 1967a) that an isolated leaf disc from a plant with salt secretory glands will support a considerable transglandular potential when it is bathed on both sides by an appropriate ionic solution.

Present addresses:
* School of Biological Science, Sydney University, Sydney, Australia.
† School of Life Sciences, N.S.W. Institute of Technology, Sydney, Australia.

The study of the permeability of plant cells and the nature of the bio-electric potentials has yielded information indicating that a connexion exists between the selective accumulation of ions in the cell and the physico-chemical properties of its subdivisions (Arisz, 1963). The purpose of the present studies is to determine the transglandular and intraglandular potential profiles pertaining to the multicellular salt gland system of *Aegiceras* using microelectrode techniques, in order to provide evidence as to the configuration of the potential and its possible origin.

II. Materials and Methods

Young leaves of the mangrove *Aegiceras corniculatum* Blanco were taken from plants grown in a greenhouse on soil watered by tapwater. Leaf discs, 2·15 cm diameter, were pretreated overnight in solutions of varying concentrations of sodium or potassium chloride in illuminated aerated plastic vials. Following pretreatment, the lower cuticle of the leaf disc was abraded and the leaf disc was then clamped horizontally in a perspex chamber so that the abraded lower surface was in contact with a slow-moving stream of bathing solution. The preparation was continuously illuminated at 4600 lx using heat-filtered light from a microscope lamp. The experiments were performed at 21°C.

Two 3 M KCl glass microelectrodes, with 1–3 μm tips, were employed in the experiment. One was inserted into the salt gland and the other was used as a reference electrode in the solution bathing the lower abraded surface of the leaf. The microelectrodes were connected to the differential input of an electrometer amplifier of input impedance 10^{14} Ω via 3 M KCl–agar salt bridges and calomel half-cells. The output from the electrometer was recorded continuously on a potentiometric chart recorder. The whole apparatus was Faraday-caged to avoid external electrical interference.

The potential difference across the leaf was determined by placing a large drop of bathing solution on the upper surface of the leaf and locating the tip of the microelectrode in the drop.

To measure the intraglandular potential, the tip of the microelectrode was first located optically about 10 μm above the gland, prior to insertion. This manipulation was performed with the upper leaf surface relatively dry as it was not possible to observe the microelectrode tip or distinguish the glandular cuticle if the upper surface was covered by solution. However, solution was added immediately after the microelectrode was positioned. The microelectrode could then be inserted almost vertically to any given depth into the gland. Initial contact between the tip and the glandular cuticle was marked by a voltage spike on the recorder. Insertion was normally performed in steps of 5 or 10 μm. The potential was recorded continuously between steps until the potential became steady. The tip potential of the microelectrode was determined before and after insertion. If the variation was greater than 15 mV the experiment was rejected.

Structure of the *Aegiceras* Salt Gland

A detailed account of the anatomy of the glands has been previously presented (Cardale and Field, 1971). The main features of the mangrove salt gland may be seen in Figs 1 and 2. Viewed from above, the gland is

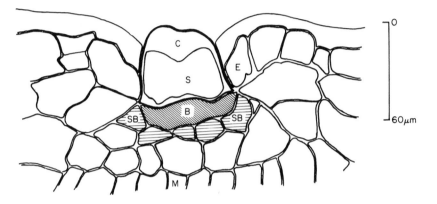

FIG. 1. Representation of a longitudinal section of the *Aegiceras* salt gland on the upper surface of a leaf, showing the main regions and the approximate dimensions. C, gland cuticle; S, secretory cell; B, basal cell; SB, sub-basal cells; E and M, the leaf epidermal and palisade mesophyll cells respectively.

roughly elliptical in shape, with a major axis ranging from 60 to 100 μm and a minor axis ranging from 50 to 70 μm. The three major regions constituting the gland are the cuticle, the secretory cells and the single basal or collecting cell.

The secretory cells in the *Aegiceras* salt gland are situated over the basal cell and are arranged such that each has a common junction with a central region (Fig. 2). The number of secretory cells varies from 24 to 40 depending on the size of the gland. Electron micrographs show the secretory cells densely packed with mitochondria and other organelles, suggesting some metabolically active function for these cells. The structure of the basal cell is far less clear, but it appears to be more vacuolated.

The thin (< 0.5 μm) junction between the basal cell and the secretory cells has been found to contain plasmodesmata. The junction between the basal cell and the sub-basal cells which form a layer above the palisade mesophyll seems to be a partially cutinized area that is a continuation of the leaf cuticle.

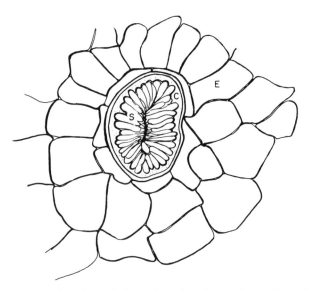

FIG. 2. A transverse section through the region of a salt gland containing the secretory cells (S), which illustrates the particular arrangement of these cells. C is the gland cuticle which extends down the sides of the gland; E is an epidermal cell.

The distance from the top of the cuticle to the lower basal cell wall of a typical gland was determined by the measurement of 12 longitudinal sections as $60.6 \pm$ S.E. 1.4 μm. The greatest variation occurs in the thickness of the basal cell which is $22.0 \pm$ S.E. 1.4 μm at the centre. Thus the mean distance from the glandular cuticle to the boundary between the basal cell and the secretory cells is approximately 38 μm. Provided the depth of insertion was known, the location of the tip could be estimated. A simple coordinate system was defined such that the direction of the positive x was into the gland, and $x = 0$ represented the point at the centre of the cuticular upper surface of the gland, corresponding to the point of microelectrode insertion.

III. Results

A. Characteristics of the Electrical Potential Profile (E.P.P.) of the Salt Gland

The results of a typical experiment involving insertion into a gland, with 100 mM NaCl as the bathing solution, are shown in Fig. 3. Initially the tip of the microelectrode was positioned within about 10 μm of the point $x = 0$ and the potential across the leaf (Ψ_L) was measured. As the microelectrode was inserted into the gland the point of initial contact $(x = 0)$ was marked by a potential spike and an immediate rise in potential thereafter. After about

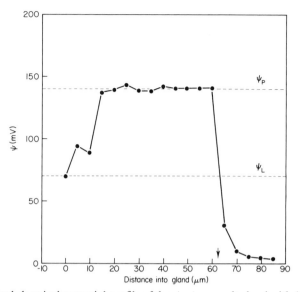

FIG. 3. A typical electrical potential profile of the *Aegiceras* salt gland with 100 mM NaCl external bathing solution. The points shown are the potentials observed as the microelectrode was advanced into the gland at 5 μM steps. Ψ_L is the transglandular potential, Ψ_P is the plateau potential within the gland. The arrow shows the value of the cut-off distance (x_g) where a depolarization was observed.

10–15 μm the potential levelled at a value of the order of twice the value of Ψ_L. Further movement of the electrode resulted in no significant change in this value, which will be termed the plateau potential (Ψ_P). Sometimes potential transients were observed in this plateau region but the potential always tended to return to within 10% of the level value. Hence Ψ_P was considered as a significant parameter, and taken as constant and single-valued for a particular experiment.

After twelve 5 μm steps had been completed, further insertion resulted in a sudden depolarization of the potential of the order of 100 mV or more. By estimating a movement of less than 5 μm, it was determined that this step occurred in a region less than 3 μm in thickness. The result suggested that a membrane had possibly been penetrated. For $x > 90\,\mu$m, the potential was effectively zero. The experiment was repeated on a number of different preparations and the characteristic "plateau" shape was always obtained, with a sharply defined drop at approximately the same distance into the gland each time. This cut-off point is denoted as x_g. The results are summarized in Table I; the results for similar experiments with KCl solution as the bathing medium are also shown. In these experiments, the characteristic profile shape was similarly observed.

Microelectrode insertions into the leaf epidermis between the glands were made under similar experimental conditions to evaluate whether the "plateau" effect in the electrical potential profile was unique to the intraglandular region. At no time was such a plateau potential observed at up to

TABLE I. Determination of the electrical potential profile of the salt gland of *Aegiceras*: effect of different concentrations of bathing solution. Potential measurements following insertion of microelectrode into the gland. The plateau potential is the steady potential measured at 20–60 μm within the gland. (The number of experiments is given in parentheses.)

Bathing solution	Concentration (mM)	Transglandular p.d.[a] mean ± S.E. (mV)	Plateau p.d.[a] mean ± S.E. (mV)	Mean cut-off distance x_g (μm) ± S.E.
NaCl	10	-112 ± 10 (8)	-169 ± 15 (5)	64 ± 1 (5)
NaCl	100	-69 ± 5 (20)	-139 ± 5 (20)	61 ± 1 (20)
NaCl	500	-36 ± 9 (6)	-155 ± 2 (4)	65 ± 2 (4)
KCl	10	-78 ± 11 (9)	-136 ± 4 (8)	60 ± 1 (8)
KCl	100	-81 ± 12 (4)	-126 ± 5 (4)	65 ± 2 (4)
KCl	500	-39 ± 7 (9)	-108 ± 8 (7)	51 ± 2 (7)

[a] p.d., potential difference.

150 μm into the leaf and the epidermal layer potential (probably of the cytoplasm of the epidermal cells) usually did not exceed -30 mV. Thus the characteristic profile could only be obtained in the salt gland.

B. The Transglandular Potential (Ψ_L) and the Plateau Potential (Ψ_P) as Functions of the Concentration of the External Solution

Leaf discs were pretreated in the appropriate solutions and a number of measurements of Ψ_L were made with 10, 100 and 500 mM external NaCl solutions. The mean results from these measurements are given in Table I. They suggest that the potential across the gland is reduced as the concentration of the solution on both sides of the leaf disc is increased. Similar measurements were made with KCl, and the results are shown in Table I. The trend appears less significant in this case.

The characteristic plateau potential was measured with both NaCl and KCl solutions. The mean values of Ψ_P for a number of experiments are given in Table I. The figures show no particular trend of Ψ_P with concentration of NaCl in contrast to the results for Ψ_L.

IV. Discussion

The salt gland of *Aegiceras* demonstrated a characteristic electric potential profile, which was retained in all experiments with different NaCl and KCl solutions. The potential variations in the observed E.P.P. corresponded well to the anatomical features of the salt gland. It can be seen from Fig. 3 that the greatest potential irregularities were found for $x < 20$ μm, corresponding to the apical gland cuticle. The potential subsequently remained virtually constant until approximately 60 μm into the gland where a well-defined potential change occurred. No significant change was found in the range 35–45 μm which can be considered to correspond to the secretory cells–basal cell junction. Such a result might be expected since the plasmodesmata observed in this region would presumably act as low resistance connexions between the basal and secretory cells, thus effectively short-circuiting any potential that might develop across the junction. The mean distances x_g to the potential cut-off point (Table I) were in good agreement with the measured distance into the gland of approximately 60 μm to the junction between the basal cell and the sub-basal cells, which forms the lower boundary of the gland. The sudden nature of the depolarization suggests that it occurred when the tip of the microelectrode penetrated a membrane, possibly the lower plasmalemma of the basal cell. This would seem to indicate that a major part of the trans-glandular potential is initiated at this site, since no potential difference was evident across the secretory cells–basal cell junction.

The significant effect on the transglandular potential of different concentrations of NaCl (Table I) compares with similar variations of the membrane potential with concentration in other biological systems (Adrian, 1956; Etherton and Higinbotham, 1960). Such a variation may possibly be partly explained by considering the effects of different ionic concentrations on the Donnan boundary potentials at fixed charge membranes (Teorell, 1953). However, the situation is complicated by the possibility of a contribution to the total transglandular potential of some metabolically coupled electrogenic ion transport mechanism (Saddler, 1970) present in the gland. This contribution may also be expected to depend on the nature and concentration of the external salt.

There appears to be a trend for the transglandular and plateau potentials obtained using KCl solutions to be less than or equal to those obtained using NaCl solutions. This effect requires further investigation, as it could be significant in determining the ion transport mechanisms present in the gland. The transglandular potential, which would be expected to be related to the movement of ions from the sub-basal leaf mesophyll through the gland, may depend on the availability of Na^+ in the mesophyll as well as in the gland. Whereas the external NaCl concentration might readily influence the concentration of Na^+ within the gland, the intraglandular exchange of K^+

for Na$^+$ with KCl solutions could be slow. One reason for this could be a relatively low permeability of the basal membranes for K$^+$. The relatively low value of the cut-off distance with 500 mM KCl (Table I) suggests a possible osmotic shrinkage of the gland which was not seen with 500 mM NaCl. This would not occur if K$^+$ were able to equilibrate freely between the mesophyll and the intraglandular space. A slow K$^+$–Na$^+$ exchange would also be expected if a certain amount of intraglandular Na$^+$ were bound in the cytoplasmic phase. In addition, the K$^+$–Na$^+$ exchange in the sub-basal mesophyll could be slow, particularly at a low KCl concentration (10 mM), thus minimizing the effect on Ψ_P. With a high concentration of KCl (500 mM), much of the mesophyll-Na$^+$ might be replaced, and an effect of high KCl concentration on Ψ_P might be expected. If this apparent dependence of the potentials on the presence of Na$^+$ is a real effect, then it would seem that both the intraglandular and the mesophyll-Na$^+$-content is important.

The present work has determined the electrical potential profile, described it as a function of the external concentration and correlated it with morphological features of the gland. It has thereby established the lower basal cell plasmalemma as the region of interest in future electrophysiological studies of the salt gland.

Acknowledgements

The work described in this paper was supported by a research grant from the Australian Research Grants Committee to one of us (C.D.F.) and a scholarship from the Rural Credits Bank of Australia to T.E.B.

References

ADRIAN, R. H. (1956). *J. Physiol.* **133**, 631.
ARISZ, W. H. (1963). *Protoplasma* **57**, 5.
ARISZ, W. H., CAMPHUIS, I. J., HEIKENS, H. and VAN TOOREN, A. J. (1955). *Acta bot. neerl.* **4**, 322.
CARDALE, S. and FIELD, C. D. (1971). *Planta* **99**, 183.
ETHERTON, B. and HIGINBOTHAM, N. (1960). *Science* **131**, 409.
HILL, A. E. (1967a). *Biochim. biophys. Acta* **135**, 454.
HILL. A. E. (1967b). *Biochim. biophys. Acta* **135**, 461.
SADDLER, H. D. W. (1970). *J. gen. Physiol.* **55**, 802.
SCHOLANDER, P. F., HAMMEL, H. T., HEMMINGSEN, E. and GAREY, W. (1962). *Pl. Physiol.* **37**, 722.
TEORELL, T. (1953). *Progr. Biophys. biophys. Chem.* **3**, 305.

Discussion

Lüttge first commented on the work of his student, Winter, in connection with Flowers' paper. He said there are clearly salt-stimulated ATPases but the enzymes of intermediary metabolism are more ambiguous in the response to salt. *Mesembryanthemum crystallinum* can be grown on NaCl concentrations from 0–500 mM; at high salt it switches from Calvin cycle metabolism to crassulacean acid metabolism. One can observe a diurnal malate rhythm with a maximum malate concentration after the 12 h dark period. Plants grown in 350 mM NaCl are able to fix CO_2 in the dark whereas in the light there is a net loss of CO_2. The dark fixed CO_2 is incorporated into malate and the malate is then metabolized in the light. Succulence does develop with this salt treatment and perhaps an explanation is (a) salt induction of an enzyme of intermediary metabolism or (b) an effect of succulence whereby malate can be stored more readily in the succulent i.e. large volume, leaves. Assays of the tissue show that Na^+ is in excess at 500 mM and Cl^- is a little less; malate may balance the charge. Flowers replied that Na^+ and Cl^- are approximately in balance in *Suaeda*; there are, however, several *Atriplex* species where organic acids are accumulated in the vacuoles apparently to balance charge.

Waisel then commented that when Flowers said *Suaeda* he meant *S. maritima*; in *S. meloica* there is an excess of Na^+. He said that Flowers had surprised him by saying that cytoplasmic Na^+ was low; the only evidence of which he was aware is for *S. meloica* by Eshel and himself and there the cytoplasmic Na^+ seems to exceed the vacuolar. All the enzymes must function in high Na^+ conditions. Waisel then asked, if there is low enzyme activity and low leucine incorporation in high salt plants, why is it that the plant does in fact grow better in high salt conditions. It is not simply the Na^+ level but the K^+:Na^+ ratio which is important, Waisel continued, and further it may be the Cl^- rather than the Na^+ which is causing the low leucine incorporation. Finally he said that growth stimulation by salt is dependent on atmospheric conditions; in *Atriplex* there is no salt stimulation of growth if the plants are in a saturated atmosphere, but in 50–60% r.h. there is salt stimulation. The salt level at which stimulation is produced varies with the age of the plant, being less in younger plants. Flowers replied that growth stimulation in *Suaeda maritima* is complicated; K^+ is necessary but Na^+ does not seem to affect growth. The K^+ response is complex and depends on the anions present.

Jennings then commented that in *Kalanchoe* crassulacean acid metabolism is remarkably sensitive to Na^+; with small amounts of Na^+ in the medium growth is markedly reduced. There is also an increased organic acid synthesis but it is thought to be a hormonal response and not a direct Na^+ effect. The ionic content of the plant does not change markedly; there is malate in the vacuoles but it does not seem to matter whether it is ionized or not. Lüttge replied that the system must be different in *Kalanchoe*; in *Mesembryanthemum* there is clearly a growth optimum at 350 mM Na^+.

Loughman then asked Kylin and others what evidence there is that the ATPases are functioning as such *in vivo*. What are the K_M values for UTP or GTP? Plant cells generally have a wide range of nucleotides available, for example in synthesizing oligosaccharides for cell wall. It is possible that some of these ATPases are used *in vivo* to regulate cell wall synthesis. Kylin said that ATP was used as substrate simply for laboratory convenience; it did not imply that ATP was necessarily the *in vivo* substrate. Hall then commented in support of Loughman's point; he said there are many ATPases in the plant cell. There may be four or five in the wall fraction and all that one can say about different substrates is that ATP is the most rapidly hydrolysed. Even in the animal cell transport ATPases it is known that other substrates can be utilized; GTP, UTP and even ITP, an unnatural triphosphate. Cram then asked why there are not tens of thousands of ATPases in a cell; ATP seems to be involved in so many biochemical reactions.

Kylin replied to Cram that the philosophy is simple. Membranes must perform active functions and the simplest way to provide energy is to hydrolyse ATP. Therefore these membrane-bound fragments which hydrolyse ATP are pragmatically identified as ion carriers etc. "If membrane-bound ATPases are specifically activated by the ion I think they are carrying, then it is my pragmatic view that these ATPases are carrying those ions." Cram retorted that one must characterize an ATPase very well before ascribing function. Kylin said that is just what he is trying to do. Jennings then suggested that proper characterization would involve comparison of K_M values for ion fluxes and for ATPase activations. He observed that it is unfortunate that the plant system is ouabain-insensitive. Kylin agreed and said that there were other important differences from the animal situation. One litre of blood provides enough erythrocytes to last a lifetime; a great deal of work is involved to obtain an equivalent amount of membrane material from plant cells. Kylin continued that flux measurements in sugar beet were now in progress, and when completed the comparison suggested by Jennings would be attempted. Eventually the evidence from ATPase studies, flux measurements and electro-potential studies will be brought together for sugar beet.

Smith commented that there was something of an obsession with Na^+-K^+ ATPases and not enough work on anion ATPases which were perhaps the major distinction between plants and animals. He continued, is it known what balances Na^+/K^+ uptake in excess of Cl^- in halophytes? It might be H^+ extrusion or possibly NO_3^- uptake; most halophytes are high protein plants. He then asked B. Hill if the anion-stimulated ATPase is specific for Cl^- or if it can operate on NO_3^- or other anions. B. Hill said that they had only worked with *Limonium* where the enzyme is specific for Cl^-. She told Cram that though there may be a large number of ATPases, there is no reason not to start to characterize them. Kylin then commented that the specific activity of Hill's isolate was small, and replied to Smith that it is easier to establish zero levels for Na^+-K^+ ATPases than it is for anion ATPases.

Kylin next devoted himself to an assessment of the respective roles of Mg^{++} and Ca^{++} in the action of ATPases. He first suggested to Hill that Mg^{++} leaching from the chloroplasts during the preparation might be a disturbing factor on their results and expressed surprise at the apparent lack of effect with Ca^{++} in *Limonium*. How much Ca^{++} is normally present in *Limonium*'s ecological niche? he asked. To demonstrate his point he gave results on ATPases from wheat roots where there is both Ca^{++} and Mg^{++} stimulation. In low-salt, low-pH material Mg^{++} and Ca^{++} give equal ATPase stimulation but in high-salt, high-pH plants Ca^{++} gives substrate stimulation of the ATPases while Mg^{++} gives substrate inhibition. If *Limonium* Cl^--stimulated ATPase is similar, one would expect it to be Ca^{++}-dependent, not Mg^{++}-dependent as has been reported. In other species these ATPase properties can be correlated with ecological observation. In acid soils oats predominate over wheat and wheat over barley and these observations can be interpreted in terms of the ATPase properties. Under acid conditions oat ATPase will function with a Mg^{++} requirement while wheat and more particularly barley ATPases have a requirement for Ca^{++} under these conditions. Thus in order to grow wheat or barley on acid soils one must apply Ca^{++} as a fertilizer. Kylin then concluded that perhaps frost tolerance, salt tolerance and drought tolerance have much in common at the membrane level. The lipid composition of the membrane may somehow protect the enzymes located there. For instance it is known that di-galacto-diglycerides decrease and mono-galacto-diglycerides increase in the membranes of frost-tolerant species following heat treatment. This change means that one layer of protective water along the membrane is lost and the permeability and ATPase activity is altered. Jennings commented that in this regard it is interesting that in two *Atriplex* species it is Mg^{++} at 300 mM which gives the best growth of all cations.

Pallaghy then asked Field how he had determined where the charge separation occurred in the mangrove salt gland. Field replied that the micro-electrode studies placed it at the basal layer. Waisel commented that salt secretion is not light-sensitive in all halophytes if the rate of water flow through the leaf is kept constant. He added that in many species substituting K^+ for Na^+ leads to a K^+ secretion although Na^+ is always preferentially secreted from mixtures. He then asked Field to elaborate on the mechanism of secretion which involves accumulation into the basal cells followed by secretion from them. Field replied that there is no net current under short circuiting with K^+; but he added that the short circuit situation is markedly different from that in *Limonium* as reported by Hill. The currents do not add up and there is a marked discrepancy between Na^+ and Cl^-. He then commented on the question of secretion; passing local currents within the gland reveals an odd situation at the cuticle. The resistance is high but the voltage drop is low. There are two possible explanations: either the cuticle shows no differential resistance to ions or there is a bulk flow of solution through the cuticle. It is likely that the secretion fluid is collected in the sub-cuticular and intra-gland spaces and is then driven outward by a standing gradient osmotic process.

Finally Walker asked what examination was made of the condition of the cells as the micro-electrode was driven through; the cells are about 3 μm in diameter and an electrode tip from 1–3 μm in diameter is likely to damage these cells extensively. Field replied that he had used long thin electrodes to minimize damage and that the voltage profile recorded as the electrode was withdrawn was similar to that recorded on entry.

Section VI

Cl$^-$ Transport and Vesicles

Chairman: J. Dainty

VI.1

The Role of Protein Synthesis in Ion Transport

J. F. Sutcliffe

School of Biological Sciences, University of Sussex
England

The remarkable ability of actively growing cells to absorb inorganic ions and to act as a "sink" towards which materials move via the phloem in intact plants is well established but the metabolic basis of these effects is still uncertain. One of the characteristic features of growing cells is rapid synthesis of protein and the purpose of this paper is to examine the extent to which the ability of cells to absorb salt is dependent on this property.

The first clear evidence of a relationship between ion absorption and protein synthesis came from studies with potato slices by Steward and Preston (1940, 1941a, b). They showed that conditions such as good aeration, a suitable pH, a supply of phosphate, nitrate, ammonium and potassium ions all of which are conducive to protein synthesis also stimulate ion accumulation, while, conversely, factors such as high carbon dioxide or bicarbonate and calcium ion concentration which suppress the synthesis of protein inhibit absorption. Later work with tissue cultures has confirmed this relationship. When cells of storage tissues are stimulated to proliferate and synthesize protein rapidly by addition of such stimulants as coconut milk, absorption of ions is enhanced (Steward and Millar, 1954). Steward and Mott (1970) showed that in rapidly growing carrot root cultures the daily increments of protein and potassium content both reach a peak at about the same time and then decline. Similarly, during the extension growth of pea root cells there is a rapid increase in both protein and potassium content (Fig. 1), and in each case the rate of increase reaches a peak at about the time the cells are enlarging most rapidly. A similar situation has been observed in young developing bean leaves (*Phaseolus vulgaris*) (Yagi, 1972).

Although the rates of increase in protein and ion content seem to change in a similar fashion during the various phases of growth there is no fixed quantitative relationship between the two processes. Steward and Millar

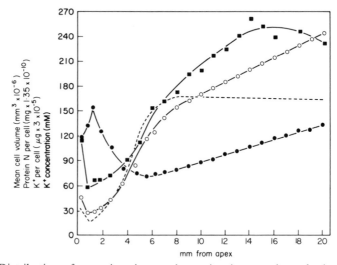

FIG. 1. Distribution of potassium ions and protein nitrogen along the longitudinal axis of young pea roots.

------ mean cell volume ($mm^3 \times 10^{-6}$)

■——■ protein N per cell ($mg \times 1.35 \times 10^{-10}$)

○——○ K^+ per cell ($\mu g \times 3 \times 10^{-5}$)

●——● K^+ concentration (mM)

For details of the techniques employed in obtaining these data see Sexton and Sutcliffe (1969).

(1954) noted that the rate of caesium accumulation in rapidly growing carrot explants was considerably lower per unit of fresh weight (and therefore per unit of protein) than in slowly growing cultures. Similarly Brown and Cartwright (1953) found that potassium absorption per unit of cell surface area and protein content was lower in the rapidly growing tip than in more slowly growing segments excised from the elongating region in the primary root of *Zea mays*. Both groups of investigators attributed the differences observed to a larger volume of vacuoles in the slowly growing cells. The presence of vacuoles probably stimulates ion absorption by providing a reservoir into which ions can be transferred from binding sites in the cyto-plasm. It has been shown (Sutcliffe, 1952) that in slowly growing cells of storage organs the rate of accumulation of potassium ions decreases markedly with time and eventually stops. This is believed to be the result of suppression of the uptake mechanism *per se* rather than to an increase in the rate of exsorption (cf. Pitman, 1963). In growing cells the concomitant absorption of water reduces the rise in concentration of vacuolar sap associated with the accumulation of solutes and this permits more sustained uptake of mineral salts. In growing pea root cells the apparent concentration of potassium ions falls markedly during the period of most rapid growth and rises thereafter as growth declines (Fig. 1).

Stimulation of protein synthesis does not always lead to increased absorption of ions as it does in tissue cultures induced to growth by coconut milk. Waisel *et al.* (1965) observed that when excised barley roots were placed in solutions of kinetin or benzyl adenine for 4 h there was a marked increase in the rate of incorporation of ^{14}C labelled leucine into protein but no effect on the absorption of rubidium over a period of 1 h. It would be interesting to know whether growth was stimulated and ion absorption increased over longer periods of time.

On the other hand, inhibitors of protein synthesis have often been found to reduce the rate of ion absorption (Sutcliffe, 1960; Balough *et al.*, 1961; Uhler and Russell, 1963; MacDonald *et al.*, 1966). By the use of specific inhibitors of protein synthesis it should be possible to decide whether there is a direct relationship between the synthesis of protein and absorption of ions in plant cells. Since protein synthesis and ion absorption are both inhibited in storage tissue slices by D-*threo*-chloramphenicol (CAM) at a concentration which did not affect oxygen absorption or protoplasmic streaming I concluded that ion absorption was affected directly by inhibition of protein synthesis rather than by an interference with energy supply. Hanson and Hodges (1963) showed that high concentrations of D-*threo*-chloramphenicol inhibit oxidative phosphorylation as well as calcium absorption in isolated maize mitochondria. This observation led these workers and others (e.g. MacDonald *et al.*, 1966) to the conclusion that CAM inhibits ion absorption by interfering with ATP synthesis. In my opinion the extrapolation of data obtained with isolated mitochondria to intact cells is, in this case, quite unjustified. It is likely that CAM penetrates much more effectively to the sites of phosphorylation in isolated mitochondria than in intact cells leading to more drastic metabolic effects. No direct evidence of uncoupling of oxidative phosphorylation from respiration by CAM in intact cell systems has yet been obtained and in the meantime I have confirmed my earlier results. While it is true that uncoupling is not necessarily (although usually) accompanied by changes in oxygen uptake (Beevers, 1961) it is difficult to understand how cytoplasmic streaming can be unaffected if ATP synthesis is inhibited. The immediate and complete reversibility of the inhibition caused by CAM when the tissue was transferred after many hours to a solution without the inhibitor is in marked contrast to the irreversible effects of uncoupling agents such as DNP under the same conditions.

A feature of cells of storage tissues is that when first cut from the dormant organ they are incapable of accumulating salt actively, but a capacity to do so develops if they are washed for a time in an aerated salt solution. The onset of absorptive ability is attributable to synthesis or activation of the transport system and the fact that it is inhibited by CAM suggests that protein synthesis

is involved (Sutcliffe, 1960). MacDonald *et al.* (1966) found that the development of ion absorption capacity in red beetroot disks, at least in its early
stages, seemed to be more sensitive to D-*threo*-CAM than is ion absorption
per se and they concluded that the inhibitor was acting differently on two
processes. They speculated that at low concentration D-*threo*-CAM inhibits
the development of ion absorption capacity by preventing the synthesis of
ATPase proteins which may form part of the absorption mechanism, and
that at higher concentration it prevents oxidative phosphorylation, and
hence ion uptake. No direct evidence was presented for either conjecture.
The differences in the apparent sensitivity of the two processes to CAM may
well be attributable to the fact that whereas in the study on development of
ion-absorption capacity the tissue was treated with the inhibitor for several
days, the absorption experiments were carried out over a period of only 3 h.

MacDonald *et al.* (1966) also examined the effects of the L-*threo* isomer of
CAM on ion absorption and protein synthesis and found that despite the
fact that this substance is not an antibiotic it inhibits both processes in red
beet tissue. They found some quantitative differences between the effects of
the different isomers on ion absorption *per se* and on the development of
absorptive capacity which they attributed to a greater influence of the
L-isomer on oxidative phosphorylation and a lesser effect on ATPase
synthesis. Hanson and Krueger (1966) showed that the L-*threo* isomer
suppresses oxidative phosphorylation in maize mitochondria more effectively
than the D-*threo* isomer. They investigated the mechanism of its action and
showed that it is not an uncoupler of the DNP type. It may act by damaging
mitochondrial membranes rather than by blocking synthesis of a phosphorylated intermediate.

The work with CAM has been extended in recent years both in my own
laboratory and elsewhere to embrace a variety of other inhibitors including
amino acid analogues, nucleic acid antagonists and cycloheximide. All have
been shown to be inhibitory to ion absorption in plant tissues and this
strengthens my belief that ion transport in plant cells is dependent on continued protein synthesis. The precise mechanism by which these various
substances inhibit ion absorption is still uncertain and may well be different
for individual compounds and under different conditions. The fact that
cycloheximide (Macdonald and Ellis, 1969) has been shown to uncouple
phosphorylation at certain concentrations does not rule out the possibility
that at others it exerts its effect on ion transport and other physiological
processes by interfering directly or indirectly with protein synthesis. Dr
Nassery, working in my laboratory, has shown recently that potassium uptake
by excised radish roots is inhibited by cycloheximide at concentrations much
lower than those employed by Macdonald and Ellis and at concentrations
up to 0·1 μM root growth is unaffected (Table I). It seems unlikely that the
inhibitor is interfering with energy supply in these experiments since this

would be expected to inhibit growth also and the observations point to a rather specific effect on ion transport. This effect merits further investigation and it remains to be seen whether there is any inhibition of protein synthesis at low concentrations of cycloheximide which inhibit ion transport, but not growth.

TABLE I. The effect of different concentrations of cycloheximide upon root growth of intact radish plants and of ion uptake by their excised roots. Figures in parentheses are percentages of inhibition of uptake in the presence of cycloheximide.

Concentration of cycloheximide (μM)	Average increase of root length of twelve plants cm/cm/20 h $\times$ 10^2	μequiv potassium/ g dry wt/4h Mean
0	7·15	58·15
0·01	5·69	28·5 (51·0)
0·05	6·86	30·1 (48·3)
0·1	7·94	22·7 (61·0)
0·5	3·29	16·7 (71·3)
1·0	1·88	14·3 (75·4)

Roots of 8-day-old intact plants were pretreated for 20 hours in 0·1 mM $CaCl_2$ with or without cycloheximide. The primary root tips were then excised and exposed to a solution with the same composition as that of pretreatment plus 0·2 mM KCl. ^{86}Rb was used as a tracer for potassium.

Protein synthesis may be implicated in ion transport in two different ways. First it is necessary for the establishment of the absorption machinery and secondly it may be involved in the operation of the mechanism itself. The importance of protein synthesis in development of absorptive capacity has been demonstrated in a variety of tissues including rejuvenated tissue slices, growing tissue cultures, developing roots and expanding leaves. Whether the proteins involved are specific ones such as membrane-bound ATPases or a range of enzymic proteins associated with the carrier mechanisms remains uncertain. This question will no doubt be settled as soon as techniques for determining the amounts and location of specific proteins within growing cells are sufficiently improved.

A need for synthesis of components of the carrier system is not of course necessarily confined to cells in which an ion absorption capacity is developing. It is likely that carrier systems like other cell components are labile and being continually broken down and resynthesized. Even if the turnover rate of carriers is high the proportion of the overall protein synthesis required to maintain it will be relatively small but might be measurable once the components of the carrier system has been identified. Some progress has been made recently in the isolation of carrier proteins from micro-organisms

(Pardee, 1968) and when this is eventually achieved it ought to be a relatively simple matter to determine the turnover rates of the proteins implicated in ion transport.

Whether synthesis and breakdown of a transport protein are involved in the operation of ion absorption *per se* remains an open question. To account for the relationship observed between absorption and protein synthesis, Steward and Street (1947) suggested that a cycle of protein synthesis and breakdown is an integral part of the carrier mechanism. They speculated that a phosphorylated, energy-rich nitrogen compound may be formed at the outer surface of the cell, which might possess amphoteric properties, combining with both anions and cations, and be capable of escorting them across the cytoplasm to the site of protein synthesis. Here the substance might be utilized in the synthesis of peptides and the ions released into the vacuole. Later Steward and Millar (1954) elaborated the hypothesis in the light of increased knowledge of the biochemistry of protein synthesis implicating nucleic acids, and identifying γ-glutamyl peptides as the carrier molecules. These ideas have in time given way to even more elaborate schemes (Steward and Lyndon, 1965) in which ion absorption is seen as an integral part of all the various activities of a growing cell.

However, an essential feature still remains (Steward and Mott, 1970) namely the binding of ions at the cell surface to newly synthesized sites on protein molecules followed by breakdown within the cytoplasm to release the bound ions which are then transferred into vacuoles. The precise mechanism by which ions are transferred across the surface membrane into the cytoplasm and subsequently transported to the vacuole is left open. On this point, I suggested (Sutcliffe, 1962) that the constituents of the surface membrane of a cell are continually being broken down and re-formed, providing sites on newly synthesized protein molecules in the membrane to which ions may be bound. These proteins with their associated ions may possibly be transferred into the bulk of the cytoplasm by micropinocytosis (Fig. 2) where the ions are released by distribution of the membrane-bounded vesicles to become involved in metabolic processes, or alternatively to be transferred to organelles such as mitochondria and vacuoles. Direct evidence supporting this hypothesis has come from the work of my colleague, J. L. Hall (1970), who has demonstrated the presence of membrane-bound vesicles apparently derived from the plasmalemma in barley root cells. Formation of these vesicles, the membrane of which is rich in ATPase, as is the plasmalemma, is stimulated by the presence of salt in the external solution. Using an entirely different approach MacRobbie (1969) has concluded from a study of the kinetics of chloride uptake in *Nitella* that salt is accumulated by a process which involves the formation at the plasmalemma of salt-filled vesicles that move through the cytoplasm, with some exchange of ions on the way, to discharge either directly at the tonoplast or by fusion with the

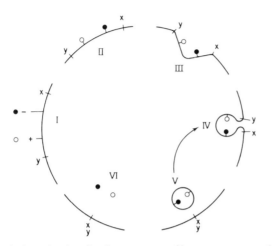

Fig. 2. A hypothetical mechanism for the transport of ions across a membrane involving membrane synthesis and pinocytosis.

endoplasmic reticulum from which a new vacuole may subsequently form. Gradually plant physiologists are beginning to discard the idea largely derived from studies with animal cells that the plasma membrane is a static barrier to passive diffusion across which a carrier mechanism operates, in favour of concepts which are more in keeping with our rapidly growing knowledge of the dynamic nature of cell ultrastructure.

References

BALOUGH, E., BÖSZÖRMENYI and CSEH, E. (1961). *Biochim. biophys. Acta* **52**, 381–382.
BEEVERS, H. (1961). "Respiratory Metabolism in Plants". Row Peterson Biological Monographs, Evanston, Illinois, U.S.A.
BROWN, R. and CARTWRIGHT, P. M. (1953). *J. exp. Bot.* **4**, 197–221.
HALL, J. L. (1970). *Nature, Lond.* **226**, 1253–1254.
HANSON, J. B. and HODGES, T. K. (1963). *Nature, Lond.* **200**, 1009.
HANSON, J. B. and KRUEGER, W. A. (1966). *Nature, Lond.* **211**, 1322–1323.
MACDONALD, I. R. and ELLIS, R. J. (1969). *Nature, Lond.* **222**, 791–792.
MACDONALD, I. R., BACON, J. S. D., VAUGHAN, D. and ELLIS, R. J. (1966). *J. exp. Bot.* **17**, 822–837.
MACROBBIE, E. A. C. (1969). *J. exp. Bot.* **20**, 236–256.
PARDEE, A. B. (1968). *Science, N.Y.* **162**, 632–637.
PITMAN, M. G. (1963). *Aust. J. biol. Sci.* **16**, 647.
SEXTON, R. and SUTCLIFFE, J. F. (1969). *Ann. Bot.* **33**, 407–419.
STEWARD, F. C. and LYNDON, R. F. (1965). *New Phytol.* **64**, 451–476.
STEWARD, F. C. and MILLAR, F. K. (1954). *Symp. Soc. exp. Biol.* **8**, 367–406.
STEWARD, F. C. and MOTT, R. L. (1970). *Int. Rev. Cytol.* **28**, 275–370.
STEWARD, F. C. and PRESTON, C. (1940). *Pl. Physiol., Lancaster* **15**, 23–61.
STEWARD, F. C. and PRESTON, C. (1941a). *Pl. Physiol., Lancaster* **16**, 85–116.
STEWARD, F. C. and PRESTON, C. (1941b). *Pl. Physiol., Lancaster* **16**, 481–519.

STEWARD, F. C. and STREET, H. E. (1947). *A. Rev. Biochem.* **16**, 471–502.
SUTCLIFFE, J. F. (1952). *J. exp. Bot.* **3**, 59–76.
SUTCLIFFE, J. F. (1960). *Nature, Lond.* **188**, 294–297.
SUTCLIFFE, J. F. (1962). "Mineral Salts Absorption in Plants". Pergamon, Oxford.
UHLER, R. L. and RUSSELL, R. S. (1963). *J. exp. Bot.* **14**, 431–437.
WAISEL, Y., NEUMAN, R. and ESHEL, Y. (1965). *Physiologia Pl.* **18**, 1034–1036.
YAGI, M. (1972). D. Phil. Thesis, University of Sussex.

VI.2

The Possible Role of Vesicles and ATP.ases in Ion Uptake

R. A. Leigh, R. G. Wyn Jones and F. A. Williamson*

Department of Biochemistry and Soil Science
University College of North Wales
Bangor, Caernarvonshire, Wales

I. Introduction

The uptake of a number of ions by plant tissues has been described by two major concentration isotherms (Epstein, 1966; Laties, 1969). In the case of potassium, the lower isotherm, hereafter referred to as System I, is operational at concentrations below 1 mM and shows a high affinity for potassium compared with sodium. In contrast, the upper isotherm (System II) which is observed above 1 mM, exhibits a slightly higher affinity for sodium than for potassium.

Despite a formidable volume of work on these and other aspects of ion relations, there is no general agreement as to the biochemical mechanisms

* Present address: Department of Botany and Plant Pathology, Purdue University, Lafayette, Indiana, U.S.A.

operating nor their location within the cell. An important point of contention is whether the two systems act in parallel at the plasmalemma or in series; System I being at the plasmalemma and System II at the tonoplast. The latter hypothesis was based originally on a comparison of uptake by vacuolated and non-vacuolated cells of corn root segments which suggested a relationship between System II and the presence of vacuoles (Torii and Laties, 1966). Further evidence has been adduced from a study of long-distance transport (Lüttge and Laties, 1967), and ion fluxes (Lüttge and Bauer, 1968; Osmond and Laties, 1969) to support this interpretation (see Laties, 1969).

The alternative concept of two parallel pathways at the plasmalemma is favoured by Epstein and co-workers (Welch and Epstein, 1968, 1969) who observed that the amount of tracer absorbed during five 1 min pulses was 50% of that absorbed during a 10 min continuous uptake; both were terminated by a 30 min wash in cold, inactive solution.

Some doubt is cast upon both of these interpretations by Pitman (see Pitman et al., 1968 and Pitman, 1970) who suggested that System II may diminish as roots reach high salt status and is therefore an artefact of the low salt status of the roots generally used in uptake experiments. He did not, however, examine the rate of uptake by System I at high and low salt status. On the other hand, Cram and Laties (1971) appear to have found that System I for chloride uptake was inhibited when root tissue is equilibrated with high salt solution.

The classical interpretations of the dual isotherm have both rested on the concept of the cell as three compartments, cell wall, cytoplasm and vacuole, between which the transport of ions must necessarily be sequential. Recently, a number of workers have suggested that a direct pathway exists, possibly by way of membrane-bound vesicles, from the external solution to the vacuole (MacRobbie, 1970; Cram and Laties, 1971; Pallaghy et al., 1970; Neirinckx and Bange, 1971; Williamson and Wyn Jones, 1972).

In this paper, we examine the kinetics of potassium uptake by maize root segments. On the basis of this and other evidence we present a number of suggestions regarding the mechanism and cellular location of the potassium uptake isotherms. There is a regrettable lack of substantial evidence for a number of points raised, nonetheless we believe that this volume provides an appropriate platform for the presentation and discussion of these ideas.

II. Materials and Methods

A. Growth of Plants

Corn seeds (Zea mays L., WF9 × M14, Crow Hybrid Corn Co., Milford, Illinois) were surface-sterilized in sodium hypochlorite solution containing 1–2% available chlorine, germinated and grown at 27°C for 5 days on anodized, expanded aluminium trays covered with filter paper. The trays were suspended over either aerated 0·5 mM $CaCl_2$ (low salt conditions) or 2·5 mM KCl + 7·5 mM NaCl + 0·5 mM $CaCl_2$ (high salt conditions). After 5 days, roots were excised into distilled water as approximately 1 cm sub-apical sections.

B. Uptake Experiments

Excised roots were placed in cheesecloth "teabags" (Epstein et al., 1963) and placed in the aerated experimental solution of KCl, at the appropriate concentration, containing 0·5 mM $CaCl_2$. ^{86}Rb (Radiochemical Centre, Amersham) was used as a tracer for potassium. Schimansky and Marschner (1971) have shown that ^{86}Rb is a suitable tracer for potassium in maize roots at both low and high salt status. Experimental solutions were held at 27°C in a thermostatically controlled water bath. Uptake was terminated by a wash in cold, aerated, unlabelled 5 mM KCl + 0·5 mM $CaCl_2$. Two different uptake–wash treatments were used:

(a) Short uptake–wash; a 10 min uptake period followed by a 5 min wash.

(b) Long uptake–wash; a 40 min uptake period followed by a 30 min wash.
Uptake of K^+ was calculated by comparison of the activity of ^{86}Rb in the tissue and the specific activity of the experimental solution.

C. Efflux Experiments

The method used was similar to that first described by Pitman (1963) and since used by several authors (Cram, 1968; Lüttge and Bauer, 1968; Pallaghy et al., 1970). Roots were equilibrated in unlabelled solution for 3–4 h, allowed to accumulate tracer for 12 h and efflux was measured over the next 12 h. All experiments were performed at 2–4°C. Plotting of the data and calculations of half-times of cellular compartments were described by Pitman (1963).

D. Radioactive Counting Procedure

^{86}Rb was determined on a Philips' automatic liquid scintillation analyser by measurement of Cerenkov radiation. The method of counting was as described by Lauchli (1969) in which 2·5 mM A.N.D.A. (7-amino,1,3,napthalene disulphonic acid, Eastman Organic Chemicals) is employed as a wavelength shifter.

E. Freeze-etching Technique

Isolated membrane fractions were prepared by the method described by Williamson and Wyn Jones (1972). Freshly isolated membrane samples were fixed in 6% glutaraldehyde, 10% sucrose in 0·025 M cacodylate buffer, pH 7·0. Fixed pellets were then kept in 28 % (w/v) glycerol until freeze-etched. The procedure for freeze-etching was basically as described by Moor and Mühlethaler (1963). An etching time of 1 min was employed. Biological material was removed from the platinum–carbon replicas as described by Power and Cocking (1970). Mounted replicas were examined in an AEI/GEC EM6M electron microscope. Micrographs shown in this paper are processed so that shadows appear white and the direction of shadowing is from bottom left to top right.

III. Results and Discussion

A. Influence of Salt Status and Uptake–Wash Regime upon Uptake Kinetics

The kinetics of (^{86}Rb)K$^+$ uptake were determined over the concentration range 0–0·3 mM (i.e. System I) using both high and low salt roots. In addition, a comparison was made of the kinetics obtained using short and long uptake–wash treatments analogous to those used by Cram and Laties (1971). High salt status severely depressed the System I isotherm independently of the uptake and wash periods used (Fig. 1), although the short uptake–wash regime gave a somewhat higher V_{max} in the low salt tissue compared with the long uptake–wash treatment. Cram and Laties (1971) reported that Cl$^-$ influx into barley and maize roots was about equal over the range 0·02–1·0 mM using both short and long uptake–wash regimes.

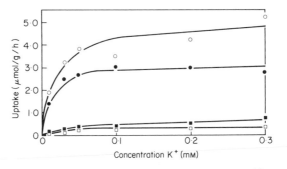

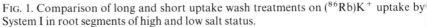

FIG. 1. Comparison of long and short uptake wash treatments on (^{86}Rb)K$^+$ uptake by System I in root segments of high and low salt status.
○—○   Low salt status, short uptake–wash.
●—●   Low salt status, long uptake–wash.
□—□   High salt status, short uptake–wash.
■—■   High salt status, long uptake–wash.

In contrast the kinetics of the upper isotherm of potassium absorption is influenced by uptake–wash treatment as well as by the salt status of the tissue (Fig. 2). The short uptake–wash regime gave an almost linear and

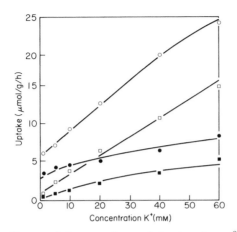

FIG. 2. Comparison of long and short uptake–wash treatments on (^{86}Rb)K$^+$ uptake by System II in root segments of high and low salt status. Symbols as for Fig. 1.

steep relationship between rate of uptake and potassium concentration up to 40 mM while the standard long uptake–wash conditions gave the characteristic shallow upper isotherm of maize (see Torii and Laties, 1966; Fisher *et al.*, 1970). High salt status displaced both isotherms towards lower uptake rates. This displacement may be accounted for by the influence of salt status upon the lower isotherm noted above. This relationship is further illustrated in Fig. 3 which shows the influence of salt status on both isotherms as

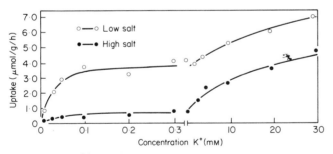

FIG. 3. Dual isotherm of (^{86}Rb)K$^+$ uptake at high and low salt status determined by long uptake–wash treatment.

measured by the long uptake–wash technique. The inhibition of the rate of uptake by System II (2–2·5 μmol g^{-1} h^{-1}) is more than accounted for by the lowering of the V_{max} of System I. To obtain further evidence on this point, the alteration of rate of K$^+$ influx from 0·1 mM KCl and 10 mM KCl solutions was measured during the transition from low to high salt status. The

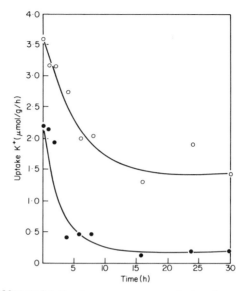

FIG. 4. Change in (^{86}Rb)K$^+$ influx into root segments during the transition from low to high salt status in 2·5 mM KCl + 7·5 mM NaCl + 0·5 mM CaCl$_2$. Influx was determined by the long uptake–wash treatment.

○—○   Influx from 10 mM KCl + 0·5 mM CaCl$_2$.
●—●   Influx from 0·1 mM KCl + 0·5 mM CaCl$_2$.

transition was brought about by transferring low salt seedlings to a solution containing 2·5 mM KCl 7·5 mM NaCl and 0·5 mM CaCl$_2$. The influx rates into root segments excised from these plants were then measured at intervals during the subsequent 30 h transition period. Influx was measured by the long uptake–wash regime. The similarity between the behaviour of the two influx curves at 0·1 and 10 mM is shown in Fig. 4 and it appears that the lowering of the rate of K$^+$ influx at 10 mM can be attributed to changes in the rate of System I. However, a very low-level activity of the lower isotherm can still be detected and the same selectivity for K$^+$ over Na$^+$ is maintained (Leigh and Wyn Jones, unpublished data). Thus it would appear that System II is not an artefact of low salt roots, as suggested by Pitman, but that its lower activity in high salt roots is apparently a reflection of the disappearance of System I.

It is considered in both the major hypotheses that System I is located at the plasma membrane (cf. Laties, 1969; Welch and Epstein, 1969). The similarity between the short and long uptake–wash isotherms at low concentrations, for both potassium in maize and chloride in barley (Cram and Laties, 1971), is consistent with these ideas. The possible significance of the small difference between the two uptake–wash treatments is under further scrutiny. In contrast the marked differences in the isotherms obtained in the range of System II would imply that an association with an internal cellular compartment must be considered in their interpretation if the short uptake–wash treatment measures influx across the plasmalemma as suggested by Cram and Laties (1971). The difficulties of applying this concept to potassium in maize are further discussed later.

B. Potassium Efflux Curves

The efflux curves for ^{86}Rb from preloaded corn root segments in 0·1 mM KCl and 5 mM KCl (both containing 0·5 mM CaCl$_2$) at 2–4°C are shown in Fig. 5. The curve obtained at 0·1 mM shows a distinct and reproducible "shoulder" between 1 h and 2 h which is not obtained at 5 mM, i.e. within

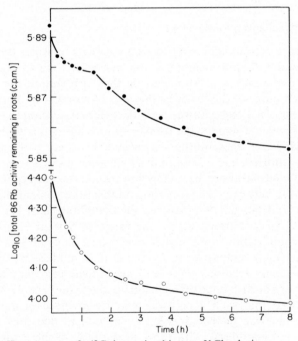

FIG. 5. ^{86}Rb efflux curves at 2–4°C determined in two KCl solutions.
O—O   5 mM KCl + 0·5 mM CaCl$_2$.
●—●   0·1 mM KCl + 0·5 mM CaCl$_2$.

the range of System II. These results confirm those of Pallaghy et al. (1970) who observed a "shoulder" in the potassium efflux curves throughout the range of the lower isotherm but not at concentrations within the range of System II. The "shoulder" is interpreted by Pallaghy et al. (1970) as a uni-directional direct pathway from the external solution to the vacuole which is tentatively identified with System I. The 5 mM curve allows estimates to be made of the half-times for ^{86}Rb efflux from the cell compartments, assuming there to be two intracellular compartments in series (Pitman, 1963). At 2–4°C the vacuole has a half-time of approximately 72 h and the cytoplasm a half-time of approximately 2 h. Difficulty was experienced in separating the two compartments with faster turnover rates (i.e. cell wall and cytoplasm) and the data presented by Heller et al. (this volume, IV.5) would suggest that a further rapidly exchanging phase associated with a portion of the cytoplasm may exist. Nevertheless, the half-time of ^{86}Rb in the major cytoplasmic phase is in reasonable agreement with the published value for beet (Osmond and Laties, 1968) and clearly significantly slower than the value for chloride in barley ($t_{\frac{1}{2}}$ cytoplasm is approximately 10 min) obtained by Cram and Laties (1971).

C. Possible Mechanism for System I

The evidence accumulated over the last ten years indicates that System I is located at the plasma membrane and our results are in line with this concept. The presence of a "shoulder" in the ^{86}Rb elution curve with the concentration range of the lower isotherm is interpreted by Pallaghy et al. (1970) in terms of a direct pathway from the external solution to the vacuole. The structure responsible for the transfer of ions along this pathway is not suggested. However, these authors associate the direct pathway with the high affinity isotherm and account for its apparent disappearance at higher external salt concentrations by changes in membrane permeability leading to equilibration between the cytoplasm and this third compartment.

Neirinckx and Bange (1971) studied the equilibration of Na^+ in barley roots at two concentrations within the range of System I and found an irreversible equilibration at the higher level, which they interpret in terms of a third intracellular compartment. As well as these studies on higher plants, work on the ionic relations of certain giant algae also may suggest the involvement of a third compartment in ion uptake to the vacuole. MacRobbie (1970) indicated that such a pathway may be involved in the quantized transport of chloride to the vacuole of Nitella, and that ions within this pathway do not equilibrate with those in the bulk of the cytoplasm. This claim has, however, been contested (Findlay et al., 1971).

Thus there is some evidence for a direct pathway from the external solution to the vacuole which may tentatively be associated with the whole of or part of the lower isotherm. Polya and Atkinson (1969) have shown that uptake in this range, unlike System II, is not directly ATP-dependent. The most obvious mechanism for such a direct pathway would be pinocytotic vesicles as indicated by MacRobbie (1970), although a number of alternatives may be considered including the concept of contact diffusion into pre-existing vesicles, which was suggested by Williamson and Wyn Jones (1972).

The inhibition of System I in high salt tissue is relatively easy to explain on the basis of pre-existing vesicles. At high internal salt status they can be envisaged as being saturated by ions from the cytoplasm and these are transported to the vacuole at the expense of those in the external solution. The evidence obtained to date appears to be compatible with, but certainly not proof of, this concept.

D. Possible Mechanism of System II

The difference in the potassium isotherms obtained by short and long uptake–wash treatments may be a major problem in the interpretation of System II. The application of a cell wall correction factor suggested by Cram and Laties (1971) indicates that the maximum error attributed to inadequate washing of the cell wall is 35% at 60 mM. Further work is in progress to determine this error more precisely, but the present evidence implies that the differences between short and long uptake–wash treatments cannot be attributed merely to a cell wall factor. The differences between the uptake rates under the different uptake–wash regimes suggests that a proportion of the cytoplasmic compartment is rapidly labelled but washed out extensively in 30 min despite the long cytoplasmic half-time observed for potassium. It therefore appears that the isotherm obtained with a long uptake–wash treatment is associated with a rate-limiting step within the cytoplasm but this data does not show that System II is specifically associated with the vacuole. The results of Cram and Laties with chloride in barley, where the cytoplasmic half-time is 9–10 min, are compatible with the specific association of System II with the tonoplast.

There is, however, no fundamental conflict if the uniform behaviour of the cytoplasmic phase is not assumed and the results of Heller et al. (this volume) cast serious doubt on the uniform behaviour of the cytoplasmic phase. These results may also invalidate the interpretation of Epstein's pulse experiment (Welch and Epstein, 1969), since using a 30 min wash in System II apparently involves a compartment internal to the plasmalemma.

There is evidence that System II is associated with a Na^+- or K^+-stimulated Mg^{2+}-dependent ATPase (Atkinson and Polya, 1967; Fisher et al., 1970; Williamson and Wyn Jones, 1972). An enzyme with characteristics similar to those of the upper isotherm was found by Williamson and Wyn Jones to be bound to vesicles, closely resembling the provacuoles previously isolated by Matile (1968). The Mg^{2+}-dependent activity was stimulated by either Na^+ or K^+ with little selectivity and K_m values in the range of 5–8 mM although the fit to the classical Michaelis–Menten formalism was poor. The ATPase was distinguishable from a non-specific acid phosphatase found associated with the vesicles. A further attempt to characterize the vesicles and especially their bounding membranes has been made using the technique of freeze-etching. Figure 6 shows structures typical

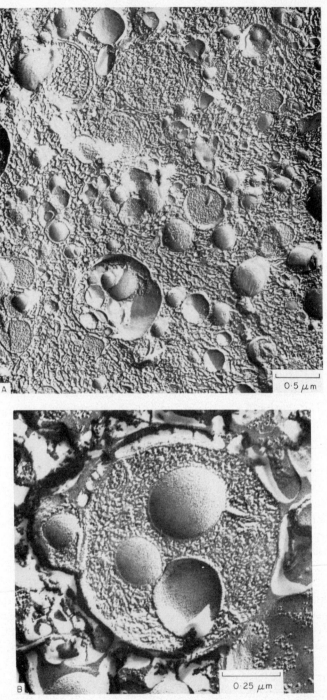

FIG. 6. Freeze-etch micrographs of the vesicular fraction isolated by Williamson and Wyn Jones (1972). A. Typical structures found within the fraction. B. A vesicle at higher magnification.

of the vesicular fraction of Williamson and Wyn Jones (1972). These structures are identical to the provacuoles observed by Matile and Moor (1968). It is therefore possible that the membrane bounding these structures is a vacuolar membrane identical to that found around vacuoles of mature tissue. However, an attempt at definitive characterization by particle size analysis, as employed by Matile and Moor (1968), failed to produce conclusive results (Leigh, unpublished data) because it has not yet been possible to show statistically valid size differences between particles on the cell membranes *in situ*. Nonetheless, the similarity between these structures and provacuoles suggests that Williamson and Wyn Jones (1972) may have been examining an ATPase associated with the tonoplast bounding either large vacuoles or smaller provacuoles within the cytoplasm. Indeed, the presence of the vesicles close to the plasma membrane (cf. Hall, 1970) may help to explain the rapidity with which System II was found to be operational by Welch and Epstein (1969).

IV. Conclusions

The examination of the kinetics of (^{86}Rb)K$^+$ influx and efflux by maize root segments leads to the formation of a tentative hypothesis. It is suggested that part of the high affinity isotherm may be a reflection of a direct pathway from the external solution to the vacuole. This mechanism becomes inoperative at high salt concentration as the vesicles become saturated with ions absorbed from the cytoplasm. System II may be due to a Na$^+$ or K$^+$-stimulated Mg^{2+}-dependent ATPase which is associated with the tonoplast of provacuoles and larger vacuoles.

Acknowledgements

We are grateful to Professor E. C. Cocking, Department of Botany, University of Nottingham, for use of the freeze-etching unit, and Mr Martin Willison, also of the Department of Botany, University of Nottingham, for assistance and advice in preparing freeze-etch replicas.

R. A. Leigh and F. A. Williamson are grateful to the Science Research Council for Research Studentships.

References

ATKINSON, M. C. and POLYA, G. M. (1967). *Aust. J. biol. Sci.* **20**, 1069–1086.
CRAM, W. J. (1968). *Biochim. biophys. Acta* **163**, 339–353.
CRAM, W. J. and LATIES, G. G. (1971). *Aust. J. biol. Sci.* **24**, 633–646.
EPSTEIN, E. (1966). *Nature, Lond.* **212**, 1324–1327.
EPSTEIN, E., RAINS, D. W. and ELZAM, O. E. (1963). *Proc. natn. Acad. Sci. U.S.A.* **49**, 684–692.
FINDLAY, G. P., HOPE, A. B. and WALKER, N. A. (1971). *Biochim. biophys. Acta* **233**, 155–163.

FISHER, J. D., HANSEN, D. and HODGES, T. K. (1970). *Pl. Physiol., Lancaster* **46**, 812–814.
HALL, J. L. (1970). *Nature, Lond.* **226**, 1253.
LATIES, G. G. (1969). *A. Rev. Pl. Physiol.* **20**, 89–116.
LAUCHLI, A. (1969). *Int. J. App. Radiat. Isotopes* **20**, 265–283.
LÜTTGE, U. and BAUER, K. (1968). *Planta (Berl.)* **80**, 52–64.
LÜTTGE, U. and LATIES, G. G. (1967). *Pl. Physiol., Lancaster* **42**, 181–185.
MACROBBIE, E. A. C. (1970). *J. exp. Bot.* **21**, 335–344.
MATILE, PH. (1968). *Planta (Berl.)* **79**, 181–196.
MATILE, PH. and MOOR, H. (1968). *Planta (Berl.)* **80**, 159–175.
MOOR, H. and MÜHLETHALER, K. (1963). *J. Cell. Biol.* **17**, 609–623.
NEIRINCKX, L. J. A. and BANGE, G. G. J. (1971). *Acta Bot. Neerl.* **20**, 481–488.
OSMOND, C. B. and LATIES, G. G. (1969). *Pl. Physiol., Lancaster* **44**, 7–14.
PALLAGHY, C. K., LÜTTGE, U. and VON WILLERT, K. (1970). *Pflanzenphysiol.* **62**, 51–57.
PITMAN, M. G. (1963). *Aust. J. biol. Sci.* **16**, 647–668.
PITMAN, M. G. (1970). *Pl. Physiol., Lancaster* **45**, 787–790.
PITMAN, M. G., COURTICE, A. C. and LEE, B. (1968). *Aust. J. biol. Sci.* **21**, 871–881.
POLYA, G. M. and ATKINSON, M. R. (1969). *Aust. J. biol. Sci.* **22**, 573–584.
POWER, J. B. and COCKING, E. C. (1970). *J. exp. Bot.* **21**, 64–70.
SCHIMANSKY, CH. and MARSCHNER, H. (1971). *Z. Pflanzenernähr. Bodenkunde* **129**, 141–152.
TORII, K. and LATIES, G. G. (1966). *Pl. Physiol., Lancaster* **41**, 863–870.
WELCH, R. M. and EPSTEIN, E. (1968). *Proc. natn. Acad. Sci. U.S.A.* **61**, 447–453.
WELCH, R. M. and EPSTEIN, E. (1969). *Pl. Physiol, Lancaster* **44**, 1301–1304.
WILLIAMSON, F. A. and WYN JONES, R. G. (1972). IAEA/FAO Use of Isotopes in Soil–Plant Relations. 69—International Atomic Energy Agency, Vienna.

VI.3

Chloride Transport in Vesicles. Implications of Colchicine Effects on Cl Influx in *Chara*, and Cl Exchange Kinetics in Maize Root Tips

W. J. Cram

School of Biological Sciences, University of Sydney
Australia

I. Introduction

Kinetic experiments on giant algal coenocytes have shown that the cytoplasm does not behave as a single compartment, the undisputed fact being that a large fraction of the Cl^- entering the cell gets to the vacuole without mixing with the bulk of the cytoplasmic Cl^- (MacRobbie, 1969). The Cl^- which gets to the vacuole rapidly might possibly be in microvesicles moving along microtubules, as in many plant and animal transport systems (Newcombe, 1969; Schmitt, 1969). This possibility can be tested by examining the effect of colchicine, which breaks down microtubules into their subunits, on Cl^- fluxes. In this paper some preliminary results are reported, showing the effects of colchicine on Cl^- influx to the cell, photosynthetic CO_2 fixation, and electrical potential difference between the vacuole and the external solution in cells of *Chara corallina*.

The structure of higher plant cells would lead one to expect kinetic complexity in their cytoplasm also. However, ion exchange in higher plant cells approaches more closely to the sum of two first order components (loosely identified with the cytoplasm and the vacuole), the closer the tissue is to a steady state and the more uniform it is. This raises the (probably false) hope that cytoplasmic organelles exchange their contents with the ground cytoplasm in a much shorter time than the cytoplasm as a whole exchanges its contents across the plasmalemma and tonoplast.

Two types of experiment have shown that the higher plant cell does not behave as two kinetic compartments.

(1) Observation of non-first-order exchange kinetics. This will often be due to non-steady-state conditions; however in some cases this may not be so (e.g. possibly in carrot tissue at low temperatures—Cram, 1968). It may then indicate relatively slow exchange between compartments within the cytoplasm. A different type of departure from first order exchange kinetics has been found by Pallaghy et al. (1970), but this may be due to stelar rather than subcytoplasmic complications (Weigl, 1971).

(2) Near a steady state the apparent contents of the cytoplasmic component of the efflux increase with increased vacuolar specific activity (Cram, unpublished); this would not occur if the cell behaved as two compartments, however arranged.

This paper also reports preliminary experiments on the kinetics of Cl^- exchange in root tips, in the hope that cytoplasmic complexity would be more amenable to investigation, but with the proviso that processes in the cytoplasm of root tip cells may be more than quantitatively different from processes in the cytoplasm of differentiated parenchyma cells.

II. Effects of Colchicine on Chloride Influx in *Chara*

A. Effects on Cl Influx, CO_2 Fixation and Electrical p.d.

Colchicine has been found to inhibit the light-dependent influx of Cl^- to cells of *Chara corallina*. Cl^- influx over 30 min from F.P.W. (Na: 2 mM, K: 0·2 mM, Cl: 2·3 mM, Ca: 0·05 mM) labelled with ^{36}Cl was inhibited 65% when 0·3% colchicine was present during the uptake period only. A 2 h pretreatment in unlabelled F.P.W. +0·3% colchicine did not significantly increase the inhibition. The effect of colchicine is therefore completed in a time short compared with 30 min.

The effect of colchicine on the electrical potential difference between the vacuole and the bathing medium is also very rapid. The normal p.d. of 170 mV fell to 80 mV within 4 min. After removing colchicine there was an initial rapid increase in p.d., followed by a slower recovery to the original value after 30 min.

In three experiments colchicine had no effect on light-dependent CO_2 fixation.

The effect of a range of colchicine concentrations on CO_2 fixation and Cl^- influx in the same cells was measured. The artificial pond water was labelled with ^{36}Cl and $H^{14}CO_3$. Batches of 10 cells were pretreated in the light in inactive FPW plus colchicine at various concentrations for several hours, and one batch was pretreated in the dark overnight. At the beginning of the uptake period the labelled solutions were substituted for the pretreatment solutions. After 1 h the cells were washed in inactive solution (with the same concentration of colchicine) and their sizes measured. The cells were then dried down on planchettes and counted twice, with and without a calibrated shield. From the two counts the activities of ^{36}Cl and ^{14}C in the individual cells, and the influx of Cl^- and HCO_3, were calculated.

In the first batch of cells the Cl^- influx was not light-dependent, although CO_2 fixation was reduced nearly to zero in the dark, and neither CO_2 fixation nor Cl^- influx were inhibited by colchicine up to 0.3%.

Figure 1 shows the effect of colchicine on a second batch of cells in which the Cl^- influx was light-dependent. Colchicine inhibited Cl^- influx to just below the dark rate, but did not inhibit CO_2 fixation.

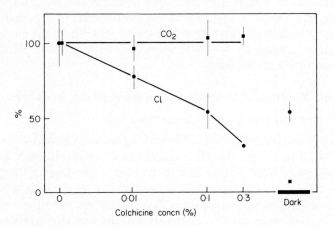

FIG. 1. The effect of colchicine on Cl^- influx and CO_2 fixation in *Chara corallina*. Bathing solution: F.P.W. Uptake time: 60 min.

B. Discussion

The critical experiment in relation to whether colchicine affects Cl^- transfer across the cytoplasm to the vacuole, is to measure the fraction of the Cl^- in the cell that is in the vacuole after various periods of loading with ^{36}Cl. The fraction would be much reduced in the early stages of loading if colchicine did inhibit a straight-through influx to the vacuole.

An inhibition of Cl^- transfer across the cytoplasm to the vacuole would not necessarily lead to an inhibition of Cl^- uptake to the cell. One possible situation in which inhibiting Cl^- transfer across the cytoplasm could lead to an inhibition of Cl^- influx to the cell would be when Cl^- entry to the cell was limited by the rate of microtubule-dependent removal of Cl^- from just inside the plasmalemma. However, the very rapid effects of colchicine on the electrical p.d. and on Cl^- influx suggest that colchicine may not only affect microtubule assembly, but may inhibit Cl^- influx more directly.

Colchicine does not appear to inhibit Cl^- influx via an effect on the generation of photosynthetic energy supply, since CO_2 fixation, which needs both energy currencies of photosynthesis—ATP and reducing power—is not affected by colchicine (cf. Smith and West, 1969). However, it is remarkable that colchicine does not inhibit Cl^- influx in *Chara* cells in which Cl^- influx is not light dependent, (cf. Findlay *et al.*, 1969) even though presumably this is still an energy-dependent process occurring by the same mechanism. Colchicine therefore appears to affect specifically the light-dependent transport. One alternative explanation would be that light-dependent transport is different from light-independent transport. For instance, microtubules might be present in one state of the cell but not in the other. The second alternative is that colchicine does interfere with the photosynthetic energy supply, but at a site outside the chloroplast. It is perhaps taking speculation too far to suggest that colchicine might inhibit the movement of microvesicles containing NADPH or with a high internal pH from the chloroplast to the plasmalemma.

III. Kinetics of Chloride Exchange in Maize Root Tips

A. Preparation and Properties of Excised Root Tips

Maize (*Zea mays* L.) root tips 1–1·5 mm long were cut from young seedlings with roots about 2 cm long. The excised tips increased in fresh weight by 0·5–1 % per hour over at least 24 h in distilled water. During this time the amount of slime on the tips decreased somewhat, so the cells would have increased in fresh weight rather faster than did the excised tips.

There was virtually no Cl^- in freshly excised tips after germination in distilled water. In 25 mM KCl + 0·5 mM $CaSO_4$ the Cl^- content increased at 1·5–2·0 μmol $g^{-1} h^{-1}$ more or less linearly with time for about 8 h. After 10–15 h the Cl^- content remained steady at about 20 μmol g^{-1}.

As judged by growth and Cl^- accumulation, therefore, these tips appear to be active.

B. The Time-course of Choride Exchange

The exchange of ^{36}Cl was followed by loading the tips with ^{36}Cl for various periods, and then measuring the loss of Cl^- into aliquots of inactive solution of the same composition as the loading solution. Figure 2 shows the shape of the wash-out curves after loading for 5 or 62 min. Log (counts remaining in the tissue) is plotted against time. The tissue was in 25 mM $KCl + 0.5$ mM $CaSO_4$.

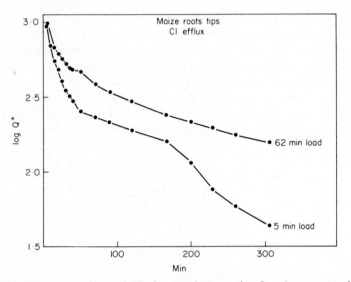

FIG. 2. The time-course of loss of Cl^- from maize root tips. Log (counts remaining in the tissue) is plotted against time. Bathing solution: 25 mM $KCl + 0.5$ mM $CaSO_4$.

In isolated maize root cortical cells (Cram, unpublished) and in many other plant tissues, the rate of loss of radioactively labelled ions falls continuously with time. The most notable aspect of the exchange of Cl^- in maize root tips is the secondary increase in the efflux beginning at about the 170th minute of wash-out after a 5 min load, and at the 50th minute of wash-out after a 62 min load. A secondary increase has previously only been found in the efflux from excised barley roots at low concentrations (Pallaghy et al., 1970). Maize root tips differ from excised barley roots in that the secondary increase in efflux appears at high (50 mM) as well as at low (0·1 mM) KCl concentrations in the external solution.

The secondary increase in efflux is not always present in maize root tips. This may be related to uncontrolled factors, such as the method of germination of the seeds or the time after excision at which the experiments were done. In one experiment roots which had accumulated KCl to a steady state (over about 24 h) showed no secondary increase in efflux, although a secondary increase did appear in the freshly excised tissue in which there was a large net influx (the tissue of Fig. 2). It is not yet known if the difference is related to the difference in net influx, or content of Cl^-, or time after excusion.

Simple compartmental analysis, assuming proportionality between flux and specific activity, cannot be used to fit a model to the kinetics of Fig. 2 because the system is far from a steady state. However, qualitatively one can say that a system of compartments in parallel would not account for the secondary increase in efflux, whereas a partly series arrangement such as that proposed by Pallaghy *et al.* (1970) would account for the secondary increase in efflux. The simplest picture, represented in Fig. 3, is that of minivacuoles accumulating Cl^- from the external solution or the cytoplasm, holding this Cl^- with little exchange with the cytoplasmic Cl^- while passing across the cytoplasm, and finally discharging the Cl^- into the main vacuole(s). The beginning of the secondary increase in efflux would then correspond to an increase in the specific activity in the main vacuoles caused by the arrival of Cl^- of higher specific activity from the minivacuoles.

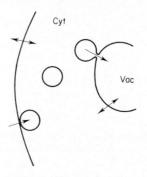

FIG. 3. A possible model for the involvement of minivacuoles in Cl transport in maize root tip cells. Arrows indicate postulated Cl movements.

C. A Test of "Linear Transit" Models

The only hypothesis tested so far is the model above with the restrictions:
  (i) that the minivacuoles take up Cl^- from the external solution exclusively, and
  (ii) that the minivacuoles then remain a certain fixed or average time (ΔT) in the cytoplasm before discharging their salt into the main vacuole(s).

(Analogies with conveyor-belts suggest the name "linear transit" for such a model.) In this case the beginning of the secondary increase would be ΔT minutes after first putting the cells into radioactive solution. The secondary increase in efflux should begin ($\Delta T - 60$) minutes after the end of a 60 min load; and ($\Delta T - 5$) minutes after the end of a 5 min load. The difference in time between the beginnings of the secondary increases in efflux should equal the difference in the loading time.

TABLE I. The relation between the time of appearance of the secondary increase in ^{36}Cl efflux and the initial ^{36}Cl loading period

Loading time (min)	Time between beginning of secondary increase and beginning of loading (min)
5	213 ± 19 (4)
59	105 ± 6 (3)

Replicated values shown as mean ± standard error of the mean (no. of replicates). A t-test shows that the times between the beginning of loading and the beginning of the secondary increase differ significantly at the 1 % level.

As shown in Fig. 2 and Table I, the secondary increase does occur sooner after a 60 min load than after a 5 min load. However, the difference between the start of the secondary increases (158 min) is significantly greater than the difference between the loading times (54 min). In other words, the time between first putting in radioactive solution and the beginning of the secondary increase is not constant with different loading times. The difference would have been even more marked if the peaks of the secondary increase in efflux (the points of inflexion of the curves) had been compared.

D. Discussion

The two next steps in investigating the Cl$^-$ compartmentation in root tips must be the construction of a quantitative model and an examination of the cellular ultrastructure.

Although the cells are not at a steady state, the contents of the cells increase more or less linearly with time over several hours. This may reduce the complexity of the non-steady-state situation to one in which a quantitative description is not impossible. The misapprehensions that have arisen concerning the kinetics of the much simpler system of two compartments in series at a steady state are a sufficient warning that a conceptual model can only be considered adequate when its kinetics have been analysed and fitted quantitatively to the measured data.

It may be possible to detect the accumulation of RbBr in minivacuoles in root tips under the electron microscope, as has been done for *Tamarix aphylla* salt glands (Thomson *et al.*, 1969). This depends, amongst other things, upon RbBr being an adequate tracer for KCl in the cells. If RbBr accumulation can be visualized, then it may prove possible to distinguish different types of vacuole if they exist, or alternatively, to locate the ions in other organelles. The electron probe microanalyser is another instrument with the potential to help in subcellular location.

The most unusual thing about the maize root tip kinetics is the secondary rise in efflux during wash-out of Cl⁻. The main experimental conclusion reached so far is that the secondary increase does not begin a fixed time after the beginning of the loading period. This may disprove what may be called "linear transit" models, in which some Cl⁻ taken up from the external solution takes a fixed time to pass through an intermediate compartment before reaching another compartment from which it can be lost. In particular, this observation is thought to disprove a particular model in which Cl⁻ was pictured as moving in a system of vacuoles which would be analogous to the pinocytotic–phagocytotic-vesicle, lysosome system of animal cells (Davson, 1970).

It would seem possible that these kinetics result from the process of vacuolation. We do not know whether vacuolation is related to the process which gives rise to the complex cytoplasmic kinetics in a mature *Nitella* or *Chara* cell. However, mature cells may still be expanding and their vacuoles increasing in size. One may therefore reasonably propose that secretion into and increase in size of vacuolate cells, or even vacuolar turnover, is by the same process as vacuolar development (MacRobbie, 1971), even though the salts accumulated in the two stages of development are different (Steward and Mott, 1969). The advantages of working with root tips may then not only be that the proportion of cytoplasm is greater than in mature cells (making possible measurements which could not be made on higher plant parenchymatous cells) but also that some aspects of ion transport may be more in evidence in root tips than in mature cells.

References

CRAM, W. J. (1968). *Abh. deut. Akad. Wiss. Berlin*, Kl. Med. 117-26.
DAVSON, H. (1970). "General Physiology", 4th Edition. Churchill.
FINDLAY, G. P., HOPE, A. B., PITMAN, M. G., SMITH, F. A. and WALKER, N. A. (1969). *Biochim. biophys. Acta* **183**, 565–576.
MACROBBIE, E. A. C. (1969). *J. exp. Bot.* **20**, 236–256.
MACROBBIE, E. A. C. (1971). *A. Rev. Pl. Physiol.* **22**, 75–96.
NEWCOMBE, E. H. (1969). *A. Rev. Pl. Physiol.* **20**, 253–288.
PALLAGHY, C. K., LÜTTGE, U. and VON WILLERT, K. (1970). *Z. Pflanzenphysiol.* **62**, 51–57.
SCHMITT, F. O. (1969). *In* "Cellular Dynamics of the Neuron". (S. H. Barondes, ed.) pp. 95–111. Academic Press, New York.
SMITH, F. A. and WEST, K. R. (1969). *Aust. J. biol. Sci.* **22**, 351–363.
STEWARD, F. C. and MOTT, R. L. (1969). *Int. Rev. Cytol.* **28**, 275–370.
THOMSON, W. W., BERRY, W. L. and LIU, L. L. (1969). *Proc. natn. Acad. Sci. U.S.A.* **63**, 310–317.
WEIGL, J. (1971). *Z. Pflanzenphysiol.* **64**, 77–79.

Discussion

Barr initiated the next discussion by reporting a finding of his student Ryan; *Nitella* cells were loaded with ^{36}Cl and after the efflux had stabilized in 3–4 h to $0.5 \, \text{pmol} \, \text{cm}^{-2} \, \text{s}^{-1}$, a hyperpolarizing current was injected. During the first 30 min of applying this current the Cl$^-$ efflux increased to twice the previous value; during the second 30 min of current flow it rose to five times the original value. This delay suggests a rapid compartment in the outer cytoplasm for exchange to the external medium. The specific activity in this compartment is low and it takes time for the current to pull Cl$^-$ from somewhere more internal. This is similar, perhaps, to the work of Costerton and MacRobbie and the possibility of cytoplasmic compartmentation may mean that flux analysis methods need revision.

Cram then offered a general comment that he would like to dispute at this point the existence of dual isotherms of uptake. Unless one knows which fluxes are at which membranes, can one properly distinguish separate mechanisms? Müller replied that Göring in Berlin has demonstrated dual isotherm uptake of one particular sugar which is not metabolized in roots, while a second rather similar sugar shows only the uptake characteristics expected of diffusion. Wyn Jones then said that it would be a great help if there were standard procedures for measuring uptake, for example how long to load with isotope, how long to free space wash, etc. Heller had said that the cytoplasm may contain compartments with different wash-out times, and the situation was very confusing.

Cram replied to this point about cytoplasmic compartmentation; he said that he had not found shoulders in efflux curves from isolated root cortex and that Weigl's criticism of Pallaghy and Lüttge's work, that there may be a time lag along the stele, might therefore have to be considered. Pallaghy replied that he agreed that Weigl's comment was justified but re-affirmed that shoulders on efflux curves had been found in *Mnium*, *Elodea*, isolated root cortex in his case and root tips. Jeschke had found a shoulder under certain conditions and not under others. Pallaghy said in reply to a further direct question from Cram, "Yes, we did try to build in a delay but now we have found a shoulder in the uptake curve as well." He commented that he

427

thought such shoulders would always be found if measurements were made on individual tissues and not on batches.

Jeschke then confirmed that he had found shoulders in the Na^+ efflux curves from *Elodea* leaves in the light but not in the dark. Pre-treatment with K^+ removes them but without the pre-treatment the inflexions are very marked. In the same material no such inflexions are found in the K^+ or Cl^- curves. Findlay then asked Jeschke what happened to the efflux after 400 min. Jeschke replied that he had not measured it for any longer. Findlay then commented that there appears to be an accelerating efflux. Could it be that the membrane permeabilities are simply changing continuously? If a *Chara* cell were effluxing in this way it would almost certainly have stopped streaming and would be discarded. Shone then said that xylem exudation of ^{36}Cl from maize roots shows a kink in the relationship after 2–4 h. The explanation may be that Cl^- is resorbed from the xylem but it seems unlikely. Cram reported that the secondary increase in efflux from root tips occurred at different times depending on loading time; a root tip loaded for a long time showed the blip earlier in the efflux.

Jennings then suggested that the increased efflux with time might be due to cell division, causing an increase in the surface area/volume ratio during the experiment. Pallaghy replied that he did not think cell division was responsible for the efflux characteristics although he had not made cell counts. In his isolated cortex tissue the meristem had been removed; furthermore he has followed efflux for 1200 min and there is no evidence that the tissue is dying. Thus, he concluded, this effect is a reflection of a compartmentation process. Anderson asked how significant these blips are with regard to the statistical uncertainties of radioactive assay, errors in experimentation and the numbers of damaged cells under observation. Bentrup then showed an efflux curve from the unicellular alga *Meugotia*, in which there is neither cell division nor damaged cells, and here too there are blips in the efflux. The algal culture is not dying, he said to Findlay, and these blips are found both in the light and the dark.

The discussion then turned to the other contributions; Heller asked both Wyn Jones and Sutcliffe if the velocity of either pinocytosis or protein synthesis is compatible with the known rates of ion transport. Are sufficient quantities of ATP produced by respiration and metabolism to allow a cell to maintain a steady state against the passive leakage of K^+ and Cl^-? Sutcliffe replied that he could not say if the rate of pinocytosis and protein synthesis is sufficient because certain vital items of information are lacking; e.g. the number of ions bound per protein molecule and the rate of protein turnover. However, it is no more difficult to explain ion transport by pinocytosis than it is by any other proposed model of transport. Wyn Jones said that again there was insufficient information to allow full comment on the question of the ATP levels. However, using the luciferase–luciferin reaction an answer might soon be forthcoming. Wyn Jones continued that he thought Neumann would agree that it is difficult to interpret ATP levels because the cell apparently has a buffer system for ATP.

Baker then asked Wyn Jones about the size of vesicles found in freeze-etched preparations; if ions in the vesicles are destined for long-distance transport then there should be an upper limit on vesicle size to allow passage through the plasmodesmata, particularly if the plasmodesmata are partially occluded. Wyn Jones replied that there are a number of different size ranges of vesicles; the smallest are several microns in diameter although there may be a population which band slightly lower on the gradient, of diameter perhaps 0·1 μm. Hall then asked Wyn Jones which cellulase he used and whether there were proteases in it. Are there oxalo-reductases in the vesicles, he continued, as well as ATPases; this would indicate their origin in the ER as Mathile believes. Finally Hall asked, if kinks in efflux curves are due to the presence and action of vesicles, are there problems in explaining ion selectivity? Wyn Jones said that the cellulase he used is a Japanese product of a pectinase–cellulase mixture; the cells are pre-treated for 12 h with the enzymes and are then thoroughly washed. After this preparation, one can isolate protoplasts that still show cytoplasmic streaming. There are still enzyme assays to be done but it is believed that these vesicles are identical with Mathile's and originate in the ER. On the selectivity problem Wyn Jones agreed that it was difficult and then called on Leigh to talk about the work presently in progress in Nottingham. Leigh reported that this work involved examining the pinocytotic uptake of latex spheres by tomato protoplasts. Two types of cytoplasmic vesicle have been identified in freeze etchings; one set have ice crystals inside, similar to those found in the external medium and it is believed that these are formed by pinocytosis at the plasmalemma. These pinocytotic vesicles fuse only rarely with the tonoplast and latex spheres are only seldom found in the vacuoles.

White then commented to Wyn Jones and Sutcliffe: he said that his experiments on phosphate uptake into whole plants at realistic levels—i.e. soil solution levels—of phosphate showed that accumulation saturated at around 60 μM external phosphate and was affected by nitrogen levels. Furthermore plant growth rate response is optimized at about this level. Why, then, this preoccupation with dual isotherms where saturation occurs a good order of magnitude higher than under field conditions? MacDonald then said that he agreed with Sutcliffe that protein synthesis might be involved with ion transport in two ways; either in the synthesis of carriers or in the actual process of uptake. The idea that carriers are proteins which must be synthesized continually is not in dispute. What is disputed is that on the basis of inhibitor studies ion transport has been shown to be directly linked to protein synthesis and turnover. It would first be necessary to establish the specificity of the inhibitor being used; until it is established that an inhibitor inhibits protein synthesis and nothing else, there is no logical basis for the type of experiment that attempts to link ion uptake directly with protein synthesis. This applies with particular force to both chloramphenicol and cycloheximide. Chloramphenicol has been tested on 80 S ribosomes from various plant and animal cells and has not been shown to have any effect on protein synthesis in these systems. On the other hand both chloramphenicol and cycloheximide have been shown to interfere with electron transport in mitochondria isolated from turnip.

Sutcliffe said that he agreed with most of what MacDonald had said; the effects of inhibitors on cells must be carefully examined. However, the same objections apply to demonstrations that chloramphenicol and cycloheximide affect oxidative phosphorylation; these demonstrations are no less equivocal than the reports of effects on protein synthesis. In his experiments cycloheximide does not affect plant growth rate, and this seems good evidence that it does not affect oxidative phosphorylation. He is further satisfied that chloramphenicol does affect protein synthesis and does not affect phosphorylation in red beet tissue. Lüttge then said that he thought it was important to realize that assumption of vesicular ion transport necessarily implied a transport of membrane material associated with the vesicular movement. If there is pinocytosis from the plasmalemma to the tonoplast, there will be a continuous movement of membrane material from the one to the other; it will thus be necessary to regenerate plasmalemma material and remove tonoplast material, either by a reverse vesicle movement from tonoplast to plasmalemma or by removal of the tonoplast as unconstituted membrane material.

Van Steveninck commented that cycloheximide will inhibit the development of ion uptake in freshly cut wheat root slices. However cycloheximide applied after the development of these mechanisms has a differential inhibitory effect; Cl^- uptake is inhibited after a lag of 1–3 h while K^+ uptake is inhibited after 6–8 h. These are not effects of poor inhibitor penetration; there is turnover of carriers and the observed delays may be due to different turnover rates in different carrier mechanisms. Sutcliffe agreed that these differential effects are more easily explained on the basis of turnover of specific proteins than on the basis of general non-specific inhibition of oxidative phosphorylation by cyclohexamide. Finally Spanswick asked Sutcliffe why the undisturbed growth rate in the presence of cyclohexamide is good evidence of lack of oxidative phosphorylation inhibition and is not also good evidence of lack of inhibition of protein synthesis. Sutcliffe replied that an answer must await future experimental results; it may be that cycloheximide is inhibiting the synthesis of proteins which are not rate-limiting on growth.

VI.4

Vacuolar Ion Transport in *Nitella*

E. A. C. MacRobbie

Botany School, University of Cambridge
England

I. Introduction

A good deal of work has been done on the ion fluxes at the plasmalemma of fresh water giant algal cells such as *Nitella*, *Hydrodictyon* and *Chara*, and the pattern of active ion transport processes responsible for the maintenance of the normal ion concentrations in the cell has been reasonably well established (reviews by MacRobbie, 1970a, 1971a; Raven, 1968a). In both *Nitella* and *Hydrodictyon* there appear to be two independent transport systems, a cation exchange pump responsible for the high K/Na of the cell interior, and a (chloride + cations) system, responsible for net salt accumulation and the maintenance of cell turgor. Both transport systems are light-dependent, but whereas the cation transport can be supported in far-red light, in which ATP is the only net photosynthetic product, the chloride transport does not function under these conditions, and appears to require some product or consequence of non-cyclic electron flow other than ATP.

Much less information is available on fluxes of ions between cytoplasm and vacuole. This paper aims to summarize results on the distribution of tracer chloride within the cell which suggest that the process of vacuolar transfer is very closely linked to the primary process of chloride entry to the cell, that the distribution of tracer between cytoplasm and vacuole is in fact controlled by the influx. We then need to consider types of entry process that could achieve this sort of control.

The work to be described starts with the assumption that it should be possible to get information on the equilibration of ions between cytoplasm and vacuole, or of ions within different phases of the cytoplasm, by the analysis of the kinetics of appearance in the vacuole of tracer ions from the outside solution. We must then test the kinetic behaviour observed against that predicted for various models of the cell, the aim being to determine the simplest hypothetical arrangement of cytoplasmic phases which is consistent with the observed kinetics. Application of this method to characean internodal cells suggests that simple models of fixed compartments in the cytoplasm, in series with the vacuole, are quite inadequate to explain some odd features of the kinetics observed (MacRobbie, 1969, 1970b, 1971b).

The experiments consist of labelling cells, primarily of *Nitella translucens* but also of *Tolypella intricata*, for varying short periods of time, and then counting a measured length of cell to determine the total activity in the cell (Q_T^*), and a weighed sample of vacuolar sap to measure the amount of tracer in the vacuole (Q_V^*). We may then calculate the total influx to the cell (M_T) and the fraction of the total activity in the cell which has reached the vacuole ($P = Q_V^*/Q_T^*$). We must then produce a model of the cell that will explain the observed time dependence of the vacuolar fraction.

Before considering the experimental results we may look at the behaviour predicted for two simple models of the cell (MacRobbie, 1969, 1971a).

(1) Two phases, cytoplasm (c) and vacuole (v) in series with contents Q_c and Q_v, and fluxes M_{oc} (outside to cytoplasm), M_{co} (cytoplasm to outside), M_{cv} (cytoplasm to vacuole) and M_{vc} (vacuole to cytoplasm). During uptake the cytoplasmic specific activity will rise as $(1 - e^{-kt})$, with the rate constant k given by $(M_{co} + M_{cv})/Q_c$. In the very early stages of uptake the specific activity of the cytoplasm will rise linearly with time, and the amount of tracer in the vacuole will therefore rise with the square of the time; the vacuolar fraction will rise linearly with time, given by $P = \frac{1}{2}kt$.

(2) Two cytoplasmic phases, a slowly exchanging cytoplasmic phase (c), and a small rapidly exchanging phase (r), with the assumption $Q_r \ll Q_c \ll Q_v$; if the vacuole receives activity from phase r, and the slow phase can exchange with r, then we predict that after a very short initial period the vacuolar fraction will take the form $P = P_0 + \gamma t$, i.e. it will rise linearly with time but with an intercept P_0. (P_0 is determined by the fluxes from the rapid phase; $P_0 = M_{rv}/(M_{rv} + M_{rc})$. γ is related to the exchange of the slow cytoplasmic phase and is equal to $M_{rc}/(M_{rv} + M_{rc}) \cdot \frac{1}{2} k$, with k the rate constant for the $(1 - \exp)$ rise of activity in the slow cytoplasmic phase, equal to

$$M_{cr} \cdot M_{rv}/Q_c(M_{rv} + M_{rc})$$

II. Experimental Results

A. Vacuolar Time Course

Time courses for the amount of tracer in the vacuole (Q_v^*) and for the vacuolar fraction (P) were determined (MacRobbie, 1969). It was found that, with ^{36}Cl, the rise of vacuolar activity Q_v^* had no initial lag (i.e. it was less than about 1 min), and that the graph of P was linear with time but did not go through the origin. Hence there appear to be two components of transfer of chloride to the vacuole, a fast component which is linear with time, and a second slow component involving transfer from a cytoplasmic phase in which the specific activity is rising linearly with time over this period of uptake. That the fast component does not arise simply from cytoplasmic contamination of the sap sample may be seen in Fig. 1, showing the results

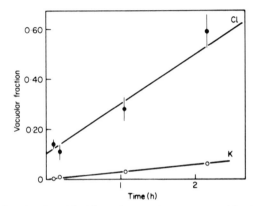

FIG. 1. Vacuolar fraction for chloride and potassium, measured in the same cells after varying times of uptake.

of a double labelling experiment using ^{42}K and ^{36}Cl in the same cells, in which the fast component is present for chloride but not for potassium (although potassium does reach the flowing cytoplasm under these conditions).

Hence the graph of vacuolar fraction against time is characterized by two quantities, the intercept P_0, which is a measure of the fast component, and the subsequent slope, a measure of the contribution of the slow component; we may write the relation as $P = P_0 + \gamma t$, and are then required to establish and interpret the behaviour of P_0 and γ under different influx conditions. It is essential to look at the separate behaviour of P_0 and γ, and in any experiment we must be clear where on the time course the experimental times fall, and whether our measured P values reflect largely the fast component, or have also a significant slow contribution. By choosing very short experimental times (and checking that they are short enough) we can look at the behaviour of the fast component P_0. By the use of double labelling with ^{36}Cl and ^{82}Br for unequal times of uptake it appears to be possible to measure both fast and slow components of vacuolar transfer in the same cell (MacRobbie, 1971b).

B. Chloride–Bromide Double Labelling

In cells loaded for equal times in ^{36}Cl and ^{82}Br the vacuolar percentages are equal for bromide and chloride, even when the influxes are very different. The regression lines of P^{Cl} on P^{Br} are shown in Table I for three separate experiments, with the ratio of influxes for the two ions in each case. This equality of vacuolar fractions holds whether the loading time is short or long, hence we must conclude that both P_0 and γ are equal for bromide and chloride.

TABLE I. Equality of vacuolar fractions of bromide and chloride for equal loading times

No.	Number of cells	Ratio of influxes M_T^{Cl}/M_T^{Br}	Regression of P^{Cl} on P^{Br}		r	$r_{0.001}$
			Slope	Intercept (P^{Cl})		
1	20	3·1	1·04 $\pm$ 0·13	0·04 $\pm$ 0·09	0·887	0·679
2	16	6·0	0·89 $\pm$ 0·05	0·038 $\pm$ 0·007	0·979	0·742
3	13	9·3	1·16 $\pm$ 0·24	$-0·003 \pm 0·01$	0·821	0·801

r: calculated correlation coefficient.
$r_{0.001}$: value of r for which $P = 0.001$.

The equality of vacuolar fractions of bromide and chloride after equal loading times suggests that if cells are loaded with bromide and chloride for unequal times we may determine both P_0 and γ on the same cell; if we label for a short time in bromide and a longer time in chloride, we get P_0 from P^{Br}, and we may determine γ from the difference between P^{Cl} and P^{Br}.

In the first experiment in which this method was used (MacRobbie, 1971b) it was shown that there was a much closer correlation between the contribution from the slow component ($P^{Cl} - P^{Br}$) and the total entry Q_T^{*Cl}, than between the slow component and time. Thus the correlation coefficient between ($P^{Cl} - P^{Br}$) and Q_T^{*Cl} was 0·926 (8 d.f.; $P < 0·001$), but r fell to 0·520 ($P \sim 0·10$) for the plot of ($P^{Cl} - P^{Br}$) against time. This confirmed the conclusion from the earlier work (MacRobbie, 1969) that the vacuolar fraction after any given time of uptake is related to the total entry to the cell during this time. The relation between vacuolar fraction and time should be written as $P = P_o + aM_T \cdot t$, i.e. the slope of the slow component is proportional to the influx M_T.

It was also shown that the cells in this experiment fell into groups with differing values of P_o but the same value of the slope a. This is shown in Table II, in which the regression lines for P^{Cl} against Q_T^{*Cl} are shown for

TABLE II. Relation between P^{Cl} and Q_T^{*Cl} in groups of cells selected on the values of the fast fraction for bromide

| No. of cells | P^{Br} | | Calculated regression $P^{Cl} = P_0^{Cl} + aQ_T^{*Cl}$ | | | |
	Range	Mean	P_0^{Cl}	a	r	P
11	$0.13 - 0.27$	0.21 ± 0.01	0.24 ± 0.006	0.125 ± 0.013	0.953	<0.001
7	$0.34 - 0.43$	0.38 ± 0.01	0.40 ± 0.03	0.132 ± 0.044	0.802	<0.05
						>0.02

P vacuolar fraction.
a fraction per nmol cm^{-2}.

cells (labelled for a short time in bromide and a long time in chloride), which have $P^{Br} \leqslant 0.27$ and P^{Br} 0.34–0.43 respectively. The distribution of values of the fast fraction will be discussed later, after consideration of the behaviour of the slow fraction in two subsequent experiments.

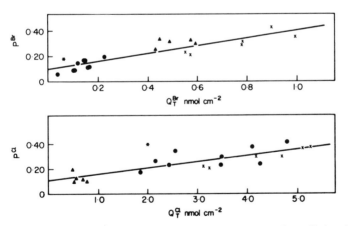

FIG. 2. Vacuolar fraction P plotted against total tracer entry Q_T for cells in which the fast fraction is $\leqslant 0.20$ (cells of experiment 2 of Table I).
● short bromide, long chloride (cell * not included in calculation of regression).
▲ long bromide, short chloride.
× long bromide, long chloride.

In the first of these, cells were loaded for either equal times in bromide and chloride, for longer in chloride than in bromide, or for longer in bromide than in chloride. The amount in the fast fraction was established by the short label, and the graph of Fig. 2 shows the plots of P^{Cl} against Q_T^{*Cl}, and of P^{Br} against Q_T^{*Br} for cells in the group in which the amount in the fast fraction is $\leqslant 0.20$. For both bromide and chloride the correlation between P and total entry Q_T^* is better than that between P and time alone. The calculated regression lines for P and Q_T^* are:

$$P^{Cl} = 0.11 + 0.050\, Q_T^{*Cl} \quad (r = 0.844; 18 \text{ d.f. } r_{0.001} = 0.679)$$
$$P^{Br} = 0.10 + 0.312\, Q_T^{*Br} \quad (r = 0.883; 18 \text{ d.f.})$$

The regression of P^{Cl} on time has a value of r of only 0.643 and that of P^{Br} on time has r = 0.849, both lower than the corresponding values above. Hence the slope of the slow component is again related to the influx to the cell. The ratio of the slopes of these two lines is not significantly different from the inverse ratio of the influxes of bromide and chloride, as indeed is required by the observation that the vacuolar fractions are equal in cells loaded for equal times. Thus we have:

$$P_o^{Cl} = P_o^{Br}$$

and

$$a^{Cl} . M_T^{Cl} = a^{Br} . M_T^{Br}$$

The values of a determined thus (from cells with small values of the fast fraction P_o) can then be used to predict values of vacuolar fraction after long loading times, irrespective of whether P_o is small or not. Thus we can predict P^{long} from P^{short} and the slope of the slow fraction, both for long chloride/short bromide loading, and for long bromide/short chloride. In Fig. 3 the predicted values of P^{long} are compared with the measured value of P^{long}. Of the cells in Fig. 3, 14 were used in the calculation of the slopes but 11 cells have high values of P_o and were not previously included; of these cells 9 out of 11 gave reasonable agreement between predicted and observed values.

Hence the conclusion is that both fast and slow components for bromide and chloride are equal. The equality of the slow components over a range of influx values of chloride and bromide, with the relation between the slope of the slow component and the influx, demands that the relation be written in terms of halide influx M_T^h (of bromide plus chloride). We should therefore write the relation as:

$$P = P_o + a^h . M_T^h . t$$

and consider it to apply to both bromide and chloride.

A second experiment (experiment 3 in Table I), done at the same time on the same batch of cells, confirms this conclusion. The chloride concentration

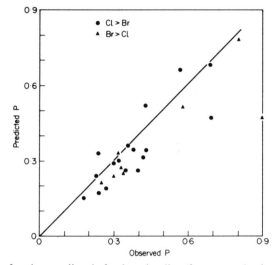

FIG. 3. Vacuolar fraction predicted after long loading (from vacuolar fraction after short loading and the slope a), compared with measured value. (Cells of experiment 2 of Table I).
● chloride.
▲ bromide.

in the bathing solution was increased, thereby increasing the chloride influx with no increase in bromide influx. The vacuolar fractions of bromide and chloride were still equal after equal loading times, although the ratio of chloride influx to bromide influx had been increased from 6·0 to 9·3. For cells loaded for longer in chloride the slope of the slow component was $0·042 \text{ nmol}^{-1} \text{ cm}^2$ chloride, when calculated from cells which had a low fraction of vacuolar bromide; the graph is shown in Fig. 4. Since P^{Cl} and P^{Br} are equal in cells loaded for equal long times in the two isotopes this implies that the slope of the curve of P^{Br} against Q_T^{*Br} must be $(0·042 \times M_T^{Cl})/M_T^{Br}$ i.e. it must be $0·39 \text{ nmol}^{-1} \text{ cm}^2$ bromide. Again we should write the relation in terms of halide. The results of these two experiments are compared

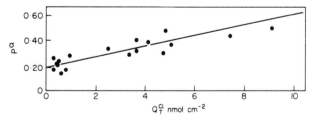

FIG. 4. Vacuolar fraction for chloride plotted against total chloride entry, for cells in which the fast fraction (bromide) is low. (Cells of experiment 3 of Table I).

in Table III. The most important feature of the behaviour of the slow component is the link with the influx. The slow component involves a cytoplasmic phase in which the specific activity is rising linearly with time, but the results show that the flux out of this phase is itself proportional to the flux to the cell. The second point to note is that once inside the cell there is no discrimination between bromide and chloride, and the ratio of tracer bromide to tracer chloride in the vacuole remains equal to that in the cell as a whole throughout the early stages of influx.

TABLE III. Comparison of experiments 2 and 3

	External concentration mM	Influx, M_T pmol cm^{-2} s^{-1}	Slope of slow component, a	aM_T
Experiment 2				
Chloride	0·6	1·08 ± 0·06	0·05 (meas.)	0·054
Bromide	0·6	0·18 ± 0·01	0·31 (meas.)	0·056
Halide		1·26		
Experiment 3				
Chloride	1·2	1·34 ± 0·09	0·042 (meas.)	0·056
Bromide	0·6	0·14 ± 0·01	0·39 (calc.)	—
Halide		1·48		

C. Distribution of Values of P_o

If the distribution of values of P_o, the fast fraction, in any single experiment is examined it is seen to be non-random, and the cells seem to fall into distinct groups (MacRobbie, 1970b, 1971b).

One such example is shown in Fig. 5, showing the cumulative distribution of values of vacuolar fraction P in a short-term influx experiment involving 19 cells of *Tolypella intricata* (i.e. the number of cells having vacuolar fractions

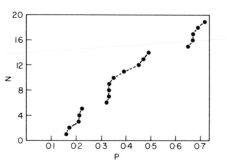

FIG. 5. Cumulative distribution of values of the fast fraction P in a short term influx experiment using cells of *Tolypella*. (N = number of cells having values of vacuolar fraction up to and including given values of P.)

less than or equal to the plotted values). It is clear that there are unoccupied ranges of P; although the values of P range from 0·16 to 0·72 there are no cells with vacuolar fractions in the range 0·23–0·31 or 0·50–0·64. We can test the probability of a distribution of this form, as a sample of a large population of cells with some hypothetical distribution. Thus we can test the probability of the measured distribution as a sample of a single Gaussian population, although since it clearly is not a Gaussian distribution it is difficult to ascribe values to the parameters required (the mean and standard deviation of the hypothetical Gaussian distribution). We could, for example, try the fit of the Gaussian distribution defined by the quartiles of the measured population (since we know that in a Gaussian distribution half the population lies within $\pm 0.6745\,\sigma$ of the mean); we could then use tables of Z, the cumulative normal frequency distribution, to predict frequencies in given sub-ranges of the total distribution. However in view of the pattern of deviation found it is also worth comparing the observed distribution with a

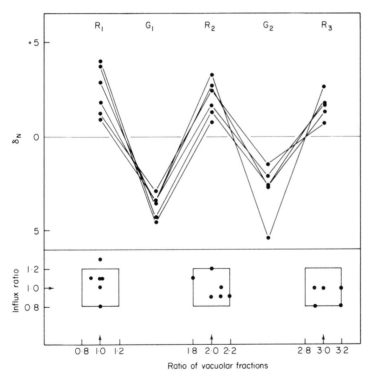

FIG. 6. (above) Deviations between observed distributions of P values in the fast fraction and a uniform population with the same total range of P. δ_N = number of cells observed in a sub-range minus the number predicted for that range. (below) Ratio of influxes and vacuolar fractions in the groups of cells in the sub-ranges R_1, R_2 and R_3.

uniform distribution, in which there is an equal probability of any value of vacuolar fraction in the total occupied range. In any single population, even if it deviates significantly from the Gaussian ideal, we expect to find fewer cells than in the uniform population at the top and bottom ends of the population, but an excess over the uniform distribution somewhere in the middle. If, therefore, we calculate the deviations from the uniform population (δ_N = number of cells observed minus number of cells predicted in given sub-ranges), we expect it to be first negative, then positive, and then again negative. What we in fact find is shown in Fig. 6 for six experiments (those of Table 1a, MacRobbie, 1971b). Two points are important, firstly that the same pattern of deviation is shown in each experiment, and secondly that the pattern is of positive, negative, positive, negative, positive deviations. Thus the whole occupied range can in every case be divided into alternating sub-ranges in which there are too many cells (positive deviations), followed by ranges without cells (negative deviations). In Fig. 6 the occupied ranges have been termed R_1, R_2 and R_3, and the unoccupied intervening sub-ranges have been termed G_1 and G_2. The probabilities for a uniform population, and the observed fraction of cells in the ranges R_1, G_1 etc., with their mean values, are shown in Table IV, from which it is clear that the gaps, the unoccupied ranges, occupy significant fractions of the total.

TABLE IV. Comparison of observed distribution and uniform distribution

Range	Probability in uniform distribution		Fractional occupation	
	Spread of values	Mean value	Spread of values	Mean value
R_1	0·10–0·23	0·16	0·22–0·54	0·29
G_1	0·13–0·36	0·24	1 value 0·13, 5 values zero	0·02
R_2	0·13–0·33	0·24	0·23–0·50	0·40
G_2	0·08–0·29	0·19	0	0
R_3	0·10–0·38	0·19	0·23–0·44	0·31

These results imply that the cells do not represent a single population, but divide into groups, centred around different mean vacuolar fractions. If we examine the mean values of influx and vacuolar fraction in the cells in the groups R_1, R_2 and R_3 in each experiment, we find that the mean influxes do not differ significantly, but that the mean vacuolar fractions are in the ratio 1:2:3. This is seen at the foot of Fig. 6 for each of the experiments in this set. Two other similar sets of experiments were discussed in the previous paper (MacRobbie, 1971b; Table 1b, c), and the pattern was the same.

The conclusion is therefore that in any given experiment the cells can be divided into groups, each covering the whole range of influx values and having the same mean influx, but in which the vacuolar fractions in the fast component are in the ratio $1:2:3$. Thus the fast component of vacuolar transfer is quantized.

The ranges R_1, G_1 etc. have been chosen by inspection for each experiment, and their mean values of the vacuolar fraction have then been compared. An alternative method is to identify only the first gap, to define the unit value from its position, and then to examine the number of cells in the ranges centred around the relative values 1.0, 1.5, 2.0, 2.5 and 3.0, i.e. in the equal sub-ranges $0.8-1.2$, $1.3-1.7$, $1.8-2.2$, $2.3-2.7$ and $2.8-3.2$. The hypothesis is then that the cells are more frequent in the ranges centred on 1.0, 2.0 and 3.0, and less frequent in those centred on 1.5 and 2.5. There is no significant difference between the unit value defined as the mean of the group of cells before the first gap, and that defined by identifying the centre of the first gap as 1.5 (or indeed from that used originally (MacRobbie, 1970), obtained as a weighted mean of the values in R_1, R_2 and R_3 in their ratios of $1:2:3$). Thus in the experiment shown in Fig. 5 the cells before the first gap (in the range $0.16-0.22$) have a mean vacuolar fraction of 0.195, whereas the first gap ($0.22-0.32$) is centred on 0.27, or 1.5×0.18. Figure 7 shows the fraction of cells occupying each of the equal length sub-ranges thus defined, for the

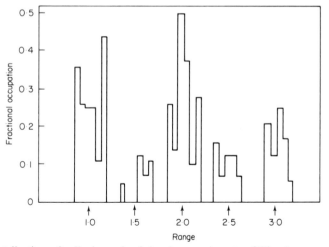

Fig. 7. Distribution of cells, in each of the six experiments of Fig. 6, among the sub-ranges specified relative to a unit of vacuolar fraction. The unit is defined by taking the centre of the first gap to be 1.5, but is not significantly different from that defined by taking the mean vacuolar fraction in cells before the first gap to be 1.0. The histogram shows the number of cells in the equal sub-ranges $0.8-1.2$, $1.3-1.7$, $1.8-2.2$, $2.3-2.7$ and $2.8-3.2$.

same set of six experiments, and it is clear that there is a grouping of cells around the values 1·0, 2·0 and 3·0, with a deficiency of cells in the intervening ranges. The separation becomes blurred at higher values as would be expected, since the spread is also multiplied, but in every case there is grouping of the cells. In the individual experiments the number of cells in the range 1·3–1·7 is only 0–0·6 of those in the range 0·8–1·2 (mean ratio 0·21), or 0–0·75 of those in the range 1·8–2·2 (mean ratio also 0·21). Hence the same pattern is present in every case, and the distribution of P_o values is multi-peaked, with a grouping of cells around values in the ratio of 1 : 2 : 3.

In the individual experiments in Fig. 6 the number of cells expected in each range is too small for a χ^2 test to be valid; we can lump together the numbers in R_1, G_1 etc. in all the experiments but it may then be argued that the absolute values of P covered by each range differ in different experiments, and that therefore the lumping is dubious. The pattern is however obviously the same in each, and this would not be expected to recur unless it had some significance.

The same pattern is seen in the double labelling experiments in which there are more cells. The cumulative distribution of values of the fast fraction for bromide for the cells of experiment 1 of Table I is shown in Fig. 8, where again we find distinct gaps in the distribution, which are much larger than we could reasonably expect to find by chance. More important, the same pattern is seen in this distribution as in those already considered. For example in Fig. 8 we have 28 cells covering a total range of vacuolar fraction 0·13–0·66, and we find a marked gap in the distribution in that there are no cells having values of 0·28–0·33 (inclusive), although there are 6 cells in the equal length sub-range before this gap and 5 cells in the equal length sub-

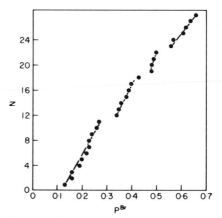

FIG. 8. Cumulative distribution of values of the vacuolar fraction in fast component for bromide. (Cells of experiment 1 of Table I).

range after this gap. If we use this gap to define the unit (either from the mean vacuolar fraction of cells before the gap taken to be 1·0, or from the centre of the gap taken to be 1·5) we can then examine the numbers of cells in the sub-ranges centred on 1·0, 1·5, 2·0, etc., and this is done in Table V. Again the same pattern of deviation from a uniform distribution is found, with alternating positive and negative deviations, and with more cells in the ranges centred around 1·0, 2·0 and 3·0 than in the intervening ranges. Although the existence of gaps at some place in the distribution is likely, a recurring pattern of such gaps would not be expected unless it had some significance.

TABLE V. Distribution of P values after short bromide loading

Range of vacuolar fractions		Number of cells
Relative values	Absolute values	
0·8–1·2	0·16–0·24	9
1·3–1·7	0·26–0·34	3
1·8–2·2	0·36–0·44	5
2·3–2·7	0·46–0·54	4
2·8–3·2	0·56–0·64	6

Cells of experiment 1 of Table I.

The existence of significant gaps in the distribution of P_o, and the quantization of its values, were questioned by Findlay et al. (1971), who reported that they could find no evidence for such effects in *Chara corallina*. However, if the behaviour of *Chara* is similar to that of *Nitella* we would not expect to find clear evidence of quantization in an experiment of their design. They did not characterize their time course, but it seems likely that both fast and slow components contributed to their measured values of vacuolar fraction, since their total entries were in the range 0·3–2·0 nmol cm^{-2} in the time chosen. This is suggested by the fact that in their results both apparent influx to the vacuole (Q_v^*/t) and vacuolar fraction seem to increase with time. Their distribution is similar to that expected for *Nitella* with a slope of 0·07 nmol^{-1} cm^2 in the slow phase, which is within the range previously found for *Nitella*. The cumulative distribution of ($P - 0.07\, Q_T^*$) in their experiment A does not show clear gaps, but it does show alternating positive and negative deviations from a uniform distribution. Hence the vacuolar kinetics in *Chara* are not well enough defined to rule out behaviour of the type found in *Nitella*—the existence of fast and slow components, of quantization in the fast component, and of an influx-dependent slope to the slow component.

III. Discussion

The experimental results show that:

(1) The vacuolar fraction after various times of uptake can be represented by the equation

$$P = P_o + \gamma t$$

We are more interested in the instantaneous rate of transfer of tracer to the vacuole ($M_v = dQ_v^*/dt$), than in the percentage of vacuolar activity, and we may therefore calculate M_v from the relation between Q_v^* and time. We have then the relation:

$$\frac{M_v}{M_T} = P_o + 2\gamma t$$

(2) The slope of the slow component (γ) is proportional to the influx to the cell, i.e. the relations should be written:

$$P = P_o + a . M_T . t$$

$$\frac{M_v}{M_T} = P_o + 2a . M_T . t$$

where a is a constant, independent of M_T.

The relations mean that the level of activity in the cytoplasm does not specify the rate of transfer to the vacuole directly, but only the rate as a fraction of the influx. Influx and transfer to the vacuole are so intimately linked that we can predict only their ratio at any given tracer content, and not the absolute value of the vacuolar flux. The process of vacuolar transfer seems to be controlled by the influx to the cell. If we double the influx to the cell, the flux to the vacuole in the slow component is also doubled; since, over the uptake times with which we are concerned, doubling the influx will also double the cytoplasmic specific activity after a given time of uptake, we shall find four times as much tracer in the vacuole.

(3) The vacuolar fractions for chloride and bromide are equal after equal loading times, even when the influxes of the two ions are very different. Thus both P_o and γ are equal for the two ions.

(4) The distribution of values of P_o is non-random; in any given experiment the cells fall into distinct groups, having the same mean influx, but in which the mean values of P_o are in the ratio $1:2:3$. Thus the fast component of vacuolar transfer is quantized.

(5) The slope of the slow component is not quantized, but has the same value in the different groups of cells selected on their values of P_o.

The problem is to interpret such behaviour of P_o and γ in terms of distinct structural components of the cytoplasm. We have to explain the existence of fast and slow components and define the relation between them, to explain the existence of quantization in the fast component but not in the slow component, to explain the dependence of the slope of the slow component on the influx to the cell.

We may first discuss the existence of quantization of the fast component. The method measures the vacuolar tracer after uptake and cutting the cell, and the question arises whether the quantization might result as an artefact in the cutting, depending perhaps on the number of action potentials generated in the handling and cutting process. It was argued in the first paper (MacRobbie, 1969) that this was unlikely for two reasons. The first was that too large a fraction of cytoplasmic tracer was involved, and that a transfer of this fraction of total cytoplasmic chloride per action potential would be detectable as a change of cytoplasmic contents (which does not appear). The second reason was that the presence of bicarbonate in the labelling solution increased the chloride transfer to the vacuole, whereas the presence of bicarbonate in a washing solution did not increase vacuolar transfer. If we accept these arguments and rule out the identification of the fast component as the transfer in the cutting process, then we must explain it as a quantized discharge to the vacuole from some small internal phase in the cytoplasm, which must be in parallel with the slow component. On the present experimental evidence no firm conclusion can be drawn, but we might notice that we can by-pass the difficulties of the explanation in terms of action potentials by making further assumptions. We could argue that transfer during the action potential was from a cytoplasmic phase containing only a very small fraction of the total cytoplasmic chloride, but nearly all of the cytoplasmic tracer after short uptake times. We could also argue that we need a well defined time course for the vacuolar fraction in the presence of bicarbonate, to establish more clearly the effect of bicarbonate on fast and slow fractions in an uptake experiment. The effects of light and inhibitors on bicarbonate transport suggest that it may be powered in the same way as chloride, and bicarbonate will inhibit chloride influx (Raven, 1968b; Smith, 1968); hence bicarbonate entry is likely to be by the same mechanism as chloride entry. In view of the results with chloride and bromide we might then expect a stimulation of the slow phase of vacuolar chloride transfer by bicarbonate, and it therefore becomes important to look carefully at the time course of vacuolar transfer in the presence of bicarbonate. In the original experiment the effect appeared at short times, but the time course should be better defined before we can be sure that the effect is on the fast component. Hence an explanation of the fast component in terms of action potentials on cutting is not ruled out at present (provided further assumptions are added), and the matter remains uncertain. Further experiments are in progress to resolve this question.

The most important feature of the results is the link between the slow phase of vacuolar transfer and the influx. It means that to reach a given rate of transfer to the vacuole from the slow phase, a given total entry is required, independent of the rate of entry. The partition of activity between cytoplasm and vacuole is specified by the total activity in the cell, independent of the time over which this was achieved. This implies that the flux from the slow phase is proportional to influx, and must be very closely controlled by the influx. It is difficult to explain this type of close link in a system of fixed cytoplasmic compartments, and it was therefore suggested that some sort of dynamic membrane system in which new vacuole was created was involved. It was originally suggested that a process of pinocytosis at the plasmalemma was consistent with the vacuolar kinetics. Provided all the tracer chloride reached the vacuole via the endoplasmic reticulum, a process involving the separation of small vacuoles from the endoplasmic reticulum, at a rate closely linked to the chloride input to this phase, would also serve the results. In view of the extensive sheets of endoplasmic reticulum just inside the plasmalemma (Costerton and MacRobbie, 1970), and the likelihood that these provide considerable barrier to ion movement inwards, it may well be that uptake by the endoplasmic reticulum is the fate of chloride ions entering the cell. The results suggest that some sort of continuous compartmentation within the cell is involved, the creation of new vacuolar volume, but the association of the kinetic features with distinct structural entities in the cell is as yet uncertain; it can only come from an understanding of the significance of the fast component and of its relations to the slow component, and this remains the aim of future work.

References

COSTERTON, J. W. F. and MACROBBIE, E. A. C. (1970). *J. exp. Bot.* **21**, 535–542.
FINDLAY, G. P., HOPE, A. B. and WALKER, N. A. (1971). *Biochim. biophys. Acta* **233**, 155–163.
MACROBBIE, E. A. C. (1969). *J. exp. Bot.* **20**, 236–256.
MACROBBIE, E. A. C. (1970a). *Q. Rev. Biophys.* **3**, 251–294.
MACROBBIE, E. A. C. (1970b). *J. exp. Bot.* **21**, 335–344.
MACROBBIE, E. A. C. (1971a). *A. Rev. Pl. Physiol.* **22**, 75–96.
MACROBBIE, E. A. C. (1971b). *J. exp. Bot.* **22**, 487–502.
RAVEN, J. A. (1968a). *Abh. dt. Akad. Wiss. Berl.* **4a**, 145–151.
RAVEN, J. A. (1968b). *J. exp. Bot.* **19**, 193–206.
SMITH, F. A. (1968). *J. exp. Bot.* **19**, 207–217.

VI.5

Intercellular Movement of Chloride in *Chara* — A Test of Models for Chloride Influx

N. A. Walker and T. E. Bostrom

*School of Biological Sciences, University of Sydney
Australia*

I. Introduction

A. General

Although the internodal cells of plants of the Characeae have been used as single cell preparations for fifty years (Osterhout, 1922), almost nothing is known of the physiology of the whole plant.

That the intact plant has effective chemical communications between its cells is apparent. The existence of colourless rhizoids, and the underground storage of starch, indicate translocation of photosynthate; the phenomenon

SYMBOLS USED

M_{AB}	flux from A to B	A_I	surface area of cell I
s_A	specific activity in A	A_N	area of junction of cell I with node N
Y_A	quantity of radioactivity in A	m	the parameter Y_V/Y_{cell}
Q_A	quantity of chemical species in A	w	the parameter $\dfrac{(Y_N + 2Y_{II})}{Y_{VI}}$

of apical inhibition of lateral buds requires the translocation of a hormone; while the rapid growth of maturing internodes, and the proliferation of the fine rhizoids underground, both suggest the intercellular transport of mineral salts. An autoradiographic study of some salts and of simazin and 2-4-D (Evrard and Chappell, 1967) has shown that they are moved around the plant of *Chara vulgaris.*

Good electrical contact between the cytoplasm of neighbouring cells—provided presumably by the plasmodesmata—was found by Spanswick and Costerton (1967), who in their surprise described the plant of *Nitella translucens* as a syncytium.

This paper represents an early phase of the attempt to make quantitative measurements of solute movement from one cell to another in plants of the *Characeae.* Attention is limited to isolated pairs of internodes with their common node, to the halide ions, and to the question of the phase of the internode from which the flux to the next cell arises.

B. Structure of the Node

Members of the family Characeae are distinguished by a somewhat simple approach to the problem of organizing a plant. The thallus consists essentially of long cylindrical internodes forming axes of limited or unlimited growth (roughly analogous to leaves and stems, respectively). The junction between each pair of internodes is referred to as a node, and consists of a number of small cells, arranged so that no two internodal cells make contact with each other. The structure of such a node in an axis of unlimited growth is shown in Fig. 1.

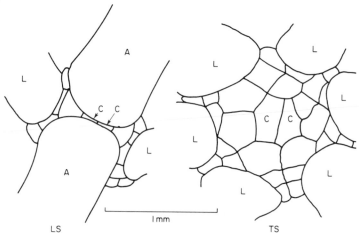

Fig. 1. Node of main axis of plant of *Chara corallina* (*australis*), in longitudinal and transverse section. A, main axis internodes; L, lateral internodes; C, central cells of node. In the longitudinal section, the thickness of cells C has been exaggerated for clarity.

Spanswick and Costerton (1967) found that in the mature node of *Nitella translucens* about 0·007 of the area of each intercellular wall was occupied by plasmodesmata. The figure for *Chara corallina* is probably not very different.

Microscopic observation of the node suggests that most of the node cells are vacuolate; their cytoplasm exhibits cyclosis.

For the purpose of the present study the experimental preparation consists of two adjoining internodal cells from a main axis of unlimited growth, with the node which separates them. All other internodes are removed, but the terminal nodes remain.

C. Models for Chloride Influx

If one considers a simple experiment in which one internode of the pair (but not the intervening node) is bathed in radioactive chloride solution, it is clear that the calculation of the flux from this cell to the unlabelled one, across the node, would be desirable. This requires a knowledge of the specific activity of the chloride in the cytoplasm of the first cell. There are reasons for believing that a direct measurement of radioactivity of the cytoplasm would be misleading, especially at short times of labelling.

Before 1969, measurements of ionic fluxes in cells of *Nitella* and *Chara* were interpreted in terms of a simple model of the cell, consisting of three compartments (Krogh *et al.*, 1944; MacRobbie and Dainty, 1958; Diamond and Solomon, 1959). Diamond and Solomon showed that the slowest compartment was kinetically in series with the intermediate one; the identification of the compartments as vacuole, cytoplasm and cell wall seemed generally agreed. The known complexity of compartments in the cytoplasm did not prevent this simple model from being adequate for some years (MacRobbie, 1962, 1964). It failed to agree with the measurements of MacRobbie (1969) on the distribution of chloride between cytoplasm and vacuole of *Nitella translucens* at short times of labelling (2–30 min). At about the same time, Larkum and Pring (unpublished) concluded that data on the radioactive chloride distribution in freeze-dried cells of *Nitella* could be best fitted to a model with three cytoplasmic compartments. MacRobbie (1969) considered a simpler model, which will be called the four-compartment model, and on the basis of a number of arguments found herself able to reject it. Figure 2, which represents data· of Findlay and Walker (unpublished) on *Chara australis*, illustrates the essence of her argument, and puts the maximum constraint on the four-compartment model.

The time course of Fig. 2 is constructed from data for a large number of cells, each cell contributing a value to only one point, since it must be sacrificed in order to determine the radioactivities of the vacuole and cytoplasm separately. The means for a number of different times can then be

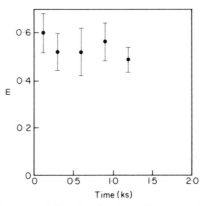

Time (ks)

FIG. 2. Time course of the vacuole fraction for chloride ($m = Y_V/Y_{\text{cell}}$) in *Chara corallina*, at 9°C (Findlay and Walker, unpublished). Points are mean and S.E.M. for batches of ten similar cells.

fitted by curves drawn by the recorder attached to an analogue computer, programmed to represent the model of Fig. 3(a). The cell wall is not represented in the model, since its chloride content is negligible after a brief wash.

In fitting the data, the "final" value of m, that fraction of the radioactive chloride in the cell which is found in the vacuole, is determined by the ratio of M_{CV} to M_{CP}. The time constant with which m rises from zero at short times is determined by ($Q_C/M_{CV} + M_{CP}$), where Q_C is the quantity of chloride in phase C. In the experiment illustrated it can be seen that the value of the half-time, 0.69 ($Q_C/M_{CV} + M_{CP}$) is likely to be less than 70 s. This is a smaller value than is generally needed to fit MacRobbie's own data on *Nitella translucens*.

In the four-compartment model of Fig. 3(a), the compartments are identified as C, watery phase of cytoplasm; P, plastids and other membrane

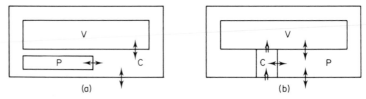

(a)   (b)

FIG. 3. Models for chloride influx in internodal cells of Characeae. (a) The four-compartment model: C, fast phase (cytoplasmic stroma); V, slow phase (vacuolar sap); P, intermediate phase (plastids and other membrane bound organelles). (b) A vesicle transport model: C, fast phase (endoplasmic reticulum and vesicles); V, slow phase (vacuolar sap); P, intermediate phase (cytoplasm, i.e. stroma and plastids and other organelles). ↔ molecular flux through membrane. ⇒ transport by exocytosis or endocytosis.

enclosed regions in the cytoplasm; V, vacuole. The movements of ions between the phases in this model are assumed to be normal membrane fluxes, via molecular mechanisms.

Arguments which tend to invalidate this model are of three kinds:

(i) that the values of m are not randomly distributed, but are "quantized" on a series of integrally related values 1α, 2α, 3α, etc. (MacRobbie, 1970, 1971) which seems inexplicable on the basis of molecular fluxes;

(ii) that the value of Q_C required by the results is too small to represent the chloride content of the stroma of the cytoplasm (MacRobbie, 1969), and

(iii) that certain correlations exist between fluxes in the four-compartment model which are inexplicable on the basis that they are independent molecular fluxes (MacRobbie, 1969).

These arguments led MacRobbie to reject the model as described and to substitute one, which she has not precisely defined, involving vesicular fluxes. She identifies C as a small membrane-bound phase in the cytoplasm,

TABLE I. Estimates of chloride concentration in cytoplasm

Value mM	Method	Species	Reference
3	Estimation of a_{Cl} from equilibrium potential in voltage clamp studies of action potential.	C. corallina (australis)	Findlay and Hope, 1964
10	Estimation of a_{Cl} by inserted chlorided silver micro-electrode.	C. corallina (australis)	Coster, 1966
35	Concentration measurement on flowing cytoplasm of perfused cell with nitrate sap.	N. flexilis	Kishimoto and Tazawa, 1965
70	As above, cell with normal sap containing 120 mM chloride.	N. flexilis	Kishimoto and Tazawa, 1965
65	Concentration measurement on flowing cytoplasm centrifuged to end of cell.	N. translucens	Spanswick and Williams, 1964
87	Estimate of a_{Cl} by chlorided silver electrode in centrifuged cytoplasm.	N. translucens	Hope et al., 1966

perhaps the endoplasmic reticulum. This model is represented in Fig. 3(b), though the details therein are ours, not MacRobbie's.

The evidence on which the model of Fig. 3(a) has been rejected is still most inadequate. Findlay *et al.* (1971) have shown "quantization" of m to reflect MacRobbie's subjective treatment of the data rather than an established phenomenon. The concentration of chloride in the stroma is not well characterized, with different methods giving widely different values (Table I). These values suggest that agnosticism may be the most healthy attitude, and they do not seem to rule out low values of stroma chloride concentration. (A thorough study of other ions has not been made, and might provide a good test here). Arguments from correlations in experiments of this kind are inherently weak, and in large experiments Findlay, Hope and Walker (unpublished and 1971) have found internal correlations unpredictable in their appearance.

The rejection of the four-compartment model (Fig. 3(a)) might be more convincing if it could be shown that a second cell attached to the cell being labelled with chloride, was labelled from a compartment different from C. In other words, if C is identified as the cytoplasmic stroma, it should be the compartment from which the second cell is labelled. This paper attempts to determine whether C is indeed the compartment from which the node is directly labelled.

II. Theory

A. The Static Model

A model for test, based on Fig. 3(a), is shown in Fig. 4. It represents the node as a single cell, which will be justified below, and its important feature is that compartment C, of the left hand cell, is the compartment from which both the vacuole of cell I, and the node and cell II, receive their flux. The

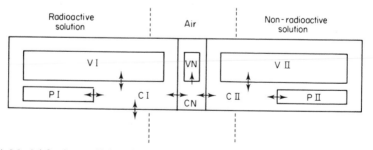

FIG. 4. Model for intercellular chloride movement in the two-cell preparation used in this paper; model based on the four-compartment model for one cell, see Fig. 3(a). VI, vacuole of internode I; VN, vacuole of node cell; VII, vacuole of internode II. Other phases labelled similarly.

test of this model will be whether, at short times after radioactive label is added to the solution bathing cell I, the fluxes CI–VI and CI–CN rise with the same time course.

The flux $M_{\text{CI–VI}}$ can be estimated from

$$Y_{\text{VI}} = A_{\text{I}} \cdot M_{\text{CI–VI}} \int s_{\text{CI}} \, dt - A_{\text{I}} \cdot M_{\text{VI–CI}} \int s_{\text{VI}} \, dt \qquad (1)$$

Since for times up to 5 ks it seems safe to neglect s_{VI}, we have

$$Y_{\text{VI}} = A_{\text{I}} \cdot M_{\text{CI–VI}} \int s_{\text{CI}} \, dt \qquad (2)$$

The radioactivity in the node and in cell II can be measured, and we have

$$Y_{\text{CN}} + Y_{\text{VN}} + Y_{\text{II}} = A_{\text{N}} \cdot M_{\text{CI–CN}} \int s_{\text{CI}} \, dt - A_{\text{N}} \cdot M_{\text{CNCI}} \int s_{\text{CN}} \, dt \qquad (3)$$

Since CN is a small compartment, it is not safe to assume that s_{CN} is small enough to neglect; and since the node is composed of cells in at least three size classes one would need a great deal of information in order to use the equality above as it stands. If, however, we can assume (i) no net fluxes, and (ii) the symmetry of the node, we can write

$$M_{\text{CNCI}} = M_{\text{CICN}} = M_{\text{CIICN}} = M_{\text{CNCII}} \qquad (4)$$

and hence from (3) and (4):

$$Y_{\text{CN}} + Y_{\text{VN}} + Y_{\text{II}} = A_{\text{N}} \cdot M_{\text{CICN}} \left(\int s_{\text{CI}} \, dt - \int s_{\text{CN}} \, dt \right) \qquad (5)$$

also

$$Y_{\text{II}} = A_{\text{N}} \cdot M_{\text{CNCII}} \int s_{\text{CN}} \, dt - A_{\text{N}} \cdot M_{\text{CIICN}} \int s_{\text{CII}} \, dt \qquad (6)$$

It should be safe to set $s_{\text{CII}} = 0$ for at least the first 2 ks of an experiment, since during this time protoplasmic streaming will bring a steady supply of fresh, unlabelled cytoplasm to the node. (The estimate of 2 ks is based on a cell 5 cm long streaming at the rate of 50 μm s^{-1}). So from (6) and (4) we have

$$Y_{\text{II}} = A_{\text{N}} \cdot M_{\text{CICN}} \int s_{\text{CN}} \, dt \qquad (7)$$

and from (7) and (5)

$$Y_{\text{N}} + 2Y_{\text{II}} = A_{\text{N}} \cdot M_{\text{CICN}} \int s_{\text{CI}} \, dt \qquad (8)$$

and from (2) and (8)

$$\frac{Y_{\text{N}} + 2Y_{\text{II}}}{Y_{\text{VI}}} = \frac{A_{\text{N}} M_{\text{CI–CN}}}{A_{\text{I}} M_{\text{CI–VI}}} = w \qquad (9)$$

and this fraction w should be independent of time, and is happily independent of the parameters Q_N which cannot easily be determined, and which in the real node differ as the sizes of the cells differ.

In some experiments Y_{VI} was not measured; a comparison is still possible if we use the relation $Y_{VI} = mY_I$, and accept that m is approximately constant from 100 to 5000 s in *Chara australis* (Findlay, Hope and Walker, unpublished). We then have the prediction that

$$\frac{Y_N + 2Y_{II}}{Y_I} = mw,$$

and will be constant to about 2 ks or so.

B. A Streaming Model

Of a number of complex problems introduced by protoplasmic streaming, only the simplest will be considered here. If the node is supposed to be square and uniform the value of s_{CII} will rise linearly as the streaming cyto-plasm passes over the node to a value ks_{CN} where k is a constant and $k < 1$. This will cause a transfer of radioactivity back to the node at a rate given by $M_{CIICN} \cdot k/2 \cdot s_{CN}$, and eqn (7) should now read

$$Y_{II} = A_N \cdot M_{CICN}\left(1 - \frac{k}{2}\right)s_{CN}\, dt \tag{10}$$

and eqn (9) becomes

$$\frac{Y_N + [(4-k)/(2-k)]Y_{II}}{Y_{VI}} = \frac{A_N \cdot M_{Cl-CN}}{A_I \cdot M_{Cl-VI}} \tag{11}$$

The factor $[(4-k)/(2-k)]$ rises from the value 2 at $k = 0$ to the value 3 at $k = 1$, with the value 2·33 at $k = 0.5$. For the present purposes, and with the preliminary data available, the value of $[(4-k)/(2-k)]$ will be taken as 2·0. No significant difference would be made if we took it to be 2·3 or 2·5.

At short times, 1 or 2 ks, we would expect the L.H.S. of eqn (11) to remain constant; at long times, when for other reasons back-fluxes are appreciable, it might be expected to fall somewhat.

III. Experiments

Pairs of cells were excised from the main axes of plants of *Chara corallina* (*australis*), maintained in clonal culture in continuous light at 22°C. They were bathed during the flux measurement in a solution containing:

NaCl	1·0 mM
KCl	0·2 mM
CaSO$_4$	0·05 mM
HEPES (Sigma)	2·0 mM
pH	7·0 (adjusted with NaOH)

Cell pairs were maintained in light and at $22°C \pm 1°C$. The solution was labelled with ^{36}Cl; and where appropriate also with ^{82}Br, in which case half the chemical chloride present was replaced by bromide. Double labelling was used to give two times of labelling for each cell, following the idea of MacRobbie (1971). During the experiment the cell pair was held in a narrow trough of polymethylmethacrylate, the node being isolated in moist air by two small pairs of stocks, while most of the surface area of each cell was bathed in the solution above.

After a period during which the solution surrounding one cell of the pair was labelled with ^{36}Cl, the pair of cells was rinsed in inactive solution, blotted gently, dried for a moment by blowing, and then cut with a new razor blade. Measured volumes of the sap of each cell, the remainder of each cell, and the whole node, were put on separate planchettes for determination of radioactivity.

In double-labelled experiments, solution labelled with ^{82}Br alone preceded solution labelled with both ^{82}Br and ^{36}Cl. The samples were counted twice at an interval of several weeks and the decay constant of ^{82}Br used to calculate the separate activities of the two isotopes.

In some experiments it was desired to calculate the chloride flux from the first cell to the second, across the node. Here the node was treated as a "membrane" and it was required to calculate M_{CICII} from Y_{II} and s_{CI}. These experiments were run for 3 h, during most of which time phases CI and PI should have very similar specific activities. At the end of the experiment the cytoplasm and sap of cell I were separated by passing an air bubble through the cell and $s_{(C+P)I}$ determined by radioactive counting followed by electrometric titration of chloride. In the same way $s_{(C+P)II}$ was determined and M_{CICII} calculated from

$$M_{CICII} = Y_{II} \cdot (s_{(C+P)I} - s_{(C+P)II})^{-1} \cdot A_N^{-1} \cdot \Delta t^{-1}$$

In 12 experiments in which Δt was 10·8 ks, the mean value of M_{CICII} was 4·1 nmol cm^{-2} s^{-1} and the standard error of the mean was 0·8 nmol cm^{-2} s^{-1}. The mean influx from the solution across the plasmalemma of cell I was 0·40 pmol cm^{-2} s^{-1}.

In other experiments it was desired to test the prediction of eqn 11. These gave the results shown in Table II and Fig. 5. Experiments 1B and 2B provide data for a direct test, since the parameter w can be evaluated. In experiments 3W–5W the value of Y_{VI} was not determined and the result quoted is equivalent to $m_I w$. Since the value of m_I in other experiments in this series has been about 0·7, a rough estimate of w can be made. In the experiments 4W and 5W, double labelling was employed, and the result quoted is the ratio found between $m_I w$ determined with ^{36}Cl to that determined with ^{82}Br. The controls 4B and 4W, in which ^{36}Cl and ^{82}Br are applied for equal times, show a moderately satisfactory approach to the ratio of 1·0 expected from

TABLE II. Experimental results to date (Tests of eqn 11)

Experiment No.	Influx time ks	Mean ratio ± S.E.M. (No. of values)
		w
1B	1·8	0·58 ± 0·17 (3)
	3·6	1·08 ± 0·76 (3)
	5·4	1·35 ± 0·64 (3)
	7·2	1·37 ± 0·53 (3)
2B	1·2	0·43 (1)
	1·8	0·59 ± 0·18 (5)
	7·2	0·54 (2)
	10·8	1·53 (2)
		$m_I w$
3W	1·0	0·20 ± 0·12 (5)
		$\dfrac{(m_I w)_{Cl}}{(m_I w)_{Br}}$
4W	Cl 1·0 Br 1·0	1·33 ± 0·20 (5)
4B	Cl 2·0 Br 2·0	1·25 ± 0·23 (6)
	Cl 10·0 Br 10·0	1·05 ± 0·18 (4)
5W	Cl 1·0 Br 2·0	0·74 ± 0·26 (2)
	Cl 1·0 Br 5·0	0·16 ± 0·06 (6)

MacRobbie's 1971 results. If one accepts their result as showing no discrimination between chloride and bromide, the result of experiment 5W indicates a real change in the value of $m_I w$ with time.

Taken together, the results of experiments 1, 2, 3 and 5, as plotted in Fig. 5, seem to show a marked increase in w with time.

IV. Discussion

These results are very preliminary. One can, however, look at them as indicators of what further work can be expected to confirm.

The flux across the node is large, some 40 pmol s^{-1}, and the immediate question is how this large amount of chloride arrives at the node, about 2 mm from the surface which is directly exposed to isotope-labelled solution.

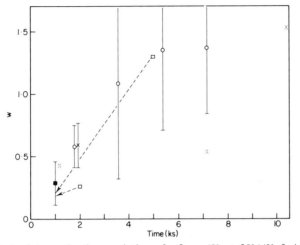

FıG. 5. Results to date on the time variation of w [$w = (Y_N + 2Y_{II})/Y_{VI}$]. All results are given as means, the tails giving the standard error of the mean. Open circles, experiment 1B. Crosses, experiment 2B; dotted crosses are points based on only one or two values. Closed square, experiment 3W. Open squares with arrows, experiment 5W. In both 3W and 5W only mw was determined experimentally; the plotted value is $mw/0.7$, based on the assumption that the mean value of m for these cells is 0·7. Open squares are mean values of $mw/0.7$ for 2·0 and 5·0 ks respectively; the arrow head represents the value of $mw/0.7$ at 1·0 ks which would give the observed mean ratio (see Table II).

Diffusion in the cytoplasm can be shown to be inadequate to account for the observed rate of arrival at the node, and diffusion across the tonoplast is likewise inadequate. Streaming of the cytoplasm seems the only possible mechanism. If we take the rate to be 50 μm s^{-1}, and the width of the stream to be 1·5 mm, we find that the minimum chloride content of the flowing cytoplasm is 50 nmol cm^{-2}. In the four-compartment model of Fig. 3(a) this has to be identified as Q_C. (It is true that on this model part of Q_P may be in the flowing cytoplasm, e.g. in mitochondria, but during the maximum of 20 s which the stream takes to cross the node, this mobile part of Q_P could not contribute much chloride to the phase C and hence to the node flux). Unpublished experiments by Findlay, Hope and Walker set an upper limit of about 70 s for the half-time of compartment C for *Chara*. If this were to be consistent with the value of 50 nmol cm^{-2}, we would require the fluxes M_{CP} and M_{CV} to be about 100 and 400 pmol cm^{-2} s^{-1}, respectively. While not quite impossible, these figures are very high.

The results of the experiments to test whether the same compartment C is the origin of both the flux to the vacuole, M_{CIVI} and that to the node, M_{CICN}, seem to show that this is not the case. The indication is that the flux to the node rises much more slowly with time than that to the vacuole. The results would not be inconsistent with the suggestion that the node flux originates in a compartment whose half-time is of the order of 2 ks. This would rule out the model of Fig. 4 and probably that of Fig. 3(a) on which it is based.

Clearly, further work is needed. The present results have little force, and serve as indicators only. They should be continued and extended. Effort should also be put into the definition of a vesicle model that will allow testable predictions to be made—it seems unsatisfactory to believe in vesicle-mediated chloride uptake because yet another simple compartmental model has failed to fit results.

References

COSTER, H. G. (1966). *Aust. J. biol. Sci.* **19**, 545.
DIAMOND, J. and SOLOMON, A. K. (1959). *J. gen. Physiol.* **42**, 1105.
EVRARD, T. O. and CHAPPELL, W. E. (1967). "Translocation of Growth Regulators in *Chara vulgaris*". Virginia Polytech., Blacksburg, Va.
FINDLAY, G. P. and HOPE, A. B. (1964). *Aust. J. biol. Sci.* **17**, 400.
FINDLAY, G. P., HOPE, A. B. and WALKER, N. A. (1971). *Biochim. biophys. Acta* **233**, 155.
HOPE, A. B., SIMPSON, A. and WALKER, N. A. (1966). *Aust. J. biol. Sci.* **19**, 355.
KISHIMOTO, U. and TAZAWA, M. (1965). *Pl. Cell. Physiol.* **6**, 507.
KROGH, A., HOLM-JENSEN, I. and WAARTIOVAARA, V. (1944). *Acta bot. Fenn.* **36**, 1.
MACROBBIE, E. A. C. (1962). *J. gen. Physiol.* **45**, 861.
MACROBBIE, E. A. C. (1964). *J. gen. Physiol.* **47**, 859.
MACROBBIE, E. A. C. (1969). *J. exp. Bot.* **20**, 236.
MACROBBIE, E. A. C. (1970). *J. exp. Bot.* **21**, 335.
MACROBBIE, E. A. C. (1971). *J. exp. Bot.* **22**, 487.
MACROBBIE, E. A. C. and DAINTY, J. (1958). *J. gen. Physiol.* **42**, 335.
OSTERHOUT, W. J. V. (1922). *J. gen. Physiol.* **4**, 275.
SPANSWICK, R. M. and COSTERTON, J. W. F. (1967). *J. Cell. Sci.* **2**, 451.
SPANSWICK, R. M. and WILLIAMS, E. J. (1964). *J. exp. Bot.* **15**, 193.

Discussion

The discussion started with a prepared comment from Walker on Mac-Robbie's work; he showed figures from the original *Tolypella* work and contended that re-plotting these data leads to the unbiased production of four straight lines through the points. Walker said, "The *Tolypella* experiment is certainly pretty groupy, but it can be grouped in different ways. It suggests to me that breaking it up in different ways is highly subjective". He next showed the same data plotted as a cumulative distribution with 95% confidence limits indicated. Walker asserted that one should not be too concerned with details within the data because of these limits, but should consider the gaps between the groupings. He later submitted the following statement on the significance of the gaps in various experiments:

Table of statistical tests on data of MacRobbie, E. A. C. (1970). *J. exp. Bot.* **20**, 236

| Experiment number | David's empty cell test | | | Run test | |
	N	E	P_E	N_R	For rejection of uniformity at 5% N_R
T	17	7	0·23	7	6
NA1	6	2	0·5	—	—
NA2	14	6	0·23	8	4
NA4	13	4	0·31	7	2
Fig. 8, MacRobbie, this volume, VI.4	22	5	most probable value of E is 6	—	—

Neither test rejects the hypothesis of uniform random distribution (used as test of uniformity of distribution).

Empty cell test
   N is the number of equal intervals into which the range of data values was divided. In an experiment yielding n values x_j, the range $x_n–x_1$ was divided into $n-2$ intervals,

and the distribution of the values $x_{n-1}-x_2$ in these intervals observed. E is the number of these intervals empty of data values. P_E is the probability of finding this number of empty cells if the values $x_{n-2}-x_2$ are uniformly randomly distributed in the interval x_n-x_1.

Run test (for clustering of empty and full cells)

N_R is the number of runs, of both full and empty cells, in the distribution; the hypothesis of uniform random distribution of data values can be rejected at the 5 % level if N_R is less than or equal to the number quoted in the next column. The Kolmogorov Statistic used to derive 95 % confidence limits for the maximum deviation of the population distribution from that of the sample in MacRobbie's Fig. 8 shows the 95 % confidence limits at ± 3.4 on the y axis.

Thus Walker asserted that until there is a good statistical test—and he indicated that MacRobbie did not have one—the peaks in the distribution are only there if they are seen to be there. Walker then showed the following slide of 5 apparently grouped distributions:

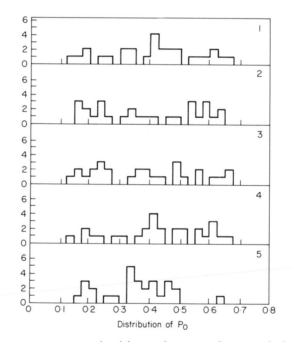

Distribution of P_0

four of these were generated with random numbers, and the fifth is the *Tolypella* data. This provided strong visual impact of the old adage that nothing is obvious to a mathematician (or statistician). Walker's conclusion, therefore, was that, apart from the original *Tolypella* results, all the tests that can be done show that the distributions asserted by MacRobbie to be grouped are in fact random, although there may be more sensitive tests than those he has used. MacRobbie replied that her argument is that the gaps do not occur at random, but that the same pattern of gaps appears in every experiment.

Müller then asked whether the appearance of the pattern of the distribution could be correlated with the acid–base bands of the cell. MacRobbie replied that *Nitella* is not as banded as *Chara*, although there can be bands in *Nitella*; nor are the pH differences along the cell as marked as in *Chara*. Cram asked what one would now know about the tonoplast fluxes. Mac-Robbie replied that, until it is known what compartmentation there is in the cytoplasm, tonoplast fluxes cannot be calculated. If the fast cytoplasm phase is due to cutting, then the slow cytoplasm phase is what should presumably be used to calculate tonoplast fluxes. However, there is still likely to be some difficulty; if the fast phase is due to an action potential on cutting, the Cl^- found in this fast phase cannot have been drawn from the cytoplasm, so that there must by implication be at least two compartments in the cytoplasm. Cram then asked if the influx measured over an hour is still the plasmalemma influx. MacRobbie did not directly answer the question; she replied that a fast phase which occurs batch-wise would be small, and would occur within the first few minutes.

Walker then returned to the discussion; he said the second part of Mac-Robbie's argument is that the slope of the slow rise to the vacuole is correlated with the influx, and that these quantities are not correlated on most simple membrane models. This claimed correlation, he asserted, is not a functional relation. One cannot alter the value of one parameter by adjusting the value of the other. It is simply a correlation in the scattered results of similarly treated, randomly selected cells. MacRobbie claimed that in some of the early papers, both by her and by Coster and Hope, there is some evidence of a functional relationship. Pitman then pointed out that in many of the experiments the distribution of results of percentage was not random but was skew, clustered towards the bottom end of the scale.

Smith asked if MacRobbie's two main recent contributions, namely that Cl^- flux is independent of phosphorylation and that vesiculation somewhere is necessary to describe the Cl^- flux, are compatible. Vesiculation may be dependent on protein synthesis and protein synthesis is dependent on ATP, or at least phosphorylation. MacRobbie replied that the evidence is only indirect that protein synthesis is involved; there may simply be membrane turnover with no net protein synthesis, and this may be powered by other high energy compounds than ATP. Barber then asked what is the co-ion for Cl^- in the fast transfer. There is, for instance, a fast transfer of Cl^- but not of K^+. MacRobbie replied it may be Na^+. There is a fast transfer of Na^+, although it is not clear if it is Cl^- linked. Cultured *Nitella* cells have a fast transfer of K^+, collected cells do not. In *Tolypella* there is a fast K^+ transport. However, in no case does the fast transfer of cations $Na^+ + K^+$ balance the Cl^-. The balance may be made up with H^+.

Section VII

Ion Transport in Roots and the Symplasm

Chairman: E. MacRobbie

VII.1

Regulation of Inorganic Ion Transport in Plants

M. G. Pitman and W. J. Cram

School of Biological Sciences, University of Sydney
Australia

I. Introduction

Our purpose in this paper is to ask what control mechanisms are needed at the level of cells and whole plants to account for observed patterns of ion uptake, and what control mechanisms are known. The ions we are particularly concerned with are the univalent cations and their balancing anions.

Uptake of ions to the plant as a whole and their subsequent distribution around the plant involve many processes. Ions taken up by the root may be retained in root cells or transported to the shoot mainly in the xylem but possibly also in the phloem. Within the leaves ions are accumulated in cells but K is also retranslocated from older leaves to younger ones. Under certain conditions both uptake to the plant and the subsequent distribution in the plant may be affected by transpiration.

In order to account for observed patterns of uptake we consider regulatory mechanisms to be necessary at two levels of organization in the plant. Control is needed to regulate transport of ions to the shoot, both at the site of secretion of ions into the xylem and during uptake from soil to root.

Plants also appear to regulate the concentration of ions in cells of the leaf and root, implying some control of fluxes into and out of individual cells. Evidence for controls operating on these processes will be discussed.

II. Regulation of Transport in Whole Plants

A. Fluxes of Ions in the Root

Plant roots have a remarkable ability to take up ions such as K, NO_3 and PO_4. Uptake of both cations and anions can take place from solutions of concentrations 10^2-10^3 times lower than in the plant. Efficiency in uptake has been expressed as a "Michaelis–Menten" constant, k_m, relating rate of uptake to external concentration. Values of k_m for K, NO_3 and PO_4 in grasses and cereals are probably about 10^{-5} M, 10^{-5} M and 5×10^{-5} M, respectively (Epstein, 1966; Asher and Loneragan, 1967; Edwards, 1969; Keay et al., 1970).

The relationship between rate of uptake and concentration for low-salt roots has been studied extensively, notably by Epstein and his collaborators. For the purpose of this paper the main feature of this relationship is that the rate of uptake is found to increase with concentration at least up to 40–50 mM. This increase is found for both tracer and net K uptake.

Fewer measurements have been made of tracer uptake to high-salt roots when external concentration was varied. Cram and Laties (1971) measured both short term (7 min) and long term (60 min) uptake. Short term uptake increased with concentration at least up to 80 mM, but long term uptake was nearly independent of external concentration above 10 mM.

To interpret these results it is convenient to refer to a model for the plant root such as that in Fig. 1. The cytoplasm of the cells of the root is assumed to form a "symplasm" into which ions move from the external solution, and from which ions move into the vacuoles of the cells in the root, or into the stele where they can be transported to the shoot.

The short term uptake tends to estimate ϕ_{oc}, while the long term uptake at higher external concentrations tends to estimate ϕ_{cv}; ϕ_{oc} appeared to increase with concentration but ϕ_{cv} was independent of external concentration. These results are similar to the response of fluxes of K to concentration in beetroot cells (Pitman, 1963) and also of Cl in carrot (Cram, 1968).

This interpretation of fluxes in high-salt roots is not inconsistent with results for low-salt roots. Low-salt roots are grown on a $CaSO_4$ solution and depend upon seed reserves for growth. These reserves are adequate for the first five days, but the level of K in the root is low (about 20 mM) and the osmotic content of the vacuoles is mainly due to hexose sugars (60–100 mM, depending upon growth conditions). When the roots are put into salt solution the sugar provides an easily accessible and unusually large source of respira-

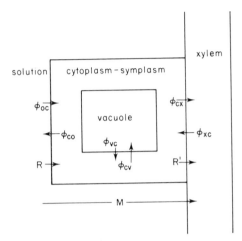

FIG. 1. Diagram of fluxes involved in uptake to the shoot from the root. Fluxes (ϕ) into and out of each phase are described by the order of subscripts; solution (o), symplast-cytoplasm (c), vacuole (v) and xylem (x). R and R' are net fluxes such that $R = \phi_{oc} - \phi_{co}$, etc. M allows for direct transport to the xylem by-passing the symplast, but appears to be small in barley. (After Pitman, 1971).

tory substrate and organic acid anions. Consequently rates of K uptake to the vacuoles of low-salt roots are large, and could be as large as ϕ_{oc} at low concentrations.

The "low-salt" and "high-salt" roots described by Hoagland and Broyer (1936) and used in many experiments since are physiological extremes of a natural continuum. Plants growing in mixtures of sand of low K content and loam of high K content show values of K content and sugar level intermediate between the extremes shown by plants growing in culture solutions (Table I). It is most likely that there is some other relation between sugar and K level than a simple energy source–energy sink relationship, or the need for K in pyruvic kinase activity.

TABLE I

Exchangeable K in sand/loam mixture (μequiv/100 g)	Level in plant root μmol (g fresh wt)$^{-1}$	
	K	Sugar (hexose)
50	26	96
106	34	66
162	42	27
275	52	17

B. Transport to the Shoot

These measurements with excised roots are concerned mainly with entry to the root cells, which can be very different from the transport that takes place into the shoot from the root.

1. Effect of external concentration on uptake to whole plants

Johansen *et al.* (1968) grew barley plants on continuous flow culture solutions and were able to make reliable measurements of uptake at very low external concentrations (down to 20 μM K). They found at 20, 200 and 2000 μM K the rate of transport of K from root to shoot was independent of external concentration. For example, we can calculate the following rates of transport from their data:

K concentration (μM)	20	200	2000
K transport relative to roots (μequiv g^{-1} h^{-1})	2·2	2·6	2·7

The relative growth rate was about 0·2 day^{-1} and root weight was approximately 40 % of total weight for each sample.

In this tabulation "transport" is calculated as the hourly average of (increase in K content of shoot/log mean root weight).

Figure 2 gives other data showing that at least up to 100 mM K (or K + Na) the rate of transport from the roots changes very little, and certainly less

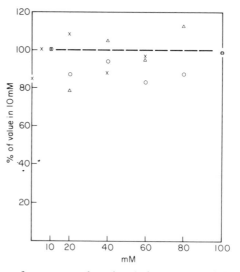

Fig. 2. Rates of transport from root to shoot in relation to external (K + Na) concentration for barley seedlings growing in culture solution containing K alone (○); containing (K + Na) (△) and for mustard seedlings (×). Rates expressed relative to value in 10 mM culture solution; (○) = 84 μequiv g dry wt^{-1} h^{-1}; (△) = 60 μequiv g dry wt^{-1} h^{-1}, and (×) = 125 μequiv g dry wt^{-1} 1 h^{-1}.

than ϕ_{oc}, which increased about 20 times between 1 and 80 mM in Cram and Laties' (1971) data. This independence of external concentration was shown for both barley and mustard, even though these plants differ in the concentrations reached by (K + Na) in the leaves.

Data for (K + Na) are included for plants growing in culture solutions containing both ions since it was found that the total level of K + Na in either root or shoot was independent of the ratio of K to Na in the solution (Pitman, 1965a, 1966).

The levels of K and of (K + Na) in barley were very nearly constant. The K concentration in the shoot increased $0.05 \pm 0.2\%$ per 1 mM change in K concentration in the solution; for K + Na the increase was $0.16 \pm 0.16\%$ per 1 mM change in (K + Na) concentration (Pitman, 1965b).

Mustard, like many other plants, has levels of (K + Na) which increase with external concentration. In this example the increase was nearly 0.5% per 1 mM change in external concentration. As external concentration increased, relative growth rate decreased and the leaves became smaller.

We conclude from these results that both ϕ_{cv} and R' are independent of external concentration over very wide ranges (0.02–100 mM). Plants growing in soil are probably exposed to effectively lower levels than these due to the slower rates of supply through soil solution to the roots. However, in experiments using culture solutions the main factor limiting the supply of K to the shoot is not concentration but some intrinsic factor connected with plant growth.

2. Uptake to whole plants in relation to growth

There are many experiments in the literature which show that uptake of nutrients is broadly related to plant growth. In some cases it is possible to show correlations between relative growth rates and rates of transport to the shoot (Greenway, 1965; Pitman, 1972b).

Plants of barley were grown in nitrate culture solution containing 10 mM K at varied relative growth rates. Transport of K to the shoot was estimated from harvests three days apart. Figure 3 shows that there was a good correlation between rate of transport and relative growth rate for the plants growing on culture solution.

Two samples of plants growing in soil and sand, and the results given for 20, 200 and 2000 μM by Johansen et al. (1968) lie well below this correlation line. Examination of the relationship between relative growth rate, K level in the shoot, and the ratio of shoot weight to root weight shows that the rate of transport = (relative growth rate × K concentration in shoot × ratio of weight of shoot to root). The ratio of weight of shoot to root ranged from about 3.5 for plants in culture solution to just less than 1.0 for plants in sand.

If rate of transport is expressed relative to shoot weight, then the ratio of weights of shoot to root is eliminated. Figure 4 shows the transport calculated

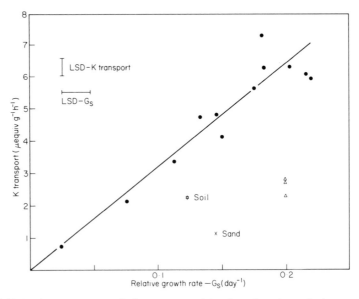

FIG. 3. Potassium transport relative to root weight plotted against relative growth rate based on shoot fresh weight (G_s). Data from Johansen *et al.* (1968) included ($\triangle$).

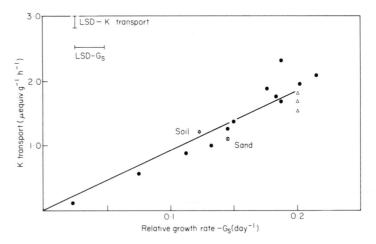

FIG. 4. Potassium transport relative to shoot weight (M_K) plotted against G_s. (Data calculated from Johansen *et al.* (1968)—$\triangle$.)

in this way plotted against relative growth rate. The good correlation was in this case a consequence of the concentration of K in the shoot being nearly constant, independent both of relative growth rate and the ratio of shoot to root (and, as we have seen in Fig. 2, to external concentration as well). This constancy suggests very strongly that some control system operates from shoot to root regulating the import to the shoot.

C. Mechanisms for Control of Transport from Root to Shoot

1. Uptake and retranslocation

A constant level in the shoot could be due to export of excess K back to the root, where it could diffuse into the solution or be recycled in the plant. To what extent is retranslocation able to account for the control of shoot content?

Import and export of K to leaves of barley seedlings at high-salt status were compared by measuring changes in total and tracer K when plants were transferred to labelled culture solution (Greenway and Pitman, 1965). The relative growth rate of the plants was 0.064 day^{-1} and total transport from root to shoot was 29 μequiv (g dry wt)$^{-1}$ h^{-1}. In the same units net K import, tracer import and estimated retranslocation were:

	Net K import	Tracer import	Retranslocation
L1	1.5	8.0	6.5 export
L2	14.0	11.5	(Not significant)
L3	13.5	7.5	6.0 import

Export from the oldest leaf (L1) was not a large proportion of total transport to the shoot, and was mainly to another part of the shoot.

In experiments of this kind where barley seedlings are transferred from unlabelled to labelled culture solution for periods of about three days, there is no evidence for large K export to the root. In the above example net transport to the shoot (29 μequiv g^{-1} h^{-1}) was not significantly different from tracer transport (27 μequiv g^{-1} h^{-1}). In other measurements the largest difference between total K transport and tracer transport from root to shoot was found for plants growing in a 2 h photoperiod, when tracer transport was 1.3 μequiv (g fresh wt)$^{-1}$ h^{-1} and net K transport 0.7 ± 0.5 μequiv (g fresh wt)$^{-1}$ h^{-1} (Pitman, 1972b). Re-translocation does not seem to be larger than 0.5–1.0 μequiv (g fresh wt)$^{-1}$ h^{-1}.

2. Sugar level and transport

The dependence of R' and ϕ_{cx} on metabolic substrates could provide a simple control relating overall growth and uptake.

Where relative growth rate is low, the sugar level in the root does seem to control rates of uptake and accumulation in the vacuoles. Figure 5 shows that there is an increase in accumulation following illumination of barley seedlings growing in a 2 h photoperiod. This increase lagged behind illumination by about 1 h. A similar lag was found in transport of radioactive sugars to the root following a pulse of labelling at the start of the light period. In this case at least 95 % of the label was present as sucrose.

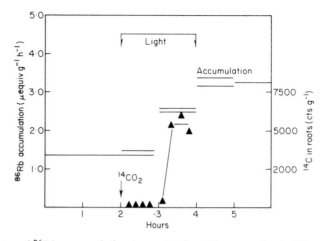

FIG. 5. Rates of ^{86}Rb accumulation by roots of seedlings growing in 10 mM K culture solution in 2 h photoperiods. $t = 0$ is 2 h before the end of the 22 h dark period. Also shown is rate of translocation of ^{14}C to the roots following a pulse of ^{14}CO$_2$ at the start of the light period (data from Pitman, 1972b).

Though sugar level may be a factor correlated with rate of transport from root to shoot at low growth rates, at higher relative growth rates sugar level increases with no further increase in transport. Figure 6 shows the rate of transport plotted against sugar level in the roots. At higher growth rates some other factor must be involved in balancing transport with growth of the shoot.

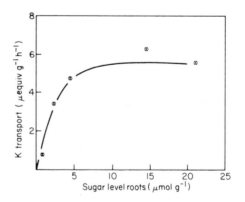

FIG. 6. Rate of K transport from root to shoot of plants growing with varied relative growth rates plotted against sugar level (hexose) in the roots. Relative growth rates ranged from 0·0 to 0·21 day^{-1}. (See also Fig. 3 for relationship between transport and relative growth rate.)

3. Abscisic acid

Abscisic acid (ABA) appears to be involved in water regulation of the plant as a whole. Mittelheuser and Van Steveninck (1969) showed that ABA inhibited stomatal opening and reduced transpiration. Wright and Hiron (1969), Most (1971) and Mizrahi *et al.* (1970) have shown that the ABA content increased in plants subjected to drought or saline conditions. Hocking *et al.* (1972) have shown that ABA can be translocated from the leaves to the roots. Tal and Imber (1971), and Glinka and Reinhold (1971) claim that ABA increases the permeability of tomato and carrot root cells to water, though no effect of ABA on permeability to water was found in maize roots in which exudation was inhibited by ABA (Cram and Pitman, 1972).

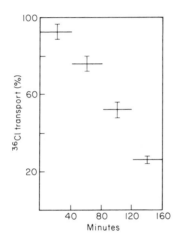

FIG. 7. Cl transport from barley roots. Means of 6 samples ± s.e.m. ABA added at $t = 0$. There was no difference in response over this concentration range. Uninhibited level was $2.0 \pm 0.2 \,\mu\mathrm{equiv}\, g^{-1}\, h^{-1}$.

ABA was found to inhibit the transport of K and Cl ions from the root to the xylem. Figure 7 shows the time course of inhibition of tracer transport through excised high-salt barley roots. Similar inhibitions were found for water and K exudation from excised maize roots, an 80–90 % inhibition of exudation occurring 2–3 h after adding ABA.

The inhibition was not detectable below 10^{-7} M ABA and the maximum inhibition was reached at about 5×10^{-6} M.

Though transport was inhibited, the uptake of tracer to the root (ϕ_{oc}) was not affected by ABA. There was in fact an increased rate of accumulation in the roots by an amount nearly equivalent to the decrease in transport from root to shoot.

This demonstration that ABA can inhibit secretion into the xylem provides a basis for a regulatory mechanism. Figure 8 shows the processes involved in the uptake and transport within the plant. It is suggested that levels of ABA change in the leaf in response to changes in levels of ions or in their osmotic effects. Translocation of ABA to the root transmits this change in ABA level which then controls the rate of secretion of ions to the xylem.

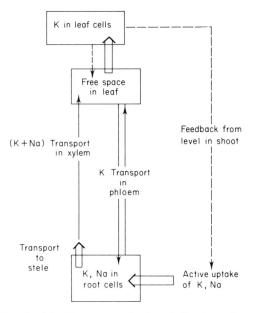

FIG. 8. Processes involved in K transport in the whole plant. (from Pitman, 1972b).

III. Regulation of Transport in Cells

A general question that can be asked is "how is a particular concentration in a cell determined?" A steady level such as in the root or leaf of a plant will be maintained when the influx equals the efflux plus that retained in an expanding cell. To investigate how a particular level is determined one can examine how fluxes change in relation to internal contents as a steady level is approached in simple tissue systems such as excised roots or storage tissue. It would seem probable that the same control mechanisms operate in cells in the intact plant.

A. Feedback in Determining Equilibrium Level

As a steady state is approached influx could become equal to efflux as a result of an increasing passive leak or a decreasing active transport. It appears probable that during accumulation of salt by plant cells there is a decrease in active influx (Cram, 1968; Pitman, 1963, 1969; Cram and Laties, 1971). This decrease must be signalled in some way. The simplest hypothesis, and therefore the one that should be examined first experimentally, is that salt transport is related to vacuolar concentration or to osmotic or hydrostatic pressure.

In *Valonia ventricosa* Gutknecht (1968) observed a 3- to 4-fold decrease in active K influx, but no change in passive fluxes, on increasing the internal hydrostatic pressure from 0 to 1×10^2 kPa (0–1 bar). In higher plants the internal hydrostatic pressure can only be changed as a result of changing the external osmotic pressure with a non-permeant solute. The equivalence of increasing the external osmotic pressure and decreasing the internal hydrostatic pressure on active K influx in *Valonia ventricosa* has been qualitatively confirmed (Cram, 1973, in press). If the increased internal hydrostatic pressure were the signal for decreased influx in salt-loaded higher plant cells, one should be able to stimulate Cl influx to its value in the unloaded state by increasing the external osmotic pressure. As shown in Table II, Cl influx to KCl-loaded carrot tissue was not affected by increasing

TABLE II. The effect of raising the external osmotic pressure on Cl influx in carrot tissue

Raffinose conc. in pretreatment and uptake solns. (mM)	Pretreatment time	Final Cl conc. (μmol g^{-1})	Cl influx (μmol g^{-1} h^{-1})
0	—	82 ± 2	1·4 ± 0·05
200	30 min	87 ± 1	1·3 ± 0·01
0	—	86 ± 6	1·0 ± 0·05
200	24 h	78 ± 4	1·0 ± 0·04

Carrot tissue washed in aerated water 6 days, 10 mM KCl 4 days, then Cl influx from 10 mM KCl^{36} ± 200 mM raffinose was measured after 30 min or 24 h pretreatment in inactive solutions ± raffinose. Loading time in radioactive solution: 45 min. Non-radioactive wash: 45–60 min.

the external osmotic pressure with raffinose. Similar results have been obtained with maize roots. In barley, the Cl influx is not related to the internal osmotic pressure determined by incipient plasmolysis (Table III). Robinson and Smith (1970) also found no stimulation of Cl influx in citrus leaf slices by increasing the external osmotic pressure. Stimulations of cation influx in chopped pea leaves (Nobel, 1969) and in beet disks (Sutcliffe, 1954) have, however, been observed.

More specific relationships between influx and internal ion concentrations have been found in *Lemna* (Young *et al.*, 1970) and in carrot and barley roots (Cram, 1973, in press). In *Lemna*, K influx (which has not been shown to be active) was found to be inversely proportional to the square of the internal

K concentration. In one experiment the Na concentration varied considerably, and K influx therefore appears to be related to the concentration of K, rather than Na or the sum of the univalent cations. This relationship does not describe all the variation in the K influx.

Table III shows values of Cl influx in carrot and barley root tissues which were allowed to accumulate a series of salts, and then washed in water for

TABLE III. Variation in Cl influx with internal contents in excised barley roots

Loading solution	H_2O	10 mM KCl	4 mM K_2SO_4 + 2 mM $KHCO_3$
Tissue contents (μmol g^{-1})			
Cl	17 ± 3	68 ± 1	11 ± 0·4
K	30 ± 0·4	94 ± 4	94 ± 3
Na	4 ± 0·1	4 ± 0·3	2 ± 0·8
Malate	7·8 ± 0·6	7·3 ± 0·5	27 ± 2
Reducing sugars	29 ± 3	11 ± 1	11 ± 1
Nitrate	2·2 ± 0·1	2·2 ± 0·1	—
Osmotic pressure (m Osm)	160 ± 7	243 ± 12	232 ± 14
Cl influx (μmol g^{-1} h^{-1})	4·1 ± 0·1	0·8 ± 0·0	3·8 ± 0·1

several hours to let the cytoplasm equilibrate with the vacuole before Cl influx was measured. As shown in Table III, Cl influx is not reduced after accumulating K malate, but is much reduced after accumulating KCl. Accumulated KNO_3 had the same effect as KCl. Cl influx was not correlated with K, Na, K + Na, malate, reducing sugar, NO_3 or Cl concentration, or with the internal osmotic pressure in barley roots. There was a highly significant correlation ($P < 0.01$) between Cl influx and log [(NO_3 + Cl) concentration] in both carrot and barley tissues. Net NO_3 influx was also inhibited by both high NO_3 and Cl concentrations in the vacuole.

The location in the cell of these effects is not certain, although it is probable that the influx to the vacuole is reduced more than the influx to the cell (Cram, 1968; Cram and Laties, 1971). This is also suggested by the fact that a larger fraction of the salt taken up is transferred to the xylem, rather than being accumulated in cortical cells, in salt-loaded roots. The site of action is obviously of importance in deciding whether this type of control is involved in uptake and metabolism of nitrate, and also in understanding regulation of influx to the xylem.

B. Energy Supply in Determining Equilibrium Level

The nature of the energy source for active transport has been extensively studied, particularly in giant algal coenocytes. Is transport ever naturally regulated by energy supply in higher plant cells?

The known relation of transport to light intensity in giant algal coenocytes and certain higher water plants suggests that at least at night, and possibly also in shaded habitats, salt transport would be limited by photosynthetic energy supply.

In non-photosynthetic cells the prevalence of salt and uncoupler-stimulated respiration shows that generally respiration is not rate-limited by respiratory substrate concentration. This has not been very extensively investigated, however, and certain observations suggesting respiratory substrate control need to be considered. Phillis and Mason (1940) observed that illuminating the shoot may stimulate uptake by the root, and suggested that this was by supplying sugars. Figure 5 in the present paper also shows light stimulation of uptake correlated with the transport of labelled photosynthate to the root.

Hoagland and Broyer (1936) first observed a parallel between salt transport and sugar concentration in cells when they discovered the difference between "high-salt" and "low-salt" barley roots. Several reports have qualitatively confirmed this, and recently Pitman et al. (1971) have observed a quantitative parallel between Cl and Rb uptake and reducing sugar concentration in barley roots during transition from low- to high-salt status. On the other hand, Hoagland and Broyer themselves showed that the high sugar–high Cl influx parallel did not hold in vigorously aerated low-salt barley roots. There is also no relationship between Cl influx and reducing sugar concentration in carrot root tissue during development or in carrot or barley root cells during accumulation of various salts (Table III and Cram, 1973).

There is therefore a parallel between Cl influx and sugar concentration under some, but not all, conditions. It is difficult to distinguish between a relationship due to cause and effect and one due to control in parallel.

C. Other Controlling Factors

1. Changes during differentiation

During differentiation of a cell, increases in size and vacuolation are accompanied by changes in transport processes. For instance, a fall in the K/Na ratio and a change from organic anion to Cl accumulation is well known (MacRobbie, 1962; Scott et al., 1968; Steward and Mott, 1970). It is almost trivial to say that these internally controlled changes are major factors in the regulated growth of the whole plant. However, they and other types of change such as occur in excised storage tissue do not give any clues as to what sort of control operates, or what sort of process is being affected. As with the use of inhibitors, changes of this sort by themselves are useful when differential or parallel changes in two processes can be followed, but are ambiguous if only a single process is followed.

2. Induction of transport systems by transported substance

The classic case of this type of control is, of course, that of β-galactosidase in *Escherichia coli* (de Crombrugghe *et al.*, 1971).

The only known case of induction of transport in higher plant cells is that of NO_3 transport in tobacco pith cultures (Filner, 1969). No case of induction of transport of inorganic ions in higher plant cells has appeared, except possibly in *Lemna*, where the rate of K transport changes with the external K concentration under conditions when the internal (vacuolar) K concentration is constant (Young *et al.*, 1970). Too few conditions and cells have been examined to draw any general conclusions about the prevalence of this type of control.

3. Cytoplasmic pH

A set of controlling factors also probably centres around the relationship of cation transport to inorganic or organic anion transport. In K malate-loaded cells Cl influx is not reduced, but net K influx is; and in $KHCO_3$ K is accumulated with malate. K and Cl influx are therefore not obligatorily linked. The fact that malic acid is not accumulated in KCl solutions has also to be explained. The suggestion has frequently been made that K and Cl influx to the cell (e.g. Poole, 1966; Smith, 1970), and malate synthesis (e.g. Jacoby and Laties, 1971) are controlled, driven, or linked by cytoplasmic pH. Such ideas are currently popular, but as yet have little experimental basis or quantitative parallels.

4. Hormonal and phytochrome control

In a few cases (e.g. Lüttge *et al.*, 1968) hormones appear to alter the rate of transport of inorganic ions. Phytochrome has been more directly implicated in the movement of K salts in pulvini of leaves of *Mimosa pudica* and some other legumes (e.g. Satter and Galston, 1971). There is no apparent phytochrome dependent transport process in *Hydrodictyon africanum* (Raven, 1969) or in excised carrot tissue (Cram, unpublished), although phytochrome has been shown to be present in carrot tissue cultures (Wetherell and Koukkari, 1967).

IV. Discussion

We have discussed factors controlling influx to cells and to the xylem of roots, but the level of ions may also be controlled by effects on efflux. This must be borne in mind when interpreting the relationship between fluxes and internal concentrations.

In the whole plant, transport processes will be regulated in relation to a combination of the factors we have considered, and probably others also.

Quantitative data are needed before the relative importance of various factors during growth and development of the cell are assessable.

The fact that the internal concentrations of several ions in roots and shoots are maintained constant suggests, by itself, the operation of regulating factors in transport. We have mainly concentrated on trying to pinpoint what the primary signals for transport regulation are, rather than speculating about mechanism.

The idea that the internal hydrostatic pressure is a control signal for transport is an attractive one. Although it does not appear to be the determining factor in the steady level of Cl reached in root cells, it would seem very likely that hydrostatic pressure might be a controlling factor in relation to cellular osmolarity in some other cells, and that it might affect the accumulation of other ions or molecules (cf. Thimann et al., 1960). In carrot and barley root cells the signal for decreased accumulation of Cl appears to be the vacuolar concentration of Cl plus NO_3, and the same is true for net NO_3 transport. It is not yet certain whether or not Cl and NO_3 are transported by the same process, but this does not alter the conclusions about the primary signal controlling their transport.

The evidence as to whether energy supply is a regulating factor is equivocal. It would seem likely that part of the relationship between sugar concentration and Cl influx is via respiratory substrate limitation of energy supply to Cl transport (Fig. 5), although the known k_m for plant hexokinases is lower than Cl transport-limiting concentrations of hexose in the cell vacuole (Saltman, 1953). It would also seem likely that there is some other relationship between Cl influx and sugar levels which shows up under certain other conditions (Pitman et al., 1971).

The results in Tables I and III in this paper show that Hoagland and Broyer's distinction between "low-salt" and "high-salt" roots must always be used with some qualification, since there is a continuum of levels possible, and since the differences depend upon what the salt involved is.

Apart from control within the cell there must be integration of the activities in different parts of the plant, in particular for regulating transport from root to shoot in relation to the activity of the shoot. In this case, again, sugar concentrations may be a controlling factor, although the evidence is not conclusive. Under other conditions, where light is not a factor limiting growth, it seems most likely that abscisic acid is an important messenger involved in the integration of the activities of the whole plant. In particular, consider the system suggested in section II C in relation to the plant's response to conditions of drought or salinity. Both conditions lead to increases in ABA levels in the leaf. Transport of ABA to the root would reduce salt secretion into the xylem of the root. In the absence of this restraint on secretion the concentration of salt would build up in the xylem and subsequently in the leaf. Unless the leaf were able to increase its accumulation of salt in the cell vacuoles, or increase its export of salts, the osmotic pressure in the extra-cellular spaces would rise and decrease the hydrostatic pressure in the leaf cells. Since photosynthesis would be reduced, by ABA-induced closure of stomata, growth of the plant would not provide a sink for any excess salt.

The same system could operate in the normally growing plant, providing a fine negative feedback to allow regular adjustment of ion transport to growth. Thus ABA, and also possibly cytokinins, could be part of a system regulating ion levels and water potentials in the plant.

References

ASHER, C. J. and LONERAGAN, J. F. (1967). *Soil Sci.* **103**, 225–239.
CRAM, W. J. (1968). *Biochim. biophys. Acta* **163**, 339–353.
CRAM, W. J. (1968). *Abh. Deut. Akad. Wiss. Berlin, Kl. Med.* Bd. a, 117–126.
CRAM, W. J. (1973). *J. exp. Bot.*, in press.
CRAM, W. J. and LATIES, G. G. (1971). *Aust. J. biol. Sci.* **24**, 633–646.
CRAM, W. J. and PITMAN, M. G. (1972). *Aust. J. biol. Sci.* **25**, 1125–1132
DE CROMBRUGGHE, B., CHEN, B., ANDERSON, W., NISSLEY, P., GOTTESMAN, M., PASTAN, I. and PERLMAN, R. (1971). *Nature New Biology* **231**, 139–141.
EDWARDS, D. G. (1969). *Aust. J. biol. Sci.* **23**, 255–269.
EPSTEIN, E. (1966). *Nature, Lond.* **212**, 1324–1327.
FILNER, P. (1969). *Develop. Biol.* Suppl. **3**, 206–211.
GLINKA, Z. and REINHOLD, L. (1971). *Pl. Physiol.* **48**, 103–105.
GREENWAY, H. (1965). *Aust. J. biol. Sci.* **18**, 249–268.
GREENWAY, H. and PITMAN, M. G. (1965). *Aust. J. biol. Sci.* **18**, 235–247.
GUTKNECHT, J. (1968). *Science* **160**, 68
HOAGLAND, D. R. and BROYER, T. C. (1936). *Pl. Physiol.* **11**, 471–507.
HOCKING, T. J., HILLMAN, J. R. and WILKINS, M. B. (1972). *Nature New Biology* **235**, 124–125.
JACOBY, B. and LATIES, G. G. (1971). *Pl. Physiol.* **47**, 525–531.
JOHANSEN, C., EDWARDS, D. G. and LONERAGAN, J. F. (1968). *Pl. Physiol.* **43**, 1722–1726.
KEAY, J., BIDDISCOMBE, E. F. and OZANNE, P. G. (1970). *Aust. J. agric. Res.* **21**, 33–47.
LÜTTGE, U., BAUER, K. and KÖHLER, D. (1968). *Biochim. biophys. Acta* **150**, 452–459.
MACROBBIE, E. A. C. (1962). *J. gen. Physiol.* **45**, 861–868.
MITTELHEUSER, C. G. and VAN STEVENINCK, R. F. M. (1969). *Nature, Lond.* **221**, 281.
MIZRAHI, Y., BLUMENFELD, A. and RICHMOND, A. E. (1970). *Pl. Physiol.* **46**, 169–171.
MOST, B. H. (1971). *Planta* **101**, 67–75.
NOBEL, P. S. (1969). *Pl. Cell. Physiol.* **10**, 597–605.
PHILLIS, E. and MASON, T. G. (1940). *Ann. Bot. N.S.* **4**, 635–650.
PITMAN, M. G. (1963). *Aust. J. biol. Sci.* **16**, 647–668.
PITMAN, M. G. (1965a). *Aust. J. biol. Sci.* **18**, 10–24.
PITMAN, M. G. (1965b). *Aust. J. biol. Sci.* **18**, 987–998.
PITMAN, M. G. (1966). *Aust. J. biol. Sci.* **19**, 257–269.
PITMAN, M. G. (1969). *Pl. Physiol.* **44**, 1417–1427.
PITMAN, M. G. (1971). *Aust. J. biol. Sci.* **24**, 407–421.
PITMAN, M. G. (1972a). *Aust. J. biol. Sci.* **25**, 243–258.
PITMAN, M. G. (1972b). *Aust. J. biol. Sci.* **25**, 905–920.
PITMAN, M. G., MOWAT, J. and NAIR, H. (1971). *Aust. J. biol. Sci.* **24**, 619–631.

POOLE, R. J. (1966). *J. gen. Physiol.* **49**, 551–563.
RAVEN, J. A. (1969). *New Phytol.* **68**, 1089–1113.
ROBINSON, J. B. and SMITH, F. A. (1970). *Aust. J. biol. Sci.* **23**, 953–960.
SALTMAN, P. (1953). *J. biol. Chem.* **200**, 145–154.
SATTER, R. L. and GALSTON, A. W. (1971). *Science* **174**, 518–200.
SCOTT, B. I. H., GULLINE, H. and PALLAGHY, C. K. (1968). *Aust. J. biol. Sci.* **21**, 185–200.
SMITH, F. A. (1970). *New Phytol.* **69**, 903–917.
STEWARD, F. C. and MOTT, R. L. (1970). *Int. Rev. Cytol.* **28**, 275–370.
SUTCLIFFE, J. F. (1954). *J. exp. Bot.* **5**, 215–231.
TAL, M. and IMBER, D. (1971). *Pl. Physiol.* **47**, 849–850.
THIMANN, K. V., LOOS, G. M. and SAMUEL, E. W. (1960). *Pl. Physiol.* **35**, 848–853.
WETHERELL, D. F. and KOUKKARI, W. L. (1967). *Pl. Physiol.* **42**, 302–303.
WRIGHT, S. T. C. and HIRON, R. W. (1969). *Nature, Lond.* **224**, 719
YOUNG, M., JEFFERIES, R. L. and SIMS, A. P. (1970). *Abh. Deut. Akad. Wiss. Berlin,*
    Kl. Med. Bd. b, 67–82.

VII.2

The Origin of the Trans-root Potential and the Transfer of Ions to the Xylem of Sunflower Roots

D. J. F. Bowling

Department of Botany, University of Aberdeen
Scotland

I. Introduction

The channelling of inorganic nutrients from the soil to the shoot is perhaps the most important function of roots. Unfortunately, despite half a century of investigation and experiment we still know very little about how this centripetal ion transport is brought about. It is generally agreed that the well known phenomenon of root exudation is the result of active transport of ions to the xylem with water moving in along a gradient of water potential which is set up as a result of the salt flux. It is the active transport of anions which appears to power the process (Bowling *et al.*, 1966). There has been much speculation about where the active transport is located. Crafts and Broyer (1938) in their well known theory placed the active step at the membranes of the cells of the outer cortex. Lundegardh (1954) as a result of his studies of the electrochemical properties of the outer surface of the root also considered that the primary accumulation takes place at the epidermis with a diffusive spread along a downward gradient across the cortex into the stele and exuded sap.

A number of investigators, however, have favoured an active step at the interface between the living cells of the stele and the xylem vessels, notably Arisz (1956) who was impressed by the ability of the salt glands of *Limonium* to secrete salt on to the leaf surface and he envisaged a similar secretory activity in the stelar cells. Secretion into the vessels is also favoured by Yu and Kramer (1969) who observed that the stelar cells of corn roots can accumulate ions to the same extent or in some cases to a greater extent than the cells of the cortex.

Lauchli et al. (1971) used an electron probe analyser to determine the pattern of potassium distribution in cross sections of corn roots. Their analysis showed that the cytoplasm and vacuoles of the xylem parenchyma cells accumulated potassium to a much greater extent than the cortical and other stelar cells. They obtained evidence from a study of the ultrastructure of the xylem parenchyma cells that they contain numerous membranes. It was concluded that the xylem parenchyma cells secrete ions into the vessels. However, it is not possible to draw such a definite conclusion from this kind of evidence which at best can only be circumstantial. Before we can make a decision about whether or not secretion into the xylem occurs we must apply more rigorous criteria for active transport to this region of the root. Unfortunately as it is buried deep in the root it is technically very difficult to make reliable measurements here and this is one of the reasons why this problem has remained unsolved for so long.

Lundegardh (1954) envisaged a gradient of ion concentration across the root with the highest concentration in the epidermal cells. A number of workers such as Lüttge and Weigl (1962), Crossett (1967), Biddulph (1967) and Shone et al. (1969) using microautoradiography, Lauchli (1967) and Lauchli et al. (1971), using an electron probe analyser and Dunlop and Bowling (1971a), using specific ion microelectrodes, have determined the profiles of various ions across the root. The shapes of the profiles obtained by these workers appear to differ. This may be due to the different methods employed, the material, or because different ions were investigated. These divergent results notwithstanding, concentration gradients alone can tell us little about the location of active transport sites across the root.

One of the most important advances made in the field of ion transport in recent years has been the application of biophysical theory to plant cells and tissues and in particular the realization by plant physiologists that ions are influenced by electrochemical potential gradients and not merely by concentration gradients. Movement of an ion against an electrochemical potential gradient appears to be a reasonable criterion for active transport. To determine the location of the active transport steps from the epidermis to the vessels we have to measure the gradients of ionic activity and electrical potential across as many membranes and phases as we can, preferably in the intact root. This is a tall order but in my opinion there is no easier alternative if we are to understand how the root brings about its polar ion transport. The same approach would also be very profitable when applied to salt glands.

Some progress has been made in this direction by the pioneering work of Higinbotham and his collaborators. They (Higinbotham *et al.*, 1967) determined the electrochemical gradients acting on the major ions accumulated by pea and oat root epidermal cells. They concluded that the cations were moved in with the gradient whilst the anions were accumulated against the gradient, indicating that they were actively transported. These conclusions were in complete agreement with the findings of Bowling *et al.* (1966) for whole castor bean roots indicating that there is an active step at the epidermis in the radial transport of ions to the xylem. Analysis of radioisotope fluxes indicates that the active anion transport is located at the plasmalemma (Pierce and Higinbotham, 1970). There is also increasing evidence that potassium is actively accumulated by plant cells (Pitman and Saddler, 1967; Macklon and Higinbotham, 1970; Bowling and Ansari, 1971). By measuring the activity of K^+ in the vacuoles of all types of cell likely to be involved in radial ion transport in corn roots, Dunlop and Bowling (1971a) showed that there was no marked gradient in K^+ activity from cell to cell across the root, and all the cells, with the exception of the xylem vessels, appeared to be able to accumulate potassium against the electrochemical gradient.

A comparison of the driving forces on the ions accumulated by root cells and on ions transferred to the vessels should provide us with some idea of the size and direction of the forces acting on the ions at the interface between the living cells of the root and the vessels. Described below are some measurements made to this end on sunflower plants (*Helianthus annuus*) grown in water culture.

II. Experimental Methods and Results

Two culture solutions were employed: a standard solution, the main constituents of which are given in Table I and a 10-fold dilution of this termed 0·1 standard (Table II). The plants were grown to the age of 5 weeks at 25°C in continuous artificial light. Analyses of the exuding sap and tissue

TABLE I. Concentrations of the major ions in the external solution and exuding sap (mM) and ion contents (μequiv g fresh wt^{-1}) of the root tissue taken from a zone between 1 and 2 cm from the root tip of 5-week-old sunflower plants grown in standard culture solution. Data are means of samples from 5 root systems

Ion	External solution	Root	Exudate
K	7·0	111	19·0
Ca	4·0	26	4·0
Mg	2·0	4·8	2·2
NO$_3$	14·0	150	14·6
SO$_4$	2·0	11·5	1·9

TABLE II. Concentrations of the major ions in the external solution and the exuding sap (mM) and ion contents (μequiv g fresh wt^{-1}) of the root tissue taken from a zone between 1 and 2 cm from the root tip of 5-week-old sunflower plants grown in 0·1 standard culture solution. Data are means of samples from 5 root systems

Ion	External solution	Root	Exudate
K	0·7	59·0	10·3
Ca	0·4	5·7	3·2
Mg	0·2	2·7	2·0
NO_3	1·4	31·0	7·7
SO_4	0·2	1·8	2·1

samples were carried out to determine the levels of the major ions in roots grown in these two media and the data are set out in Tables I and II. In order to calculate electrochemical potential differences, electrical potential differences in the vacuoles of epidermal cells and the exuding sap relative to the outside solution were measured, the former with the use of microelectrodes. It can be seen from the data set out in Table III that for roots in both media the epidermal cell potentials were larger (i.e. more negative) than the trans-root potentials. Therefore there must be a potential gradient of magnitude $\Delta E_1 - \Delta E_2$ located somewhere between the epidermis and the vessels.

TABLE III. Vacuolar potential differences (mV) of epidermal cells (ΔE_1) and trans-root potential differences (ΔE_2) of 5-week-old exuding roots of sunflower (*Helianthus annuus*) grown in two strengths of complete nutrient solution. ΔE_1 means ±95% confidence limits for 20 or more cells. ΔE_2 means ±95% confidence limits for 5 or more root systems. ΔE_2 was measured 6 h after roots were excised. ΔE_1 was measured in a zone between 1 and 2 cm from the root tip

Culture solution	ΔE_1	ΔE_2	$\Delta E_1 - \Delta E_2$
Standard	-39 ± 4	-28 ± 5	11
0·1 standard	-72 ± 4	-48 ± 6	24

This potential difference appears to be in the stele at the interface between the pericycle cells and the xylem vessels. This was determined by measuring the potential profile from cell to cell across the root. A potential profile across maize roots was determined using longitudinally sectioned material (Dunlop and Bowling, 1971a); however, the roots of *Helianthus* are thinner than those of corn and so were very difficult to section without causing considerable damage. It was observed that some roots were fine enough to allow most of the cells to be seen under the microscope with the root intact. Consequently vacuolar potentials of most of the cells from epidermis to pericycle could be measured without too much difficulty in whole rootlets.

A microscopic magnification of × 700 was employed and the microelectrode could be positioned and observed in almost every cell between epidermis and pericycle. The results for roots in standard culture solution are shown in Fig. 1. They indicate that there is no statistical difference in the vacuolar potentials of the living cells likely to be involved in radial ion transport. The potential difference $\Delta E_1 - \Delta E_2$ must therefore be at the pericycle–vessel interface.

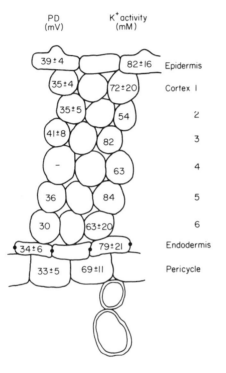

FIG. 1. Diagram of part of a transverse section of a fine root of *Helianthus annuus* showing the trend in electrical potential relative to the outside solution, and the potassium activity in the vacuole of the cells between the external solution and the xylem. The determinations were made on the intact root in standard culture solution. Confidence limits of 95 % are given where five or more determinations were made.

Figure 1 also gives the K^+ activity in the vacuole of each of the cells across the root, measured in intact rootlets in a similar way to the potential except that K^+ specific microelectrodes were used. The construction, calibration and use of these electrodes in roots has been fully described by Dunlop and Bowling (1971a). It can be seen that there was no gradient in K^+ activity from vacuole to vacuole in the intact root.

III. Discussion

We can work out the magnitude and direction of the driving forces on potassium ($\Delta\bar{\mu}_k$) at several of the interfaces across the intact root from the data in Fig. 1 using the following formula:

$$\Delta\bar{\mu}_k = zF(\Delta E_{obs} - \Delta E_n)$$

where z = valency, F = the Faraday, ΔE_{obs} = the measured p.d. and ΔE_n the Nernst potential. It is assumed that potassium was in flux equilibrium and we have some evidence for this (Bowling and Ansari, 1972). The profiles of electrical and electrochemical potential are shown in Fig. 2. Nitrate is

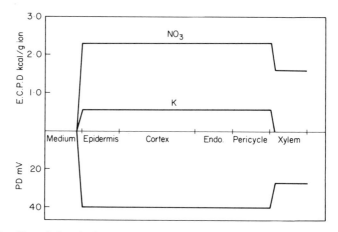

FIG. 2. Profiles of electrical potential difference (P.D.) and electrochemical potential difference (E.C.P.D.) for potassium and nitrate in sunflower roots grown in standard culture solution.

the anion which approximately balances potassium and is therefore likely to show the same activity profile and hence its electrochemical potential profile is also given in Fig. 2. These profiles are very similar to those found in maize (Dunlop and Bowling, 1971b). Clearly there is active transport of both ions from the outside solution into the vacuoles of the epidermal and cortical cells. Movement from the endodermal and pericycle cells into the vessels is down the electrochemical gradient. The net gradient on potassium between epidermis and vessels is zero as the fall in electrochemical potential at the vessels equals the uphill step at the outer surface of the root. However, nitrate is moving to the xylem from the outside solution against a net electrochemical gradient.

Unfortunately we do not have such detailed activity profiles for the other major ions investigated, nevertheless it is still possible to come to some conclusions about the driving forces acting upon them. Figure 3 shows that sulphate behaves in a similar way to nitrate at the outside solution–root and root–xylem interfaces and moves to the xylem against a net adverse gradient. Calcium and magnesium both move into the vacuoles and to the exudate with the gradient and the data provide no evidence of active transport of these two cations at any stage in their path across the root.

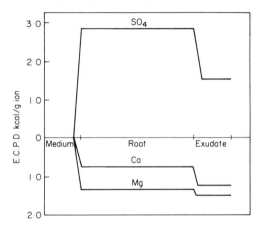

FIG. 3. Profiles of electrochemical potential difference (E.C.P.D.) for sulphate, calcium and magnesium in sunflower roots in standard culture solution.

The driving forces on the ions entering the vacuoles and the exudate of roots in 0·1 standard culture solution are given in Table IV. They broadly agree with the results for the roots in standard solution except that there appears to be a net active transport of potassium to the vessels; the downward step in the gradient from pericycle to xylem not quite cancelling out the uphill step at the epidermis.

Lavison showed as early as 1910 that coloured salts could penetrate the root cortex by way of the cell walls as far as the endodermis. It is only in

TABLE IV. Electrochemical potential differences for the major ions distributed between the external solution and the cells ($\Delta\bar{\mu}_1$) and between the cells and the exudate ($\Delta\bar{\mu}_2$) of sunflower plants grown in 0·1 standard solution kcal g ion^{-1}. $\Delta\bar{\mu}_1 - \Delta\bar{\mu}_2$ gives the net driving force on the ion between medium and exudate. Asterisk indicates active transport

Ion	$\Delta\bar{\mu}_1$	$\Delta\bar{\mu}_2$	$\Delta\bar{\mu}_1 - \Delta\bar{\mu}_2$
K	*0·92	0·46	*0·46
Ca	1·79	0·64	2·43
Mg	1·79	0·64	2·43
NO$_3$	*3·45	1·36	*2·09
SO$_4$	*4·60	2·34	*2·26

recent years that we have obtained evidence that transport of ions from cell to cell through the cytoplasmic continuum (symplasm) of roots also occurs (Jarvis and House, 1970; Ginsburg and Ginsburg, 1970; Dunlop and Bowling, 1971b) although the possibility of such transport had been suggested many years before (Crafts and Broyer, 1938). It seems virtually certain that both types of transport occur in the cortex although it is an open question as to which pathway plays the most important role in radial transport. There can be little doubt that movement across the endodermis from cortex to stele is symplasmic as the Casparian bands appear to prevent transport in the cell walls at this point. It is probable that the Casparian band acts as a resistor preventing a short-circuit of the trans-root potential by restricting ion flux to the cell membranes where the potential is generated.

It may be argued that these present results do not bear directly upon the problem of radial transport as they deal largely with accumulation in the vacuoles. Certainly an activity profile, similar to that in Fig. 1 for potassium in the symplasm, would be invaluable. However there is increasing evidence that the sites of active transport of ions into the cell are at the outer membrane. The results of Macklon and Higinbotham (1970) for instance, indicate an inwardly directed potassium pump at the plasmalemma whilst the electro-chemical potential difference on potassium at the tonoplast is zero. Therefore these results are likely to provide us with a reasonably accurate picture of the forces on ions entering and leaving the symplasm.

They show, without exception, that both cations and ions move into the vessels from surrounding cells down the electrochemical potential gradient; therefore there is no necessity to postulate active transport or secretion into the xylem at this point. Indeed, as potassium, nitrate and sulphate are actively accumulated from the outside solution it seems reasonable to postulate that the cells surrounding the vessels also actively take up these ions from the xylem sap. This provides the plant with a rather precise control over the transfer of certain ions to the xylem as exemplified by the behaviour of potassium (Fig. 2 and Table IV) and sodium (Bowling and Ansari, 1972).

Thus these results suggest that radial transport is driven by an active step at the outer surface of the root whilst movement from the living cells into the vessels is passive.

References

ARISZ, W. H. (1956). *Protoplasma* **46**, 5–62.
BIDDULPH, S. K. (1967). *Planta* **74**, 350–367.
BOWLING, D. J. F. and ANSARI, A. Q. (1972). *J. exp. Bot.* **23**, 241–246.
BOWLING, D. J. F., MACKLON, A. E. S. and SPANSWICK, R. M. (1966). *J. exp. Bot.* **17**, 410–416.
CRAFTS, A. S. and BROYER, T. C. (1938). *Am. J. Bot.* **25**, 529–535.
CROSSETT, R. N. (1967). *Nature, Lond.* **213**, 312–313.

DUNLOP, J. and BOWLING, D. J. F. (1971a). *J. exp. Bot.* **22**, 434–444.
DUNLOP. J. and BOWLING, D. J. F. (1971b). *J. exp. Bot.* **22**, 453–464.
GINSBURG, H. and GINZBURG, B. Z. (1970). *J. exp. Bot.* **21**, 593–604.
HIGINBOTHAM, N., ETHERTON, B. and FOSTER, R. J. (1967). *Pl. Physiol.* **42**, 37–46.
JARVIS, P. and HOUSE, C. R. (1970). *J. exp. Bot.* **21**, 83–90.
LAUCHLI, A. (1967). *Planta* **75**, 185–206.
LAUCHLI, A.; SPURR, A. R. and EPSTEIN, E. (1971). *Pl. Physiol.* **48**, 118–124.
LAVISON, J. (1910). *Rev. gen. Bot.* **22**, 225–240.
LUNDEGARDH, H. (1954). *Symp. Soc. exp. Biol.* **8**, 262–296.
LÜTTGE, U. and WEIGL, J. (1962). *Planta* **58**, 113–126.
MACKLON, A. E. S. and HIGINBOTHAM, N. (1970). *Pl. Physiol.* **45**, 133–138.
PIERCE, W. S. and HIGINBOTHAM, N. (1970). *Pl. Physiol.* **46**, 666–673.
PITMAN, M. G. and SADDLER, H. D. W. (1967). *Proc. natn. Acad. Sci. U.S.A.* **57**, 44–49.
SHONE, M. G. T., CLARKSON, D. T. and SANDERSON, J. (1969). *Planta* **86**, 301–314.
YU. G. H. and KRAMER, P. J. (1969). *Pl. Physiol.* **44**, 1095–1100.

VII.3

Some Evidence that Radial Transport in Maize Roots is into Living Vessels

N. Higinbotham,* R. F. Davis,† S. M. Mertz,* and L. K. Shumway*

Department of Botany, Washington State University
*Pullman, Washington, U.S.A.**

and

Department of Botany, Rutgers University, Newark, N.J., U.S.A.†

I. Introduction

Root xylem vessels have generally been considered to be dead and devoid of living protoplasm in the zone of absorption; this is a premise of most concepts of radial transfer to the xylem of roots. However some older and some more recent studies suggest that vessel elements may retain a lining of intact protoplasm throughout most of the absorption zone and that this may constitute the important barrier to solute conduction up the root. The purpose of this paper is to summarize this evidence and to present some of our own preliminary observations which tend to corroborate the conclusion.

In 1938 Crafts and Broyer summarized their views of radial transport into the xylem of roots. The major steps were as follows: (1) active absorption of solutes at the epidermis; (2) passive movement through the symplasm to the stelar parenchyma; and (3) leakage of solutes into dead xylem elements. The endodermis was thought to be an important barrier, particularly via cell walls, and a possible site of active solute transfer into the stele. A lower O_2 supply in the stelar parenchyma, resulting in a lower respiratory rate, was considered to be a likely cause for leakage of ions into xylem vessels. Laties and Budd (1964), using maize root cortical sleeves and stelar tissue isolated from one another, found that the stele had less capacity for accumulation of Cl and was leaky to ions as required by the Crafts and Broyer model. However Laties and Budd suggest that the stele may have lower metabolic activity because of the presence of a volatile metabolic inhibitor rather than low O_2 tension; steles aged for 24 h show ion uptake comparable to that of the cortex.

Yu and Kramer (1967) have contested the view that stelar parenchyma is leaky since they found that cortex and stele, separated after previous accumulation of PO_4 by intact roots, showed equal uptake.

Läuchli and Epstein (1971) also assume that xylem vessel elements are dead in the absorption zone. They show that the uncoupler carbonylcyanide m-chlorophenylhydrazone, m-CCCP, and oligomycin, which blocks phosphorylation, quickly depress Cl transport to the vessels. They conclude that transfer to the vessels is an active process driven by adenosine triphosphate (ATP); and, on the basis of electron probe X-ray profiles of K distribution across the root, it is postulated that the stelar parenchyma actively secrete ions into the xylem (Läuchli et al., 1971).

In contrast to the premise that conducting vessels are dead in the absorptive zone are the observations of Anderson and House (1967). They observed that cytoplasm may persist in vessels at levels of about 6 cm from the corn root tip. Measurements of absorption, translocation, and exudation of water and K suggest that the power for absorption and long distance transfer decreases from the tip toward the base (10 cm); this suggested the involvement of living xylem elements in the conduction pathway. Davis (1968) extended the observations on the persistence of intact protoplasm in xylem vessels, finding by electron microscopy that the plasmalemma and tonoplast may still be intact in at least some vessels at the 10 cm level of corn roots.

Scott (1949), in a study that appears to have been widely overlooked by plant physiologists, reported that in roots of *Ricinus* all of the xylem elements were alive to at least the 8 cm level and possessed plasmodesmata. She stated that the problem of transfer to non-living xylem does not arise in the root hair zone. More recently she has reaffirmed the conclusion that solute movement in the root hair zone is into living nucleated xylem elements (Scott, 1965 and private communication).

Although the case for active transport into living xylem seems strong there is also good evidence in corn roots that solutes may be transferred radially into fully mature xylem lacking protoplasm. Burley *et al.* (1970) made a careful study of the capacity of 2 mm zones along the root to absorb and translocate $^{32}PO_4$. They found that whereas absorption was quite uniform up to 23 cm, translocation to the shoot began at about 5 cm and reached a maximum at about 23 cm, the highest level measured. However the shoots were attached and consequently the translocation at the higher levels may have been due to transpiration pull rather than to root pressure movement. In addition these authors used a pressure system to perfuse eosin Y solution through the root from below, at different levels, to determine where vessels became open to conduction. They found that the late metaxylem showed dye flow at 28 cm or above. The early metaxylem showed some dye flow at about 10 cm but flow was not complete in the ring below approximately 13 cm; however, in older plants, the early metaxylem showed flow at 7–9 cm. They state that mature suberized endodermis is found at 20 cm or above.

Water uptake by roots of intact maize plants has been reported to reach a maximum at about 10 cm from the apex (Hayward *et al.*, 1942); this contrasts somewhat with PO_4 uptake discussed above but suggests that most absorption occurs at the 0–10 cm zone.

It seems clear from the discussion above that some of the critical information is not yet available, namely: (1) What is the course of anatomical differentiation? At what level do the walls separating xylem vessel elements become perforated? How long after this perforation does the protoplasm continue to line the wall? (2) How long are the individual vessels? Are they connected to one another by perforations through abutting or overlapping end walls?

II. Electropotential Studies

Root pressure exudate is electrically negative to the solution bathing roots by 20–70 mV, the magnitude being inversely related to the concentration of cations (chiefly K or Na) in the external medium (Davis and Higinbotham, 1969). The potential is believed to arise from active radial transport of ions from the outer solution into the xylem. If root segments of different lengths are cut, and exudate electropotentials measured, it should be possible to obtain an approximate measure of average vessel length provided the vessels are free to conduct; i.e. open vessels should short-circuit and the potential should drop or become zero. Likewise the electrical resistance should be lowered in the presence of open conduits. This experiment was done by Davis (1968) and the results in Table I show that over a 10 cm length of corn roots excision of 1 cm lengths in either direction had no significant effect on either the electrical potential or the resistance, except for the 11–13 cm zone.

TABLE I. Typical results from experiments relating the electrical properties to distance from basal and apical ends of maize roots

Distance of cut from root tip (cm)	Segments cut in acropetal direction[a]	Segments cut in basipetal direction[a]	
	p.d. (mV)	p.d. (mV)	Resistance MΩ
11	−30	−15	0·7
10	−31	−29	1·9
9	−31	−31	1·8
8	−29	−31	1·8
7	−35	−31	1·8
6	−47	−30	1·8
5	−43	−20	1·8
4	−31	−28	1·7
3	−24	−25	1·6
2	—	−23	1·6
1	—	−23	2·0

[a] Separate roots were used for the acropetal and basipetal cutting experiments. Root lengths were approximately 13 cm.

These results suggest that on average, below the 10 cm level, the vessels are occluded or are less than 2 cm long. This might arise from the persistence of vessel element end walls in this zone, or perhaps from occlusion due to the presence of protoplasm.

III. Microscopic Studies

A. Maceration of Steles

In order to measure vessel lengths the cortex was stripped from the stele with a wire stripper (Laties and Budd, 1964) and treated in the maceration mixture CrO_3/HNO_3 (each 5% in H_2O by volume) for 8 h. They were then stained in safranin. After staining they were placed on a slide for microscopic observation or they were run through a dehydration series and permanently mounted in balsam. With careful treatment a partial maceration was achieved with the stele being flattened out so that most xylem vessels were visible yet remained in vertical alignment.

By definition a vessel consists of a linear series of vessel elements; in early development between adjacent elements there is a wall which later becomes perforated. By implication the walls at each end are not perforated although the anatomical sources seem not to be explicit on this point. In our observations on maize root xylem some vessels were distinguished by having a series of rather clearly transverse cross walls with quite oblique end walls.

In other vessels the end walls could not be distinguished with certainty. In those cases in which we thought vessel end walls were distinguishable each vessel consisted of about 8–10 elements, the elements averaging less than 1 mm in length. In the 10 cm zone all of the late metaxylem, the large inner ring, were clearly immature, showing nuclei and having little or no secondary wall thickening. The early metaxylem vessels often showed nuclei in the elements but had clear secondary wall formation.

B. Light Microscopy

Segments were fixed in formalin-acetic-alcohol (5/5/90, v/v) and, using usual procedures, imbedded in paraffin, sectioned, and stained in Conants quadruple stain procedure.

In longitudinal sections at the 4–5 cm level, which is above the elongation zone, 50 metaxylem vessel elements had an average of 0·82 mm in length

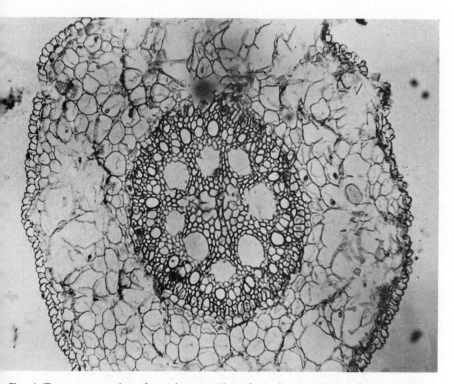

FIG. 1. Transverse section of a maize root 10 cm from the apex. Taken from a seedling 7 days old grown in aerated nutrient solution. The ring of 8 later-maturing metaxylem vessels is surrounded by a ring of earlier-maturing metaxylem vessels one of which, in the 7 o'clock position, contains a nucleus (also see Figs 2 and 3). Just outside the early metaxylem elements and adjacent to them lie 1–4 protoxylem vessels. × 120.

and 0·053 mm in diameter. There is a ring of about 17 early metaxylem vessels and 8 late metaxylem vessels (Fig. 1). In a 10 cm segment of maize root this would provide a volume of 4·7 μl; however, it is probable that the terminal 5 cm is not conducting (Burley et al., 1970) so that the functional volume would be about 2·8 μl, close to other estimates reported (Anderson and House, 1967). In another series of measurements the estimated volume of xylem was 3·5 μl.

At the 9–10 cm level many of the early metaxylem vessel elements can be seen to have intact nuclei although the lateral walls are quite well differentiated (Figs 2 and 3). At this preliminary stage in our investigations we are inclined to think that most if not all of the early metaxylem vessel elements retain living protoplasm at, or near, the 10 cm zone although roots of this length show quite vigorous exudation (Davis, 1968), and, as will be seen below, there is good evidence that the perforation plates are fully developed. At this zone the inner ring of late metaxylem vessels elements retain intact protoplasts and there is little or no secondary wall formation; the cross walls

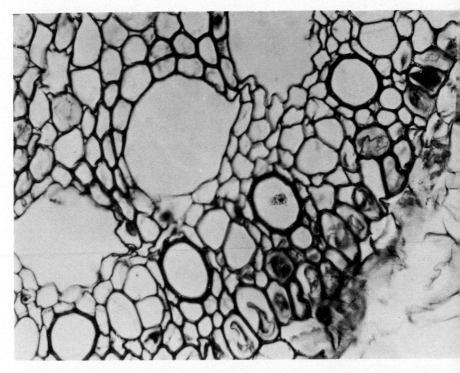

FIG. 2. Portion of Fig. 1 enlarged. This shows an early metaxylem vessel with an intact nucleus; two protoxylem elements lie to the outside of it. Note the absence of secondary wall thickening in the large vessels. × 500.

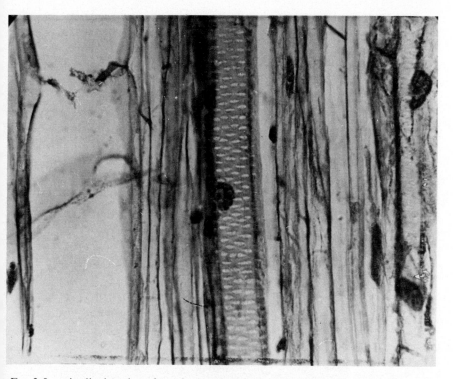

Fig. 3. Longitudinal section of a maize root in the 9·5–10·0 cm zone. An early metaxylem element with an intact nucleus may be seen in a central position. A large immature vessel may be seen on the left. × 500.

at this stage are imperfect, in agreement with the observations of Burley *et al.* (1970).

We consider these results inconclusive simply because we have not been able, at this stage, to account for each vessel at each level and are inclined to the view that only by electron microscopy of whole stelar cross-sections can it be determined where the protoplast is intact and where it has broken down.

C. Electron Microscopy

Material for electron microscopy was obtained by cutting out small pieces of tissue and placing them in ice-cold 6% glutaraldehyde in 0·2 M phosphate buffer, pH 7·0. The solution containing the specimens was subjected to vacuum for 5–10 min, then left on ice for 90 min. Following a 5 min wash with buffer the tissues were placed in 1% osmium tetroxide, buffered at pH 7·0, for 90 min and, following dehydration in ethanol propylene oxide, embedded in Araldite. Sections, 600–900 Å thick, were stained 5–10 min with 1% aqueous $Ba(MnO_4)_2$ and then examined in a Zeiss EM 9A electron microscope.

The electron microscope observations were initially undertaken to confirm the report of Anderson and House (1967) that some vessels retained cytoplasm up to the 10 cm level. This was confirmed for all of the late metaxylem of the inner ring of large vessels (Davis, 1968), and it now appears that these elements may not become perforate below 28 cm (Burley *et al.*, 1970). (No doubt this is provisional since in slowly growing roots full differentiation would proceed closer to the apex.) At that time we failed to realize the extent to which cytoplasm persisted in other metaxylem vessels. Consequently all that we can say at present is that some membranes persist in some of these vessels

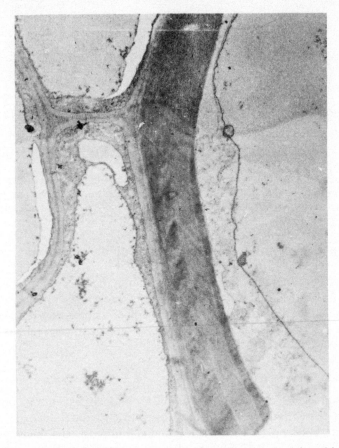

FIG. 4. Electron micrograph of an early metaxylem element, on right; although not clearly visible intact mitochondria were observed. The section was made at the 10 cm level. × 8700.

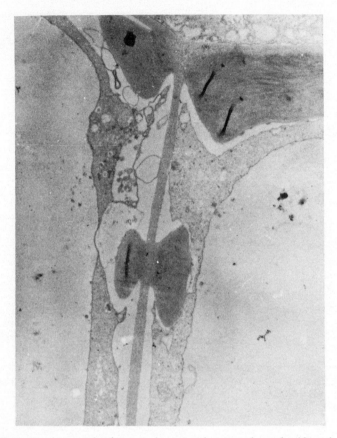

FIG. 5. Electron micrograph of two early metaxylem vessels at the 10 cm level. Both plasmalemma and tonoplast are intact. × 8800.

at the 10 cm level (Fig. 4). We believe that excision of stelar tissue in our preparations involved cutting open long vessels with consequent poor preservation of the cytoplasm; better methods will be used in future work. The failure of others to observe protoplasm at higher levels is probably due to the same cause.

Figures 4–6 of the electron micrographs clearly show the persistence of protoplasm in early metaxylem vessels at the 10 cm level. Figure 7 shows two large, late metaxylem vessels with protoplasm.

IV. Ink Perfusion Experiments

Since the microscopic studies had convinced us that cytoplasm lined most vessel walls in the 5–10 cm zone the next step was to ascertain whether

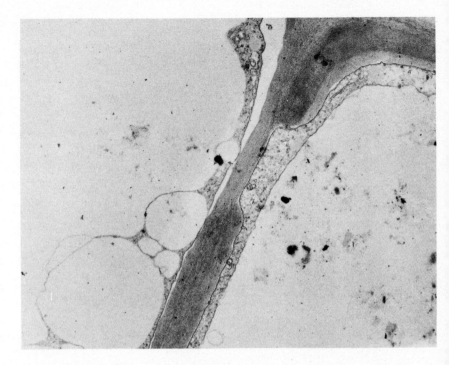

FIG. 6. Electron micrograph of a vessel, on right, at the 10 cm level. It is believed to be early metaxylem but was not clearly identified. Intact cell membranes and mitochondria are evident. × 11,400.

the perforation plates were developed at this level. Roots were excised under water at 5, 10, 15, and 20 cm from the apex. To the cut stump of the basal portion of the root a small polyethylene tube 2–3 cm long was sealed with a lanolin–paraffin mixture (95/5 w/w) and filled with a 1/50 dilution of India ink. Bathing the root in 0·3 M sucrose solution causes a rapid movement of ink down the stele. The cut end of the root to which the shoot is attached may be dipped in ink resulting in perfusion of the stele in the upward direction by transpiration pull. The steles were studied microscopically by the maceration technique described above.

The results are shown in Table II. At the 5 cm level only the outermost protoxylem elements showed extensive perfusion. At the 10 cm level and above, both upward and downward movement was in the early metaxylem since in the roots the large late-maturing metaxylem vessels were never perfused. The downward movement indicates that perforations have developed in the 5–10 cm region, since the average element length is less than 1 mm but perfusion extends from 3 cm downwards.

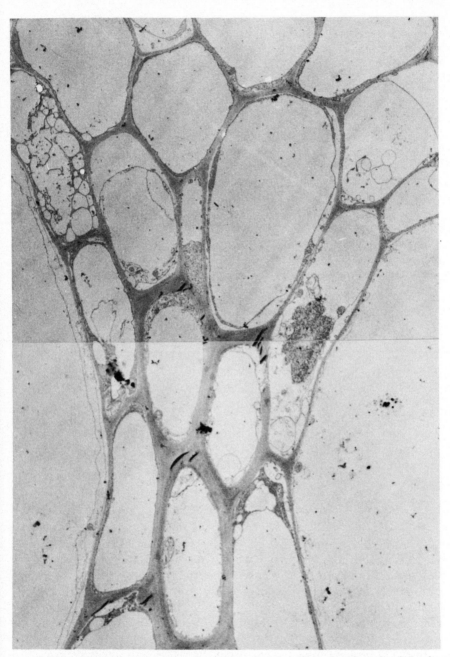

Fig. 7. Electron micrograph of two late metaxylem vessels at the 10 cm level. In the vessel on the left, intact plasmalemma and tonoplast are visible; mitochondria persist in each. × 3600.

TABLE II. Distance of perfusion by India ink upward and downward in maize roots of seven day seedlings. These represent the four longest perfusions in each of four roots

Level of cut	5 cm[a]	10 cm	15 cm	20 cm
Upward				
Range, cm	0·4–1·2	1·2–21·0	4·8–6·4	3·1–6·3
Average, cm	0·72	8·0	5·6	4·3
Downward				
Range, cm	0·4–1·3	0·9–4·3	1·5–3·2	3·2–7·0
Average, cm	0·7	3·1	2·1	5·3
Average root length, cm	25	23	22	25

[a] Protoxylem vessels only.

It will be recalled that anatomical observations of partially macerated flattened stele microscopic sections suggested, assuming that vessel ends were correctly identified, that a single vessel consists of 8–10 elements averaging 0·82 mm and thus the vessel would be somewhat less than 1 cm long. Since India ink is a suspension of fine carbon particles it cannot pass through pits but does move rapidly through perforation plates. It seems clear that in corn root the vessels are either much longer than 1 cm or that abutting vessel end walls become perforated. The latter possibility is not unlikely since perfusion of air through some stems has been shown to extend over several metres (Greenidge, 1952). F. M. Scott (private communication) has informed us that in *Ricinus* stems the vessels extend the full length of an internode.

It is believed that much of the perfusion observed here at the 5–10 cm zone is in vessels containing cytoplasm; it is likely, of course, that cutting and perfusion disrupts the cytoplasm. One likely consequence is that protoplasmic occlusions would increase the resistance to perfusion and to electrical current.

V. Interpretation

It has been established that all but the protoxylem vessels retain protoplasts well above the 5 cm level where the evidence from perfusion shows that perforation plates are developed; thus the protoplast persists for several centimetres after perforations of cross walls. It is in this region that translocation begins (Burley *et al.*, 1970; Kramer, 1969) and root pressure may reach its maximum here. The results are in accord with the test-tube hypothesis of Hylmo (1953).

In radial profile across the corn root the cell electropotentials, p.d.s, are about 100 mV negative when the bathing solution contains 1 mM K (Davis, 1972; Dunlop and Bowling, 1971). The p.d. of the epidermis is slightly lower than other cells, but the endodermis and stelar tissue are similar to the cortex. The uniformity of the p.d.s is in accord with the concept of the symplast; it also suggests that the endodermis may not constitute a secretory layer and that stelar tissue is not leaky. It is likely that intact developing vessel elements have a similar electropotential and a turgor pressure of about 7 atm. However with the dissolution of the last cellulosic microfibrils of the cross walls destined to become perforation plates, the vacuolar pressure is released upward as an exudation pressure. If the protoplasm persisting in the vessel elements with perforation plates continues to pump ions, the maximum exudation pressure which could develop would be that found in an intact cell. The largest root exudation pressure reported is 6 bars in tomato roots (White, 1938).

According to our interpretation the electropotential of the root exudate is that of a long cell which becomes short-circuited where the membranes become disrupted. By this view the drop in p.d. of the exudate compared to intact cells then should constitute a measure of the resistance through the cell wall pathway. However K outward diffusion via wall material having cation exchange properties could tend to augment the p.d. of the exudate.

As vessels lose their protoplasm the inner walls become lined with suberin in the plants carefully studied so far (Scott, 1965) and this would increase the resistance to lateral solute movement. Parenchyma cell surfaces also become suberized. It is not known whether the plasmodesmata become plugged or whether they could continue to serve as a pathway into the symplast. The endodermis walls become fully thickened at about the 20 cm level (Burley et al., 1970) and above this point could constitute an effective barrier tending to retain solutes in the stele. Unfortunately, so far as we are aware, all the studies to evaluate the functioning of the endodermis as a barrier have been with roots below this level, although, perhaps, the Casparian strip may have been detectable.

One critical test of the model supported here, Hylmo's test-tube model, would be to determine whether living vessels can exude. Smith (1970) has demonstrated that maize segments at 3·5 and 5·0 cm levels show appreciable pressure exudation; also inverted segments exuded although at about one-third the rate. Certainly our observations suggest that all of the metaxylem vessels contain protoplasm at this level but perhaps not the protoxylem vessels. It might be argued that all of the exudate arises from the protoxylem, but this seems unlikely to us; we construe these results as being entirely consistent with the test-tube hypothesis.

VI. Summary

Most of the current theories on transport into the xylem are based on the premise that there is a transfer into dead elements (Crafts and Broyer, 1938;

Laties and Budd, 1964; Läuchli and Epstein, 1971). However, various studies (Scott, 1949, 1965; Anderson and House, 1967) have indicated that in the region of greater absorption (about 5 cm from the apex) many or all of the metaxylem elements retain living protoplasm lining the walls. This has been confirmed for maize roots in our own studies by light and electron microscopy up to the 10 cm level. These observations suggest that radial transfer into the xylem is via living protoplasm and that root pressure exudate represents an extension of vacuoles much in accord with Hylmo's (1953) proposal. The anatomical and physiological evidence for this view is presented. The critical observation is the occurrence of protoplasm with intact membranes and nuclei in metaxylem vessels in the 9–10 cm zone, a level at which vigorous root pressure exudation occurs.

Acknowledgements

This work was supported in parts by Grants GB5117X and GB19201 to N.H. and GB29615X to L.K.S. from the National Science Foundation. We are also grateful to Mr. Robert Murray for preparing some of the microscope slides.

References

ANDERSON, W. P. and HOUSE, C. R. (1967). *J. exp. Bot.* **18**, 544–555.
BURLEY, J. W. A., NWOKE, F. I. O., LEISTER, G. L. and POPHAM, R. A. (1970). *Am. J. Bot.* **57**, 504–511.
CRAFTS, A. S. and BROYER, T. C. (1938). *Am. J. Bot.* **25**, 529–535.
DAVIS, R. F. (1968). Ph.D. Thesis, Washington State University, Pullman, U.S.A.
DAVIS, R. F. (1972). *Pl. Physiol.* **49**, 451–452.
DAVIS, R. F. and HIGINBOTHAM, N. (1969). *Pl. Physiol.* **44**, 1383–1392.
DUNLOP, J. and BOWLING, D. J. F. (1971). *J. exp. Bot.* **22**, 434–444.
GREENIDGE, K. N. H. (1952). *Am. J. Bot.* **39**, 570–574.
HAYWARD, H. E., BLAIR, W. M. and SKALING, P. E. (1942). *Bot. Gaz.* 104, 152–160.
HYLMO, B. (1953). *Physiol. Pl.* **6**, 333–405.
KRAMER, P. J. (1969). "Plant and Soil–Water Relationships". McGraw-Hill, New York.
LATIES, G. G. and BUDD, K. (1964). *Proc. natn. Acad. Sci. U.S.A.* **52**, 462–469.
LÄUCHLI, A. and EPSTEIN, E. (1971). *Pl. Physiol.* **48**, 111–117.
LÄUCHLI, A., SPURR, A. R. and EPSTEIN, E. (1971). *Pl. Physiol.* **48**, 118–124.
SCOTT, F. M. (1949). *Bot. Gaz.* **110**, 492–495.
SCOTT, F. M. (1965). *In* "Ecology of Soil-borne Pathogens" pp. 145–153. (K. F. Baker and W. C. Snyder, eds). University of California Press, Berkeley.
SMITH, R. C. (1970). *Pl. Physiol.* **45**, 571–575.
YU, G. H. and KRAMER, P. J. (1967). *Pl. Physiol.* **42**, 985–990.
WHITE, P. R. (1938). *Am. J. Bot.* **25**, 223–227.

Discussion

First, Thellier made a statement on the use of the hyperbolic model of ion uptake. There are many assumptions underlying the Michaelian model. It must be a steady-state reaction, with only one enzyme, only one active centre, only one substrate. Using this model, one can of course characterize enzyme kinetics in terms of two parameters K_m and V_{max}. Under properly constituted conditions as in the assumptions, K_m and V_{max} have properly constituted significance. Thus Michaelian kinetics serve two useful functions; firstly they summarize data from experiment, and secondly they allow interpretation of two important parameters. It may be possible to draw other curves than hyperbolae through experimental points, thus generating parameters with which results can also be compared. This is useful and it would not lead to the temptation of attributing mechanistic significance to these parameters. In some ways it is unfortunate that the data are normally described by hyperbolae because too much may be read into the supposed significance of K_m and V_{max}. Pitman responded to Thellier's statement; he said we must remember that ATP or some other high energy compound is also involved. It is an assumption of much of the kinetic work that the metabolic contribution is constant throughout.

Van Steveninck then produced evidence on the effect of ABA which lends support to some of Pitman's and Cram's contentions. In freshly sliced beet disks bathed in 0·5 mM KCl/0·5 mM NaCl with ABA added, there is marked stimulation of Cl^- uptake over many hours; Na^+ uptake is also stimulated, but K^+ uptake is inhibited. The response time to ABA in root tissue is about 2 h, while in guard cells it is known to be of a few minutes only. In beet disks selective for K^+, uptake of Na^+ is stimulated by ABA while K^+ uptake is inhibited. As soon as all available Na^+ is used up in the external solution, K^+ uptake is strongly stimulated in turn. Thus, uptake into the vacuoles in beet tissue is stimulated by ABA, and ion transport to the shoot may be reduced. Lüttge then asked Pitman and Cram if they had any quantitative information on the possibilities of transport from root to shoot in regulation of ion transport. They had suggested shoot to root movement of ions, of sugars, of hormones perhaps as a regulatory mechanism. Lüttge asked for experimental information, in order to provide a little more on phloem transport feedback than just the idea, but Cram gave the short answer that there is no such information. Pitman said, "The only thing we know is that there is an enhanced ion uptake by a root following a sugar pulse."

Pallaghy then described what he considers to be a positive feedback mechanism for ion transport in guard cells. An enzyme involved in the CO_2 response of guard cells—the stomata open at high CO_2 levels—is PEP carboxylase. PEP carboxylase is known to be inhibited by high levels of Na^+ and K^+. Thus, there is a positive feedback, because high levels of K^+ in the guard cells inhibit the PEP carboxylase which in turn further increases K^+ in the guard cells. Cram asked Pallaghy, "Why, then, doesn't the stoma blow up?" Pallaghy replied that there are other feedback controls; for example, permeability may increase so that an equilibrium value between 500–800 mM K^+ is established.

Dainty then corrected Pitman's implication that Gutnecht's work in *Valonia* involved osmotic pressure variations. It was hydrostatic pressure that Gutnecht found to influence ion transport. Dainty continued that turgor pressure should be more carefully considered as a means of regulating ion transport in higher plants. Cram said he thought the evidence available in barley roots was that Cl^- transport was not affected by turgor.

Raven then made a plea that metabolic terms were considered in interpreting kinetic experiments. He said that there is little difference in tissue ATP levels in light and dark, so we may have to assume that there is a change in enzyme activity induced by light–dark switching. Light may affect the amounts of ATP consumed rather than the amounts of ATP available. It is unlikely to be an effect on levels of ATP, although possibly by an allosteric effect there could be magnification of very small changes in ATP level. He then made a specific comment on a statement by Pitman and Cram that in shady habitats ion transport was limited by light. He said that growth as measured by carbon fixation is more limited. Thus the ion transport to carbon fixation ratio is in fact increased. It is misleading to look at ion transport effects by themselves in light–dark situations.

Davis asked Bowling why he did not consider the electric potential between exudate and stelar tissue rather than the potential from exudate to bathing medium. Bowling said that, on the basis of the measured electrochemical gradients, active transport would go from xylem to stelar cells. Spanswick then said he was sceptical about Higinbotham's suggestion of open-ended xylem vessels. He asked three questions: first, what happens to the cytoplasm during transpiration? Second, can you tell the difference between living cytoplasm and disintegrating cytoplasm? Third, does Davis's evidence really mean that the vessels are closed or does it mean that the longitudinal resistance of the root segment is large in relation to its radial resistance? Higinbotham replied by stressing the difficulty of cytological preparation of xylem vessels. In many segments the xylem vessel is cut open before it is in fact fixed. In reply to Spanswick's first point he said that, during transpiration, water permeability through the cytoplasm layer of the xylem is adequate to maintain the radial water flow into the vessels without the cytoplasm being torn apart. Higinbotham replied to the second question that there is a great gap in our knowledge of what happens to the disintegration products during xylem vessel ontogeny; many of the xylem vessels in maize root have intact nuclei and intact cytoplasm at many centimetres from the tip. This length includes the region normally considered to be the region of maximum ion absorption. To the third point Higinbotham said, if the elements are dead and filled with exudate fluid, they will act as open conduits, and will therefore have a high conductivity, and in fact we find this when the segments are very short. Davis then gave further details; a 5 mm segment taken 25 cm from the tip has a longitudinal resistance of 46 kΩ. A 5 mm segment 10 cm from the tip has a resistance of 192 kΩ. At 5 cm from the tip, a 5 mm segment has a resistance of 172 kΩ. These segments are measured after coating the external surface and the current is driven along the root axis. Bowling then said that he has measured electric potentials in the protoxylem compared to the exudate; thus it seems that appreciable potential gradients can develop within the dead protoxylem elements.

Läuchli asked Bowling what difference would be made to his conclusions if the measurements referred to the cytoplasm of the cells and not to the vacuoles. We assume that lateral transport occurs in the symplasm. Läuchli then asked a question of Higinbotham. He said that his information was that the xylem parenchyma contained high K^+ concentrations located principally in the cytoplasm. The ontogeny seems complicated and development may be occurring over many centimetres in the same fashion as Higinbotham claims for the xylem vessels. Is it possible that perhaps these parenchyma cells later develop into xylem vessels and release their high K^+ and presumably Cl^- content into the exudation stream?

Bowling replied first; he said that our information is that the tonoplast potential in higher plant cells is small so that his estimates of potential apply well to the cytoplasm. As regards the K^+ activity, it may well be different but he believes the shape of the K^+ activities profile to be similar for cytoplasm to what he has measured for the vacuoles. Higinbotham said in reply to Läuchli that he had no information on possible interrelations between xylem parenchyma and vessels. Higinbotham then corrected Bowling's earlier assertion that the protoxylem was dead, although ink had been found in protoxylem in ink perfusion experiments. If protoxylem contains living cytoplasm, there is no reason to be surprised that potential gradients can be found there.

Jeschke then asked if the xylem vessels with living protoplasts are functional, if the xylem protoplasts are connected with the rest of the symplasm, and if one could simply imagine a leakage from the symplasm to supply the exudate stream. Higinbotham replied that Flora Murray Scott has shown many plasmodesmata connecting the parenchyma with the xylem vessels, and that presumably the xylem protoplasm is part of the symplasm. As to the leakage, he said we must remember there will be a pressure drop into the xylem lumen and that solution flow may occur. However, ions supplied through the free space perhaps, may be pumped across the xylem membranes in the normal fashion. Higinbotham said further that the possibility of active transport across the plasmodesmata should not be too lightly dismissed. Were we to find significant potential differences in different layers of the cortex, we would then have to assume that some ionic regulation occurred within the symplasm, presumably at the plasmodesmata.

MacRobbie then referred to unpublished work by Bowen which suggests there are longitudinal symplastic connections of some importance in the root cortex. Bowling then reported his recent experiments on the effect of transpirational flow on trans-root potential, which becomes more negative as transpirational flow increases. Measurement of a root epidermal cell potential in the same experiments shows it too goes more negative, but with a lag of two minutes. Bowling claimed this lag indicates ion movement from one vacuole to the next along the root cortex, with a rate of $3 \, cm \, h^{-1}$, the rate of cytoplasmic streaming.

Pitman then spoke to the one pump–two pump proposal. He said to Bowling, "Is it fair comment that you believe there is one pump operating?" Bowling said, "Yes, the active step is uptake into the cortical cytoplasm." Pitman then suggested there is a conflict of evidence from potential measurements and inhibitor studies. Barley root segments loaded with ^{36}Cl were allowed to exude and the ^{36}Cl efflux was collected from both the xylem exudate and into the bathing medium. On the addition of CCCP, the xylem flux of ^{36}Cl was much reduced, while the flux through the whole surface of the root was increased. The total efflux remained approximately constant. The contention is that CCCP not only stops tracer entering the root, but also stops it entering the xylem. One might argue that leakage into the xylem is stopped simply because uptake into the symplasm at the exterior site is stopped, but Pitman believes this point can be negated by the fact that zero external Cl^- concentrations, which also reduce symplasm uptake to zero, do not affect the xylem flux. Thus two pumps are required; one at the external root surface and one at the xylem. This latter pump may be located at the xylem vessel protoplasm or at the xylem parenchyma cells.

VII.4

The Radial Transport of Ions in Maize Roots

D. A. Baker

School of Biological Sciences, University of Sussex
England

I. Introduction

The nature of the process by which ions are transported from the surrounding medium to the xylem of plant roots has been the subject of controversy for a number of years (see Epstein, 1972). Most modern researchers subscribe to the view that ions which are accumulated in the cytoplasm of the cortical cells diffuse centripetally through the symplasmic continuity to the stele where they are released into the xylem vessels. The back diffusion of ions from stele to cortex through the cell walls is thought to be prevented by the endodermis (see Baker, 1971). The major contention is whether the release of ions within the stele is a passive leakage or an active secretory pumping process.

Passive leakage of ions within the stele was an important premise of the hypothesis formulated by Crafts and Broyer (1938) and experimental support for this has been provided by the observation that steles freshly isolated from maize roots are leaky and only a limited uptake of ions occurs (Laties and Budd, 1964; Lüttge and Laties, 1966, 1967). The observation that such steles acquire the capacity to accumulate ions upon ageing has been found to reflect contamination by micro-organisms and not the development of an ion uptake capacity by the stelar tissue *per se* (Hall *et al.*, 1971). The postulated leaky nature of the stelar tissue was attributed by Crafts and Broyer to a diminished metabolism within the centre of the root caused by low oxygen and high carbon dioxide tensions. Fiscus and Kramer (1970) have provided evidence that the stelar tissues of roots operate at oxygen tensions which are suboptimal for respiration while freshly isolated stelar tissue has been shown to have a lower oxygen uptake than cortical tissue (Hall *et al.*, 1971), results which are consistent with the Crafts–Broyer hypothesis.

511

The observations and conclusions outlined in the preceding paragraph are contradicted in a number of reports in which data are presented which do not support the Crafts–Broyer concept. Yu and Kramer (1967) report that the stele is as active as the cortex for the accumulation of ^{32}P, measured either when the stele is isolated prior to or isolated following an absorption period. In a further study (Yu and Kramer, 1969) they obtained similar results for ^{36}Cl and ^{86}Rb, but only when accumulation preceded separation of the stele and cortex. In some instances the uptake by the stele was greater than that for the cortical tissue. These workers also reported a consistently higher oxygen uptake by isolated stele than by isolated cortex and a marked dehydrogenase activity in the stele, results which suggest that the stele is metabolically as active as, or more active than, the cortex.

Further evidence in support of an active role for the stelar tissues in radial transport has been inferred from studies using the electron probe analyser to reveal the distribution of various elements in root sections (Läuchli, 1967, 1968; Läuchli et al., 1971). These studies indicate that potassium and phosphorus are present at a higher concentration in the stelar parenchyma cells than in the cortical cells of the root.

In the present report experiments are described on the capacity of steles, both freshly isolated and in situ, to accumulate ^{86}Rb. The pattern of release of this ion from stelar and cortical tissues has also been investigated in preloaded freshly isolated material.

II. Materials and Methods

Seeds of maize (Zea mays L. Kelvedon 33) were germinated and grown for 6 days over an aerated solution of 0·1 mM $CaCl_2$. Uptake measurements were performed on 8 cm lengths of primary root which were excised and placed overnight in a large volume of aerated 0·1 mM $CaCl_2$ solution at 25°C. Roots were then either attached to 100 μl micropipettes, after re-cutting 0·5 cm from the end, or separated and excised into 2 cm lengths of isolated stele and isolated cortex. It has been previously verified that the break between stele and cortex occurs at the endodermis in maize roots (Hall et al., 1971) as shown in Fig. 1. In all cases samples were taken from the region 2–8 cm from the root tip, a zone which in these roots at this stage of development is devoid of root hairs and lateral root primordia. The material was divided into 3 replicate samples per determination, each sample composed of five 2 cm pieces of isolated stele. The material was then placed into an aerated solution of 0·2 mM KCl + 0·1 mM $CaCl_2$ labelled with 0·05 μCi ml^{-1} ^{86}RbCl at 25°C. After a measured uptake period the exuding roots were rinsed for 30 s in unlabelled uptake solution, rapidly separated and excised into similar 2 cm lengths of isolated stele and isolated cortex. All material was subsequently washed for 30 min with unlabelled 20 mM KCl + 0·1 mM

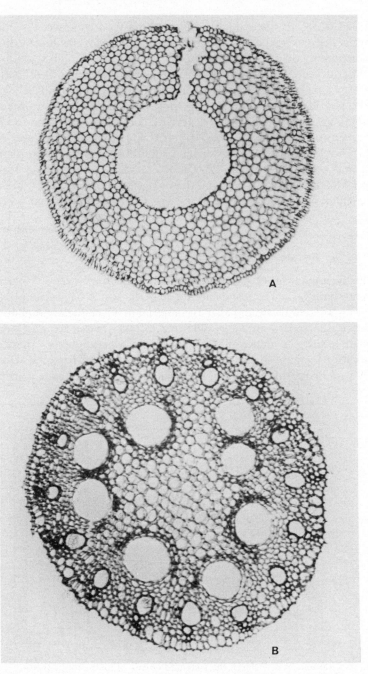

FIG. 1. Transverse frozen sections of isolated cortex (A) and isolated stele (B) from maize roots.

$CaCl_2$ at 0–2°C to remove labelled ions from the free space. The material was then dried, weighed and arranged on planchets for the direct estimation of radioactivity on a gas flow counter.

Efflux measurements were made on material which had been preloaded with the same uptake solution labelled with 0·5 μCi ml^{-1} ^{86}RbCl in a 3 h uptake period at 25°C. After a 30 s rinse the material was transferred to 10 ml of an unlabelled solution at 10°C or 25°C for 10 min and then subsequently to other washing solutions for further 10, 30 and 60 min washes. 1 ml aliquots of these washing solutions were placed on planchets, dried and counted on a gas flow counter. Measurements were made on material loaded as intact roots and then separated into stele and cortex prior to the efflux period. At the end of the efflux period the material was dried, weighed and arranged on planchets for gas flow counting.

III. Results

The accumulation of ^{86}Rb by the stele and cortex over a 5 h period is presented in Fig. 2. The results are for material which has accumulated this ion as an isolated tissue and also for material in which the accumulation has preceded the separation into stele and cortex. These are referred to as isolated tissue and intact tissue respectively. For all tissues there was a continuous increase in the uptake of ^{86}Rb over the 5 h period. However, there were major differences between isolated and intact tissue, particularly

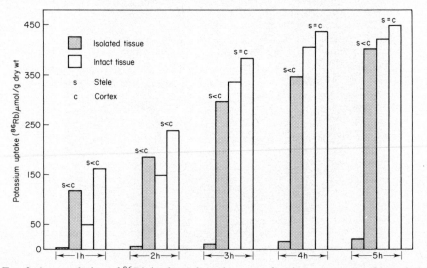

FIG. 2. Accumulation of ^{86}Rb in the stele and cortex of maize roots over a 5 h period. Uptake was measured on isolated tissue and also when uptake preceded isolation. Differences are significant at the 5% level and are indicated as s < c or s = c.

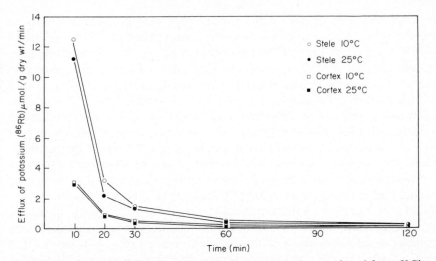

FIG. 3. The efflux of ^{86}Rb from preloaded isolated stele and cortex into 0·2 mM KCl + 0·1 mM CaCl$_2$ at 10°C and 25°C.

in the capacity of the stelar tissues to accumulate ions. The isolated steles had an extremely low level of accumulation when compared with the steles accumulating ions in the intact tissue. Also the level of ^{86}Rb in isolated stelar tissue was considerably lower than that in isolated cortex. This difference between stele and cortex in their ability to accumulate ions was less marked in the intact tissue and after 3 h the difference in the level of ^{86}Rb in these two tissues was not significant at the 5% level of probability.

The efflux of ^{86}Rb from stelar and cortical tissue which had been isolated following a 3 h uptake period by intact roots is presented in Fig. 3. Over a 2 h period the pattern of efflux from the stelar tissues was consistently lower at 25°C than at 10°C. Similarly the pattern of efflux from cortical tissue was marginally lower at the higher temperature. The large initial rapid efflux of ^{86}Rb from the free space of the tissues is complete after the first 20–30 minutes and is followed by a much smaller, steadier efflux. Both of these phases are of a greater magnitude in the stelar than in the cortical tissue, a demonstration of both the greater free space of the stelar tissue and the larger efflux of ions from the inner or non-free space compartments of the stelar tissues.

IV. Discussion

This investigation was instigated as a result of the conflicting evidence in the literature as to the ability of the stele to accumulate and release ions. The results obtained here indicate that whereas isolated steles have a low capacity for ion uptake, steles in the intact root can readily accumulate ions to a level as high as that in the cortex. This low ion uptake capacity of isolated steles is in agreement with the observations of Laties and Budd (1964), Lüttge and Laties (1967) and Hall *et al.* (1971). The relatively high uptake capacity of steles in the intact root reported here is consistent with the observations of Yu and Kramer (1969) and Läuchli *et al.* (1971).

This marked difference in the behaviour of isolated and *in situ* steles is open to a number of interpretations. Läuchli *et al.* have suggested that the isolation of steles may lead to structural and physiological alterations and that the low uptake capacity of the steles is a reflection of this damage. However, this interpretation would require a similar situation to apply in isolated cortical tissue. There is evidence in the present work that the uptake capacity of isolated cortex is slightly lower than that of the cortex of the intact root, but after 5 h this difference is only about 10 % whereas the difference in the stelar uptake is nearly 2000 %. It therefore seems unlikely that damage is the major cause of this reduced uptake capacity of the isolated stelar tissue. Another possibility is that the plasmalemma of the stelar tissues is unable to maintain ion uptake while the tonoplast can transport ions into the vacuole. Thus in the intact root ions can be transported into the stele via the symplasm and these ions can be accumulated in stelar vacuoles or released across the plasmalemma into the stelar free space. This interpretation is consistent with both the present results and with the seemingly conflicting results in the literature referred to above.

It has been suggested (Läuchli *et al.*, 1971) that excised steles may be active in ion transport. In support of this the results of Anderson and Reilly (1968), obtained with shaved roots, are quoted. However the shaved roots had only the outer cortical cells removed and the inner cortex, endodermis and stele were all intact, whereas when steles are isolated as in the present study the endodermis is ruptured and there is no cortical tissue attached. It is therefore misleading to attempt comparisons of the behaviour of such different materials.

The results obtained here for the efflux of ^{86}Rb from stelar and cortical tissues do not show any evidence of an active component. On the contrary they support a passive release of ions from the stele. The stimulation of the efflux at 10°C is attributed to an increased passive efflux at this lower temperature. As the active uptake system is inhibited so the net passive efflux is stimulated. When an active efflux pump is present low temperature inhibits the efflux of ions, as demonstrated for sodium efflux from barley roots (Nassery and Baker, 1972).

The conclusions that may be drawn from the present observations are that accumulation of ions by steles of intact roots is not evidence for an active efflux pump located at the plasmalemma as proposed by Läuchli *et al.* (1971). The anomalous behaviour of isolated steles in that they do not accumulate ions suggests that the plasmalemma of stelar parenchyma is not functional in ion accumulation and that only vacuolar uptake is reflected in the intact system. The eflux of ions from the stele is thought to be a passive leakage of ions across the plasmalemma of the stelar parenchyma cells. Thus ions which are accumulated by the cortical cells are transported centripetally across the cortex, possibly aided by cytoplasmic streaming (Baker, 1968), and into the stelar tissues. These ions may be accumulated within the stelar vacuoles or may leak passively into the free space of the stele. Experiments with metabolic inhibitors are in progress to characterize further the nature of the ion release from stelar tissues.

Acknowledgements

I wish to thank the Agricultural Research Council for financial support for this investigation and Miss C. Munck for skilled technical assistance.

References

ANDERSON, W. P. and REILLY, E. J. (1968). *J. exp. Bot.* **19**, 19–30.
BAKER, D. A. (1968). *Planta (Berl.)* **83**, 390–392.
BAKER, D. A. (1971). *Planta (Berl.)* **98**, 285–293.
CRAFTS, A. S. and BROYER, T. C. (1938). *Am. J. Bot.* **25**, 529–535.
EPSTEIN, E. (1972). "Mineral Nutrition of Plants: Principles and Perspectives". Wiley, New York.
FISCUS, E. L. and KRAMER, P. J. (1970). *Pl. Physiol., Lancaster* **45**, 667–669.
HALL, J. L., SEXTON, R. and BAKER, D. A. (1971). *Planta (Berl.)* **96**, 54–61.
LATIES, G. G. and BUDD, K. (1964). *Proc. natn. Acad. Sci. U.S.A.* **52**, 462–469.
LÄUCHLI, A. (1967). *Planta (Berl.)* **75**, 185–206.
LÄUCHLI, A. (1968). *Vortr. Gesamtgeb. Bot. N.F.* **2**, 58–65.
LÄUCHLI. A., SPURR, A. R. and EPSTEIN, E. (1971). *Pl. Physiol., Lancaster* **48**, 118–124.
LÜTTGE, U. and LATIES, G. G. (1966). *Pl. Physiol., Lancaster* **41**, 1531–1539.
LÜTTGE, U. and LATIES, G. G. (1967). *Planta (Berl.)* **74**, 173–187.
NASSERY, H. and BAKER, D. A. (1972). *Ann. Bot.* **36**, 881–887.
YU, G. H. and KRAMER, P. J. (1967). *Pl. Physiol., Lancaster* **42**, 985–990.
YU, G. H. and KRAMER, P. J. (1969). *Pl. Physiol., Lancaster* **44**, 1095–1100.

VII.5

Potassium and Chloride Accumulation and Transport by Excised Maize Roots of Different Salt Status*

A. Meiri

*Volcani Center
Bet Dagan, Israel*

I. Introduction

Ion uptake by roots is divided into accumulation in the root tissue and transport to the xylem. Accumulation in tissue reaches saturation level within 24 h (Hiatt, 1969; Johansen *et al.*, 1970), while transport to the xylem and to the shoot (Pitman, 1971) or in the xylem exudate of excised roots, continues at a steady rate for a longer time, after an initial period of adjustment (House and Findlay, 1966). The relationship between accumulation and transport will depend on the root's salt status (Hodges and Vaadia, 1964) and on root characteristics (Smith, 1970). The use of excised and exuding roots enables differentiation between the accumulation and transport processes (Greenway, 1967; Pitman, 1971).

This work investigated the salt accumulation and exudation by maize roots grown at different concentrations for 3–4 days and exposed to exudation

* Contribution from the Agricultural Research Organization, The Volcani Center, Bet Dagan, Israel. 1972 Series, No. 2120-E.

media of different salinity levels for a period of one day. Concentrations up
to 50 mM were tested to investigate the effect of high salt levels.

II. Materials and Methods

Seeds of *Zea mays* (var. White Horse Tooth) were rinsed with large volumes
of tap water and finally washed with distilled water. They were then set out
to germinate on filter paper in covered trays and kept in the dark at 25°C $\pm$ 1°.
The paper's margins were dipped in water continuously to maintain its
moisture. After two days, the germinating seeds with roots approximately
1 cm in length were arranged on nylon netting so that the roots were in 10 l
containers of solutions of different concentrations. The solutions were
aerated continuously and the seedlings were allowed to grow for a further
2–4 days under the same conditions.

Three types of experiments were carried out:

(a) In one experiment, root elongation, chloride and potassium accumulation in the root tissue were measured for roots grown in solutions of 0·1 mM,
1·0 mM, 10·0 mM, 25·0 mM, and 50·0 mM KCl, with 0·1 mM $CaCl_2$ in each.
Groups of ten seedlings were taken at random at different time intervals
after transfer to these solutions. The primary roots were cut and dipped
for 30 s in 0·01 M HNO_3 to replace absorbed ions on the root surface, and
washed three times by dipping and stirring for 30 s in distilled water. The
roots were then blotted with absorbent paper, their length measured to an
accuracy of 1 mm, and their fresh weight determined to an accuracy of
10^{-4} g. Five roots were oven-dried at 75°C, weighed and extracted by
shaking for 12 h in a 5 ml acid solution (6·4 ml conc. HNO_3 + 100 ml conc.
CH_3COOH + 900 ml H_2O). The extract was analysed for chloride. The
other five roots were wet-digested for potassium analysis (Chapman and
Pratt, 1961).

(b) In a consequent set of five experiments, seedlings were grown in the
KCl–$CaCl_2$ solutions as above, and when most primary roots reached a
length of 100–130 mm, they were cut into lengths of 93–103 mm and sealed
in capillary tubes of 200 mm length and 0·47 mm radius for exudate collection,
as described by Anderson and Reilly (1968). Five bathing media of the five
different concentrations were used in each experiment. In each medium,
14–22 roots were used. The exudate of every root was collected after 27 h,
weighed to an accuracy of 10^{-4} g, and analysed for chloride and potassium.
The exuding roots were washed only in distilled water, their lengths determined, and 3–4 roots were combined to form a composite sample before
the fresh and dry weights were determined. The dry matter was extracted,
and the extract analysed for chloride and potassium. Ion concentrations
obtained by extraction and by wet digestion were similar for similar growth
medium concentrations.

(c) In another experiment, repeated three times, seedlings were grown for 3 days in a 0·1 mM solution, or for 4 days in a 50·0 mM solution. Then the primary roots were excised and arranged for exudation in bathing media of 0·1, 10·0 and 50·0 mM KCl plus 0·1 mM $CaCl_2$, for consecutive time intervals up to 1400 min from the beginning of the exudation period. Roots were divided into groups of 3–4 roots. At each sampling time, roots of one group were collected, dipped into 0·01 M HNO_3, washed, and their lengths and fresh weight determined. These roots were wet-digested and analysed for potassium. Their exudate was combined to form a composite sample. The exudates of the other groups of roots were collected with the aid of a fine-tip micropipette and combined to give a composite sample. The exudate was weighed and analysed for chloride and potassium content. Chloride was determined with a chloride-meter, and potassium with a flame photometer.

III. Results

A. Root Growth and Ion Accumulation

Root elongation during the first day after transfer to solutions of different concentrations was very similar to all concentrations. During the second day, differences in growth rate were found. These differences were small, up to 10 mM, and increased considerably at higher concentrations. Chloride and potassium contents in the roots increased more rapidly during the initial period and more slowly later, and their contents were greater for roots grown at higher concentrations (Table I).

B. Effect of Growth Medium and Exudation Medium Concentrations on Ion Transport and Tissue Ion Content

Mean exudate volumes, for a period of 27 h, of roots grown and exuded at the different concentrations (Experiment Series B) are presented in Table II. Exudate volumes were affected mainly by growth medium concentration. Roots grown at higher concentrations had a smaller exudate volume. A marked effect of exudation medium was found in the 50 mM solution.

Mean exudate concentration, for a period of 27 h, of these roots is presented in Figs 1A and B. Exudate concentration depended mainly on the concentration of the exudation medium. Increase in exudation medium concentration resulted in an increase in concentrations of potassium and chloride. Differences due to growth conditions were small and random. This indicates that the change in exudate concentration after transfer of the roots to the tested exudation medium should be rapid.

Tissue ion content in the root tissue was affected by growth medium or exudation medium concentrations. When transferred to the exudation medium the roots grown at higher concentrations had higher ion contents

TABLE I. Root length and ion content of corn roots after different growth periods in media of different KCl concentrations (length values are means for 10 roots ± S.E.; ion content was measured on composite samples of 5 roots)

Parameter	Root length (mm)			Ion content mM kg⁻¹ fresh weight					
				Chloride			Potassium		
Period, h:	6	26	49	6	26	49	6	26	49
Medium concn, mM KCl[a]									
0.1	20.8 ± 1.7	41.6 ± 3.0	81.3 ± 4.0	4.7	22.3	30.4	77.1	82.7	82.4
1.0	21.4 ± 1.8	47.3 ± 2.7	79.1 ± 4.4	6.0	27.0	34.4	80.0	95.6	94.5
10.0	26.4 ± 4.2	44.0 ± 2.5	77.3 ± 4.5	12.3	38.8	49.5		109.8	111.2
25.0	25.7 ± 3.4	43.5 ± 3.0	64.9 ± 3.6	24.9	50.6	58.7	99.3	117.1	134.1
50.0	21.7 ± 2.8	41.1 ± 2.4	47.2 ± 2.4	36.9	67.6	77.3	112.0	146.4	150.5

[a] 0.1 mM CaCl₂ was added to all media.

TABLE II. Exudate volume as affected by salt levels of the growth and exudation media (Results in μl/root for 27 h ± S.E.)

Exudation medium mM KCl[a]	Growth concentration, mM KCl				
	0.1	1.0	10.0	25.0	50.0
0.1	185 ± 14	163 ± 18	129 ± 9	116 ± 16	85 ± 11
1.0	188 ± 17	182 ± 16	155 ± 14	133 ± 9	74 ± 9
10.0	114 ± 19	177 ± 21	153 ± 12	109 ± 9	68 ± 7
25.0	182 ± 16	134 ± 13	127 ± 10	137 ± 14	53 ± 5
50.0	110 ± 12	98 ± 7	81 ± 6	84 ± 6	45 ± 4

[a] 0.1 mM CaCl₂ was added to all media.

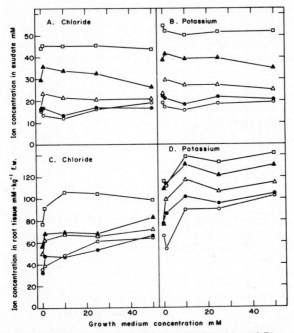

Fig. 1. Potassium and chloride concentration in exudate (A and B) and in root tissue (C and D), as related to growth media concentrations. Exudation media: 0·1 (○), 1·0 (●), 10·0 (△), 25·0 (▲), and 50·0 (□) mM KCl plus 0·1 mM CaCl$_2$. (Exudate—mean for 27 h; tissue—after 27 h of exudation).

(Table I). After 27 h of exudation roots transferred for exudation to a more concentrated medium had higher ion contents and roots transferred to a medium of lower concentration had lower ion contents than the roots which remained at a constant concentration (Fig. 1C and 1D). Therefore, we assume that these roots are at a steady state and their salt content is constant. This assumption was supported by the results reported in Fig. 2. Later on we used the concentration of these roots as a reference value to calculate salt balance in the root tissue.

Ion contents of the roots grown in concentrations of 10–50 mM were similar, and higher than in roots grown in 1 mM or 0·1 mM solutions, for any exudation medium. Therefore, the strong effect of the exudation medium did not overcome the differences due to growth conditions, indicating that either accumulation or depletion in tissue salt content requires a longer period.

Changes with time in root potassium content and exudate potassium and chloride contents were followed for roots grown at low or high external concentration and exuding from different concentrations (Experimental Series C). It was found that roots remaining in a constant concentration show

only small changes in ion content with time (Fig. 2). Roots transferred to a
higher concentration accumulate ions in the tissue, and roots transferred to
a lower concentration lose ions (Fig. 2). Accumulation rate seems to be
faster than loss rate. In roots grown in 0·1 mM solution, the concentration

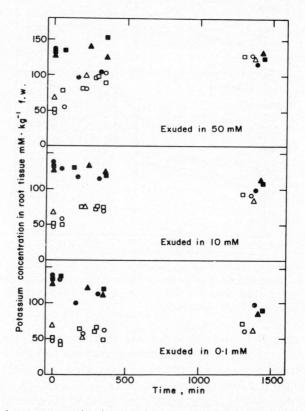

FIG. 2. Potassium concentration in root tissue as related to period of exudation from
media of different KCl concentrations. Open symbols—roots grown at 0·1 mM; closed
symbols—roots grown at 50 mM. Results of experiments 1 (○), 2 (△), and 3 (□).

reached the level of that of roots grown in 50 mM, while in roots grown in
50 mM, the concentration was higher than in roots grown in the 0·1 mM
solution, even after 22 h. Also, the concentration in roots was higher when
grown in 50 mM than in 0·1 mM, when both were exuded from a 10 mM
solution.

Potassium concentration in the exudate (Fig. 3) was about the same after
a short period for the two types of roots, when exuding from a similar
exudation medium. Results for chloride were similar.

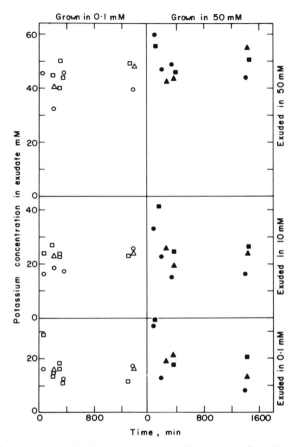

Fig. 3. Potassium concentration in exudate as related to period of exudation from media of different concentrations. Results of experiments 1 ($\bigcirc$), 2 ($\triangle$) and 3 ($\square$).

IV. Discussion

By exudation experiments we were able to study the effect of root salt status and medium concentration on ion accumulation by roots and transport to the xylem.

Transfer of roots to the exudation media of a different concentration caused a drastic change in the concentration gradient from medium to root. Considering the faster change in exudate concentration and the slower change in tissue concentration, the transfer also changes the concentration gradient from tissue to exudate. Transfer of roots from a lower growth medium concentration to a higher concentration of exudation medium reduces these gradients and results in an increase in ion uptake, transport or accumulation; and transfer of roots from a high concentration of growth medium to a low bathing medium which increases the gradients, results in lower rates of transport or accumulation.

Since the mean exudate concentration for 27 h was independent of growth conditions (Fig. 1) and roots grown in higher concentrations had volume fluxes of lower rates (Table II), the product of the two—salt fluxes in the exudate (Fig. 4A and B)—also depended on growth conditions. Roots grown at higher concentrations transported less ions to the exudate from any exudation medium.

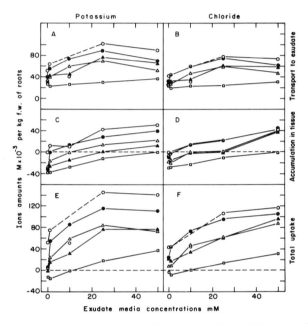

FIG. 4. Potassium and chloride transported to exudate (A and B), accumulated in root tissue (C and D), and total uptake (E and F) during 27 h of exudation for roots grown at different salinity levels as related to exudate medium concentration. Growth media: 0·1 (○), 1·0 (●), 10·0 (△), 25·0 (▲) and 50·0 (□) mM KCl plus 0·1 mM CaCl$_2$.

The change in tissue ion content (Fig. 4C and D), calculated as the difference between measured values after 27 h of exudation and the content of the reference roots show, that in all cases when the gradient from medium to root increased, accompanied by an increase in gradient from root tissue to exudate, there was a negative accumulation in the root tissue. This depletion of root tissue content could be toward the exudation stream (Meiri and Anderson, 1970; Pitman, 1971) or could be leached out to the medium (Pitman, 1971; Cram and Laties, 1971).

Comparing tissue potassium and chloride contents in root tissue or amounts that were depleted from tissue with amounts transported to exudate, indicates that the tissue may be a reservoir for ions for exudation. In roots grown in higher concentrations, that source may be more important as their ion contents were larger and the amounts of ions transported to the exudate were smaller.

Accumulation in the tissue, exudation, and the sum of the two-total uptake (Fig. 4E and F) depended on the concentration gradient from medium to root, and increased with increasing exudation medium concentration. Total uptake of potassium reached saturation at an exudate medium concentration of 25 mM. It seems that for the exudation, saturation was achieved at a somewhat lower concentration, after which it levelled off or was slightly lower (Fig. 4A and B). Chloride exudation was similar to that of potassium, while accumulation did not show complete saturation. Similar results for exudate of low salt roots were reported by House and Findlay (1966).

The shape of the curve is similar for all growth media except for roots grown in 50 mM solution. Roots grown in higher concentration have lower levels of accumulation and transport.

In extreme cases, when depletion of tissue content was greater than total transport to exudate, uptake became negative, indicating a net leaching of ions to the medium.

As indicated by the results, ion accumulation in root tissue seems to be faster than ion depletion. Calculation of the rate of change in ion concentration in the tissue of roots of experiment Series C, as a linear function of time for the first 6 h (Table III), shows clearly the differences in rates between accumulation and depletion after the transfer. Thus, the slow ion depletion of the tissue could contribute to exudation for a long period of time.

TABLE III. Rate of change in potassium tissue content for high and low salt roots during the first six hours of exudation from media of different concentrations. (Results in M $\times$ 10^{-3} . kg^{-1} fresh weight . min^{-1} $\pm$ S.E.)

Exudate medium concn, mM KCl	Growth medium concn, 0·1	mM KCl 50·0
0·1	0·020 ± 0·019	−0·059 ± 0·021
10·0	0·061 ± 0·016	−0·024 ± 0·015
50·0	0·124 ± 0·021	−0·008 ± 0·045

During the transient period, "high-salt" roots transported more ions and "low-salt" roots less ions from the same exudation medium (Table IV, short period). For long periods, ion exudation was greater from the low-salt roots, except for the 0·1 mM medium (Table IV, long period). Likewise, total ion transport during 27 h of exudation (Fig. 4A) was not related to concentration gradient from tissue to exudate. Roots grown in higher

TABLE IV. Potassium transport to exudate by high and low salt roots after short and long periods of exudation from media of different concentrations (results in $M \times 10^{-9}$/root)

Exudation medium concn, mM KCl		0.1		10.0		50.0	
Growth medium concn, mM KCl		0.1	50.0	0.1	50.0	0.1	50.0
Roots exuding for short periods	Expt 1	86 (60)[a]	534 (55)	64 (65)	396 (60)	203 (70)	117 (65)
	Expt 2	453 (210)	526 (235)	324 (215)	755 (240)	343 (220)	430 (245)
	Expt 3	114 (60)	191 (70)	63 (65)	217 (141)	76 (200)	364 (80)
Roots exuding for long periods	Expt 1	1533 (1360)	719 (1390)	2298 (1360)	1246 (1390)	3602 (1360)	1159 (1390)
	Expt 2	1169 (1375)	1258 (1415)	1804 (1375)	1117 (1423)	3371 (1375)	1338 (1428)
	Expt 3	849 (1295)	1676 (1435)	1967 (1300)	1212 (1440)	4102 (1305)	2062 (1445)

[a] Number in parentheses indicates exudation time in minutes.

concentrations and having larger gradients from tissue to exudate, exude ions at a lower rate.

All these pieces of evidence fit nicely into the compartmentation model (Epstein, 1966; Laties, 1969), where the symplast concentration and transport in the symplast will determine the exudation process. The vacuoles may be considered as large reservoirs that accumulate ions or are depleted according to concentration gradients with the cytoplasm. At constant salinity, the ion content of both compartments will be constant and the exudation rate will be constant, as a result of constant rates of influxes and effluxes across the plasmalemma, the tonoplast and toward the xylem elements (Hodges and Vaadia, 1964; Pitman, 1971; Cram and Laties, 1971). A change in external concentration will result in a transient change toward a new steady state with different rates in the system. The symplast which has a low ion content (Pitman, 1971) rapidly reaches a new level determined by the medium concentration. In roots transferred to a higher concentration, the faster influx across the plasmalemma than across the tonoplast (Cram and Laties, 1971) was sufficient to supply ions for accumulation and transport. In roots transferred to lower concentration, faster rates of efflux from symplast than from vacuoles (Cram and Laties, 1971) and the rapid transport of ions to the xylem in the exudation process (Table IV, short period results) (Pitman, 1971), result in the new concentration level. Net efflux from vacuoles continues over long periods at a relatively rapid rate (Table IV) and could contribute large proportions of the exuded ions. In extreme cases, net uptake was negative over the entire exudation period of 27 h (Fig. 4).

After achievement of steady state in the symplast, the transport to xylem became saturated at a certain concentration and not at a certain flux. Roots of very different salt fluxes had similar exudate concentrations. Considering the very small re-absorption of chloride from exudate (Hodges and Vaadia, 1964), this saturation level should be a result of the limited transport of ions to xylem. Reductions in water and ion fluxes were indicated also as the exudation medium effect for roots exuded in 50 mM medium.

Low-salt roots had faster water fluxes than high-salt ones (Table II). These faster fluxes could increase ion fluxes by accelerating symplast transport (Laties, 1969), or by removing the exuded ions in the xylem. As secretion to xylem was limited by exudate concentration, higher rates of water fluxes should result in higher rates of ion transport to the exudate (Broyer and Hogland, 1943; Brouwer, 1965).

The similar behaviour of water and ion fluxes might coincide with root tissue characteristics. Elongation of roots grown at higher salt levels was lower than in lower concentrations (Table I). Therefore, root segments of the same length grown at higher concentrations were older and the root zone of the highest accumulation and transport rates for ions (Smith, 1970) or water (Anderson et al., 1970) was shorter. Other explanations for the reduced uptake and transport rates for roots grown in saline media were reported. Under high salinity, influx rates to the symplast were reduced (Cram and Laties, 1971). Thus, limited ion supply to the symplast may result in a reduction in transport. Accumulation and transport of ions over long periods are active processes (Brouwer, 1965). High-salt roots have lower sugar contents and thus lower energy sources and lower rates of ion uptake (Pitman *et al.*, 1971). The external concentration may have a direct effect on root properties. In 50 mM solution both water and ions transport were reduced. All these factors may operate simultaneously and may interact and, as a result, roots grown under lower salt conditions will have a higher capacity for both accumulation and transport of ions.

Acknowledgements

I wish to thank Dr J. Putter and Dr A. Genizi for their help in the analysis of the data, Mrs Margot Shuali for excellent technical assistance, and Mr A. Mantell and Dr E Shmueli for their critical reading of the paper.

References

ANDERSON, W. P. and REILLY, E. J. (1968). *J. exp. Bot.* **19**, 19–30.
ANDERSON, W. P., AIKMAN, D. P. and MEIRI, A. (1970). *Proc. R. Soc.* B **174**, 445–458.
BROUWER, R. (1965). *A. Rev. Pl. Physiol.* **16**, 241–266.
BROYER, T. C. and HOAGLAND, D. R. (1943). *Am. J. Bot.* **30**, 261–273.
CHAPMAN, H. D. and PRATT, P. F. (1961). "Methods of analysis for soils, plants and waters". University of Calif., Division of Agricultural Science.
CRAM, W. J. and LATIES, G. G. (1971). *Aust. J. biol. Sci.* **24**, 633–646.
EPSTEIN, E. (1966). *Nature, Lond.* **212**, 1324–1327.
GREENWAY, H. (1967). *Physiol. Pl.* **20**, 903–910.
HIATT, A. J. (1969). *Pl. Physiol., Lancaster* **44**, 1528–1532.
HODGES, T. K. and VAADIA, Y. (1964). *Pl. Physiol., Lancaster* **39**, 109–115.
HOUSE, C. R. and FINDLAY, N. (1966). *J. exp. Bot.* **17**, 344–354.
JOHANSEN, C., EDWARDS, D. G. and LONERAGAN, J. F. (1970). *Pl. Physiol., Lancaster* **45**, 601–603.
LATIES, G. G. (1969). *A. Rev. Pl. Physiol.* **20**, 89–116.
MEIRI, A. and ANDERSON, W. P. (1970). *J. exp. Bot.* **21**, 908–914.
PITMAN, M. G. (1971). *Aust. J. biol. Sci.* **24**, 407.
PITMAN, M. G., MOWAT, J. and NAIR, H. (1971). *Aust. J. biol. Sci.* **24**, 619–631.
SMITH, R. C. (1970). *Pl. Physiol., Lancaster* **45**, 571–575.

VII.6

Heterogeneity of Ion Uptake Mechanisms Along Primary Roots of Maize Seedlings

A. Eshel and Y. Waisel

Department of Botany, Tel Aviv University
Israel

I. Introduction

Variations in uptake of minerals along roots are well known and were initially described a long time ago (Scott and Priestley, 1928; Steward *et al.*, 1942; Overstreet and Jacobson, 1946; Kramer and Wiebe, 1952; Canning and Kramer, 1958; Scott and Martin, 1962; Handley *et al.*, 1965; Torii and Laties, 1966; Bowen, 1968, 1970; Bowen and Rovira, 1967, 1968; Rovira and Bowen, 1968, 1970; Grasmanis and Barley, 1969). Although differences in uptake were recorded, little attention was given to the mechanisms involved in such phenomena and they had made little impact on our views of ion uptake by roots. Evidence for variations in uptake mechanisms along roots was obtained in our previous work (Eshel and Waisel, 1972), and this line of investigation has been continued and extended.

II. Materials and Methods

Uptake of sodium and rubidium by the apical 8 cm portions of primary roots of maize seedlings was investigated. Plants were grown for 6 days on $\frac{1}{2}$ strength Hoagland's nutrient solution to which 3 mM NaCl was added. Roots were excised, rinsed in water and immersed for an uptake period of

10 min in solutions of NaCl, RbCl, or NaCl + RbCl. Sodium was labelled with ^{22}Na and rubidium with ^{86}Rb. Following an uptake period, roots were rinsed in deionized water and desorbed for 1 h. Desorption was conducted at 0°C in a non-radioactive solution, similar to the uptake solution but containing 10 mM $CaCl_2$. Roots were then sectioned by a guillotine into 20 four millimeter slices and the isotope content of each slice was determined by liquid scintillation counting.

III. Results and Discussion

Except under low-temperature conditions (0°C), uptake rates varied along the roots, yielding a typical pattern for sodium and rubidium ions (Figs 1 and 2). The patterns of rubidium uptake were similar to those described previously for sodium (Eshel and Waisel, 1972). Higher rates of uptake were usually recorded near the root tip as well as within the 2–5 cm portion. Relatively lower rates of uptake were observed at the 6–8 cm portion. Thus, the curve describing absorption rates as a function of distance from the root apex exhibited a hump.

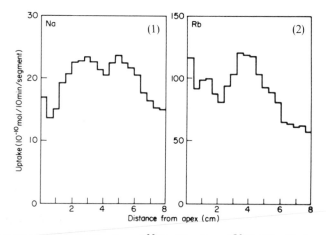

FIGS 1 and 2. Patterns of distribution of ^{22}Na (Fig. 1) and ^{86}Rb (Fig. 2) along the apical 8 cm portion of primary roots of maize seedlings.

The dependence of uptake on temperature, in the 0–30°C range can generally be described by an S-shaped curve. In cases of a chemical process the exponentially increasing part of such a curve will follow the Arrhenius equation [$U = A \exp(-Ea/RT)$]. Thus, the logarithm of the rate of uptake (ln U), plotted against the reciprocal of absolute temperature ($1/T$) will yield a straight line. Such a line has a slope of $-Ea/R$, where Ea is the activation energy of the process and R the gas constant. A is a pre-exponential constant. Examination of such relationships at 5°C intervals over the 0–30°C range indicated that dependence of uptake of sodium and rubidium on temperature was not uniform. The lines obtained for ln U vs. $1/T$ can be divided into two sections: one with a steep slope in the 0–10°C range and another one in the 10–30°C range with a relatively shallow slope (Figs 3 and 4). Such changes in slope were more pronounced in the first 4 cm portion

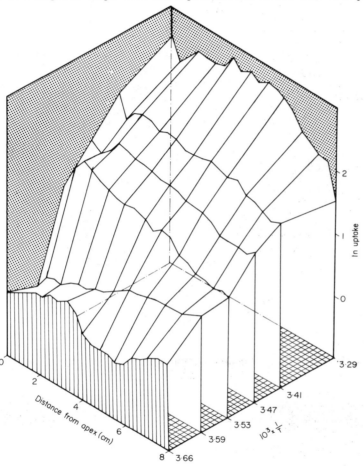

FIG. 3. Variations in the dependence of sodium uptake (ln uptake) on $1/T$ along the apical 8 cm portion of primary roots of maize seedlings.

whereas the temperature relationships of the 6–8 cm portion were very close to linearity. Such a break in the line can be interpreted as a shift from one rate limiting step with a relatively high activation energy, to another one, with a lower Ea.

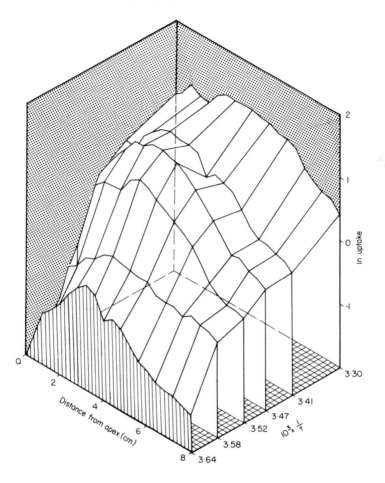

FIG. 4. Variations in the dependence of rubidium uptake (ln uptake) on $1/T$ along the apical 8 cm portion of primary roots of maize seedlings.

The uptake of either sodium or rubidium was inhibited by the presence of the other ion. However, the degree of inhibition of uptake of one cation by the other one also varied along the roots. Difficulties were encountered in obtaining distinct and replicable patterns of such inhibition. Nevertheless, the common trends are presented in Figs 5 and 6.

Inhibition of the uptake of sodium by rubidium varied with temperature. It was uniform along the roots at 0°C (approx. 50% inhibition) but showed distinct variations along the roots at 30°C. At such temperature conditions inhibition was smaller near the apex as well as within the 3–5 cm portion. Inhibition of sodium uptake at 30°C in the 6–8 cm portion was relatively high and was similar to that observed at 0°C.

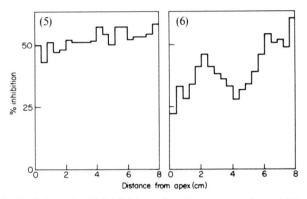

FIGS 5 and 6. Variations in % inhibition of sodium uptake by rubidium along the apical 8 cm portion of primary roots of maize seedlings. Na concentration, 5 mM; Rb concentration, 5 mM; Fig. 5—0°C; Fig. 6—30°C.

Uptake of rubidium was less affected by sodium than vice versa. However, variations along the roots were observed also for this characteristic. In some experiments inhibition was uniform along the roots, whereas in others it varied (Fig. 7). In such cases addition of sodium to the treatment solution caused a change in the pattern of rubidium uptake. Rates of rubidium uptake by the 2–4 cm root portion were affected more than those by any other

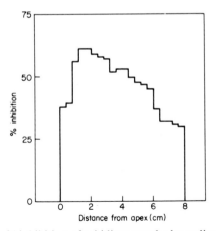

FIG. 7. Variations in % inhibition of rubidium uptake by sodium along the apical 8 cm portion of primary roots of maize seedlings. Rb concentration, 3 mM; Na concentration, 5 mM; temperature, 30°C.

portion along the roots. Consequently, the "hump" observed for rubidium uptake in this portion of the root was removed by the presence of sodium. Such changes of patterns were observed only at 20°C and 30°C treatments

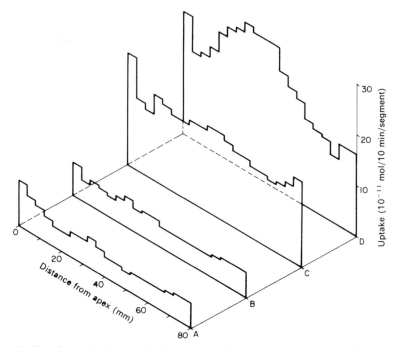

FIG. 8. Sodium-induced changes in the pattern of rubidium uptake. A, NaCl + RbCl; 0°C. B, RbCl; 0°C. C, NaCl + RbCl; 30°C. D, RbCl; 30°C.

but not at 0°C (Fig. 8). Rubidium had no such effects on the patterns of sodium uptake.

Concentration isotherms were followed either in 0·1–1·0 mM range or in the 1–20 mM range. The fitness of the data to Michaelis–Menten equation was examined either directly by non-linear regression, or by the Lineweaver–Burke transformation.

Variations along roots occurred in the Km values of sodium or rubidium uptake (cf. Hanson and Kahn, 1957). However, no full description of such changes along roots could be obtained because in a number of experiments the uptake by segments within the apical 4 cm portion did not fit the Michaelis–Menten equation. Also the values of V_{max} varied along the roots, but the patterns of such changes have not yet been defined.

The conclusions from such data seem to us to be self-evident. As various segments along primary maize roots differ in their responses to temperature, concentration, and ion composition of the medium, it is suggested that different uptake mechanisms operate at various sites along each root. Thus, analysis of uptake mechanisms using data obtained with entire roots, must include the relative contribution of each root portion to the total uptake capacity.

Changes in uptake cannot be directly correlated to the degree of vacuolation (cf. Torii and Laties, 1966) because large variations occur in that portion of the roots (2–8 cm) which are fully vacuolated.

We know of no precise physiological parameters which might cause such changes in uptake mechanisms. The identification and characterization of such parameters are essential for the description and understanding of ion uptake processes either by single roots or by whole root systems.

Acknowledgement

This investigation was supported in part by a grant from the U.S. Department of Agriculture under P.L. 480.

References

Bowen, G. D. (1968). *Nature, Lond.* **218**, 685–686.

Bowen, G. D. (1970). *Aust. J. Soil. Res.* **8**, 31–42.

Bowen, G. D. and Rovira, A. D. (1967). *Aust. J. biol. Sci.* **20**, 369–378.

Bowen, G. D. and Rovira, A. D. (1968). *In* "Root Growth" (W. J. Whittington, ed.), pp. 170–199. Butterworth, London.

Canning, R. E. and Kramer, P. J. (1958). *Am. J. Bot.* **45**, 378–382.

Eshel, A. and Waisel, Y. (1972). *Pl. Physiol., Lancaster* **49**, 585–589.

Grasmanis, V. O. G. and Barley, K. P. (1969). *Aust. J. biol. Sci.* **22**, 1313–1320.

Handley, R., Metwally, A. and Overstreet, R. (1965). *Pl. Physiol., Lancaster* **40**, 513–520.

Hanson, J. B. and Kahn, J. S. (1957). *Pl. Physiol., Lancaster* **32**, 497–498.

Kramer, P. J. and Wiebe, H. H. (1952). *Pl. Physiol., Lancaster* **27**, 661–674.

Overstreet, R. and Jacobson, L. (1946). *Am. J. Bot.* **33**, 107–112.

Rovira, A. D. and Bowen, G. D. (1968). *Nature, Lond.* **218**, 686–687.

Rovira, A. D. and Bowen, G. D. (1970). *Planta* **93**, 15–25.

Scott, B. I. H. and Martin, D. W. (1962). *Aust. J. biol. Sci.* **15**, 83–100.

Scott, L. I. and Priestley, J. H. (1928). *New Phytol.* **27**, 125–140.

Steward, F. C., Prevot, P. and Harrison, J. A. (1942). *Pl. Physiol., Lancaster* **17**, 411–421.

Torii, K. and Laties, G. G. (1966). *Pl. Physiol., Lancaster* **41**, 863–870.

VII.7

Multiphasic Ion Uptake in Roots

Per Nissen*

Department of Microbiology,
Agricultural College of Norway and
Botanical Laboratory, University of Bergen
Norway

I. Introduction

The uptake of sulphate by roots and leaf slices of barley can be described (Nissen, 1971) by single, multiphasic isotherms, i.e. by a series of phases following Michaelis–Menten kinetics and separated by discontinuous transitions. These kinetics are taken to indicate the involvement of a structure that changes characteristics (kinetic constants) at certain discrete salt concentrations (transition points).

A comprehensive re-examination (Nissen, 1973, and in preparation) of published data has shown the concept of multiphasic mechanisms to be valid for the uptake of most, if not all, inorganic ions by plant tissues. The recently proposed (Nissen, 1973) multiphasic series model of ion uptake (Fig. 1) involves a single, multiphasic uptake mechanism at each membrane, seems consistent with all observations, and reconciles seemingly contradictory evidence for the parallel and the series models of ion uptake.

In the present paper, previous examples of multiphasic ion uptake (Nissen, 1971, 1973) will be supplemented with data for chloride uptake by barley roots, rubidium uptake by tips and proximal sections of maize roots, potassium uptake and long-distance transport in maize seedlings, and phosphate uptake and long-distance transport in wheat seedlings.

* Present address: Botanical Laboratory, University of Bergen, Allégaten 70, N-5000 Bergen, Norway.

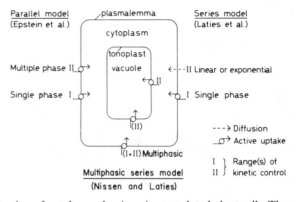

FIG. 1. Localization of uptake mechanisms in vacuolated plant cells. The parallel model consists of a homogeneous (single phase) high affinity or type 1 mechanism (system I) and a heterogeneous (multiple phase) low-affinity or type 2 mechanism (system II) in parallel at the plasmalemma, i.e. $v = v_1 + v_2$ (Epstein *et al.*, 1963; Epstein, 1966; Welch and Epstein, 1968, 1969). Rates of uptake from an external solution provide, according to this model, no direct evidence concerning mechanisms at the tonoplast. In the series model, system II is placed at the tonoplast and a diffusive component has been postulated across the plasmalemma at external salt concentrations above 5×10^{-4} M–10^{-3} M (Torii and Laties, 1966; Lüttge and Laties, 1966; Laties, 1969). The multiphasic series model involves a single, multiphasic mechanism at the plasma-lemma and a single, probably multiphasic tonoplast mechanism which may become rate-limiting above 10^{-2} M. No diffusion across the plasmalemma is required. Multi-phasic kinetics are taken to reflect all-or-none phase changes in a single structural entity or site. In contrast, the so-called low-affinity mechanism or system II has been considered to be multiple or polyvalent, i.e. to reflect simultaneously operating carrier sites which differ somewhat in their affinities. Energy requirements are appreciably lowered at high external salt concentrations (in preparation), and it is not clear whether uptake mediated by the higher phases should be termed active.

II. Re-analysis of Data

Original data were kindly provided by Dr O. E. Elzam for the chloride experiments and by Dr D. G. Edwards for the phosphate experiments. Data for the rubidium and potassium experiments were obtained by measur-ing points in the original figures to the nearest 0·05 mm with a precision ruler. Kinetic constants were calculated by a Fortran program minimizing $\log v$ (kindly provided by Dr W. W. Cleland—cf. Cleland, 1963 and 1967). Kinetic constants based on more than 2 points are given with 95% con-fidence intervals provided the entire interval remains positive. Adjacent phases have been tested for fit to one or two phases.

Data for chloride uptake by excised barley roots, in the presence of 5×10^{-4} M $CaSO_4$, are presented directly in a double reciprocal plot in Fig. 2 (see also Fig. 3 and Table I). Uptake of chloride in the range 5×10^{-6} M–5×10^{-2} M is accurately represented by 7 phases separated by discontinuous

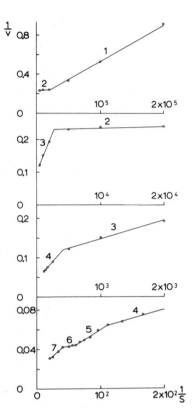

FIG. 2. Double reciprocal plot for uptake of chloride by barley roots. (FIG. 1 in Elzam *et al.*, 1964.)

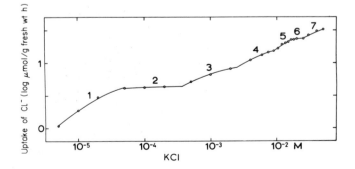

FIG. 3. Uptake of chloride by barley roots. (FIG. 1 in Elzam *et al.*, 1964.)

TABLE I. Kinetic constants and transition points for uptake of chloride by barley roots (Fig. 1 in Elzam et al., 1964)

Phase	V_{max} μmol h^{-1} (g fresh wt)$^{-1}$	K_M M	Transition M	
1	7.3 ± 0.4	$2.8 \pm 0.2 \times 10^{-5}$		
			3.6×10^{-5} (calc.)[b]	**[a]
2	4.3 ± 0.4	2.6×10^{-6}		
			3.6×10^{-4} (calc.)	**
3	10.0 ± 2.4	$4.8 \pm 3.2 \times 10^{-4}$		
			2.5×10^{-3} (calc.)	***
4	22.5 ± 0.8	$4.1 \pm 0.4 \times 10^{-3}$		
			9×10^{-3} (chosen)	**
5	51.3 ± 6.3	$2.1 \pm 0.4 \times 10^{-2}$		
			1.65×10^{-2} (chosen)	**
6	26.6 ± 1.6	$2.9 \pm 1.4 \times 10^{-3}$		
			2.5×10^{-2} (chosen)	*
7	52.5 ± 7.7	$3.0 \pm 0.9 \times 10^{-2}$		

[a] Significance of fit to adjacent phases (tested against a single phase): n.t.—not testable (insufficient points), n.s.—not significant ($P > 0.05$), *—$0.01 < P \leqslant 0.05$, **—$0.001 < P \leqslant 0.01$, ***—$P \leqslant 0.001$. Common (chosen) transition points have been assigned first to the lower, and then to the higher phase.
[b] 5×10^{-5} M used as chosen transition in Figs 2 and 3.

TABLE II. Kinetic constants and transition points for uptake of chloride by barley roots in the presence of 10^{-2} M $CaSO_4$ (Experiment referred to in Elzam et al., 1964)

Phase	V_{max} μmol h^{-1} (g fresh wt)$^{-1}$	K_M M	Transition M	
1	11.0	4.7×10^{-5}		
			2×10^{-5} (chosen)	n.s.[a]
2	4.5 ± 0.3	$7.6 \pm 2.8 \times 10^{-6}$		
			3.4×10^{-4} (calc.)	**
3	9.4 ± 4.0	3.8×10^{-4}		
			2.4×10^{-3} (calc.)	***
4	17.2 ± 0.1	$2.7 \pm 0.1 \times 10^{-3}$		
			1.2×10^{-2} (chosen)	***
5	25.6 ± 1.5	$1.0 \pm 0.1 \times 10^{-2}$		
			2×10^{-2} (chosen)	n.s.
6	18.6	1.5×10^{-3}		
			2.5×10^{-2} (chosen)	n.s.
7	24.7 ± 0.5	$1.0 \pm 0.1 \times 10^{-2}$		

[a] cf. Table I.

transitions. Originally (Elzam *et al.*, 1964), phases 1 and 2 were taken to represent a homogeneous high-affinity mechanism, while phases 3 and 4, 5 and 6, and 7 were interpreted as 3 carrier sites of a heterogeneous low-affinity mechanism. The same 7 phases are evident for chloride uptake in the presence of 10^{-2} M $CaSO_4$ (Fig. 4 and Table II).

Uptake of rubidium from solutions of 5×10^{-5} M–10^{-2} M RbCl or Rb_2SO_4 by tips or proximal sections of maize roots can be represented by 5 phases with virtually identical transition points (Figs 5–7, Tables III and IV). Phases 1–3 were originally (Fig. 5 in Torii and Laties, 1966) represented by a hyperbolic isotherm, while phases 4 and 5 were interpreted as a second hyperbolic isotherm for proximal sections and as an essentially straight isotherm for tips.

Uptake and long-distance transport of potassium from solutions of KCl or K_2SO_4 by maize seedlings can be represented by 4 phases in the range 10^{-5} M–4×10^{-2} M (Figs 8–11, Tables V and VI). Originally (Fig. 10 in

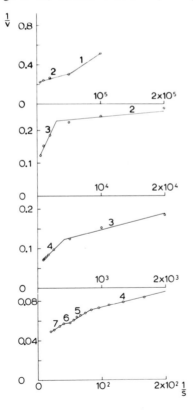

FIG. 4. Double reciprocal plot for uptake of chloride by barley roots in the presence of 10^{-2} M $CaSO_4$. (Experiment referred to in Elzam *et al.*, 1964.)

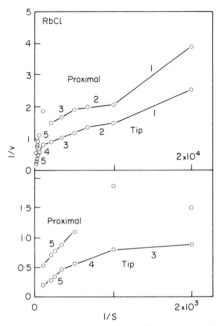

FIG. 5. Double reciprocal plot for uptake of rubidium from solutions of RbCl by tips and proximal sections of maize roots. (Fig. 5 in Torii and Laties, 1966.)

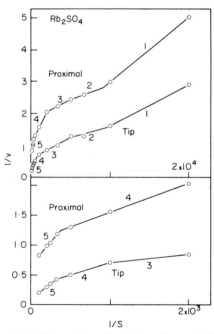

FIG. 6. Double reciprocal plot for uptake of rubidium from solutions of Rb_2SO_4 by tips and proximal sections of maize roots. (Fig. 5 in Torii and Laties, 1966.)

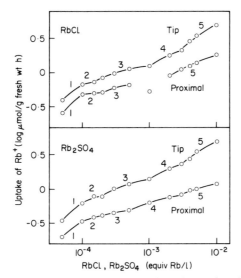

FIG. 7. Uptake of rubidium by tips and proximal sections of maize roots. (Fig. 5 in Torii and Laties, 1966.)

Table III. Kinetic constants and transition points for uptake of rubidium by root tips of maize (Fig. 5 in Torii and Laties, 1966)

Phase	V_{max} μmol h^{-1} (g fresh wt)$^{-1}$	K_M M	Transition M	
	RbCl			
1	2·28	$2·4 \times 10^{-4}$		
			10^{-4} (chosen)	n.t.[a]
2	0·89	$3·2 \times 10^{-5}$		
			$1·5 \times 10^{-4}$ (chosen)	* or n.t.
3	$1·44 \pm 0·02$	$1·4 \pm 0·0 \times 10^{-4}$		
			10^{-3} (chosen	***
4	$3·32 \pm 0·90$	$1·7 \pm 0·9 \times 10^{-3}$		
			3×10^{-3} (chosen)	* or n.t.
5	$11·4 \pm 4·7$	$1·2 \pm 0·7 \times 10^{-2}$		
	Rb$_2$SO$_4$			
1	3·16	$4·1 \times 10^{-4}$		
			10^{-4} (chosen)	n.t.
2	1·10	$7·4 \times 10^{-5}$		
			2×10^{-4} (chosen)	n.s.
3	$1·77 \pm 0·17$	$2·4 \pm 0·6 \times 10^{-4}$		
			10^{-3} (chosen)	*
4	$3·47 \pm 0·23$	$1·5 \pm 0·2 \times 10^{-3}$		
			$3·2 \times 10^{-3}$ (calc.)[b]	**
5	$10·0 \pm 4·9$	$1·0 \pm 0·8 \times 10^{-2}$		

[a] cf. Table I.
[b] 3×10^{-3} M used as chosen transition in Figs 6 and 7.

TABLE IV. Kinetic constants and transition points for uptake of rubidium by proximal sections of maize roots (Fig. 5 in Torii and Laties, 1966)

Phase	V_{max} μmol h^{-1} (g fresh wt)$^{-1}$	K_M M	Transition M	
	RbCl			
1	4·20	$7·6 \times 10^{-4}$		
			10^{-4} (chosen)	n.t.[a]
2	0·55 ± 0·10	$1·4 \times 10^{-5}$		
			2×10^{-4} (chosen)	n.s.
3	0·83 ± 0·09	$1·2 ± 0·4 \times 10^{-4}$		
4	—[b]	—[b]		
5	2·48 ± 0·15	$3·6 ± 0·5 \times 10^{-3}$		
	Rb$_2$SO$_4$			
1	1·04	$2·1 \times 10^{-4}$		
			$1·1 \times 10^{-4}$ (calc.)	n.t.
2	0·51	$4·8 \times 10^{-5}$		
			2×10^{-4} (chosen)	n.t.
3	0·57 ± 0·03	$7·5 ± 1·8 \times 10^{-5}$		
			5×10^{-4} (chosen)	*
4	0·94 ± 0·06	$4·5 ± 0·8 \times 10^{-4}$		
			$2·7 \times 10^{-3}$ (calc.)	***
5	1·47 ± 0·06	$2·2 ± 0·3 \times 10^{-3}$		

[a] cf. Table I.
[b] Cannot be calculated because of erratic value for 10^{-3} M (cf. Figs 5 and 7).

Lüttge and Laties, 1966), phases 1 and 2 were represented by a hyperbolic isotherm, while phases 3 and 4 were represented by two hyperbolic isotherms for uptake and by a linear isotherm for long-distance transport.

Uptake and long-distance transport of phosphate by wheat seedlings can, similarly, be represented by 5 and 4 phases, respectively, in the range 10^{-7} M–5×10^{-2} M (Figs 12 and 13, Table VII). The lower phases were originally (Edwards, 1970) represented by hyperbolic isotherms, while phases 3–5 were represented by a linear isotherm.

III. Discussion

The concept of multiphasic uptake mechanisms and the multiphasic series model of ion uptake have been presented elsewhere (Nissen, 1971, 1973, and in preparation), and the present data will be only briefly discussed.

The isotherms for uptake and long-distance transport of chloride, rubidium, potassium, and phosphate as analysed in this paper are clearly multiphasic. The phases are invariably separated by discontinuous transitions which become especially marked when accompanied by "jumps" as between phases 2 and 3, and 3 and 4 for potassium uptake from solutions of KCl

TABLE V. Kinetic constants and transition points for uptake (absorption) of potassium by maize roots (Fig. 10 in Lüttge and Laties, 1966)

Phase	V_{max} μmol h^{-1} (g fresh wt roots)$^{-1}$	K_M M	Transition M	
	KCl			
1	2.5 ± 0.7	$4.7 \pm 2.2 \times 10^{-5}$		
			6×10^{-5} (chosen)	n.s.[a]
2	1.6 ± 0.1	$1.2 \pm 0.8 \times 10^{-5}$		
			5×10^{-4}–10^{-3} (jump)	***
3	3.1 ± 0.1	$1.5 \pm 1.1 \times 10^{-4}$		
			1.5–2×10^{-2} (jump)	***
4	7.8 ± 1.2	$9.5 \pm 5.8 \times 10^{-3}$		
	K$_2$SO$_4$			
1	9.3	3.3×10^{-4}		
			6×10^{-5} (chosen)	n.t.
2	—[b]	—[b]		
			5×10^{-4}–10^{-3}	n.t.
3	2.1 ± 0.0	$2.9 \pm 0.5 \times 10^{-4}$		
			10^{-2} (chosen)	***
4	3.3 ± 0.1	$7.2 \pm 0.6 \times 10^{-3}$		

[a] cf. Table I.
[b] Not calculated because of erratic values for phase 2 (cf. Figs 8 and 9).

TABLE VI. Kinetic constants and transition points for long-distance transport of potassium in maize seedlings (Fig. 10 in Lüttge and Laties, 1966)

Phase	V_{max} μmol h^{-1} (g fresh wt roots)$^{-1}$	K_M M	Transition M	
	KCl			
1	—[b]	—[b]		
			6–8×10^{-5}	n.t.[a]
2	0.051 ± 0.008	$7.0 \pm 3.5 \times 10^{-5}$		
			5×10^{-4}–10^{-3} (jump)	***
3	0.30 ± 0.13	6.3×10^{-4}		
			8×10^{-3} (chosen)	**
4	11.3	2.2×10^{-1}		
	K$_2$SO$_4$			
1	—[b]	—[b]		
			4–6×10^{-5}	n.t.
2	0.035 ± 0.006	$6.6 \pm 3.3 \times 10^{-5}$		
			5×10^{-4} (chosen)	* or **
3	0.17 ± 0.07	$2.5 \pm 1.7 \times 10^{-3}$		
			5×10^{-3}–10^{-2}	n.t.
4	—[b]	—[b]		

[a] cf. Table I.
[b] Cannot be calculated (negative values obtained for V_{max}) because of erratic data (cf. Figs 10 and 11).

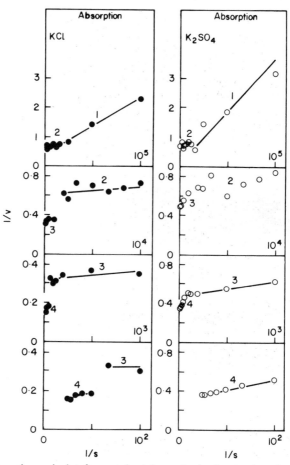

FIG. 8. Double reciprocal plot for uptake (absorption) of potassium by maize roots. (Fig. 10 in Lüttge and Laties, 1966.)

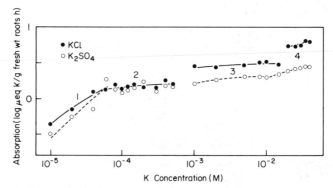

FIG. 9. Uptake (absorption) of potassium by maize roots. (Fig. 10 in Lüttge and Laties, 1966.)

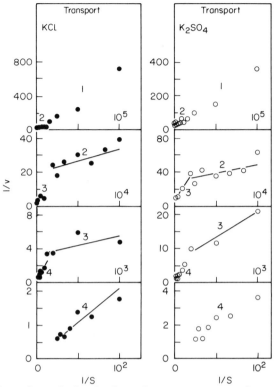

FIG. 10. Double reciprocal plot for long-distance transport of potassium in maize seedlings. (Fig. 10 in Lüttge and Laties, 1966.)

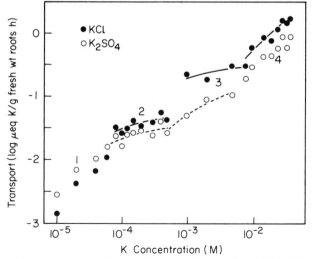

FIG. 11. Long-distance transport of potassium in maize seedlings. (Fig. 10 in Lüttge and Laties, 1966.)

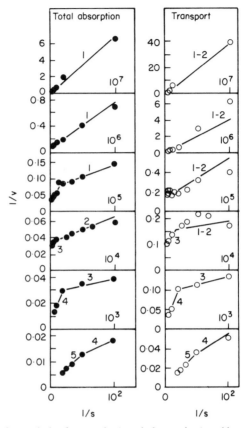

FIG. 12. Double reciprocal plot for uptake (total absorption) and long-distance transport in wheat seedlings. (Fig. 1 in Edwards, 1970.)

(Figs 8 and 9) and between phases 2 and 3 for long-distance transport of potassium from the same solutions (Figs 10 and 11).

The fit to Michaelis–Menten kinetics is impressively precise for most of the data, indicating that only one phase functions at any one concentration. Extrapolation and subtraction of any one phase from adjacent phases result in meaningless kinetics (cf. Nissen, 1971). At no concentration can the data be represented as the sum of two (or more) mechanisms, i.e. $v \neq v_1 + v_2$.

Uptake in the so-called low concentration range is, with the exception of sodium, usually biphasic for alkali cations and halide ions (Nissen, 1973, and in preparation). The kinetic constants for phase 2 are usually lower than those for phase 1. This is evident ($P < 0.01$) for chloride uptake in the presence of 5×10^{-4} M $CaSO_4$ (Figs 2 and 3, Table I) and is also indicated in the presence of 10^{-2} M $CaSO_4$ (Fig. 4, Table II). Uptake of rubidium (Figs 5–7,

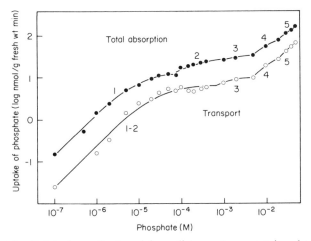

FIG. 13. Uptake (total absorption) and long-distance transport in wheat seedlings. (Fig. 1 in Edwards, 1970.)

TABLE VII. Kinetic constants and transition points for uptake (total absorption) and long-distance transport in wheat seedlings (Fig. 1 in Edwards, 1970)

Phase	V_{max} $\mu mol\ h^{-1}$ (g fresh wt roots)$^{-1}$	K_M M	Transition M	
	Total absorption			
1	14.0 ± 9.8	$9.9 \pm 1.0 \times 10^{-6}$		
			7.5×10^{-5} (chosen)	***[a]
2	29.9 ± 2.7	$9.7 \pm 2.4 \times 10^{-5}$		
			10^{-3} (chosen)	n.s.
3	35.7 ± 10.6	3.8×10^{-4}		
			5×10^{-3} (chosen)	n.s. or *
4	139 ± 18	$1.5 \pm 0.3 \times 10^{-2}$		
			2×10^{-2} (chosen)	*
5	491 ± 132	$1.0 \pm 0.4 \times 10^{-1}$		
	Transport			
1–2[b]	6.6 ± 0.7	$2.7 \pm 0.5 \times 10^{-5}$		
			6.4×10^{-4} (calc.)	n.s.
3	10.7 ± 1.0	$4.4 \pm 2.0 \times 10^{-4}$		
			5×10^{-3} (chosen)	* or n.s.
4	69	2.9×10^{-2}		
			2×10^{-2} (chosen)	n.s. or *
5	971	6.7×10^{-1}		

[a] cf. Table I.
[b] Calculated as one phase, but may possibly represent two phases (cf. Figs 12 and 13).

Tables III and IV) and uptake and long-distance transport of potassium (Figs 8–11, Tables V and VI) follow this pattern in all experiments.

The patterns for chloride uptake at the two different calcium levels are strikingly similar (Figs 2 and 4) and are fairly typical for uptake of alkali cations and halide ions in general (Nissen, in preparation). The lowering of kinetic constants upon transition from phase 1 to phase 2 may be repeated once or twice at higher concentrations, in this case upon transition from phase 5 to phase 6. The patterns for phosphate uptake and long-distance transport (Figs 12 and 13) are more regular and are similar to those for sulphate uptake (Nissen, 1971). The similarities between uptake patterns for various ions, plant tissues, and experimental conditions indicate that the transitions reflect changes of state in the membrane. There is nothing in these or other patterns to indicate a duality as implied in "system I" and "system II", and the use of these and similar terms should be abandoned.

The multiphasic series model (Fig. 1) combines the concept of multiphasic uptake mechanisms with the series model of ion uptake and obviates the need for a diffusive component across the plasmalemma at high external salt concentrations. This postulate is, furthermore, not supported by measurements of uptake across the plasmalemma alone.

Isotherms for chloride uptake by tips and proximal sections of maize roots (Torii and Laties, 1966) have thus been shown (Nissen, 1973) to be multiphasic and qualitatively very similar. Rubidium uptake by the same tissues is also characterized by multiphasic isotherms (Figs 5–7, Tables V and VI).

As previously pointed out (Nissen, 1973), data for uptake and long-distance transport are also qualitatively similar. The isotherms for uptake and long-distance transport of potassium in maize seedlings (Lüttge and Laties, 1966) can, despite some scatter, be resolved into 4 phases (Figs 8–11, Tables V and VI). The isotherms for total uptake and long-distance transport of phosphate, presumably as $H_2PO_4^-$ (Edwards, 1970), are, furthermore, both multiphasic (Figs 12 and 13, Table VII).

Compelling evidence for single, multiphasic uptake mechanisms at the plasmalemma is provided by the precise multiphasic kinetics exhibited by vacuolated and non-vacuolated plant cells. Rate-limitation at the tonoplast may, however, still occur at high external salt concentrations and after "filling" of the cytoplasm as predicted by the series model and supported by a variety of observations (Laties, 1969). Quantitative differences between the isotherms for tips and proximal sections and between the isotherms for uptake and long-distance transport (Figs 9 and 11) indicate that uptake across the plasmalemma may be faster than uptake across the tonoplast at high concentrations. Above 10^{-2} M, the influx of chloride across the plasmalemma has been shown, by the use of short uptake and still shorter wash periods (Cram and Laties, 1971), to increase much more rapidly than the quasi-steady over-all influx. The kinetics of tonoplast influx cannot, however, be established from available data.

References

CLELAND, W. W. (1963). *Nature, Lond.* **198**, 463–465.
CLELAND, W. W. (1967). *Adv. Enzymol.* **29**, 1–32.
CRAM, W. J. and LATIES, G. G. (1971). *Aust. J. biol. Sci.* **24**, 633–646.
EDWARDS, D. G. (1970). *Aust. J. biol. Sci.* **23**, 255–264.
ELZAM, O. E., RAINS, D. W. and EPSTEIN, E. (1964). *Biochem. biophys. Res. Commun.* **15**, 273–276.
EPSTEIN, E. (1966). *Nature, Lond.* **212**, 1324–1327.
EPSTEIN, E., RAINS, D. W. and ELZAM, O. E. (1963). *Proc. natn. Acad. Sci. U.S.A.* **49**, 684–692.
LATIES, G. G. (1969). *A. Rev. Pl. Physiol.* **20**, 89–116.
LÜTTGE, U. and LATIES, G. G. (1966). *Pl. Physiol., Lancaster* **41**, 1531–1539.
NISSEN, P. (1971). *Physiol. Pl.* **24**, 315–324.
NISSEN, P. (1973). *Physiol. Pl.* **28**, 113–120.
TORII, K. and LATIES, G. G. (1966). *Pl. Physiol., Lancaster* **41**, 863–870.
WELCH, R. M. and EPSTEIN, E. (1968). *Proc. natn. Acad. Sci. U.S.A.* **61**, 447–453.
WELCH, R. M. and EPSTEIN, E. (1969). *Pl. Physiol., Lancaster* **44**, 301–304.

VII.8

Symplasmic Translocation of α-Aminoisobutyric Acid in *Vallisneria* Leaves and the Action of Kinetin and Colchicine*

E. Müller and E. Bräutigam

*Institute for Biochemistry of Plants
Halle/Saale, Weinberg, G.D.R.*

I. Introduction

It was shown by Arisz (1960, 1969) that symplasmic translocation of substances can be identified in plant tissues; once a substance has entered a parenchyma cell it may move from cell to cell within the symplast without intermediate efflux and reabsorption. A valuable experimental system, introduced by Arisz, is represented by the leaves of *Vallisneria spiralis* L. Apoplastic translocation can be excluded in this system. The symplasmic transport through the leaf parenchyma resembles phloem transport (Mac-Robbie, 1971) in some respects as has been shown with parenchyma- and phloem-bridges by Arisz (1960).

Although much is known about symplasmic transport (Helder, 1967) our knowledge is not sufficient to give us a model of the translocation mechanism. As symplasmic transport is of fundamental interest for an understanding of the physiology of plant tissues we have started experiments for further elucidation of this process. (For details of the method

* Part of the doctoral dissertation of E.B.

see Arisz, 1960, 1969 and Bräutigam, 1971. All our experiments are evaluated statistically.)

II. Results and Discussion

A. Characterization of Symplasmic Translocation

1. Uptake of AIB* into the cells through the plasma membrane

The first step of the sequence of translocation processes in the leaf is uptake into the cells through the plasmalemma. We have investigated the uptake of α-aminoisobutyric acid into pieces of *Vallisneria* leaves, 4 cm long, immersed in a large volume of experimental solution. Changes in the concentration of the solution are negligible.

Uptake into the leaf parenchyma proceeds proportionally with time over a period of 120 h in the low concentration range and during 24 h in a 2×10^{-2} M solution, the highest concentration used in our experiments. The concentration dependence of the uptake velocity shows a saturation curve with a half saturation at 10^{-3} M. An intracellular concentration of about 2×10^{-2} mol kg^{-1} fresh weight is reached after 120 h in a 5×10^{-3} M experimental solution. Although we have not investigated the mechanism of this transport in detail it can be concluded by analogy that it is an active transport through the plasma membrane. For the following discussion it is important to know the uptake kinetics.

2. Symplasmic translocation

In these experiments the amino acid is applied to a leaf strip (5–15 cm long) in a limited zone at one side of the strip (zone I) and translocation is investigated by cutting the whole strip into pieces and measuring radioactivity of each piece, or by evaluation the leaf-strips with a scanner. Results can be summarized as follows:

(a) The velocity of translocation is higher than would be predicted by a simple diffusion mechanism (for calculations see Bräutigam, 1971).

(b) The concentration profile differs from a logarithmic form in most cases. Some fraction of the amino acid moves "faster" than would be expected from a logarithmic curve (see e.g. Fig. 4). This may be explained by different velocities in the parenchyma and in the phloem strands as is evident from autoradiographs.

(c) If the amino acid is applied at the middle of a short leaf strip, it moves in the apical as well as the basal direction with a comparable velocity (see Fig. 3).

(d) The efflux of the amino acid is low and can be neglected at a first approximation in our uptake experiments. Under extreme conditions (93 h washing in tap water of a leaf strip, previously incubated 50 h in a

---

* α-aminoisobutyric acid.

10^{-3} M solution of AIB) the tissue loses 4% of the radioactivity and this fraction can be further reduced by kinetin to practically zero (see II.B(e)).

One experiment must be described in detail since it allows a further characterization of symplasmic transport:

Radioactive AIB (0·1 μCi 7×10^{-3} M solution) is applied in droplets from a pipette at one side of the leaf strip (zone I) and starts to move through the middle part (zone II) of the leaf strip to the opposite side (zone III). Zone II is immersed into a solution of non-radioactive AIB 9 h after the start of translocation from zone I. Under these conditions the experiment is continued for another 15 h. The concentration of the non-radioactive AIB, supplied in zone II (2×10^{-2} M) is as high as possible without osmotic damage of the leaf cells.

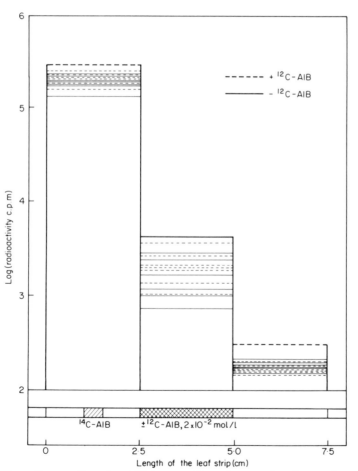

FIG. 1. Effect of non-radioactive AIB in zone II on the symplasmic translocation of (^{14}C)-AIB from zone I to zone III through zone II (details see text).

The result is illustrated in Fig. 1: No influence of the non-radioactive AIB on the translocation of the radioactive AIB from zone I to zone II and III can be detected.

This experiment again emphasizes the symplasmic nature of the translocation process. The amino acid supplied in zone II would compete with the radioactive AIB, if a transport step at the plasmalemma were involved. The uptake through the plasmalemma is saturated at an outside concentration of 5×10^{-3} M, whereas the concentration of the AIB solution, surrounding the cells in zone II is about 2×10^{-2} M.

Apparently another conclusion can be drawn: The process of symplasmic translocation does not exhibit saturation kinetics comparable with those of the uptake process into cells. The intracellular concentration of the non-radioactive AIB in zone II is higher than 5×10^{-3} M (using kg fresh weight instead of litre solution as a reference quantity). At this concentration the inactive amino acid should compete with the radioactive AIB if the translocation process had the same saturation characteristics as the uptake process (see II.A,1).

B. The Action of Kinetin

Mothes (1966) has shown that application of kinetin to plant tissue may lead to a "directed translocation" of substances towards the treated part of the tissue under certain conditions. In our experiments with kinetin we have tested whether this hormone alters the source–sink relationship or whether it acts on the translocation mechanism directly. The concentration profile of AIB under the influence of kinetin, expressed as radioactivity, is indicated by Figs 2–4. The results are, in summary:

(a) Kinetin leads to an "attraction" of AIB if applied to zone III of the leaf (Fig. 2). The promoting action on the amino acid accumulation in zone II, of kinetin applied in zone III, is connected with a transport of kinetin (or the "active" substance). No long-distance action of kinetin is demonstrable (Bräutigam, 1971).

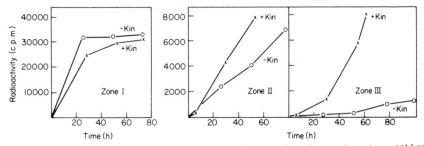

FIG. 2. Action of kinetin, applied in zone III, on the symplasmic translocation of [^{14}C]-AIB, applied in zone I.

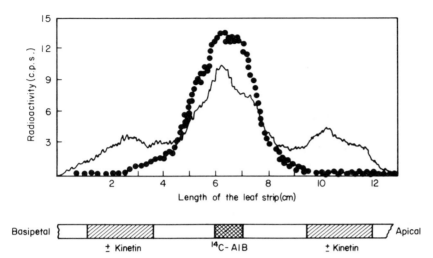

FIG. 3. Basipetal and apical symplasmic translocation of [^{14}C]-AIB (49 h) and the action of kinetin at the two sides (preincubation with a 10^{-6} M solution during 24 h).

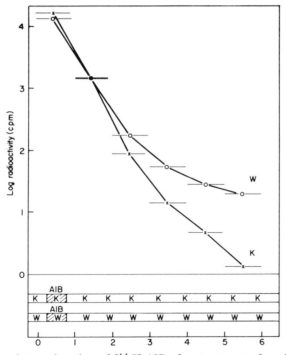

FIG. 4. Symplasmic translocation of [^{14}C]-AIB after treatment of a whole leaf strip with kinetin (K), compared with a water control (W). 48 h incubation in a 10^{-6} M solution of kinetin, after this 20 h translocation of AIB.

(b) The "attractive power" of kinetin can develop at both sides of the place of application of AIB. In such an experiment (Fig. 3) an attraction in opposite directions takes place at the same time.

(c) Application of kinetin to zone I or to the whole leaf strip restricts the translocation (Fig. 4).

(d) Kinetic stimulates the uptake of AIB into the cells (Fig. 5).

(e) Kinetin reduces the efflux of intracellular AIB (see II.A, 2(d)).

(f) The kinetin action on all the processes mentioned has an identical concentration dependence (see (d) and Bräutigam, 1971).

From these results some conclusions can be drawn as to the action of kinetin on translocation. One hypothesis was that the development of an "attractive power" may be explained by the ability of the hormone to improve the retention properties of the plasmalemma so that the treated cells successfully compete for material with the more leaky, untreated cells.

Our experiments demonstrate that the plasmalemma is not involved in the kinetin-induced accumulation of AIB in zone III: The amino acid reaches this zone through the symplast without crossing a plasmalemma. The kinetin-treated part of the leaf is a stronger "sink" and attracts AIB

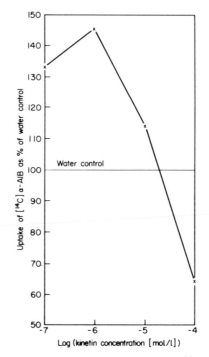

FIG. 5. Action of kinetin simultaneously applied with $[^{14}C]$-AIB on the uptake of the amino acid (54 h).

from a "source" through the symplast. If the AIB "source" or the whole leaf is treated with kinetin, the "source" cells, or all cells respectively, are improved in their ability to attract AIB and the translocation decreases.

This source–sink relationship inside the symplast may be explained hypothetically by compartmentation of AIB. Under the influence of kinetin a fraction of AIB would be trapped in a compartment in such a way that it is no longer available for translocation. Although we have not been successful in proving this hypothesis in our experiments with tobacco leaves (Rothe and Müller, 1970) further work has to be done before a final discussion is possible. We are now developing techniques for better ultrastructural localization of water-soluble substances.

C. Action of Colchicine

Symplasmic transport operates with a higher velocity than can be explained by diffusion. Tyree (1970) has developed a model which combines diffusion through the plasmodesmata and intracellular mixing by protoplasmic streaming. A possible connection between protoplasmic streaming and translocation has been in discussion since the paper of de Vries (1885), but the experimental evidence has never been conclusive. We have started

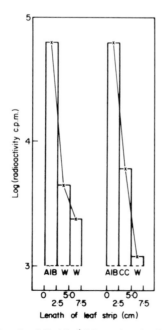

Fig. 6. Action of colchicine ($=$CC, 10^{-4} M, preincubation 2 h) on the symplasmic transport of $[^{14}C]$-AIB from zone I to zone III.

a programme to test the interdependence between protoplasmic streaming or other processes of intracellular movement with translocation. Some preliminary results have been obtained.

As indicated by Fig. 6, colchicine inhibits translocation of AIB into zone III if applied in zone II. As has been shown in other experiments, colchicine has no effect on the uptake of AIB into the cells (Bräutigam, 1971). This is further evidence that the two processes are different.

Colchicine is known to combine with the proteins of microtubuli. Apparently as a consequence of this, the microtubuli disappear from electron microscope images and some processes of cellular movement which are considered to be driven by microtubuli are stopped (Borisy and Taylor, 1967).

In our experiments colchicine does not stop protoplasmic streaming even if applied in a 10^{-2} M solution for 48 h.

Further experiments and application of other inhibitors (e.g. cytochalasin) are planned. Meanwhile, as a working hypothesis, two questions can be formulated:

(a) Do other processes of movement exist, besides protoplasmic streaming, which are connected with symplasmic translocation and which are sensitive to colchicine?

(b) Is the action of colchicine to be interpreted as an action on the plasmodesmata? Some electron microscopists interprete the tubular structures, sometimes visible inside the plasmodesmata, not as strands of the endoplasmatic reticulum but as derivatives of the spindle apparatus. Accordingly these structures are called desmotubuli. They can be compared with the microtubuli of the spindle apparatus (Robards, 1968).

Acknowledgements

We thank K. Mothes and F. Jacob for valuable discussions.

References

ARISZ, W. H. (1960). *Protoplasma* **52**, 309–343.
ARISZ, W. H. (1969). *Acta bot. Neerl.* **18**, 14–38.
BORISY, G. G. and TAYLOR, E. W. (1967). *J. Cell. Biol.* **34**, 525–554.
BRÄUTIGAM, E. (1971). Mscr. Dissertation, University of Halle.
HELDER, R. J. (1967). *In* "Encylopedia of Plant Physiology" (W. Ruhland, ed.) Vol. XIII, pp. 20–43. Springer, Berlin.
MACROBBIE, E. A. C. (1971). *Biol. Rev.* **46**, 429–481.
MOTHES, K. (1966). *Biol. Rdsch.* **4**, 211–244.
ROBARDS, A. W. (1968). *Protoplasma* **65**, 449.
ROTHE, U. and MÜLLER, E. (1970). *In* "Transport and Distribution of Matter in Cells of Higher Plants" (K. Mothes, E. Müller, A. Nelles and D. Neumann, eds) Vol. b, pp. 185–187. Akademie-Verlag, Berlin.
TYREE, M. T. (1970). *J. theor. Biol.* **26**, 181–214.
VRIES, H. DE (1885). *Bot. Z.* **43**, 1–6 and 17–26.

Discussion

Shone opened the discussion by commenting on Meiri's paper; he described exudation experiments with maize grown under high- and low-salt conditions. Under low-salt conditions ^{36}Cl applied to the external medium appears only slowly in xylem exudate but reaches a specific activity ratio of 0.6; in high salt roots ^{36}Cl appears almost immediately in xylem exudate but only reaches a specific activity ratio of 0.3. Therefore there seems to be a direct throughput of Cl^- in the symplasm of high-salt roots but in low-salt roots the vacuoles are somehow involved in symplasmic movement. Meiri replied that these results are similar to those of Hodges and Vaadia.

Cram then introduced the topic of exudation by isolated steles. He said that in their experiments they did not find exudation except in iris, where the endodermis remains after separation; isolated cortex, with the root tip removed and sealed over, does exude. Baker commented that this finding demonstrates the endodermis to be the osmotic barrier, a point he wished to raise with Higinbotham. Presumably iris steles exude at only a limited rate because the large collecting area of the cortex has been removed. Cram replied that it is difficult to compare with intact excised root exudation in iris; the intact root does not exude properly because the cortex is very thick. Lüttge then asked Cram what he meant by saying "the isolated cortex exudes". Cram then described the usual preparation for isolated cortex exudation attempts. Anderson then said that he could not detect exudation in such a preparation; he asked the rate of exudation and the contents of the exudate fluid in the light of Ginzburg and Ginsburg's claim that it is isotonic and represents active water movement. Cram said he did not yet have this information but the exudation rate was probably between 10 and 25% of the intact excised root rate.

Läuchli then referred to the autoradiographic work of Clarkson and Sanderson which showed ^{45}Ca in the living parenchyma cells around the xylem after a short pulse. After a short chase with cold Ca^{++} the ^{45}Ca appears rapidly in the xylem, which seems to indicate that it is readily transported from the parenchyma cells to the xylem. Pallaghy changed the direction of the discussion at this point by asking Cram if the cortex preparation he finds to exude is the one he uses for efflux studies, because *his* cortex preparation is properly isolated. Cram replied that in his efflux preparation the isolated cortex is sliced along its length so that there are no problems with dead volume in the internal cavity. MacRobbie then asked if exudation from isolated steles was necessarily a test that there is a pump there. Might not exudation be radially outward because the constraint of the endodermis is lost? Baker said that he did not know about the pump but it does seem to him to be a test of where the osmotic barrier is. MacRobbie replied that if vessels have intact protoplasts and membranes this would then constitute an osmotic barrier, and that she still thought a stele might exude radially outward and not upward into the collecting pipette. Cram said that they had sealed the broken ends of their steles so that exudation should be up into the pipette. Anderson then commented that preparing an isolated stele causes an extension of its length by 10 % or more and that many cells are dislocated; it is not surprising that there is no exudation. Furthermore, reported exudation rates for both isolated steles and isolated cortex, between 10 and 25 % of the rate for intact excised roots, mean that the level of exudate collected in 24 h, at which time one would normally assume the treated material to be dying, is only about the level expected by capillary rise. Davis then adduced the evidence on polar ^{45}Ca movement to show that the vessels are not open pipes. Spanswick retorted that differences in the number of vessels opening at the upper and lower ends of the root segment might give an apparent polar movement although the vessels were dead.

Waisel then opened the discussion on Nissen's paper by saying that he agreed with Nissen but would probably increase the number of phases to the number of cells in the tissue. Further he disagreed with the maize root data; different segments along the root behave differently. Some do not obey Michaelis kinetics, others do. When one sums up all these different isotherms one will get some sort of multi-phasic curve depending on the contribution of each segment to the total uptake measured. Nissen replied that what he was trying to impress on people was the precise nature of the Michaelis kinetics in each phase; this would not be produced by summing up heterogeneous contributions. The same kinetics are found in *Chlorella* or with amino acid uptake by bacteria where this type of variation is not possible; the precise kinetics at any given concentration demonstrates a very basic mechanism.

Cram then said that Nissen must distinguish individual transport processes at individual membranes. One cannot sum fluxes on the assumption they are single processes; active and passive components might be involved. He continued, one graph of 23 points is fitted by an equation which has 14 arbitrary parameters. How can this be strong evidence for agreement with Michaelis–Menten kinetics? Does curve-fitting ever prove anything about mechanism? Nissen replied first to the query about fluxes; he asserted that it can be shown from Cram's data on carrot or from Higinbotham's data on roots that the values he uses are very good estimates, no more than 10% over, of the plasmalemma influx. [Thus he infers that his model refers to the plasmalemma—Editor's comment or query.] In regard to the point on curve-fitting, Nissen commented that he agreed there were many parameters but asserted that the chance of finding similar patterns in different tissues by different workers at different times is very small. He did not indicate how this chance was calculated. Finally Nissen said that his sulphate data, which are designed to test the model, allow no possibility of drawing a continuous curve through the experimental points. MacRobbie broke the discussion at this point and offered a final comment which went perhaps to the heart of the matter; it does not matter how many phases there are but is it clearly established that the transition from one phase to the next is discontinuous?

Bentrup then asked Müller how many plasmodesmata there are connecting the *Vallisneria* cells and had he calculated the fluxes on this basis? Müller replied that his information would be available soon; he thought that the calculation would show the flux to be very high. Cram asked how symplasmic movement was distinguished from phloem movement in the *Vallisneria* leaves. Müller agreed that it was not easy and said that they probably observed a combination of movement in phloem and symplasm. This may also be the reason why logarithmic profiles are not observed. However phloem movement is similar in nature to symplasmic movement although probably a little faster. Pitman then asked why the efflux from the middle segment in Fig. 1 is so low; is it because the membranes have low permeability or is the activity sequestered in vesicles or tubules? Müller agreed the efflux was low but could not give any reason. He added that it can be further reduced to an undectable level by the addition of kinetin which probably affects intercellular compartmentation. Abscisic acid and kinetin act synergistically, not in opposition, but the overall effect is complex.

The discussion finally turned to Waisel's paper and Shone asked if the results could be interpreted on the basis of variations in the trans-root potential, which are known for different segment lengths. Since Bowling has shown that the trans-root potential originates at the outer root cortex, this may suggest that some cortical cells have a greater ability to take up ions. Shone continued that they had similar results for Na^+, K^+ and Cl^- in maize but in barley the pattern of uptake is much more uniform along the root. Waisel replied that they had not made potential measurements but it could not be the only reason; different cations show different uptake patterns along the root. Sutcliffe asked if growth pattern varied along the root length. Waisel replied that they had no information on this point.

Pallaghy said that intracellular potential is uniform along a root even although different regions of the root are growing at different rates.

Pitman then said that he agreed with Shone that there is much less variation in uptake in barley; he then pointed out inter-species differences in temperature response and submitted the following statement:

Lyons and Raisen in various publications have investigated the effects of temperature on mitochondrial activity and on membrane properties. They found in general that plants could be divided into two classes on the basis of their response to temperature. In one group, rate of mitochondrial activity was linearly proportional to the reciprocal of absolute temperature; in the other group, the rate of mitochondrial activity plotted in the same way was distinctly biphasic with a break at about 13–15°C, and usually with a steeper slope at the lower temperatures. Plants in the first group were less sensitive to chilling (barley) than plants in the second group (maize, sweet potato). Since their earlier investigation, many other organisms have been investigated and the temperature response has been shown to be related to the state of lipid in the membrane.

The measurement of uptake to apical segments of maize roots shows behaviour characteristic of the "chilling-sensitive" plants, possibly showing that uptake is limited by mitochondrial activity. In contrast, uptake of Rb and Cl by short lengths of barley, and the efflux of these ions, are linearly proportional to the reciprocal of absolute temperature, as might be expected for a "chilling-insensitive" plant.

Pitman concluded that the effects on transport may be due to the varying activity of the mitochondria along the root. Waisel agreed but said he had no information on mitochondrial activity. Loughman also spoke of the temperature response in barley; he said that phosphate uptake is not temperature sensitive and that phosphorylation is operating almost as rapidly at 0°C as at 25°C. West then commented that breaks in the Arrhenius curve in bacteria are correlated with freezing of the membrane lipids and this may equally apply to mitochondria. Waisel replied that this is the point he wished to stress; various regions of the root have various responses to temperature, to ion concentration, to pH and so on.

Läuchli then asked if Ca^{++} would affect the uptake patterns. Waisel replied that in these short-term experiments Ca^{++} is not important; the material was grown in a medium with Ca^{++} supplied. Läuchli next described an experiment on plants grown on complete Hoagland solution and then transferred to a KCl solution with no Ca^{++}; there is an efflux of ^{86}Rb-labelled K^+. In the presence of Ca^{++} no such efflux is found. Waisel replied that he too had done this experiment but did not find the effect.

Nissen then commented that the poor fit to Michaelis kinetics may be due to the operation of a single multiphasic isotherm and submitted the following statement:

I submit that the uptake of inorganic ions is mediated by single, multiphasic mechanisms which remain fundamentally unchanged along the roots or even from one plant or tissue to another. Reanalysis of data of Torii and Laties (*Pl. Physiol., Lancaster* **41**, 863–870 (1966)) thus yields 5 phases with virtually identical transition points for uptake of rubidium by tips and proximal sections of maize roots in the range 5×10^{-5}–10^{-2} M (Nissen, this volume, VII.7). The isotherms for chloride uptake are also multiphasic and qualitatively similar for tips and proximal sections (Nissen, *Physiol. Pl.*, in press, and in preparation). For concentrations up to 5×10^{-2} M, reanalysis yields 5 phases for uptake of phosphate by roots of wheat (Nissen, this volume), and barley (in preparation), and leaves of *Elodea densa* (in preparation). Sulphate uptake by leaf slices of barley was similarly mediated by 5 phases of a single, multiphasic mechanism (Nissen, *Pl. Physiol., Lancaster*, **24**, 315–324 (1971)). The patterns for alkali cations and halide ions tend to be similar, as do the patterns for phosphate and sulphate. These similarities indicate that the transitions reflect changes of state in the membrane. The heterogeneity observed by Eshel and Waisel may be due to differences in energy metabolism, membrane composition etc. along the roots.

The finding that uptake in certain segments did not fit the Michaelis–Menten equation in the 1-20 mM range may well be because there is more than one phase of the single, multiphasic mechanism in this range (cf. Nissen, this volume), and more detailed experiments would be required to bring this out.

Waisel replied that if there are in different regions of the root uptake mechanisms that differ in the activation energy, differ in the kinetics, differ in the response to medium composition, then he would call them different mechanisms.

White then returned to the absence of Ca^{++}; he said that it is known that Ca^{++} affects ion selectivity, so what is the value in calculating K_M values? Waisel replied that to his mind the evidence of the Ca^{++} effect is equivocal, but anyhow some segments fit Michaelis kinetics while others do not; this cannot be a Ca^{++} effect. White then commented on an experiment of Fe^{+++} uptake from an agar plate by a wheat root attached to the plant. He said that accumulation was greatest 3 cm from the tip but that the Fe^{+++} was found in greatest concentration *at* the tip. Thus the region of greatest uptake may not be the region of maximum concentration in the root. The desorption period at 0°C, White continued, will not necessarily stop longitudinal transport and the apparent ion accumulation pattern may be thus altered. Waisel replied that he was sure that transport was negligible at 0°C.

Lion (Liverpool) asked if Waisel had considered the variation in total cell surface area for the same segment length at different levels of the root. The maximum zone of uptake seems to coincide with the elongation region and with the region where root hairs start to develop. Waisel replied that these roots were grown in water culture and had no hairs. Shone then commented that the Letcombe studies have shown that ion uptake assayed in the manner used by Waisel may be associated with microbial contamination on the root surface. He then pointed out that Waisel had not demonstrated which regions are effective in xylem supply. They had evidence that K^+ and Cl^- are transported uniformly along the whole root length, whereas Ca^{++} is not; in older regions Ca^{++} is readily accumulated but is not apparently available for xylem transport. Waisel agreed that here he is measuring uptake into the cortex and is not concerned with cortical supply to the xylem. He continued that although they have not yet used sterile roots the results cannot be explained as uptake by bacterial colonies. That would require two populations on different regions, one preferring Na^+ and the other K^+, and it seems unlikely.

Section VIII

Ion and Water Movement
Chairman: D. H. Jennings

VIII.1

A Comparison of the Uptake and Translocation of some Organic Molecules and Ions in Higher Plants

M. G. T. Shone, D. T. Clarkson, J. Sanderson and Ann V. Wood

Agricultural Research Council, Letcombe Laboratory
Wantage, Berkshire, England

I. Introduction

The increasing use of soil-applied pesticides over the past few years has encouraged interest in the manner in which organic molecules are absorbed and translocated by plant roots.

Much of the work carried out on this subject suggests that transport of these molecules across the root is a passive process; however the multi-compartmental nature of the root can lead to difficulties in distinguishing unequivocally between active and passive transport of a given solute. For ions these difficulties might be increased by gradients of electrical potential and varying rates of ion transport in different parts of the root. An associated question which has aroused considerable discussion is the extent to which radial movement of solutes to the xylem vessels occurs via the symplast as proposed originally by Crafts and Broyer (1938).

The present article is concerned with the uptake and transport of organic molecules which are not usually present in the natural environment of the plant root, and to see what features there are in common with the absorption of solutes normally found in the soil solution. The triazines, some of which are important herbicides, and calcium, have been investigated with the comparison in mind. The choice of calcium has been based on observations that, under some conditions, its movement across the root appears to be by diffusion.

II. The Triazines

A. Properties

The triazines comprise a number of compounds which may be represented by the general formula:

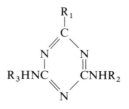

The formulae and characteristics of the three compounds considered here are given in Table I.

TABLE I. Structure and physico-chemical characteristics of triazines

Compound	R_1	R_2	R_3	Molar solubility in water at pH 7 ($\times 10^{-4}$)	pK_a
Simazine	Cl	C_2H_5	C_2H_5	0·25	1·85
Atrazine	Cl	C_2H_5	C_3H_7 (iso-)	1·61	1·85
Hydroxyatrazine	OH	C_2H_5	C_3H_7 (iso-)	0·30	4·85

Except under strongly acid conditions simazine and atrazine are essentially non-polar, but the non-phytotoxic hydroxyatrazine is positively charged under mildly acid conditions. The pK of hydroxyatrazine approximates to those of the methoxy- and methylthio-triazines ($R_1 = OCH_3$ or SCH_3) which are extensively used as herbicides.

B. Behaviour of Triazines in Plants

In common with calcium, simazine and atrazine appear to move very little in the phloem (Crafts, 1964) and when calcium and simazine are applied to segments of the roots of intact plants movement is almost entirely towards the base of the shoot (unpublished data). A number of workers (Minshall, 1954; Sheets, 1961; Wax and Behrens, 1965; van Oorschot, 1970) have noted a broad correlation between rate of transpiration and uptake of triazines and other soil-applied herbicides.

We have recently (Shone and Wood, 1972) investigated the relationship between rate of transpiration and transport of simazine to the shoots of barley under a range of different conditions in water culture. Although simazine is a potent inhibitor of photosynthesis there was no evidence in the short-term experiments described here that the herbicide affected the active uptake of nutrient ions. The relationship between transport of simazine and water is conveniently expressed as a Transpiration Stream Concentration Factor (TSCF) (Russell and Shorrocks, 1959) where:

$$\text{TSCF} = \frac{\mu\text{g simazine in shoots per ml water transpired}}{\mu\text{g simazine per ml of uptake solution}}$$

For a non-polar solute which should not be affected by gradients of electrical potential, values of the TSCF greater than unity would imply a direct dependence of transport of the solute on metabolism, since the concentration in the transpiration stream is greater than that in the ambient medium. We found that, in all treatments, the TSCF was less than 1, ranging from 0·63 to 0·98 (Shone and Wood, 1972). There was a significant effect on the TSCF of temperature, light intensity and humidity; the magnitude of the TSCF appeared to be positively correlated with the concentration of simazine in the roots. Neither 10^{-3} M sodium azide nor 5×10^{-5} M 2–4 dinitrophenol significantly affected the TSCF. We therefore concluded that the radial movement of the herbicide was largely a passive process, depending to a marked extent on movement of water.

The role of the root in the overall uptake and transport of tissues has been examined in further experiments, using labelled simazine ($4·9 \mu\text{Ci mg}^{-1}$), atrazine and hydroxyatrazine ($7·6 \mu\text{Ci mg}^{-1}$). Hydroxyatrazine was prepared from atrazine by the method of Castelfranco et al., (1961). These compounds were supplied in aqueous solutions at concentrations of $0·5 \times 10^{-6}$ or $1·0 \times 10^{-6}$ M (0·1 or 0·2 ppm). Previous work had established that for simazine there was no effect of the concentration in the uptake solution on the TSCF. Details of experimental methods are given in a previous publication (Shone and Wood, 1972).

1. Time course of uptake and loss of triazines by roots

When plant roots are placed in a solution there is usually an initial rapid entry of solute, followed by a slower process of accumulation. While the second process appears, in many circumstances, to be under metabolic control, the rapid initial uptake is generally considered to be largely due to the entry of solutes by diffusion and, if they are charged, to exchange on sites within the tissue.

Figure 1 shows the results of an experiment in which roots of intact plants were placed in labelled simazine solution for different periods of time. The roots were then blotted with tissue and the simazine content determined by

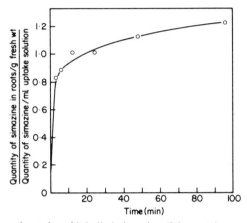

FIG. 1. Time-course of uptake of labelled simazine (0·2 ppm) by roots of intact barley plants.
The concentration of simazine in the roots in samples taken at successive intervals of time is expressed relative to the concentration in the uptake solution.

counting. The concentration of simazine in the roots is expressed relative to that in the uptake medium. The initial rapid uptake appears to be virtually complete in about 3 min; thereafter the uptake continues at a very slow rate. After 10 min the concentration in the root exceeds that in the ambient medium; this could be due to physical adsorption onto the root tissues. Comparable results were obtained with atrazine and hydroxyatrazine.

The ordinate in Fig. 1 has the dimensions of the Apparent Free Space which has been used to describe the volume within the root which ions can enter freely by diffusion (Briggs *et al.*, 1962), thus:

$$AFS = \frac{\text{quantity of readily diffusible solute/g fresh weight root}}{\text{quantity of solute/ml uptake solution}}$$

The magnitude of the AFS depends on the method of measurement. If for non-polar molecules it is assumed that the free space is occupied by water containing nothing which reacts with the solute, and the density of the root be taken as unity, then the AFS as defined above will represent the fraction of the total volume of the root which is occupied by the solute at the same concentration as in the ambient medium. The term AFS is used in the present article, although for some of the solutes considered this quantity would seem to represent a volume greatly in excess of the intercellular spaces implied in the original definition.

The magnitude of the AFS was examined by placing roots of intact plants in labelled solutions for 24 h. The roots were gently blotted and transferred to water; aliquots of this were sampled at successive time intervals after the start of the elution. The data were examined on the lines suggested by previous work (MacRobbie and Dainty, 1958; Diamond and Solomon, 1959; Dainty and Hope, 1959; Pitman, 1963); the logarithm of the percentage of triazine originally present in the root was plotted against time of elution. For all three compounds there was a rapid loss to the water with a half-time of less than 1 min; this was followed by much slower losses which were linear on the semi-logarithmic plot for up to 160 min. Extrapolation of the straight lines to zero time gave a value for the fraction of triazine most rapidly lost by the roots; this was used as a basis for calculating the AFS. The concentration remaining in the roots relative to that in the uptake solution has for convenience been termed the non-free space (NFS). Estimates of the AFS were also made from the quantity of triazine taken up rapidly in an arbitrary period of 3 min (Fig. 1).

2. Relationship between TSCF and AFS for triazines

Preliminary experiments showed that whereas uptake and translocation of simazine and atrazine were unaffected by the pH of the ambient medium, or by increasing the concentration of calcium chloride from 10^{-4} to 10^{-2} M, both these treatments affected the absorption of hydroxyatrazine. At pH 4 hydroxyatrazine will carry a positive charge and it was found that both uptake by the roots and translocation to the shoots was markedly reduced by calcium chloride, presumably as a result of competition by the calcium ion for negatively charged exchange sites.

Figure 2 show the relationship between the TSCF and the AFS, determined both by elution and by uptake for a period of 3 min, for the three triazines; for hydroxyatrazine data are given for three different pH values of the uptake solution and at pH 4 in the presence and absence of 0·01 M calcium chloride. The regression line was calculated from the values for the AFS obtained by elution; this has been assumed to pass through the origin on the basis that the concentration in the transpiration stream should be zero when there is no freely diffusible triazine in the roots. Figure 2 strongly suggests that the concentration of freely diffusible triazine in the roots determines the concentration in the transpiration stream. In Fig. 3 the TSCF is plotted against the NFS. Although there appears to be a correlation between the TSCF and the NFS for hydroxyatrazine, there is no evidence of any correlation when all three triazines are considered.

When intact plants, which had taken up labelled simazine for 23 h, were transferred to unlabelled solution for a further 24 h, there was no significant

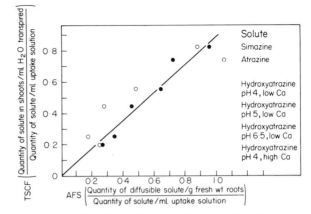

FIG. 2. Relationship in barley plants between Transpiration Stream Concentration Factor (TSCF) and Apparent Free space (AFS) as defined in the text, for simazine, atrazine and hydroxyatrazine under different conditions of pH and calcium concentration in the uptake solution.
Solid symbols: values for AFS obtained by elution after uptake for 24 h. Open symbols: values obtained from quantity taken up by roots in an arbitrary period of 3 min.

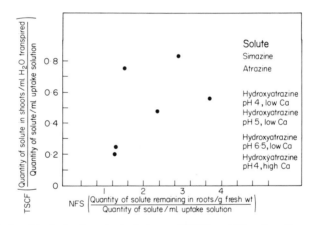

FIG. 3. Relationship in barley plants between TSCF and Non-free Space (NFS). The NFS is the concentration in the roots, relative to that in the uptake solution, after elution of the AFS as shown in Fig. 2.

increase in the simazine content of the shoots, implying that the simazine in the NFS was largely unable to move in the transpiration stream.

When plants are placed in an apparatus consisting of two compartments, so that the apical portions of the roots are in labelled simazine and the basal portions in unlabelled solution, simazine continues to move to the shoots although there is some loss to the solution surrounding the basal zone. Preliminary experiments seem to indicate that this loss largely represents material which has moved in the cortex and in consequence the endodermis would appear to act as a barrier to the outward movement of simazine.

If the concentration of freely diffusible simazine and atrazine in the root is the same as that in the ambient medium, as might be expected for non-polar molecules, it must be concluded that the AFS comprises not only the intercellular spaces but also the total volume of the tissues of the cortex including the cell vacuoles. Simazine and atrazine must then be freely diffusible across the cell membranes; this accords with the lipophilic nature of both these compounds.

At pH 4 in the presence of 0·01 M calcium, $AFS_{hydroxyatrazine}$ corresponds closely to estimates made of the intercellular, or water-free, space in barley roots. Thus Epstein (1955) using sulphate and selenate as solutes found a value of 23 % of the root volume and use of the non-polar monosilicic acid gave a value of 17 % (Shone, 1964). Hydroxyatrazine is strongly lipophobic and consequently might be expected to cross cell membranes with difficulty. It therefore seems likely that the readily diffusible fraction of this compound lies largely in the negatively charged intercellular spaces, where its concentration will depend on the extent to which it acquires a positive charge and on the concentration of cations competing for exchange sites. This is borne out by the decrease in $AFS_{hydroxyatrazine}$ with increasing pH and calcium content of the uptake solution.

These results suggest that the triazines largely move with water across the root to the transpiration stream rather than by the symplastic pathway. By contrast, there is good evidence that 2–4 dichlorophenoxyacetic acid and related compounds are mobile in the phloem (Crafts, 1964) and it is possible that for these the symplasm represents an important pathway for transport across the root.

III. Calcium

Although calcium is an element essential to plants, the dependence of absorption of this ion on concurrent metabolism appears to vary between different species and different experimental conditions. Thus in beans (Drew and Biddulph, 1971) and detached barley roots (Moore et al., 1961) metabolic inhibitors and lowering the temperature were found to have little effect on absorption. On the other hand in maize roots (Handley and Overstreet, 1961; Maas, 1969) these treatments brought about substantial reductions in absorption. Measurements of transpiration stream concentration factors under conditions of contrasting humidity in beans (Drew and Biddulph, 1971) and barley (Barber and Shone, unpublished data) indicate that, by contrast with the situation in the triazines, at low concentrations of the uptake medium the TSCF can be greater than unity and that transport of calcium to the leaves does not depend greatly on the rate of transpiration, although at higher concentrations, transport of both calcium (Drew and Biddulph, 1971) and phosphate (Russell and Shorrocks, 1959) shows a marked dependence on transpiration. Furthermore, the concentration of calcium in the xylem sap of detached roots of barley (Moore *et al.*, 1965) castor oil plants (Bowling *et al.*, 1966) and maize (Shone, 1968) can exceed that of the ambient medium. However, if the Nernst equation is accepted as valid when applied to electrochemical measurements on exuding root systems, transport of calcium can appear to be passive (Bowling *et al.*, 1966) since there should be a very marked effect of the gradient of electrical potential on the divalent cation, leading to a lower electrochemical potential of this ion in the sap than in the ambient medium. This approach can lead to difficulties of interpretation (Shone, 1968) and in calculating differences in electrochemical potential for divalent cations, appreciable errors may arise when ionic activities are approximated by concentrations, not only because for a given concentration, activity coefficients of divalent salts are lower than those of monovalent solutes, but also because there may be organic substances in the sap which substantially further reduce the activity of calcium. Assessments based on electrochemical considerations of whether transport across the root is active or passive, are therefore inherently more subject to uncertainties for calcium than for monovalent ions.

A. Pathways of Movement of Calcium across the Root

There is increasing evidence (Arisz, 1969; Läuchli and Epstein, 1971; Läuchli *et al.*, 1971) that long-distance active transport of some ions occurs through the symplast. In the older portions of cereal roots, the endodermis

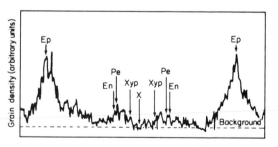

FIG. 4. Densitometer scan showing the distribution of ^{45}Ca in a transect (10 μm wide) across the diameter of a section cut transversely 1 cm from the tip of a seminal axis of barley. The root, while still part of an intact plant, was treated for 1 min with a solution containing 0·1 mM CaCl$_2$ with 5·25 μCi ^{45}Ca ml^{-1}, then washed with water for 30 s. The section (10 μm thick) was cut from frozen, unfixed root at −25°C and exposed to Kodak AR 10 emulsion for 14 days at −25°C. Density measurements made with scanning microspectrophotometer (Leitz) using a slit area of 25 μm^2.
Ep = epidermis; En = endodermis; Pe = pericycle; XyP = xylem parenchyma and X = central xylem vessel.

can become heavily suberized and thus appears to present a barrier to the radial movement of solutes and water. Nevertheless, the observations (Clarkson *et al.*, 1971) that the thickened endodermis is penetrated by plasmodesmata accords with the fact that phosphate, and to a lesser extent potassium and water, can cross the endodermis in this region of the root (Clarkson *et al.*, 1968, 1971). Calcium does not cross the suberized endodermis to any great extent, and this casts doubt on whether this ion can move into the stele in the symplast.

When younger segments of the roots of intact plants are exposed to solutions labelled with ^{45}Ca, the ion reaches the protoxylem poles of the pericycle and the xylem parenchyma within 1 min (Fig. 4); this movement to the stele is unaffected for at least 2 h, and probably for longer, by 10^{-5} M DNP or 10^{-3} M KCN (unpublished data). The data in Table II, which come from time-course labelling of barley roots, seem to suggest that the vascular and cortical tissues fill up in parallel rather than in series, as if ^{45}Ca reaching the stele bypassed the cortex.

TABLE II. Accumulation of labelled calcium in tissues of the seminal axis of barley roots 1 cm from apex

	Duration of labelling (min)			
	1	5	120	180
	Concentration of calcium (mM)			
Epidermis	0·25	0·28	0·26	0·39
Mid-cortex	0·08	0·05	0·17	0·27
Endodermis	0·12	0·11	0·32	0·49
Pericycle	0·11	0·24	0·37	0·69
Xylem parenchyma	0·09	0·24	0·46	0·97
Xylem vessel (centre)	0·03	0·06	0·17	0·40

Assessments of the concentration of ^{45}Ca in tissues were made by comparison of the grain density of microautoradiographs with calibrated standard sources. The external concentration of calcium was 0·1 mM.

Further evidence to support this interpretation comes from experiments with the polyvalent cations aluminium and scandium which profoundly reduce both the entry of calcium into the free space and its translocation to shoots (Clarkson and Sanderson, 1971). This effect is thought to be due to the reversal of the electrical sign of the entry points into the free space, as a result of the deposition of positively charged precipitates of hydroxide and hydroxyphosphates. Aluminium does not, however, reduce the quantity of calcium accumulated in a non-exchangeable form in the root, although a much higher proportion of the calcium in the cortex is present in the peripheral layers. These results show that the amount of calcium in the NFS of the cortex is not directly related to the amount translocated so that the primary pathway to the transpiration stream is through the free space and its associated exchange sites.

Some evidence consistent with the idea that calcium does not move readily to the stele via the symplast comes from other autoradiographic studies which are summarized in Fig. 5. This figure shows that a short period of

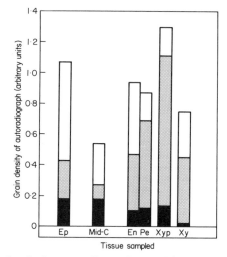

FIG. 5. Average gain density in autoradiographs overlying selected tissues in transverse sections of barley seminal root axes, 1 cm from the tip.
Values represent means of 10 replicate samplings from 2 separate roots.
The roots of intact plants were treated for 2 h with a solution containing 0·1 mM CaCl$_2$ with 4·5 μCi ^{45}Ca ml^{-1} and then were either (A) blotted to remove surplus uptake medium; radioactivity of this sample $\equiv$ whole height of column, or (B) exchanged for 10 min with unlabelled 0·1 mM CaCl$_2$; radioactivity of this sample $\equiv$ solid + stippled portions of column, or (C) kept in unlabelled 0·1 mM CaCl$_2$ for 6 h; radioactivity of this sample $\equiv$ solid portion of column.
Mid-C = mid-point of cortex, other abbreviations as in Fig. 4.

washing with an unlabelled solution of calcium chloride rapidly removed ^{45}Ca from the cortex, presumably by ion-exchange in the free space; this exchange has a half-time of about 7 min (Clarkson and Sanderson, 1971). Proportionally much less ^{45}Ca was removed from the pericycle and xylem parenchyma during this period; the marked decline in ^{45}Ca in the central xylem vessel would have more probably been due to displacement in the transpiration stream than to ion exchange with the external solution. During the following 6 h the radioactivity in the stelar tissues declined much faster than in the cortex so that the latter became the most radioactive tissues rather than the least radioactive—the situation after 10 min "exchange" treatment. This differential becomes even more pronounced after 24 h treatment with unlabelled solution (unpublished data). Other experiments, to be reported elsewhere, confirm that ^{45}Ca lost from the stele is recovered quantitatively in the shoot.

IV. Conclusions

In barley, the transport of triazines and calcium have certain features in common. There is little evidence that the symplasm plays an important role in their long-distance transport; the major pathway seems to be by mass flow with water or by diffusion. Owing to the low solubility of the triazines, and the low specific activity of available labelled materials, it has not yet been possible to localize them autoradiographically. Nevertheless, it seems likely that the AFS for the lipophilic compounds comprises the vacuoles of the cortical cells, whereas for the lipophobic hydroxyatrazine, the freely dif- fusible fraction which can readily pass to the vascular tissues is, as for calcium, confined to the extracellular spaces. Transport of ions, including calcium, is related to rate of transpiration when the concentration is above a given level. However, this relationship does not apply at low concentrations. We have no evidence that for the triazines the broad correlation between trans- port and transpiration rate is affected by concentration. It is possible that for calcium at low concentrations the rate at which the ion moves across the root as far as the stele is largely controlled by its rate of diffusion along ex- change sites in the free space, as, indeed, is its upward movement in the xylem vessels themselves (Bell and Biddulph, 1963).

The existence of difference in electrical potential between the xylem sap and the ambient medium implies that the free space cannot be a continuous pathway across the root. The possibility that the potential difference could arise from electro-osmotic potentials due to movement of salts and water through the free space (Briggs, 1968) seems unlikely, since the magnitude of the potential difference does not necessarily depend on the rate of exudation of the sap. Furthermore, recent work by Dunlop and Bowling (1971) suggests that the potential difference largely arises across the plasmalemmae of the root cells.

The work described above suggests that for calcium in barley the endoder- mis acts as a boundary between the AFS in the cortex and the vascular tissues and the same conclusion may be tentatively reached for the triazines.

Acknowledgements

We are grateful to Dr R. Scott Russell for helpful comments, to Mrs P. H. Bedford for assistance with experimental work and to Ciba-Geigy Limited for supplying labelled triazines.

References

ARISZ, W. H. (1969). *Acta bot. neerl.* **18**, 14–38.

BELL, C. W. and BIDDULPH, O. (1963). *Pl. Physiol., Lancaster* **38**, 610–614.

BOWLING, D. J. F., MACKLON, A. E. S. and SPANSWICK, R. M. (1966). *J. exp. Bot.* **17**, 410–416.

BRIGGS, G. E. (1968). *J. exp. Bot.* **19**, 486–488.

BRIGGS, G. E., HOPE, A. B. and ROBERTSON, R. N. (1962). *In* "Electrolytes and Plant Cells" (W. O. James, ed.) pp. 73–99. Blackwell, Oxford.

CASTELFRANCO, P., FOY, C. L. and DEUTSCH, D. B. (1961). *Weeds* **9**, 580–591.

CLARKSON, D. T., ROBARDS, A. W. and SANDERSON, J. (1971). *Planta* **96**, 292–305.

CLARKSON, D. T. and SANDERSON, J. (1971). *J. exp. Bot.* **23**, 837–851.

CLARKSON, D. T., SANDERSON, J. and RUSSELL, R. S. (1968). *Nature, Lond.* **220**, 805–806.

CRAFTS, A. S. (1964). *In* "The Physiology and Biochemistry of Herbicides" (L. J. Audus, ed.) pp. 75–110. Academic Press, London.

CRAFTS, A. S. and BROYER, T. C. (1938). *Am. J. Bot.* **25**, 529–535.

DAINTY, J. and HOPE, A. B. (1959). *Aust. J. biol. Sci.* **12**, 395–411.

DIAMOND, J. M. and SOLOMON, A. K. (1959). *J. gen. Physiol.* **42**, 1105–1121.

DREW, M. C. and BIDDULPH, O. (1971). *Pl. Physiol., Lancaster* **48**, 426–432.

DUNLOP, J. and BOWLING, D. J. F. (1971). *J. exp. Bot.* **22**, 453–464.

EPSTEIN, E. (1955). *Pl. Physiol., Lancaster* **30**, 529–535.

HANDLEY, R. and OVERSTREET, R. (1961). *Pl. Physiol., Lancaster* **36**, 766–769.

LÄUCHLI, A. and EPSTEIN, E. (1971). *Pl. Physiol., Lancaster* **48**, 111–117.

LÄUCHLI, A., SPURR, A. R. and EPSTEIN, E. (1971). *Pl. Physiol., Lancaster* **48**, 118–124.

MAAS, E. V. (1969). *Pl. Physiol., Lancaster* **44**, 985–989.

MACROBBIE, E. A. C. and DAINTY, J. (1958). *J. gen. Physiol.* **42**, 335–353.

MINSHALL, W. H. (1954). *Can. J. Bot.* **32**, 795–798.

MOORE, D. P., JACOBSEN, L. and OVERSTREET, R. (1961). *Pl. Physiol., Lancaster* **36**, 53–57.

MOORE, D. P., MASON, B. J. and MAAS, E. V. (1965). *Pl. Physiol., Lancaster* **40**, 641–644.

OORSCHOT, J. L. P. VAN (1970). *Weed Res.* **10**, 230–242.

PITMAN, M. G. (1963). *Aust. J. biol. Sci.* **16**, 647–668.

RUSSELL, R. S. and SHORROCKS, V. M. (1959). *J. exp. Bot.* **10**, 301–316.

SHEETS, T. J. (1961). *Weeds* **9**, 1–13.

SHONE, M. G. T. (1964). *Nature, Lond.* **202**, 314–315.

SHONE, M. G. T. and WOOD, ANN V. (1972). *J. exp. Bot.* **23**, 141–151.

SHONE, M. G. T. (1968). *J. exp. Bot.* **19**, 468–485.

WAX, L. M. and BEHRENS, R. (1965). *Weeds* **13**, 107–109.

VIII.2

Solute and Water Transport in the Bladders of *Utricularia*

P. H. Sydenham and G. P. Findlay

School of Biological Sciences, Flinders University, Bedford Park
South Australia.

I. Introduction

Utricularia, the largest genus of insectivorous plants, has about 250 terrestrial, epiphytic and aquatic species. Insect-catching bladders, which are modified leaves, are borne on submerged stems of the plant. The oval bladders, about 2–3 mm in length, have a hole at one end, which is covered by a valve-like trapdoor, attached by its top and sides to the periphery of the hole. Attached to the trapdoor and projecting outwards from it are a number of two-celled filaments. A mechanical deformation of one of the sensitive filaments (by a passing insect, for instance) opens the trapdoor, and causes a sudden increase in bladder volume, produced by a water flow into the bladder. After this rapid movement the bladder resets comparatively slowly to its original or "set" condition, with its side walls convex to the inside. Any insect caught by the bladder remains inside, eventually dies, and is absorbed by the bladder.

Apart from the sensitive hairs attached to the trapdoor, at least four other types of specialized cells can be identified: (a) two-armed and (b) four-armed hairs attached to the inside of the bladder, (c) cells borne on narrow-celled stalks around the periphery of the trapdoor on the outside of the bladder, and (d) small cells in pairs attached to the external surface of the bladder and distributed fairly uniformly over it. Glandular functions have been assigned to each of the groups of cells by various authors (Lloyd, 1942).

The mechanism of action of the bladder can conveniently be considered in three parts: (a) the triggering process, (b) the rapid expansion of the bladder and (c) the slow resetting process. This paper will deal with the resetting of the bladder, since it is this process that is caused by transport of solutes and water from the inside of the bladder to the outside solution.

II. Results

We have made the following experimental observations on the *Utricularia* bladder in its set condition.

(a) The volume of the luminal solution is 1–2 μl.

(b) The hydrostatic pressure in the lumen is about -0.2 atm compared with the outside.

(c) In the luminal fluid, $[K] = 7.8 \pm 0.2$ mM, $[Na] = 10.9 \pm 0.5$ mM, $[Cl] = 8.2 \pm 0.7$ mM, while in the outside solution $[K] = 0.2$ mM, $[Na] = 2.0$ mM and $[Cl] = 2.3$ mM.

(d) The lumen is at an electric potential of $+133 \pm 5$ (31) mV with respect to the outside solution.

(e) The electric resistance from the lumen of the bladder to the outside is 118 ± 6 kΩ cm^2 (8 bladders).

(f) In bladders which had not been triggered for at least 24 h, influxes, in pmol cm^{-2} s^{-1}, were: Na 1.0 ± 0.11 (25), K 1.18 ± 0.16 (11), and Cl 0.76 ± 0.06 (5).

Following stimulation of a sensitive hair, the trapdoor opens briefly, and solution flows from the outside into the lumen.

(g) The trapdoor opens for approximately 20 ms. This time was determined by applying a voltage clamp to the p.d. between the lumen and the outside solution, and measuring the duration of the surge in clamp current following stimulation of the bladder.

(h) The volume of the bladder increases by $43 \pm 2\%$ (12) and the pressure difference between the lumen and the outside goes to zero.

(i) Fluid is then transported from the lumen to the outside solution, and within 20–60 min the bladder is again in its set state.

(j) The radioactive isotopes ^{42}K, ^{22}Na and ^{36}Cl were introduced to the luminal solution only, by placing a bladder in radioactive solution, triggering it and then immediately transferring it to a non-radioactive solution. During its subsequent resetting, the bladder loses about 9% of total ^{42}K, 20% of total ^{22}Na and 6% of total ^{36}Cl.

(k) If the bladder is kept under paraffin oil during the resetting process the exudate appears to emerge from the mouth of the bladder; we presume from the stalk-like cells surrounding the trapdoor. In the exudate $[Na] = 12–16$ mM, $[K] = 2–5$ mM and $[Cl] = 5–7$ mM.

(l) The p.d. between the lumen and the exudate is about $+134$ mV.

(m) The rate of resetting of the bladder, and hence the volume flow from lumen to outside is temperature dependent. At 2°C the initial resetting rate (measured as the change of width of the bladder through its centre) is about 10% of that at 21°C.

(n) The resetting rate is reduced to about half that of normal in the presence of 1 mM sodium azide in the external solution, and is almost completely stopped when the inhibitor is introduced into the lumen. Iodoacetamide (3 mM) in the external solution has no affect on the resetting rate, but in the lumen almost completely stops the resetting.

(o) Addition of sodium azide (1 mM) to the external solution reduces the luminal p.d. to about +20 mV in about 20 min. The p.d. returns to its normal value in about the same time when the sodium azide is removed.

(p) The resetting rate is light-independent.

(q) The bladder, when reset, will remain reset for many hours after the temperature is lowered to 2°C or when sodium azide (1 mM) or iodoacetamide (3 mM) are present in the external solution.

The resetting rate is unaltered by the addition, at a concentration of 100 mM, of sucrose to the external solution.

III. Discussion

The *Utricularia* bladder, during its resetting phase, transports about 30% of the water in the lumen to the outside solution, setting up at the same time a negative pressure in the lumen. Although there is essentially isotonic transport of solution, the result described in Section II(j) shows that not all the ions in the exudate come from the luminal solution. Potassium and chloride ions in particular must come from cells in the walls of the bladder. Thus one can exclude the possibility that the outward transport of luminal fluid is an overall pinocytotic process in which luminal fluid is engulfed and transported, perhaps by vesicles, eventually to be released to the outside solution.

A more likely hypothesis to account for the transport of luminal fluid is one in which an ion, or ions, is actively transported from the lumen to the outside solution, in a specialized structure where the ion or salt transport can set up a local osmotic gradient (Diamond and Bossert, 1967), causing in turn a passive flow of water. To sustain such a hypothesis at least three important questions have to be answered.

1. Do specialized structures exist for setting up a local osmotic gradient?

As mentioned in the Introduction, at least four types of cells in the bladder might function as glands. But if the bladder resets under oil the exudate appears at the mouth and thus the large cells (supported on thinner cellular stalks) surrounding the trapdoor could be glands for the transport of salt and water. Although these cells do not appear to provide the channels as the standing gradient model of Diamond and Bossert requires, a possible site in which a local concentration of salt might be set up could be a subcuticullar space in the gland cell (Findlay and Mercer, 1971).

2. Is the water flux closely dependent on the ion flux?

Although there is no direct answer, the evidence does not conflict with the idea of a linked solute–water transport system. That the resetting process is inhibited by low temperature, iodoacetamide and sodium azide, but unaffected by dark, indicates that the energy for the outward solute transport comes from respiration and not photosynthesis. In animal systems (Keynes, 1969) and in the salt glands of *Limonium* (A. E. Hill, personal communication) salt transport is dependent on ATP production.

If the main barrier to the diffusion of ions is similar to that in single plant cells the p.d. between the lumen of the *Utricularia* bladder and the outside is of the wrong sign and magnitude to be produced by passive diffusion of ions, but is of appropriate sign to arise from an electrogenic transport of chloride ions. The magnitude of this p.d. and the rate of efflux of water from the bladder are both diminished by the addition to the external solution of inhibitors of respiration.

3. Does the ratio of water molecules transported per ion have a reasonable value?

In the exudate from a bladder resetting under oil $[K] = 5$ mM, $[Na] = 16$ mM and $[Cl] = 7$ mM. The identity of the other anions is not known, but assuming that the total ionic concentration is equivalent to 40 mOsm, and the concentration of water is about 55 M, then for each ion transported, 1375 molecules of water move from the lumen to the outside. This value, although at least four times that reported for animal transport systems (Keynes, 1969) and for the plant systems *Limonium* (Hill, 1967) and *Aegialitis* (Atkinson *et al.*, 1967), may indicate that the solute and water transport system in *Utricularia* is a very efficient one. On the other hand, there may be other solutes in the exudate, and a measure of the osmotic pressure of the exudate is needed.

A bladder can be triggered at least six times. If we suppose that about the same amount of salt is transported out of the bladder each time it resets, the loss of salt from the system after six resetting cycles is approximately 40% of the total in the lumen and the cells of the bladder wall. This calculation assumes that the luminal volume increases by 40% on each triggering, and that the total salt in the cells of the bladder wall is roughly the same as the amount in the luminal fluid of an untriggered bladder. Thus a continuous pumping of ions into the bladder in order to supply ions for the outward solute–water transport is not absolutely necessary. It may occur, but the steady-state influxes alone are not, in any case, sufficient to replace the ions lost during each resetting phase.

We conclude that a reasonable model to account for the resetting of the bladder of *Utricularia* is one in which, in the specialized gland cells, ions are pumped from the cell, possibly to a subcuticular space. These ions provide an osmotic gradient across the cell membrane, down which water moves passively. From the subcuticular space salt and water flow to the outside solution. During the resetting phase with the bladder under oil, there is a net flux of sodium, potassium and chloride ions from the bladder. The electrochemical equilibrium p.d.s between the luminal solution and the exudate are $\psi_K = -7\,\text{mV}$, $\psi_{Na} = +7\,\text{mV}$, $\psi_{Cl} = +4\,\text{mV}$. The observed p.d. is $+134\,\text{mV}$, and so the net fluxes of sodium and potassium ions are down the electrochemical gradient. The net flux of chloride ions, however, is against the gradient, and thus chloride must be actively transported. We conclude that the chloride efflux contributes to the apparent electrogenic p.d. between the lumen and the outside solution (II(o)).

The results described in II(q) indicate that when the salt transport system is stopped the water permeability is extremely low, since water does not leak back into the bladder. This observation could be explained if it is assumed that minimum hydrostatic pressure is required in the subcuticular space before solution can flow to the outside. If this flow were through large pores, sucrose in the external solution would not be expected to be osmotically effective (II(r)).

References

ATKINSON, M. R., FINDLAY, G. P., HOPE, A. B., PITMAN, M. G., SADDLER, H. D. W. and WEST, K. R. (1967). *Aust. J. biol. Sci.* **20**, 589.

DIAMOND, T. M. and BOSSERT, W. H. (1967). *J. gen. Physiol.* **50**, 2061.

FINDLAY, NELE. and MERCER, F. V. (1971). *Aust. J. biol. Sci.* **24**, 647.

HILL, A. E. (1967). *Biochim. biophys. Acta* **135**, 454.

KEYNES, R. D. (1969). *Q. Rev. Biophys.* **2**, 177.

LLOYD, F. E. (1942). "The Carnivorous Plants". Waltham.

VIII.3

Hormonal Control of Ion Movements in the Plant Root?

J. C. Collins and A. P. Kerrigan

Botany Department, University of Liverpool, England.

Physiological investigation into the membrane processes of excised roots has proceeded apace over the last decade. But, as is usually realized, there is a need to integrate these findings into the workings of the whole plant. There has been a similar expansion in plant hormone research, and there are now at least four major hormones, or hormone groups, to be considered. It is the point of convergence of these two research fields that we wish to discuss here.

Evidence has accumulated in the literature to show that cytokinins and abscisic acid (ABA) may have an important role to play in plant transport phenomena. Itai (1967) has shown that kinins are produced in the plant root and transported to the shoot, and he showed this transport to be affected markedly by water stress of the root. Kinin transport decreased under stress conditions. Conversely Most (1971) showed that ABA is accumulated in the leaf as a consequence of stress conditions; in addition Hocking *et al.* (1972) have shown that there is some transport of ABA from the leaf to the root.

In this context two other reports are pertinent; Tal and Imber (1971) purport to show that "cytokinins decrease stomatal resistance by opening stomata and increase the resistance of the root to absorption of water. The general effect of the plant hormone is to reduce turgor". On the other hand they found that ABA increased stomatal resistance but decreased the root resistance. This work on whole tomato plants is complemented to some extent by that of Glinka and Reinhold (1971) who showed that ABA increases water permeability in tissue cylinders of carrot and stem pith of *Pelargonium*; kinetin was found to have the opposite effect.

The work we wish to present shows the *in vitro* workings of hormones (Kinetin and ABA) in the excised root and we also report *in vivo* effects which we tentatively attribute to these hormones.

It is a relatively easy matter to study the exudation processes of excised maize roots when they are bathed in a medium containing added hormone, in this case either kinetin or ABA. Table I gives some preliminary data obtained from such experiments. In each case a stock solution of hormone dissolved in dimethyl sulphoxide (DMSO) was used, an equivalent amount of DMSO was added to the controls.

TABLE I. Volume flows of 3-day old maize roots treated with hormone
(Jv in $\mu l\ cm^{-2}\ h^{-1}$)

Treatment	Time from excision	
	4–5 h	24–25 h
ABA in DMSO	2.56 ± 0.26	2.68 ± 0.29
DMSO control	1.79 ± 0.42	1.96 ± 0.36
Kinetin in DMSO	1.13 ± 0.17	0.37 ± 0.07
DMSO control	1.73 ± 0.27	1.63 ± 0.20

Ionic concentrations in the exudate after hormone treatment (all in mM)

Treatment	[K]	[Ca]	[Cl]
ABA in DMSO	20.5 ± 0.7	1.10 ± 0.03	17.2 ± 0.7
Kinetin in DMSO	16.1 ± 1.1	2.90 ± 0.27	15.4 ± 1.2
DMSO control	22.1 ± 0.7	1.84 ± 0.09	23.1 ± 1.7

Hormone levels were both at 2×10^{-6} M.
DMSO in all solutions was $2.5 \times 10^{-3}\%$ v/v.
All values are mean $\pm$ standard error.

TABLE II. Ionic concentrations in the root tissue of maize after 24 h treatment with hormone

Treatment	Ion concentration (mM/Kg fresh wt)		
	[K]	[Ca]	[Cl]
Kinetin in DMSO	94.2 ± 2.9	2.48 ± 0.18	69.3 ± 4.1
ABA in DMSO	91.5 ± 2.3	3.74 ± 0.31	84.0 ± 3.2
DMSO control	69.4 ± 3.8	2.88 ± 0.25	41.5 ± 2.7

Both kinetin and ABA had a final concentration of 2×10^{-6} M in the medium.
DMSO was always at $2.5 \times 10^{-3}\%$ v/v.

The effect of kinetin and ABA on the net ion fluxes into the root tissue was studied and Table II shows values for the ionic concentrations in root tissue after 24 h immersion in solutions containing added hormones.

This seems to show that both kinetin and ABA exert profound effects on the ionic movements in isolated maize roots. Kinetin drastically reduces the volume flow, though it takes some 20 h for the full magnitude of this inhibition to become apparent; there is also a reduction in the concentration of exudate produced, and it can be concluded that Lp is decreased in the presence of kinetin. Addition of ABA to the bathing fluid increases the rate of volume flow, but has only a negligible effect on the concentration of fluid produced; here it can be concluded that Lp increases. Obviously the effect of kinetin is to reduce drastically fluxes into the exudate, whilst ABA has the opposite effect; this is true for water and ion fluxes.

The net flux into the root is increased by both hormones. This can only be an effect on the influx for the normal chloride efflux is negligible. The major part of the influx is into the cortex, as the cortex is about 10 times the volume of the stele; that is the flux is into the symplasm. It is noteworthy that it is the Ca fluxes that are least affected by hormone treatment and that Ca is very poorly transported by the symplasm.

We can conclude from these results that kinetin will increase the ion fluxes into the root whilst reducing the ion and water transport from the root; on the other hand ABA increases the ion flux into the root and also the ion and water transport from the root. ABA exerts its effect rapidly, within an hour or so, whilst for kinetin there is a time lag of some 20 h before it exerts its total effect.

What of the workings of these hormones *in vivo*? We have conducted two experiments the results of which we believe to be indicative of *in vivo* hormone control of transport processes.

The parameters of the exudation process in maize are well known, but if the isolated roots of light-grown maize be examined, then certain differences appear, as compared with the usual dark-grown material. These are set out in Table III. Further, if the ionic contents of these roots are measured (Table IV) then further differences are seen.

TABLE III. Volume flows and exudate concentrations from light and dark grown roots

Dark-grown roots Medium Concentration—KCl	Jv (μl cm^{-2} h^{-1})	osmolarity of exudate (mOsm)
0·1	1·98 ± 0·24	58·9 ± 1·1
1·0	1·41 ± 0·34	61·5 ± 1·1
10·0	1·75 ± 0·18	80·8 ± 2·2
50·0	0·88 ± 0·08	129·4 ± 3·2
Light-grown roots		
0·1	0·95 ± 0·15	61·2 ± 1·4
1·0	0·73 ± 0·14	58·9 ± 1·7
10·0	0·98 ± 0·12	90·6 ± 1·8
50·0	0·85 ± 0·06	140·3 ± 1·4

All values expressed as mean ± standard error.
All media contained 0·1 mM $CaCl_2$.

TABLE IV. Ionic concentrations in the root tissue of 3 day-old maize seedlings

	[K]	[Na]	[Ca]	[Mg]	[Cl]
Light	81·3	1·87	1·99	1·72	30·1
grown	±2·1	±0·22	±0·22	±0·11	±1·3
Dark-	74·5	0·91	1·70	1·51	38·7
grown	±2·1	±0·08	±0·14	±0·04	±1·2

All values are in mM/Kg fresh wt ± standard error.
Growth medium was 1·0 mM KCl + 0·1 mM CaCl$_2$.

The decrease in volume flow from light-grown roots and the increase in chloride concentration of the tissue is in part reminiscent of the effect of kinetin. Obviously it is possible to view these results from several different standpoints but there is good evidence that the light regime affects the hormonal balance of plants and it is not difficult to explain these results using the earlier quoted hormonal effects, or a combination of them. There is an obvious need for a study to be made of the hormonal balance of these tissues.

An obvious role for plant hormones is in the integration of root–shoot functions, and in an attempt to investigate this we have studied the chloride fluxes in both excised mustard roots and whole mustard plants. One reason why we used mustard is that the whole plant is so much easier to handle than a whole cereal plant. The exudate fluxes from young isolated mustard roots were measured, and chloride transport to the shoot of mustard seedlings of a similar age was also measured; the data are presented in Table V.

TABLE V. Comparison of chloride fluxes in excised mustard roots and whole mustard plants

Age of plant (days)	Flux from excised roots	Flux into shoot
3–4	5·92	31·6
4–5	4·95	68·7
5–6	5·11	117·5
6–7	7·15	—

All flux values are in mol × 10^{-9} plant^{-1} h^{-1}.

The chloride flux is some 5 times higher in the whole plant as compared with the excised mustard root. This does seem to give some evidence for a strong root–shoot interaction in the chloride transport processes. However, care must be taken in using results from excised roots, for not all excised roots may function in the same way as the maize roots, where there is some evidence that fluxes in the excised root and the whole plant are equivalent (Anderson and Allen, 1970).

We should perhaps conclude this miscellany of data by first stressing their preliminary nature and secondly by underlining the necessity for information on the *in vivo* hormone levels in such tissues and plants, for most plants contain their own specific cytokinins, not kinetin (6-furfuryl amino purine) and these may well elicit a much greater response.

Acknowledgements

We wish to thank the A.R.C. for financial support.
An ABA sample was kindly supplied by Shell, and we wish to thank Dr D. R. Robinson of Shell, Sittingbourne, for his help and advice.

References

ANDERSON, W. P. and ALLEN, E. (1970). *Planta* **93**, 227–232.
GLINKA, Z. and REINHOLD, L. (1971). *Pl. Physiol., Lancaster* **48**, 103–105.
HOCKING, T. J., HILLMAN, J. R. and WILKINS, M. B. (1972). *Nature New Biology* **235**, 124–125.
ITAI, C. (1967). Ph.D. thesis, Hebrew University, Israel.
MOST, B. H. (1971). *Planta* **101**, 67–83.
TAL, M. and IMBER, D. (1971). *Pl. Physiol., Lancaster* **47**, 849–850.

VIII.4

Ferric-EDTA Absorption by Maize Roots

J. T. Beckett and W. P. Anderson*

Department of Botany, University of Liverpool, England.

I. Introduction

Iron chlorosis is observed in plants in many regions throughout the world. This micronutrient deficiency is not confined to any one type of environment nor is the cause–effect relationship simple. Chlorotic symptoms may be evident on acid sandy, citrus-growing areas (Smith and Specht, 1952) and on calcareous agricultural land (Lindsay and Thorne, 1954). However, it is on the more calcareous soils that deficiency symptoms are most obvious, because at high soil pH values iron is precipitated as ferric hydroxide and is then unavailable to the plant. But soil pH is only one factor and others (e.g. the presence of other heavy metals, soil phosphate status, soil sodium levels, CO_2 and bicarbonate concentrations) all may interact to limit the supply of iron to the plant.

The chief symptom of iron chlorosis is always obvious, the yellowing of the green tissues implying a lack of chlorophyll; but the mechanism by which iron deficiency depresses chlorophyll biosynthesis remains somewhat obscure. Protoporphyrins are produced through a series of intermediate porphyrinogen compounds from δ-amino levulinic acid (ALA) which is

* *Present address*: Department of Botany, Washington State University, Pullman, Washington, U.S.A.

synthesized from the precursors glycine, pyridoxal phosphate and succinyl-CoA by the action of ALA-synthesase. Incorporation of iron into proto-porphyrins results in the formation of haem-related compounds, while incorporation of magnesium yields the chlorophylls. There is no direct requirement for iron at any stage of chlorophyll biosynthesis. An indirect requirement may be imagined in that succinyl-CoA, a precursor of ALA, is produced by the Krebs cycle which is in turn dependent on several iron-containing enzymes (Granick, 1965). Alternatively it may be that ferredoxin levels are limiting and that chloroplast biogenesis is depressed. Again, it has been suggested (Wilkie and Miller, 1960) that iron deficiency may inter-fere with riboflavin incorporation with a resultant decrease in the levels of flavoprotein enzymes. It is therefore apparent that there is, as a result of iron deficiency, a plexus of interacting biochemical consequences which together produce the observed chlorotic symptoms.

Jacobson (1951) first showed that iron deficiency in several higher plant species could be alleviated by an application of ferric iron chelated by the synthetic compound ethylene diamine tetraacetic acid (EDTA). Since this demonstration much effort has gone into field trials and laboratory investi-gations to establish practical rules for Fe-EDTA treatment in plant hus-bandry, and it is now the most common and effective method of providing iron in potentially deficient situations. Yet our understanding of the basic physiological processes upon which this effective practical treatment is founded is slight indeed. The remainder of this paper will deal with prelimin-ary attempts in this laboratory to elucidate the mechanism of Fe-EDTA uptake by maize roots.

A review of the literature relevant to our immediate aim at once reveals a most pointed difference of opinion. One group of workers (see Brown et al., 1961; Brown and Jones, 1962; Tiffin, 1970; Tiffin et al., 1959; among others) believe that Fe^{+++} is displaced from the Fe-EDTA complex at the root surface immediately prior to uptake across the cell membrane, by a process analogous to competitive chelation; the competing chelating agent is presumably the Fe^{+++} carrier molecule in the membrane. In these authors' contention, any uptake of EDTA is incidental and is not directly coupled to the Fe^{+++} uptake. A second group of workers (see Hale and Wallace, 1960; Hill-Cottingham and Lloyd-Jones, 1965; Weinstein et al., 1954; among others) may be broadly categorized as believing that the metal–ligand com-plex is transported across the cell membrane as a single entity. We have attempted to resolve this difficulty by observing the uptake of Fe-EDTA labelled with ^{55}Fe and ^{14}C(U)-EDTA, by the primary roots of 5–7 days old seedlings of Zea mays (L.). We believe from our experience in studying the uptake of other ions that there are advantages in using such relatively homo-geneous material; most other workers have used the whole root systems of week-old plants.

II. Materials and Methods

Zea mays (cultivar White Horse Tooth) seeds were briefly rinsed in $HgCl_2$ solution (0·5 %) to surface-sterilize and were then soaked in water for several hours before being set out on damp filter paper in seed trays to germinate in the dark at $25^\circ C$. When the germinating radicles were about 2 cm long the seeds were carefully arranged on an open mesh plastic net suspended at the surface of an aerated culture solution containing 1·0 mM KCl, 0·1 mM $CaCl_2$ and a range of equimolar Fe-EDTA concentrations from 0·07 mM to 1·0 mM, at a pH of 4·5. In the final series of experiments described here (see Fig. 4), 0·5 mM equimolar Fe-EDTA was used with the further addition of a range of concentrations of Na_4EDTA to give various excesses of EDTA above the Fe concentration. In all cases various amounts of Na^+ were present in solution because the Na salts of EDTA and Fe-EDTA were used. The volume of the culture solution was sufficient to ensure negligible depletion of any ion during the growth period. The seedlings were kept in continuous darkness at $25^\circ C$ until the average primary root length was *ca.* 11 cm; total growth time to this stage was in the range 5–7 days. For the last 48 h of growth (see later) Fe^{55}-EDTA-$C^{14}(U)$ was added to the culture solution.

At the end of the growth period the seedlings were removed from the culture solution and 10·2 cm apical segments were cleanly cut from the primary roots. The cut ends were inserted and sealed into 100 μl Microcap capillaries and these excised roots were set to exude (for details see Anderson and Collins, 1969) from a bathing solution identical in all respects to the growth solution. The fluid exudation rate was monitored at regular intervals and the exudates collected over 24 h were pooled in a 5 ml stoppered flask. The volume collected was determined by weighings. The exudate fluid was stored in a refrigerator in the dark. The roots used in exudation were then washed in unlabelled solution, blotted dry, weighed and digested in concentrated nitric acid. The digestate was then dry-ashed carefully to avoid charring and suspended in 25 ml distilled water. This preparation too was stored in a refrigerator in the dark.

Fe^{+++} was assayed after suitable dilutions with distilled water by comparison with standards on the absorption mode of a Unicam SP 90A flame spectrophotometer. Fe^{55} and ^{14}C were assayed by adding 10 ml of Bray's scintillation fluid to 0·1 ml aliquots of exudate and digestate and counting in a Packard 3320 "TriCarb". It was found that ^{55}Fe counted adequately in a channel set for ^3H, with only 2·7 % of the counts appearing in the channel set for ^{14}C. Digestate samples prepared without dry-ashing could not be counted; the presence of HNO_3 greatly distorts the β-energy spectra of the radionuclides as seen by the photomultipliers.

The radiochemicals were obtained from The Radiochemical Centre, Amersham, Bucks, England. All other reagents were AnalaR grade.

III. Results

Since it is convenient to assay iron by radioactive counting and since radioactive counting is possibly the only method of assay of the quantities of EDTA with which we are concerned here, we conducted a preliminary trial to determine the length of time required for these radioactive labels to reach tracer equilibrium in our system. We could chemically assay only the Fe^{+++} and so we are forced to assume that the EDTA specific activities level off at about the same time as we have observed for the Fe^{+++}. We can only provide fairly crude information of the rate of tracer build-up because our present techniques require the collection of exudate over many hours

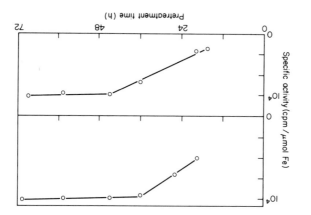

FIG. 1. (above) Plot of the specific activity of Fe in the tissue of excised roots at the end of the exudation period, against the time of pretreatment (see text). (below) Similar plot for exudate samples collected over 24 h, against pretreatment time (see text).

and hence we have relatively unresolved time-average estimates of exudate composition. Nevertheless it should be realized that our objective was simply to determine a time of pretreatment exposure to radioactive tracer after which tracer equilibrium could be safely assumed, all external conditions remaining constant both before and after root excision.

Figure 1 shows the rise in specific activity of Fe^{+++} in both root tissue and exudate, plotted against time of pretreatment exposure to tracer prior to root excision. The external solution contained 0·1 mm Fe-EDTA (equi-molar) at an iron specific activity of 10^4 cpm μmol^{-1} at pH 4·5. Note that 24 h (the exudation period) must be added to all the times indicated here if the total time of tissue exposure to tracer is required.

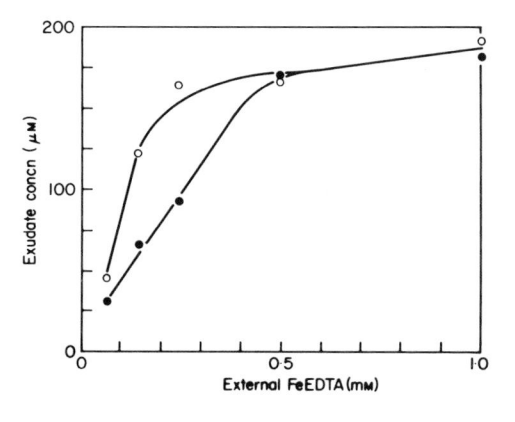

FIG. 2. Exudate concentrations of Fe (μM) and EDTA (mM) at various external Fe-EDTA concentrations.

It is clear that the specific activity of iron in both exudate and tissue has equilibrated with that of the external solution, within experimental uncertainty, if the roots are exposed to tracer for 48 h prior to excision. We took this finding to hold for EDTA specific activity, although we have no method of directly verifying this, and thenceforth we adopted a 48 h pretreatment time as standard, and used our knowledge of external specific activities to allow direct conversion of the radioactive assays of Fe^{55} and C^{14} to chemical concentrations.

Figures 2 and 3 show the concentrations of Fe^{+++} and EDTA in exudate and root tissue at the end of the exudation period, at different levels of external

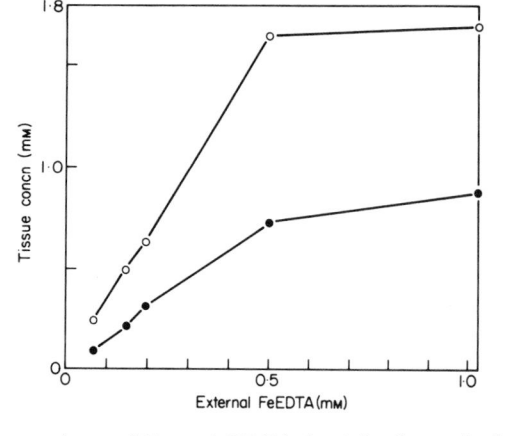

FIG. 3. The concentrations of Fe and EDTA (mM) in the excised root tissue at the end of exudation, at various external Fe-EDTA concentrations.

Fe-EDTA equimolar concentrations. It can be seen from these data that we have clear and unequivocal evidence that (a) the Fe^{+++} concentration in the xylem exudate exceeds the EDTA when the external Fe-EDTA concentration is low, and (b) at higher external Fe-EDTA concentrations there are approximately equimolar concentrations of Fe^{+++} and EDTA in the xylem stream.

Thus we appear to have evidence to support the contentions of both opposed schools of thought referred to in the Introduction; it seems, from our observations with young primary maize roots, to depend on external Fe-EDTA concentration. The tissue Fe^{+++} always greatly exceeds the EDTA concentration. We shall return to this point in the discussion.

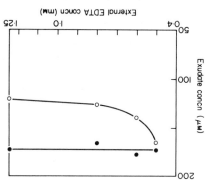

Fig. 4. Effect on exudate concentrations of Fe and EDTA (μM) of changing the external EDTA concentration at constant *external* Fe-EDTA concentration of 0·5 mm.

Finally in Fig. 4 we show the concentrations of Fe^{+++} and EDTA in the exudates of roots bathed in 0·5 mm Fe-EDTA with added $EDTA^{4-}$ in varying amounts. In this case it is clearly seen that an excess of EDTA over Fe is present in the xylem stream, again clear evidence that EDTA in some form is absorbed into and translocated by the xylem.

In all that has been said above, the assays that were assumed to be of EDTA were in fact of ^{14}C. We have evidence (Beckett and Anderson, to be published) that most of the ^{14}C-labelled compound in xylem exudate runs as EDTA in paper chromatograms (Whitman: *n*-butanol:acetic acid:water 120:30:50) with R_f values in the range 0·125–0·2 although there are considerable fractions with ^{14}C activity which show R_f values of 0·375 and 0·5. It is of interest that these fractions also have ^{55}Fe activity. They are the first degradation products of EDTA (both metabolically and photo-induced) and as such can be treated, in the present study of transport from the medium to the xylem, as though they are unaltered EDTA. Thus the ^{14}C activity is a good measure of total EDTA transport through the tissue into the xylem stream.

(a)

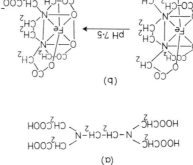

(b)

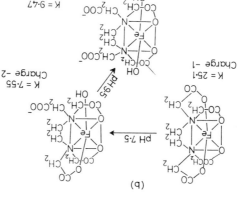

Fig. 5. (a) Structure of EDTA. (b) Structures of Fe-EDTA with increasing pH.

IV. Preamble to the Discussion

Ethylene diamine tetraacetic acid (EDTA) is well known for its ability to form very stable chelates with the transition metals. It dissociates to a tetravalent anion, the pK values being 2.00, 2.67, 6.16 and 10.26. The first two protons dissociate from the carboxyl groups of two of the acetic acids, one on each nitrogen. The third and fourth dissociations are more complex because of the intermediate formation of betaine structures with protonated nitrogens (for details see Chapman, 1955). It is the completely dissociated $EDTA^{4-}$ which forms the very stable metal chelates. The ionization state of EDTA is therefore of vital importance in any discussion of experimental data, and we should like to point out that, so far in this text, we have made no explicit reference to the ionization state in presenting the results. We shall return to this matter shortly.

$EDTA^{4-}$ acts as a hexadentate ligand for Fe^{+++}, forming an octahedral structure as shown in Fig. 6 (Hoard et al., 1961). The chelated compound usually has a net charge of -1. The theory of metal–ligand bonding is not easy, but a good account is given by Orgel (1960). The simplest explanation, which is unfortunately inadequate to describe much that has been observed, is that the five unpaired electrons in the 3d shell of Fe^{+++}, and one from the

3p shell, undergo transitions and hybridization due to the effect of the ligand field, to form six d^2sp^3 orbitals, octahedrally arranged. These orbitals then form bonds with the nitrogen atoms of $EDTA^{4-}$ and with the oxygens of the four carboxyl groups, in the double bonds of which there is now resonance; see Figs 5b, 6. The inadequacy of this simple idea is immediately apparent from Fig. 6 where we may see that it is now held that Fe^{+++} is seven

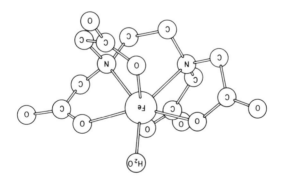

Fig. 6. Octahedral structure of Fe-EDTA.

co-ordinate in Fe-EDTA⁻, not six; the seventh position is filled by a molecule of H_2O. Nevertheless this simple explanation has the advantage that it can be visualized; a more satisfactory theory can be produced by a molecular orbital approach.

The effect on the Fe-EDTA⁻ chelate of increasing pH is shown diagrammatically in Fig. 5b. As the OH^- concentration increases, OH^- groups replace the carboxyl group bonds to the iron atom with consequent alterations to the stability constant of the complex, and to the net charge carried; these values are indicated in Fig. 5b. There is the further effect of the extreme insolubility of $Fe(OH)_3$, solubility product $\simeq 10^{-36}$, which effectively removes "all" non-chelated iron as a precipitate. Thus as pH is increased, the amount of Fe^{+++} sequestered per unit amount of $EDTA^{4-}$ will decrease for these two reasons. Consideration must also be taken of the effect of low pH on the stability of a metal–ligand complex. Firstly it should be obvious that the pK values given earlier for free EDTA will not apply to a metal–ligand complex because the protonation sites on $EDTA^{4-}$ are not freely accessible, being involved in bonding to the metal. However, as the H^+ ion concentration increases there will be increasing competition for the protonation sites, and as these are occupied by protons, the ligand field experienced by the metal will diminish. For example it is known that a $CaHEDTA^{1-}$ does exist but has a stability constant of $10^{3 \cdot 5}$ compared to the normal $CaEDTA^{2-}$ value of $10^{10 \cdot 5}$ under identical conditions.

V. Discussion

The data in Fig. 2 are clear evidence that Fe^{+++} concentration in xylem exudate exceeds EDTA concentration at external Fe-EDTA$^-$ levels up to 0.5 mm. It is quite clear that the difference $\{[\text{Fe}] - [\text{EDTA}]\}$ is dependent on external Fe-EDTA$^-$ concentration. Such observations cannot be accounted for on the basis of an uptake of Fe-EDTA$^-$ alone, nor can they be plausibly explained on a model that requires the Fe^{+++} to be displaced from the chelate immediately prior to uptake, as the only Fe^{+++} transport system. We propose to interpret these data in terms of three fluxes ϕ_1, ϕ_2 and ϕ_3 of Fe^{+++}, Fe-EDTA$^-$ and EDTA respectively. We shall return later to what precisely we mean by "EDTA" as just used. Note that these fluxes will measure only the transport from the external medium to the xylem stream, as determined by collecting xylem exudate over 24 h. They do not measure total uptake by the root.

As defined, the concentration of Fe^{+++} in the exudate at any external Fe-EDTA concentration C_x (mm) will be given by

$$C^{\text{Fe}}_{C_x} = \frac{\int_0^T \phi_1(C_x)\,dt + \int_0^T \phi_2(C_x)\,dt}{V} \qquad (1)$$

where $V(1)$ is the total volume of exudate collected in time $T(h)$; if C^{Fe} has units μm, then $\phi_1(C_x)$ and $\phi_2(C_x)$ will have units of $\mu mol\ h^{-1}$ per 10 cm root length. If we assume that our roots are in steady state exudation throughout the time T, eqn (1) is simplified to

$$\frac{VC^{\text{Fe}}_{C_x}}{T} = \phi_1(C_x) + \phi_2(C_x) \qquad (2)$$

where all quantities on the right-hand side are known.

In equivalent fashion we may obtain for the EDTA transport

$$\frac{VC^{\text{EDTA}}_{C_x}}{T} = \phi_2(C_x) + \phi_3(C_x) \qquad (3)$$

Now we note from Fig. 2 that $C^{\text{EDTA}} = C^{\text{Fe}}$ when $C_x \geqslant 0.5$ mm so that by combining (2) and (3), V and T being identical in both, we obtain

$$\phi_1(C_x \geqslant 0.5) = \phi_3(C_x \geqslant 0.5)$$

In words therefore, the flux of free Fe^{+++} is equal to the flux of "free EDTA" when the external concentration of Fe-EDTA$^-$ is equal to or exceeds 0.5 mm. Consider now the data in Fig. 4; we employ the same symbolism with the replacement of C_x, the external Fe-EDTA$^-$ concentration, by C_x, the external EDTA concentration in excess of Fe. In this case Fe-EDTA$^-$ is held constant at 0.5 mm, a level which produces equimolar concentrations of Fe^{+++} and EDTA in xylem exudate, from the equimolar external solution. We notice that the exudate EDTA concentration in Fig. 4 remains substantially constant so that we may write

$$\phi_2(C_x) + \phi_3(C_x) = \text{constant}$$

for all C_x in Fig. 4. It seems to us unlikely that ϕ_2 and ϕ_3 are not separately constant; Fe-EDTA$^-$ remains constant in the external solution because additions of excess EDTA do not much alter the equilibrium concentration of so stable a chelate. Thus the variation in exudate Fe^{+++} concentration in Fig. 4 can be all ascribed to variation in ϕ_1, the free Fe^{+++} flux. In none of our experiments did we find significant variation in the rate of volume (water) exudation. We now assume that $\phi_1 \to 0$ as $C_x \to \infty$ and that ϕ_1 is approximately zero when $C_x = 0\cdot3$ mм; thus we can find ϕ_2 directly from the level value of the iron concentration curve in Fig. 4 as $\phi_2 = 126V/T$. The mean value of V/T in these experiments was $1\cdot25$ μl h^{-1} so that ϕ_2, the Fe-EDTA$^-$ flux, has the value $157\cdot25$ pmol h^{-1} per 10 cm root length.

Let us now consider the situation when $C_x = 0$ in Fig. 4, i.e. at the equi-molar Fe-EDTA$^-$ concentration of $0\cdot5$ mм. Eqn (2) will hold of course, with C_x replacing C_x and by inserting the measured values for the RHS of (2) and the value of ϕ_2 just determined, we find $\phi_1(C_x = 0) = 55$ pmol h^{-1} per 10 cm root length. Under the conditions of the experiment of Fig. 4 this is of course also the value of ϕ_1 $(C_x \geqslant 0\cdot5)$. We have earlier shown that ϕ_3 $(C_x \geqslant 0\cdot5) = \phi_1(C_x \geqslant 0\cdot5)$ so that we now have the maximal values of all three fluxes ϕ_1^{max}, ϕ_2^{max} and ϕ_3^{max}.

The next obvious step would be to try to characterize the concentration dependence of these fluxes, deriving K_m or S_{50} values for example. Although it is not easy to separate the fluxes ϕ_1, ϕ_2 and ϕ_3 anywhere other than at the concentration independent maximal values as has just been done, it would nevertheless be possible to ask a computer for an optimal fit, based either on polynomial expansions for ϕ_1, ϕ_2 and ϕ_3, or on the assumption that these fluxes are described by some enzyme kinetics equation; for example

$$\phi_1 = \frac{\phi_1^{max}[\text{Fe}]}{K_{m,1} + [\text{Fe}]}$$

or

$$\phi_2 = \frac{\phi_2^{max}[\text{Fe-EDTA}]^{n_2}}{K_2' + [\text{Fe-EDTA}]^{n_2}}$$

depending on whether we chose the Michaelis-Menten or the Hill equation. We know the ϕ^{max} values, and the ns should be integers so that parameter values could be generated from fitting to the data in Figs 2 and 4. However, we have decided not to attempt to evaluate any such parameters for these fluxes for two reasons. The first and more important reason is that we think it is most unlikely that these exudation fluxes ϕ_1, ϕ_2 and ϕ_3 are produced by transport mechanisms at a single membrane. There is an increasing body of opinion (Lauchli et al., 1971; Pitman, 1972; Anderson, 1972) which holds that there are two membrane transport mechanisms, the first in the cortical cells and the second in the parenchyma cells of the stele, involved in the transport of ions such as K^+ and Cl^- to the xylem exudate. We believe that at least two membranes are similarly involved in the present situation and although we think there is some meaning and use in calculating the ϕ^{max} values as we have done on a minimum of assumptions directly from the data; we further think that any attempt to calculate any other parameters to characterize the flux-external concentration dependence would be quite without profit.

The second reason is that we would have difficulty in deciding what concentrations we should use, particularly in dealing with the data in Fig. 4. Let us examine the external situation in the light of our knowledge of the chemistry of Fe-EDTA$^-$. In equimolar solutions, as used in the experiments of Fig. 2, we can compute the free Fe^{+++} concentration in the external solution from

$$\frac{[Fe - EDTA]}{[Fe][EDTA]} = 10^{25.1}$$

and since the free EDTA has dissociated from Fe^{+++} we may write

$$[Fe] = \sqrt{[Fe\text{-}EDTA]} \times 10^{-12.5}$$

a truly minute level of free Fe^{+++}. One may reasonably express surprise at any suggestion of a carrier situated at an exterior membrane, which operates maximally at a substrate concentration of the order 10^{-15} M, as may seem to be the case for our free Fe^{+++} flux, ϕ_1. However it should be pointed out that a recently isolated Fe^{+++}-deferoxamine chelate from *Streptomyces pilosus* (Bock and Lang, 1972) has a measured stability constant of 10^{32}, so that naturally occurring carriers of extreme affinity for Fe^{+++} may be found. On the other hand there is an alternative to having a carrier pick up free Fe^{+++} as it dissociates from Fe-EDTA$^-$ in the external solution; this model will be discussed shortly.

It is very difficult to know the chemical state of excess EDTA in the experiments of Fig. 4. The excess was supplied as Na_4EDTA which will dissociate to 4 Na^+ and EDTA^{4-}. In the solution we have used there will be competition for this EDTA^{4-} by H^+ (the solution pH is 4.5-5.0), by Ca^{2+} (0.1 mm Ca^{2+} present) and by Fe^{+++} (the traces present as a result of mass action dissociation from Fe-EDTA$^-$). The protons will occupy the pK 10.26 and 6.16 sites of free EDTA^{4-} forming H_2EDTA^{2-} which will no longer act as a ligand. The Ca^{2+} can be chelated as Ca-EDTA^{2-} or conceivably as the much less stable chelate Ca-HEDTA$^-$. In the event of all these possibilities any attempt at evaluating the variation in Fe^{+++} concentration in the external solution with excess EDTA present, is fruitless. All we can say is that the free Fe^{+++} will decrease to less than the value at equimolar 0.5 mm Fe-EDTA$^-$. Fe^{+++} will effectively compete with Ca^{++}, both for Ca-EDTA^{2-} and for Ca-HEDTA$^-$. Hence possibly, the decrease in ϕ_1 observed in Fig. 4.

There is an alternative model to explain the observed fluxes which does not require uptake of Fe^{+++} and "free EDTA" from the minute traces of each in the external solution. We imagine that Fe-EDTA$^-$, externally present in mm concentrations, crosses the plasmalemmata of the cortical cells either by diffusion or in association with a carrier, and thence migrates in the sym-plasm. During this passage through the cell cytoplasm the Fe-EDTA$^-$ will be subjected to possible metabolic breakdown. We know that total tissue Fe much exceeds total tissue EDTA (Fig. 3), and that EDTA is degraded to the eventual evolution of $^{14}CO_2$ (Beckett and Anderson, to be published). Thus, Fe^{+++} released from a partially degraded EDTA, as yet unaltered Fe-EDTA$^-$ and various partially degraded EDTA types possibly still chelating Ca^{++} which is present in mm quantities, will be present in the stele for transport to the xylem. The proportions of these chemical species may be somewhat different from those we expected for the external solution, and may also be dependent on the external Fe-EDTA$^-$ concentration. In this manner one can imagine, perhaps more straightforwardly, how the observed exudation fluxes have arisen.

It is apparent that these are only preliminary experiments and that a great deal more detailed chemical information is needed. However we think we have delineated the problem to some extent. At low external Fe-EDTA levels the xylem exudate contains a greater concentration of Fe^{+++} than of "EDTA". At external Fe-EDTA concentrations in excess of 0.5 mm the exudate contains equimolar concentrations. The exudation flux is principally an Fe-EDTA$^-$ flux at high concentrations, but Fe^{+++} and "EDTA" fluxes must also be involved, the Fe^{+++} flux requiring a high affinity carrier, since it is responsible for the excess exudate Fe^{+++} concentration at low external Fe-EDTA levels.

References

ANDERSON, W. P. (1972). A. Rev. Pl. Physiol. (in press).

ANDERSON, W. P. and COLLINS, J. C. (1969). J. exp. Bot. 20, 72–80.

BOCK, J. L. and LANG, G. (1972). Biochim. biophys. Acta 264, 245–251.

BROWN, J. C., HOLMES, R. S. and TIFFIN, L. O. (1961). Soil Sci. 91, 127–132.

BROWN, J. C. and JONES, W. E. (1962). Soil Sci. 94, 173–179.

CHAPMAN, D. (1955). J. chem. Soc. 1766–1769.

GRANICK, A. (1965). In "Biochemistry of Chloroplasts" (T. W. Goodwin, ed.), Vol. II, 373–411, 391–392. Academic Press, London (1967).

HALE, V. Q. and WALLACE, A. (1960). Soil Sci. 89, 285–287.

HILL-COTTINGHAM, D. G. and LLOYD-JONES, C. P. (1965). J. exp. Bot. 16, 233–242.

HOARD, J. L., LIND, M. and SILVERTON, J. V. (1961). J. Am. chem. Soc. 83, 2770–2776.

JACOBSON, L. (1951). Pl. Physiol., Lancaster 26, 411–413.

LAUCHLI, A., SPURR, A. R. and EPSTEIN, E. (1971). *Pl. Physiol., Lancaster* **48**, 118–124.

LINDSAY, W. L. and THORNE, D. W. (1954). *Soil Sci.* **77**, 271–279.

ORGEL, L. E. (1960). "An Introduction to Transition Metal Chemistry". Wiley, New York.

PITMAN, M. G. (1972). *Aust. J. biol. Sci.* **25**, 243–257.

SMITH, P. F. and SPECHT, A. W. (1952). *Pl. Physiol., Lancaster* **28**, 371–383.

TIFFIN, L. O. (1970). *Pl. Physiol., Lancaster* **45**, 280–283.

TIFFIN, L. O., BROWN, J. C. and KRAUSS, R. W. (1960). *Pl. Physiol., Lancaster* **35**, 362–367.

WEINSTEIN, R., ROBBINS, W. R. and PERKINS, H. F. (1954). *Science* **120**, 41–43.

WILKIE, G. W. and MILLER, G. W. (1960). *Pl. Physiol., Lancaster* **35**, 516–520.

Discussion

Highotham began the final discussion by asking Shone if the difference in Ca^{++} distribution in endodermis, cortex, epidermis, pericycle etc. might be due to the differing cross sections of wall area from one region of root to the next. Shone agreed that it might be a possibility, but reaffirmed that the distribution is denser in the stelar regions. Penetration into the cortical cell vacuoles is possibly also slow, but it is not feasible to resolve this by auto-radiography. Jennings then asked how one might correlate Shone's early work on silica with his present work on organic molecules. Shone replied that the water free space estimate, using soluble silica, $Si(OH)_4$, was found to be independent of uptake time and of elution period, and also of silica con-centration. The silica free space is 0.2 of root volume in agreement with the present estimate of 0.25 and with other estimates; the main difference is that silica is apparently transported. In bean the xylem silica can rise to as much as 20 times the external medium concentration. It is unlikely that the silica is either polymerized or ionized, and it must be assumed that silica is one of the rare examples of non-polar compounds that are accumulated into root xylem.

Pitman then asked if the free space estimates included the trivial space at the exterior of the root. Shone said that in all the experiments the roots were merely surface-blotted. Pitman then said it may be more appropriate to subtract the surface-adhering volume, so reducing the estimates to about 5% of the root volume. Shone replied that he thought Hylmo's experiment with India ink was the only attempt to distinguish the surface-adhering volume from the internal free space volume, and that the correction factor he obtained was rather less than that suggested by Pitman.

Pallaghy asked Findlay in which direction the trap-door opens. If the internal pressure is released by inserting an open-ended capillary, does the trap-door still function on stimulating the hairs? Findlay said, "My reply to the first part is that the insects go in. Yes, it still works; it has two meta-stable states like a piece of flexed steel and it flips from one to the other". Smith then commented on Findlay's use of the "archaic inhibitor sodium azide" and then said in reference to his paper, "It is rather like watching Hamlet without the Prince of Denmark, or like eating tomato sauce without the meat pie.'' He was referring to the ''festering fly flux; in other words, what concentration of ions may be provided by the unfortunate insect?''

Findlay replied that a bladder will fire six to eight times without an insect having gone in.

Spanswick then said the hairs are rather reminiscent of the Venus fly-trap; are there action potentials associated with stimulation? Findlay replied that it is not possible to stimulate the opening by electrical means; varying the bladder potential over several hundred millivolts has not produced it. Umrath did fire one by using 1000 volts across a big coil, but a mechanical stimulus seems necessary. Old bladders are not usually stimulated by moving the hairs. Mechanical stimulus of the hair presumably distorts the myriad of cells near the trap-door and causes a collapse of turgor pressure. As the pressure collapses, the door will flip inward because there is 0.2 atm pressing it that way. The pressure is thus released, the insect has gone in, and the trap-door flips back to its second metastable state. Smith then asked about the terrestrial *Utricularia* species. Findlay said they are always found in moist conditions, and the epiphytic ones are mainly tropical. The ones known as ''Fairy Aprons'' grow in swamps under high humidity, and in fact the bladders are in water.

Cram then commented on Anderson's paper and asked if EDTA was normally present in plant cells. Anderson replied that it is a synthetic product, but is not dissimilar to naturally occurring organic anions which are taken up by plant cells. Many such anions—e.g. citrate, oxalate—will chelate iron inside the plant. If the iron is not chelated, it is sure to be precipitated as ferric hydroxide since the solubility product is 10^{-36}. Heller commented that root segments fed Fe^{+++} as $FeCl_3$ and then effluxed show an efflux of Fe^{+++}. Root segments fed chelated iron and then effluxed show an efflux of chelated iron.

Cram then commented on Collin's paper in regard to his finding of 80% stimulation and Cram and Pitman's findings of 80% inhibition in volume exudation by ABA. He said there is an agreement that ABA stimulates ion accumulation into root cells; Van Steveninck also confirms this. Cram then asked Collins if he had any idea what caused this stimulation and Collins said he had no idea. Pitman then spoke on the two pump idea for root-xylem transport. He said ABA does not affect ion entry into the root but may inhibit the secretory process to the xylem fluid; this would cause an increase of activity in the root tissue. Thus, the ABA effect on transport into the xylem is different from the effect on uptake into the tissue, which supports the idea of there being two pumps involved in ion uptake and xylem exudation, whatever the effect is at the stele. Collins then said that there is evidence over many years that volume exudation flow is stimulated by a factor 2 with ABA.

Pitman then agreed that after their original publication they discovered a great deal of variability in the ABA effect depending on the pre-treatment of the plants.

Loughman commented that in a series of studies on uptake and transloca-tion of phosphate in intact barley plants they had found no effect of kinetin, gibberellin and IAA, except at high concentration where there is some inhibi-tion of phosphate; there is also some evidence of a selective effect in that transport to the shoot seems more affected than uptake by the root. There were no synergistic effects of any of these. Pallaghy then asked about the endogenous levels of ABA in the roots. How significant is ABA in the whole plant in regulating ion transport and what is known about the rates of ABA transport from shoot to root? Collins replied that the only work he knew was by Hocking, Hillman and Wilkins in which transport from shoot to root had been demonstrated. Pitman pointed out that one of the problems is knowing where an effective concentration is. There may be a very small export from the leaf to a small volume in the root which will give a reasonably high local and effective concentration. In non-stressed plant leaves, the levels are 10^{-8}, rising to approximately 10^{-6} M in stressed plants. Collins finally remarked that in avocado, a supposedly rich tissue, the ABA levels are approximately $10 \mu g \, kg^{-1}$, a miniscule concentration.

Author Index

Numbers in italics indicate those pages where references are given in full.

Subject Index

Bold numbers show major reference; *italics* indicate discussion section